For the common-base amplifier circuit with emitter feedback:

$$A_v = \frac{R_C}{r'_e + R_E} \quad (7\text{-}23) \qquad r_{in} = (1+\beta)(r'_e + R_E) \quad (7\text{-}25)$$

$$r'_{in} = \frac{r_{in} R_B}{r_{in} + R_B} \quad (7\text{-}26)$$

For the common-emitter amplifier circuit with collector-to-base feedback:

$$r_{in} = (1+\beta) r'_e \quad (7\text{-}27) \qquad A_v = \frac{R_C}{r'_e} \quad (7\text{-}30)$$

Miller theorem:

$$R_{in} = \frac{R_B}{1 + A_v} \quad (7\text{-}28) \qquad R_{out} = \frac{A_v}{1 + A_v} R_B \quad (7\text{-}29)$$

Chapter 8 Field Effect Transistors

For JFET and depletion-type MOSFET:

$$I_D = I_{DSS}\left(1 - \frac{V_{GS}}{V_P}\right)^2 \quad (8\text{-}1) \qquad g_m \equiv \frac{\Delta I_D}{\Delta V_{GS}} \text{ for a constant } V_{DS} \quad (8\text{-}3)$$

$$g_m = -\frac{2I_{DSS}}{V_P}\left(1 - \frac{V_{GS}}{V_P}\right) \quad (8\text{-}4) \qquad g_m = g_{mo}\left(1 - \frac{V_{GS}}{V_P}\right) \quad (8\text{-}6)$$

For enhancement-type MOSFET:

$$I_D = K(V_{GS} - V_T)^2 \quad (8\text{-}7) \qquad g_m = 2K(V_{GS} - V_T) \quad (8\text{-}8)$$

Chapter 9 FET Bias, Load Lines, and Amplifiers

$$A_v = A_e = g_m R_D \quad (9\text{-}10) \qquad r'_s \equiv \frac{1}{g_m}\,\Omega \text{ or } g_m = \frac{1}{r'_s}\,\text{S} \quad (9\text{-}11)$$

$$A_e = A_v = g_m R_D = \frac{R_D}{r'_s} \quad (9\text{-}12)$$

$$A_e = A_v = g'_m R_D = \frac{R_D}{r'_s + R_S} \quad (9\text{-}13) \qquad g'_m = \frac{1}{r'_s + R_S} = \frac{1}{\dfrac{1}{g_m} + R_S} \quad (9\text{-}14)$$

Source follower:

$$A_v = \frac{g_m R_S}{1 + g_m R_S} < 1 \quad (9\text{-}17)$$

Chapter 10 Stability and Compensation

For beta stability:

$$K \equiv \left(\frac{\Delta I_C}{I_C}\right) \Big/ \left(\frac{\Delta\beta}{\beta}\right) \text{ where } 0 \le K \le 1 \quad (10\text{-}1)$$

$$K = \frac{1}{1 + (\beta + \Delta\beta)\dfrac{R_E}{R_E + R_B}} \quad (10\text{-}5)$$

For temperature sensitivity:

$$I_{CEO} = (1+\beta) I_{CBO} \quad (10\text{-}7)$$

I_{CBO} doubles for each 10°C rise in germanium transistors. I_{CBO} doubles for each 6°C rise in silicon transistors.

$$I'_{CBO} = 2^N I_{CBO} \quad (10\text{-}8c) \qquad I'_{CEO} = (1+\beta) 2^N I_{CBO} \quad (10\text{-}9)$$

$$S \equiv \frac{\Delta I_C}{\Delta I_{CBO}} \text{ where } 1 \le S \le (1+\beta) \quad (10\text{-}11a)$$

$$\Delta I_C = S \times \Delta I_{CBO} \quad (10\text{-}11b) \qquad S = \frac{R_E + R_B}{R_E + \dfrac{R_B}{1+\beta}} \quad (10\text{-}12)$$

Chapter 11 Decibels

$$dB = 10 \log_{10} \frac{P_2}{P_1} \quad (11\text{-}1) \qquad dB = 20 \log_{10} \frac{V_2}{V_1} + 10 \log_{10} \frac{R_1}{R_2} \quad (11\text{-}3)$$

$$dB = 20 \log_{10} \frac{V_2}{V_1} \quad (11\text{-}4) \qquad dB = 20 \log_{10} \frac{I_2}{I_1} + 10 \log_{10} \frac{R_2}{R_1} \quad (11\text{-}5)$$

Chapter 12 Special Amplifiers

The Darlington Amplifier:

$$r_{in} \approx (1+\beta)^2 (r'_e + R_E) \quad (12\text{-}5)$$

$$A_i = (1+\beta)^2 \quad (12\text{-}6) \qquad A_v = \frac{R_E}{r'_e + R_E} \le 1 \quad (12\text{-}7)$$

The differential amplifier:

$$A_v \equiv \frac{V_{out}}{V_{in_1} - V_{in_2}} \quad (12\text{-}9)$$

with balanced output:

$$A_v = \frac{R_C}{r'_e} \quad (12\text{-}10) \qquad A_v = \frac{R_C}{r'_e + R_E} \quad (12\text{-}11)$$

Fundamentals of Electronics

Third Edition

Fundamentals of Electronics

Third Edition

E. Norman Lurch

Professor, Electrical Technology
The State University of New York at Farmingdale

John Wiley & Sons,
New York Chichester Brisbane Toronto

Library of Congress Cataloging in Publication Data

Lurch, E Norman.
Fundamentals of electronics.

Includes index.
1. Electronics. I. Title.

TK7815.L84 1981 621.381 79-18696
ISBN 0-471-03494-0

Printed in the United States of America

10 9 8 7 6 5 4 3 2

To my daughters,
Mary Ann Johnson and
Victoria Ruth Bellias

Preface

The following paragraphs that appeared both in the original edition and in the second edition still apply to this, the third edition:

> This book is planned to meet the needs of the technician who is to work in the field of electronics. It is intended to provide the firm, solid background in fundamentals that is necessary for the study of the more specialized aspects of electronics.... A student using this book should have a working knowledge of dc fundamentals, and he or she should be studying ac circuits concurrently with Chapters 1 through 7, if ac circuits have not been studied already.
>
> The level of the text material enables the student to solve such problems as gain calculations, power outputs and graphical solutions... It is not my intention, however, in this book, to prepare the user to handle the design calculations and original derivations that are the premise of the electronic engineer. A student with a good working knowledge of algebra and right-angle trigonometry should not have difficulty with the solution of the problems.

Electronics has expanded over the last two decades and particularly over the last decade. Fortunately, we have developed simplified approaches to understanding its new technology. For example, at the technician level, it is no longer necessary to become involved in the very complex algebraic approach of the hybrid parameter; we can take a simplified approach that applies to both the transistor and the field-effect transistor.

I have tried to keep the number of necessary equations to a minimum. Each important equation is boxed. These boxed equations, listed by chapter and reference title, also appear on the inside covers of the book where the student can find them easily. The individual instructor may decide which of these equations the student must commit to memory.

Although this edition is technically the third, it has been almost completely rewritten. A new approach has been taken in most chapters. The subject of communications is omitted from this edition. Instead, these "saved" pages are used for an expansion of FET concepts and for several chapters on the operational amplifier.

I have sectionalized the material to provide for as much flexibility as possible:

Chapters 1 through 4 introduce electronic concepts, the diode, and its application as a rectifier.

The transistor (BJT) as a device and as an amplifier is discussed in Chapters 5 through 7. Similar coverage for the field-effect transistor is accomplished in fewer pages in Chapters 8 and 9. Stability and compensation concepts that apply both to the transistor and to the FET are treated in Chapter 10.

I believe that the decibel is too important a topic to be considered in a few paragraphs. The chapter on decibels (Chapter 11) is placed before the chapter on special amplifiers (Chapter 12), in which decibels are used to establish common-mode rejection ratios.

The power amplifier requires two chapters (Chapters 13 and 14). I have stressed heat-sink calculations because I believe that the technician should be able to protect a solid-state device against its dissipation.

Fall-off in gain at low frequencies and at high frequencies is considered in Chapter 15. The approach in this book does not consider the hybrid-pi circuit required for very high frequencies.

The general considerations of feedback (Chapter 16) lead directly to the operational amplifier. Three chapters are devoted to this topic to assure students' thorough assimilation of the material.

The ideal operational amplifier (Chapter 17) yields the basic amplifier equations and concepts. In Chapter 18, the details of the specifications for an operational amplifier are examined to show how we must modify our approach to the ideal operational amplifier. In Chapter 19, some specialized applications of the operational amplifier are considered such as the integrator, the differentiator, and the Schmitt trigger. At this point, the instructor can easily supplement the discussion with a consideration of other circuits that he or she would like to introduce. A full chapter (Chapter 20) is devoted to the voltage regulator, which is actually an extension of the operational amplifier.

The last two chapters (Chapters 21 and 22) discuss breakdown devices and controlled rectifiers (the SCR and the triac). I have included this material because, although it represents a large segment of the electronics industry, it is often neglected in current textbooks. I agree that the material properly belongs to the area of industrial electronics but, realistically, few programs offer anything in industrial electronics today. Individual instructors can decide whether to include this material in their courses. Study material, however, is provided for the interested student.

At each appropriate point in the text I have included a worked-out example to provide a numerical explanation of the material. The units for each number are given to show students exactly how the numerical solution is obtained.

Most of the sections within the chapters have associated problems for assignment. At the ends of most chapters there is a group of supplementary problems that cover the whole chapter. I have selected problems for the supplementary sets that are similar to the ones I use for testing purposes in my own classes. Most of the problems in the book have been used in at least two of my classes.

I thank Dr. Irving L. Kosow, formerly my editor for engineering technology at John Wiley, for his many helpful suggestions during the writing of this edition.

E. Norman Lurch
Stony Brook, New York, 1980

Contents

Fundamentals of Electronics

Chapter 1 Semiconductor Material

This chapter serves as an introduction to electronics by considering those properties of atomic particles that contribute to current flow in metals (Section 1-1). In the study of electronics we must refine our concepts of conduction and insulation by considering energy bands (Section 1-2). The structure of a crystal (Section 1-3) is very important because it is the building block for intrinsic semiconductor material (Section 1-4). *N*-type material (Section 1-5) is formed by introducing donor atoms and *P*-type material (Section 1-6) is formed by introducing acceptor atoms into the intrinsic material. Since there are many optical applications of semiconductors, there is a short discussion on the properties of light (Section 1-7).

Section 1-1 The Structure of the Atom

At the start of any study of dc circuits, the concept of current flow in a conductor is introduced. In a copper conductor one electron from each copper atom is free to move from atom to atom. Since there is a vast number of copper atoms in a conductor, the "free" electrons form a "cloud." Consider a small section of a conductor. When an emf is applied between the ends of the conductor, electrons from this "cloud" move out one end. An equal number of electrons enters the other end to keep the total quantity of electrons in the "cloud" constant. This net flow of electrons is called an *electron current.*

To explain how electronic devices work, we must examine the atom in greater detail and also study how atoms form semiconductor materials.

A diagram widely used to show the physical form of the atom is the Bohr model (Fig. 1-1). It is a three-dimensional model in which the electrons orbit in elliptical paths about a central core called the *nucleus.* For clarity, the three-dimensional concept is simplified to the two-dimensional forms of Fig. 1-2.

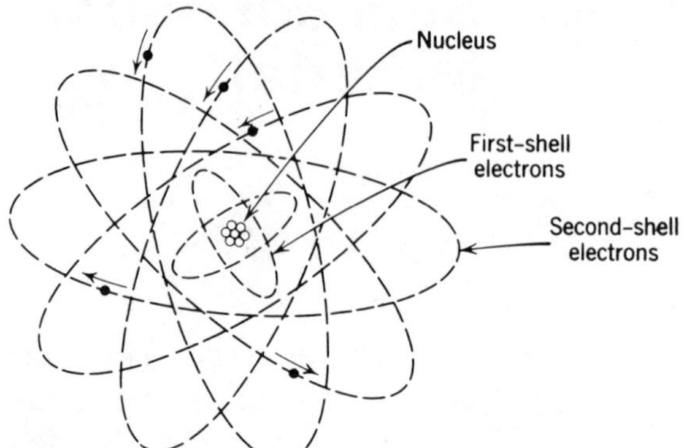

Figure 1-1 The Bohr model of an atom.

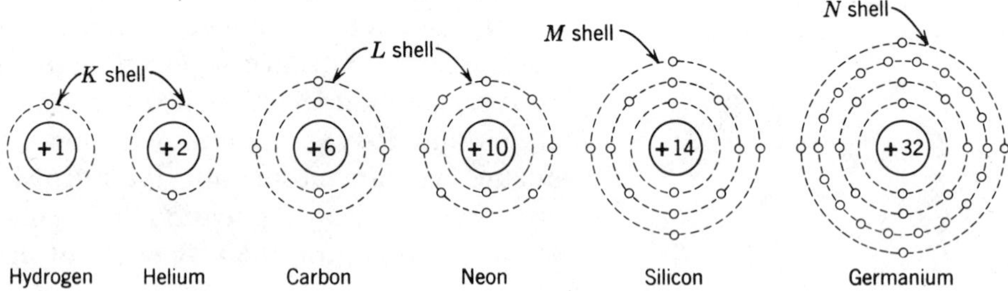

Figure 1-2 Models of various atoms.

An atom in its natural state is electrically neutral. Since each electron represents a certain fixed amount of negative charge, sufficient positive charges are necessary to balance the negative charges of the electrons. These positive particles, located in the central region of the atom, are called *protons.* For example, if a particular atom has 10 electrons in orbit (neon in Fig. 1-2), the nucleus must contain 10 protons to bring the total electrical charge to zero. The nucleus also contains neutral components (without a specific charge) called *neutrons.* Both the proton and the neutron weigh about 1850 times as much as the electron. Thus the weight of an atom is determined primarily by the total weight of its protons and neutrons. A substance or material is made up of many atoms of one or more different types.

Hydrogen (Fig. 1-2) has the simplest atomic structure. It has one proton with a single electron in orbit. We note in Fig. 1-2 that as electrons are added the number of protons (and neutrons) increases, making a heavier atom. The periodic table used in chemistry and physics is laid out according to the orderly increasing of the number of electrons and protons.

We find that electrons in orbit are confined to specific finite distances from the atomic center. Additionally, the orbital distances are arranged in groups called *shells* (Fig. 1-2). The innermost shell, the *K* shell, may contain up to 2 electrons but no more than 2. The next shell, the *L* shell, may contain up to 8 electrons. The third shell, the *M* shell, may contain up to 18 electrons. Succeeding shells have, as maximum numbers of electrons, 32 and 50 in that order. Figure 1-2 shows that there are four electrons in the *M* shell of the silicon atom. The lower shells of the silicon atom, the *K* shell and the *L* shell, must be completely filled before electrons can exist in the *M* shell.

The outer shell, the *valence* shell, determines the chemical activity of the element. If the outer shell is filled in completely, the substance is inert and does not react chemically. Examples are neon, argon, and krypton. If the outer ring is incomplete, it may join in chemical bonds with other atoms to produce the effect of filled outer shells. This action produces molecules of stable chemical compounds, such as water and salt.

The shells are finite in character and for an electron to exist within an atom the electron must exist within one of the specific shells. By this we mean that an electron cannot exist between shells. To permit an electron to move from one shell to another, definite *discrete* amounts of energy called *quanta* are required. A quantum of energy is the least unit amount of energy that can be considered in the process. Furthermore, quanta must exist as whole numbers; fractions of a quantum do not exist. If the energy required to shift an electron from one shell to another were three quantum units, an energy level of two quanta would produce no shift. If the energy level were gradually increased, then suddenly at a particular instant the necessary three quanta would be available and the electron would abruptly shift from one shell into the next.

We have stated that the shells are described as finite orbits in which electrons may travel. If we examine a particular shell carefully, we find that the *L* shell consists of two very close subshells, the *M* shell has three subshells, and the *N* shell has four subshells.

Since the subshells are very close to each other in a particular shell, the energy required to move an electron between subshells is small compared to the energy required to shift an electron from one shell to another. All this leads to the description of electrons as existing in atoms at definite, *discrete*, or *permitted* energy levels. External addition or removal of energy may move an electron from one permitted level to another, but it cannot change the permitted levels.

Section 1-2 Energy Bands

Semiconductor materials do not have electrons occupying or filling all the permitted energy level states in the outer subshells of the atom. The *Pauli exclusion principle* of modern physics theory requires that no two electrons within a *system* can have exactly the same energy content. Therefore, if there are eight permissible energy-level states, they all *must* have different values (Fig. 1-3*a*). Consider an atom that has four electrons in the outer two subshells and eight possible permissive energy-level states. When the least possible external energy is supplied to the system, the four electrons are contained in the lowest four energy levels (Fig. 1-3*b*). As energy is put into the system, the electron in the highest level (level 4) moves into the next higher level (level 5, Fig. 1-3*c*). As the energy input to the atom is increased, the four electrons eventually move into levels 5, 6, 7, and 8 with *vacancies* in levels 1 through 4.

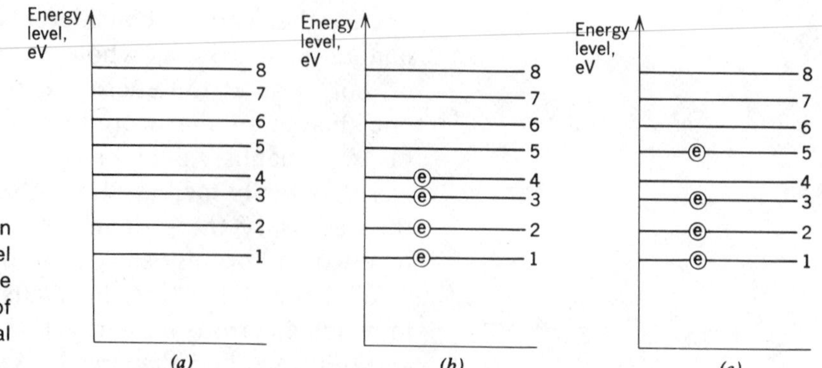

Figure 1-3 Energy level states in an atom. (*a*) The different energy-level states. (*b*) Four electrons filling the lower energy-level states. (*c*) Effect of addition of energy from an external source.

Energy levels in atoms are measured in units of the *electron-volt, eV*, which is the amount of increase in energy that an electron acquires when it is accelerated by the field created by one volt. An electron-volt is equivalent to 1.60×10^{-19} joule or watt-second.

Each molecule of a gas is a separate system. Therefore the energy-level diagram for any one molecule is the same for all other molecules.

A solid consists of a great many molecules that are bound together physically. Now by the Pauli exclusion principle all the levels for all the molecules must be different. Consequently, we can no longer show all the different individual lines because there are so many of them. These many lines now merge into *bands* as shown in Fig. 1-4.

The total possible number of energy levels in the outer subshells are divided into two classes: those that are grouped as the *valence band* and those that are grouped as the *conduction band.* The electrons that are in the valence band do not move readily from atom to atom. The electrons that are in the conduction band can move freely and are, therefore, free to serve as *current carriers.* In a conductor such as copper, the valence band and the conduction band merge (Fig. 1-4*a*). A small external emf applied to a conductor is sufficient to produce a flow of current.

In an insulator all the electrons in the outer subshells are in the valence band. Additionally, there is a large separation between the permissible lower energy levels of the valence band and the permissible higher energy levels of the conduction band (Fig. 1-4*b*). This separation is called the *forbidden band* in which no permissible energy-level states exist. Con-

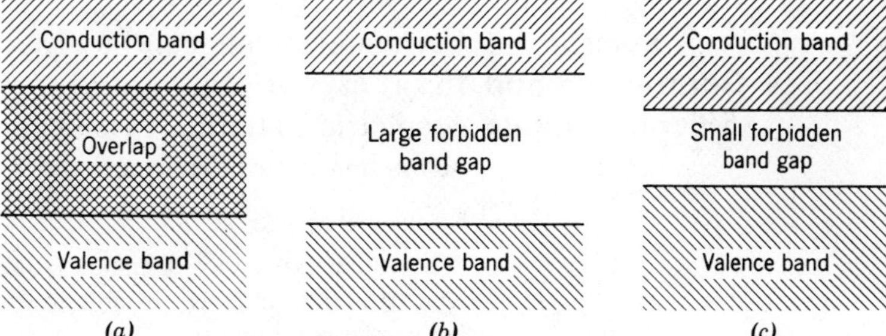

Figure 1-4 Energy levels in the outer shells of different solids. (*a*) Conductor. (*b*) Insulator. (*c*) Semiconductor.

sequently, under normal conditions, there is negligible movement of electrons. If sufficient voltage is applied to an insulator, electrons acquire enough energy (of the order of 6 eV or more) to cross the forbidden gap into the conduction band. When this occurs, current flows in the insulator. We say that the insulator has broken down under high-voltage stress.

The basic materials used for semiconductors are elements from Group IV of the periodic table. Group IV elements have eight permissible energy states in the outer subshell and a total of four electrons in the outer subshells. Examples of Group IV materials are silicon and germanium (Fig. 1-2). When a solid is formed from Group IV atoms, we obtain the two energy bands separated by a narrow energy gap of the order of 1.0 eV, (Fig. 1-4*c*). Four of the permissible energy states of each atom are in the conduction band and four of the permissible energy states are in the valence band. All four electrons enter into the valence band. In order to obtain a current flow in the material, sufficient external energy must be applied to cause electrons to cross the forbidden band into the conduction band.

For a broad classification of electrical materials, let us assume that a cubic-centimeter block of each is available for tests and measurements. If the resistance between opposite faces of each cube is measured, we find that the *insulator* yields a resistance value of many megohms. The *conductor's* resistance value will be measured in millionths of an ohm or micro-ohms. The *semiconductor* is a cross between the two, and we can expect to find its resistance value in the order of ohms. This method of classification is quite loose and very general because the boundaries between the materials often tend to be obscure.

Section 1-3 Germanium and Silicon Crystals

Certain elements can be processed synthetically into *crystals*. In a crystal, the atoms are positioned into symmetrical geometric patterns. In silicon and germanium crystals* each atom has a bond to each of four other atoms resulting in the

* The other Group IV elements from the periodic table are carbon, tin, and lead. It is theoretically possible to use these elements as the basic materials for semiconductors.

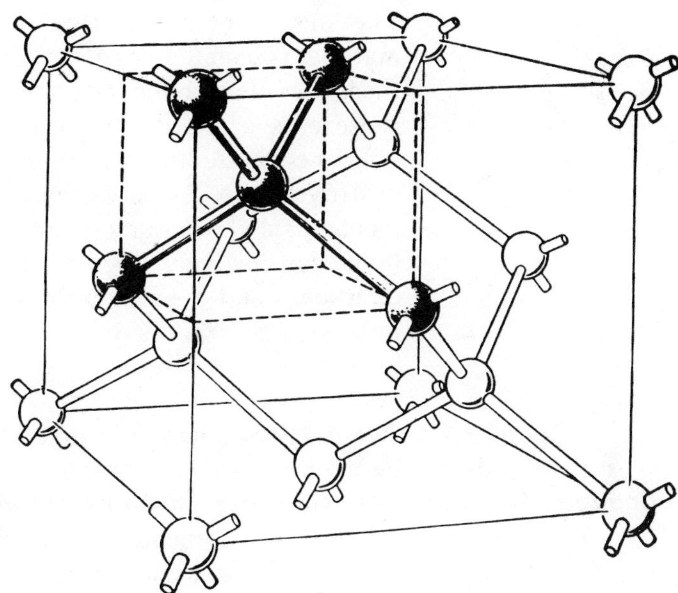

Figure 1-5 Three-dimensional model of a face-centered crystal lattice. (*Courtesy Kittel, Introduction to Solid State Physics.*)

tetrahedral crystalline structure shown in Fig. 1-5. Each atom has four electrons in the outer shell. The outer shell has eight permissive places for a completely filled shell. Thus, if the atoms in this crystal share electrons in their outer shells with their neighbors, the outer shells of all the atoms will be complete, since its own four electrons plus the four shared from the four neighbors add up to eight electrons, the maximum number allowed in the outer shell.

The three-dimensional model of Fig. 1-5 can be reduced to the two-dimensional representation of Fig. 1-6 for simplicity.

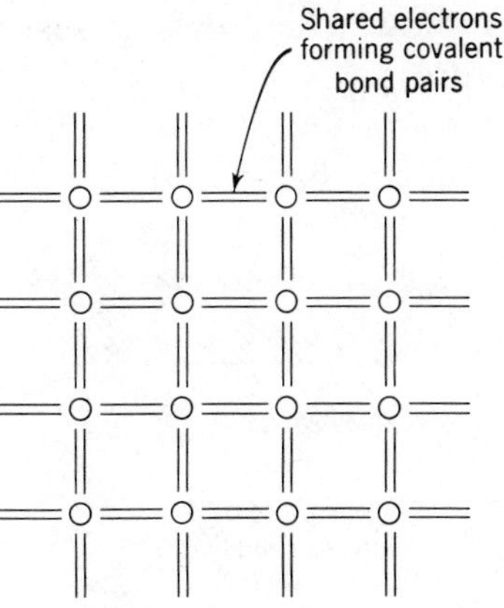

Figure 1-6 Two-dimensional model of a face-centered crystal lattice.

The sharing of an electron between two atoms of the *same* material is called a *covalent bond.* In semiconductor materials, these covalent bonds exist in pairs. Each line in Fig. 1-6 represents a shared electron thus forming the bond.

If the crystalline structure of a semiconductor device is destroyed, the device will not work in an electronic circuit. A mechanical shock such as occurs when the device falls on the floor can fracture the crystalline structure. If the device overheats and a part of the crystal melts, its properties are lost.

Section 1-4 Intrinsic Semiconductor Material

The crystalline material made from pure silicon or from pure germanium and illustrated in Fig. 1-5 and Fig. 1-6 is called *intrinsic semiconductor material.* The energy band pattern for this crystal is repeated in Fig. 1-7.

When the material is at the temperature of absolute zero (0°K), all electrons are contained in the valence-bond energy band; none are in the conduction band. Therefore, the material is an ideal insulator. The energy gap E_G is 0.785 eV for germanium and 1.21 eV for silicon at absolute zero. When energy is applied to the system in the form of heat, electrons will leave the valence band and jump across to the conduction band. The energy gap becomes smaller as the ambient temperature increases. At room temperature (300°K), E_G is 0.72 eV for germanium and 1.10 eV for silicon. Further application of heat causes more and more electrons to cross the energy gap into the conduction band. This means that the resistivity characteristics of germanium and silicon (and carbon) show a decreasing resistance with an increase in temperature.

At room temperatures, then, electrons have crossed from the valence band into the conduction band. The absence of

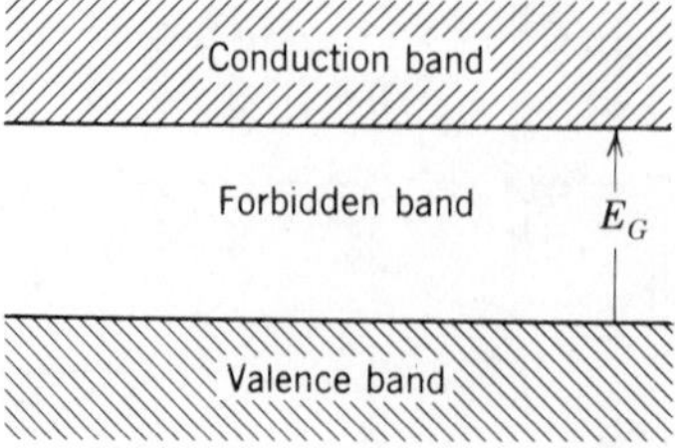

Figure 1-7 Energy bands in the outer shell.

the electron in the valence band creates a *hole*. Hence, in pure crystals, the number of electrons in the conduction band is balanced by an equal number of holes in the valence band. A pure crystal above absolute zero must contain these combinations, which are called *electron-hole pairs*. Electrons are free to move within the conduction band. Correspondingly, if an electron in the valence band moves to fill a hole, it leaves a hole where it was before. Thus, we can have not only current in the conduction band but also an independent current within the valence band that is the result of holes' "jumping" from one atom to another.

An electron in the conduction band has a higher energy level than an electron (or hole) in the valence-bond band. If an electron requires a certain total energy to be moved by an external field, the electron in the conduction band requires less additional energy than the one in the valence band. Accordingly, the largest portion of the current is due to the conduction-band carriers. However there will be a small value of *hole* current at the same time. Accordingly, in this case the electrons are the *majority current carriers* and the holes are the *minority current carriers*. Both carriers contribute to the *total* current.

A *thermistor* is a device that has a negative temperature coefficient of resistance. Thermistors are widely used as temperature sensing devices or as devices used to counteract the changes caused by *ambient temperature* changes on other electronic devices. The *ambient temperature* is the temperature of the surrounding air, or room temperature. A thermistor is made from a piece of intrinsic material. Metallic (ohmic) contacts are placed at the ends of the material. The resulting device is equivalent to a resistor. However, as the ambient temperature of the thermistor is increased, covalent bonds are broken and current flow in the thermistor increases. Accordingly, the resistance of the thermistor decreases with an increase in temperature. The circuit symbol is shown in Fig. 1-8.

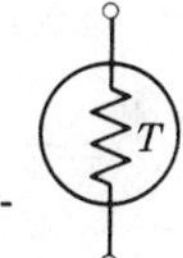

Figure 1-8 Circuit symbol for a thermistor.

Section 1-5 *N*-Type Material

The elements that have five electrons in the outer ring (Group V elements in the periodic table) are called *pentavalent* elements. Pentavalent elements used in the manufacture of semiconductors are phosphorus, arsenic, antimony, and bismuth. When one of these elements is introduced at the carefully controlled rate of the order of one part in 10 million into pure germanium or pure silicon crystals by a process called *doping*, we form *N-type material.* These deliberately introduced atoms are often called *impurity atoms* to distinguish them from the predominant germanium or silicon atoms of the crystalline structure.

Each impurity atom replaces a germanium or a silicon atom within the structure of the crystal. The *N*-type material is a true crystal with the structure explained for the intrinsic material in the last section. We call the new crystalline structure *extrinsic.*

We represent intrinsic germanium or silicon by the two-dimensional model of a face-centered lattice shown in Fig. 1-6. Now we must change the model to that shown in Fig. 1-9. Four of the five electrons of the pentavalent atom are bound within the four covalent bonds. The fifth electron is not confined to the parent atom but can "drift" within the crystal. Since the pentavalent atoms add a "free" electron to the crystal, they are called *donor atoms.*

As an aid to the memory, the student should associate the *N* in "donor" with the *N* in the "*N*-type material." *N*-type material is so called because of the associated negative-charge carriers provided by the donor atoms.

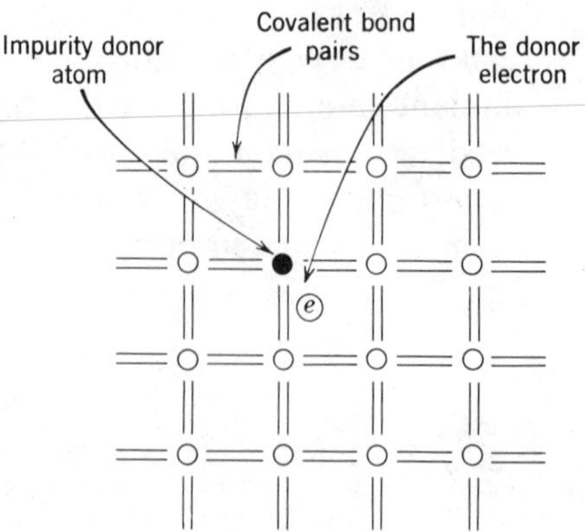

Figure 1-9 Electron affinities within *N*-type material.

N-type material inherently has more electrons than the pure intrinsic crystal, but it must be remembered that a block of this material is *definitely neutral* as far as overall net charge is concerned. It has no net charge any more than a piece of copper could have. The concept of *space-charge neutrality* states that a block of material has an equal number of positive and negative charges. If there is an "extra" free electron, the characteristic of the donor atom is such that it has in its structure, one more positive charge in its nucleus than the atom of intrinsic material has. This is true because the number of electrons in an atom equals the number of positive charges in its nucleus. The electron of the donor atom is only "free" because the donor atom is part of an orderly crystalline structure. If this crystalline pattern were lost or destroyed, this electron would no longer be "free."

The energy required to break covalent bonds in intrinsic semiconductor material is of the order of 0.7 to 1.1 eV. In the formation of *N*-type material, the energy level of the extra electrons is only about 0.01 eV below the conduction band. Since the average energy supplied to these electrons at room temperature by heat is about 0.025 eV, they shift into the conduction band.

The extra electrons contributed by the donor atoms must be in the fifth permissible energy level. Since the eight permissible levels divide equally between the valence band and the conduction band, these extra electrons must be in the conduction band (Fig. 1-10).

An applied external electric field can easily move these donor electrons. Accordingly, current in *N*-type material is due to the net movement of these donor electrons in the conduction band.

Since the material is normally at room temperature, we also have a breaking of covalent bonds caused by heat. Therefore,

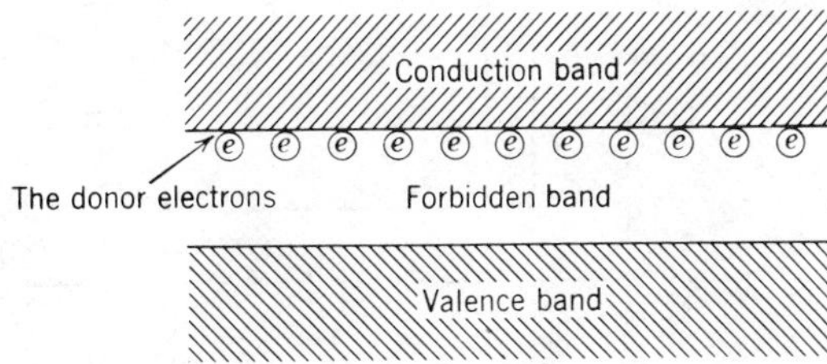

Figure 1-10 Energy levels in *N*-type material.

in addition to the donor electrons, there are also the electrons from the electron-hole pairs in the conduction band and the holes from these pairs in the valence band. These, too, contribute to current flow with the result that electrons are the *majority current carriers* and holes are the *minority current carriers* in N-type material.

At normal temperatures, there are many more electrons than holes. If the temperature of the crystal is raised, more and more electron-hole pairs are formed. At some temperature level, the number of donor electrons becomes negligible with respect to the number of electron-hole pairs, and the crystal is then, for all practical purposes, intrinsic. If the source of heat is removed, the material returns to its normal extrinsic state. However, there is a limit to the amount of heat that a crystalline substance can take without losing the basic lattice structure. If lattice pattern is destroyed by excessive temperatures, the device ceases to have the desired properties and must be replaced.

Consider a block of N-type material that has metallic contact plates at the ends. A source of emf, V, is connected to the metallic plates (Fig. 1-11). The electron in the conduction band moves through the N-type material to the positive metallic plate into the external conductor. At the same time, another electron enters from the negative metallic plate into the N-type material. Thus there is no change in the total number of electrons in the block of N-type material.

The net current that flows is I amperes. The ratio V/I gives the *bulk resistance* value of the block of N-type material. This resistance value is a function of the amount of doping during manufacture. Consequently, a light doping can be used to produce resistors with high resistance values and a heavy doping can be used to produce resistors with low resistance values.

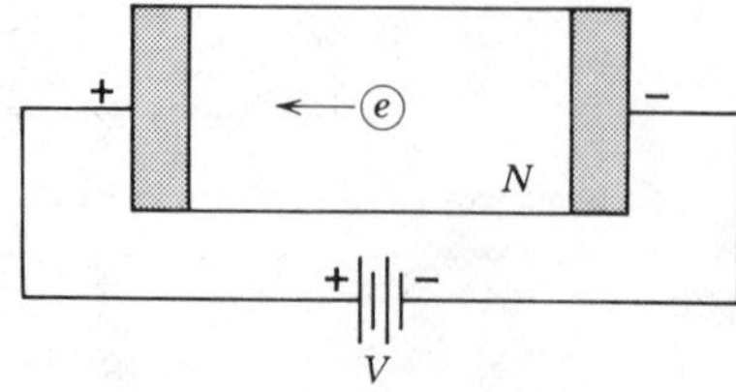

Figure 1-11 Electron flow in N-type material.

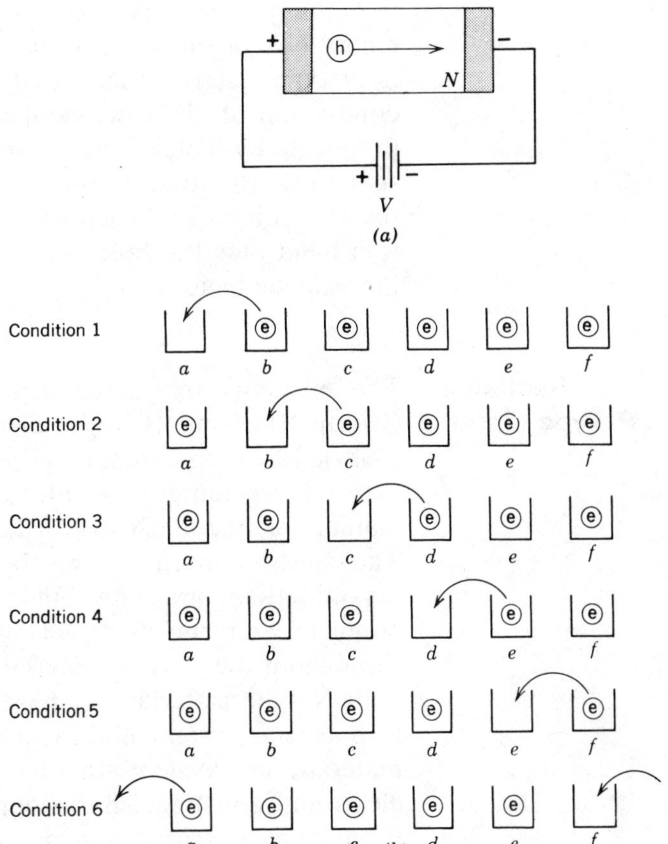

Figure 1-12 Hole movement in *N*-type material. (*a*) Hole movement. (*b*) Mechanism of hole transfer.

If a hole exists in the *N*-type material (Fig. 1-12*a*) the hole movement is toward the negative supply terminal because the hole represents a positive charge. To show the mechanism of current generated by holes, consider the six boxes, *a* to *f*, shown in the top row of Fig. 1-12*b*. These boxes represent the covalent bonds of adjacent atoms in the crystal. Box *a* is empty and boxes *b* through *f* have an electron in each. A box is used to show that a specific amount of energy is required for an electron to get out of the box.

An electron "jumps" from box *b* to box *a*. Now the empty box (the hole) is box *b*, condition 2. Successively, the empty box is box *c*, box *d*, box *e*, and box *f* as shown in Fig. 1-12*b*. In the bottom row, condition 6, if an electron jumps to the left out of box *a*, an electron must come in from the right to fill in box *f* and the process starts over.

Actually electrons are moving but the effect is that an empty box or "hole" moves to the right. The normal current in N-type material comes from electrons freely moving in the conduction band. Now we also have some additional current produced by holes moving in the opposite direction. Consequently, the total current in N-type material is the sum of the electron current (majority current carriers) in the conduction band plus the hole current (minority current carriers) in the valence band.

Section 1-6 *P*-Type Material

The elements that have three electrons in the outer ring (Group III elements in the periodic table) are called *trivalent elements*: boron, aluminum, gallium, indium, and thallium. When these impurity elements are introduced into the germanium or silicon crystal by doping, we form *P-type material.* The added impurity atoms become an integral part of the crystal-lattice structure but leave certain of the covalent bonds missing one electron (Fig. 1-13). Thus, holes are formed throughout the P-type material.

In N-type material, the pentavalent atom adds a free electron to the system and is called a *donor atom.* In P-type material, the trivalent atom leaves the system deficient by one electron. To distinguish this opposite effect, the trivalent atom is called an *acceptor atom* to contrast it with the donor atom.

As an aid to the memory, the student should associate the *P* in "accep*tor" with the *P* in "*P*-type material" and with the *P*

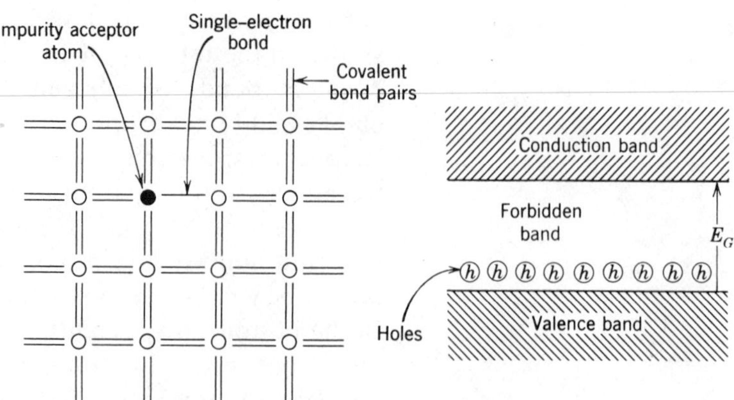

Figure 1-13 Electron affinities within *P*-type material.

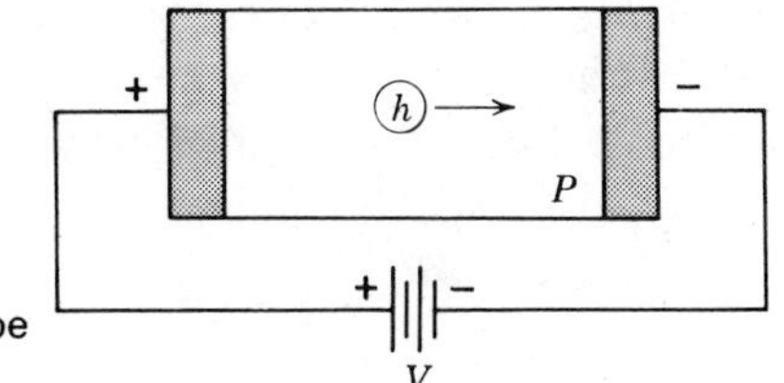

Figure 1-14 Current flow in *P*-type material.

in "*p*ositive charge carrier." *P*-type material is so called because of the positive-charge effect of the holes.

When an emf is placed across a block of *P*-type material (Fig. 1-14) the majority current carriers are the holes that shift toward the negative terminal of the voltage supply. When a hole finally reaches the negative metal plate, an electron enters from the external circuit to fill the hole. Simultaneously an electron is injected into the positive metal plate and a new hole is created. Consequently, in this circuit electrons are the current carriers in the lead wires and in the metallic plates but holes are the majority current carriers inside the *P*-type semiconductor itself.

Thermal effects can break covalent bands. When this occurs, there are free electrons within the *P*-type material that form minority current carriers. As with *N*-type material, additional heating can cause the number of thermally created electron-hole pairs to overshadow the effect of the hole-producing acceptor atoms. Now the substance will behave like an intrinsic material provided the crystalline structure is not destroyed.

Section 1-7 Light

Light, radio waves, heat, and electromagnetic radiation have similar properties. All travel through free space at the velocity of light

$$\begin{aligned} c &\approx 300{,}000 \text{ kilometers per second} \\ &\approx 300{,}000{,}000 = 3 \times 10^8 \text{ meters per second} \\ &\approx 186{,}000 \text{ miles per second} \end{aligned}$$

A particular energy has a specific frequency, f, measured in hertz. By frequency we mean that the energy is being propagated at f pulsations of energy per second. The physical distance between corresponding points of two successive pulsations of energy is called the *wavelength* λ (Greek letter lambda). The relationship between c, f and λ is

$$c = f\lambda \qquad \text{or} \qquad \lambda = \frac{c}{f} \tag{1-1}$$

The wavelength is measured in meters or in centimeters when we are concerned with radio communications. Since the wavelengths of heat and light are much shorter, we use the unit *micrometer* (μm) or *nanometer* (nm) in SI units. The micrometer is commonly called the *micron.* The former unit, which will be replaced in time by the standardized use of SI units, is the *Angstrom unit* (Å). These units are defined as

$$1 \text{ micrometer } (\mu\text{m}) = 10^{-6} \text{ meter}$$
$$1 \text{ nanometer (nm)} = 10^{-9} \text{ meter}$$
$$1 \text{ Angstrom unit (Å)} = 10^{-8} \text{ centimeter} = 10^{-10} \text{ meter}$$
$$= 0.1 \text{ nanometer}$$

Therefore

$$1 \text{ Å} = 10^{-4}\ \mu\text{m} = \frac{1}{10{,}000}\ \mu\text{m} = 0.1 \text{ nm}$$

and

$$1\ \mu\text{m} = 10^{4} \text{ Å} = 10{,}000 \text{ Å}$$
$$1 \text{ nm} = 10 \text{ Å}$$

If we use the wavelength as the horizontal scale, we can show the relationship between infrared (heat), visible light, and ultraviolet (Fig. 1-15). Sunlight is *white light*; that is, it contains light of all wavelengths and produces the continuous spectrum shown in Fig. 1-15. A special type of visible light is a *monochromatic light* which is the light of a *single* wavelength. For example, if the wavelength of a monochromatic source is 0.68 μm (6800 Å), we see a very intense red. A laser beam is a source of monochromatic light. Most electronic devices that produce light (Section 2-6) yield a light that is made up of a number of monochromatic values.

Now let us perform a thought experiment. A monochromatic source of light near 0.550 μm (5500 Å) is used as a

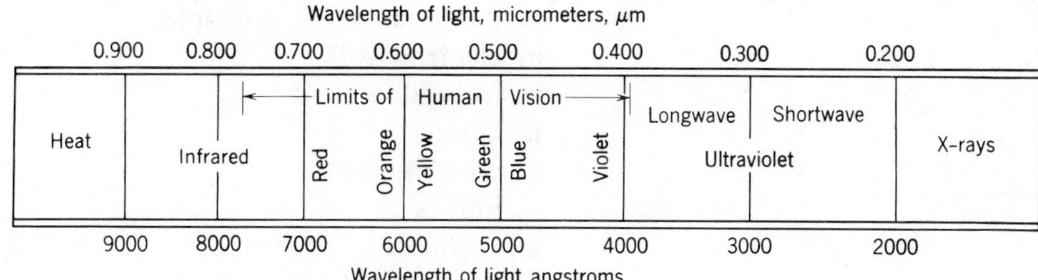

Figure 1-15 The spectral distribution of light energies.

reference. A person is asked to adjust a second source of monochromatic light at a different wavelength until he believes that the two light sources have equal intensities. After this experiment has been performed at many different wavelengths by many different people, we can obtain an average value for the *spectral response of the human eye* (Fig. 1-16). The human eye is most sensitive to light emission in the yellow–green region.

Questions

1-1 Define or explain each of the following terms.

a. molecule
b. atom
c. electron
d. shell
e. proton
f. permitted energy level

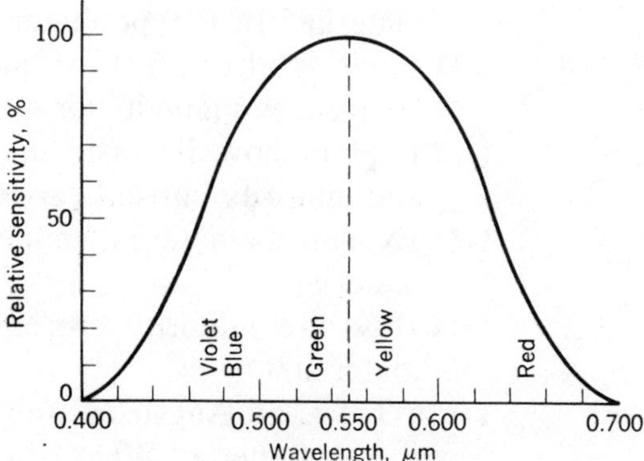

Figure 1-16 The spectral response of the human eye.

g. Pauli exclusion principle
h. electron volt
i. forbidden band
j. lattice
k. covalent bond
l. intrinsic
m. extrinsic
n. majority current carrier
o. minority current carrier
p. hole
q. donor atom
r. acceptor atom
s. wavelength.

1-2 Name the component parts of an atom.

1-3 What happens to an electron if it attempts to remain at an energy level between two adjacent permitted energy levels?

1-4 Explain how a conduction band is formed.

1-5 Explain why one material is an insulator and another a conductor.

1-6 Compare the resistive values of conductors, semiconductors, and insulators.

1-7 If a covalent bond is broken, how many current carriers are produced?

1-8 How does a thermistor function?

1-9 Assume a thermistor is used as a thermometer. If the device used to measure the resistance of the thermistor puts too much current through the thermistor, does the thermometer read high or low? Why?

1-10 What carrier is a majority current carrier in *N*-type material? In *P*-type material?

1-11 What is the effect of an increase in temperature on intrinsic semiconductor material?

1-12 Explain how intrinsic material can have both majority and minority current carriers.

1-13 Explain the action of minority current carriers in *N*-type material.

1-14 How are minority current carriers produced in *P*-type material?

1-15 What is an Ångstrom unit? What is a micrometer? What is a nanometer? What is a micron?

1-16 What colors are present in a monochromatic light?
1-17 What is a white light?
1-18 What does a spectrum of the human eye represent?
1-19 Why can a dress change color when taken from a store into sunlight?

Chapter 2 Diodes

When a *P-N* junction (Section 2-1) is formed, we create a diode that has the property of passing current in one direction only. The Zener diode (Section 2-2) is a general-purpose diode that is modified to establish a specific reverse breakdown voltage. Diodes can be used to change or to modify signal waveforms (Section 2-3). The characteristics of a diode circuit used to modify a signal waveform can be displayed on an oscilloscope as a *V-I* characteristic (Section 2-4). When we use diodes in ac signal circuits, we require the ac model (Section 2-5). Optical properties of the diode are examined (Section 2-6). The varactor diode (Section 2-7) makes use of the depletion region under reverse bias conditions.

Section 2-1 The *P-N* Junction

Consider the special case illustrated in Fig. 2-1*a*. We are using this approach as a "thought experiment" discussion to give the concept of a *P-N* junction, although it is not the actual way the junction is formed. A single continuous crystal is formed in Fig. 2-1*a*. The first part of the crystal is formed with a doping of donor atoms (*N*-type material) and the last part of the crystal is formed with a doping of acceptor atoms (*P*-type material). The resulting unit shown in Fig. 2-1*a* has a *metallurgical junction* or *barrier* between the *N*-type material and the *P*-type material. The majority current carriers are electrons in the *N*-type material and holes in the *P*-type material.

The electrons in the *N*-type material at the barrier cross over the barrier to fill the holes on the other side of the barrier (Fig. 2-1*b*). Now there are regions on either side of the barrier in which there are no current carriers (Fig. 2-1*c*). This total shaded region in Fig. 2-1*c* has no free current carriers and is called the *depletion region* or *transition region.*

It should be stressed that the depletion region is formed during the manufacturing process and is basic to the *P-N* junction. We cannot take a crystal of *N*-type material and a

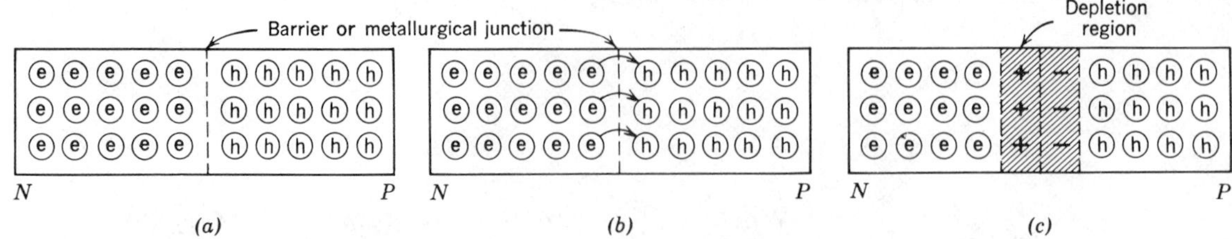

Figure 2-1 The *P–N* junction. (*a*) Construction. (*b*) Movement of charge carriers across the junction. (*c*) The depletion region.

crystal of *P*-type material and place them in contact to form a *P-N* junction.

When an electron moves across the barrier, it leaves behind an atom that is one electron short of its normal state. This atom is now *ionized* and has a positive charge. We can also say that there is an *uncovered positive charge* in this part of the depletion region. Similarly, the transfer of the electron across the barrier into the hole brings an extra electron into that atom to give it a negative charge. This atom is now a negative ion and there is an *uncovered negative charge* in this part of the depletion region. The uncovered charges in the depletion region are shown as + and − signs in Fig. 2-1*c*.

When a *P-N* junction is packaged as a semiconductor device it is called a *diode.*

Let us connect a power or voltage supply to the diode (Fig. 2-2*b*). The positive terminal is connected to the *N* side of the diode and the negative terminal is connected to the *P* side of the diode. The positive terminal of the supply pulls electrons out of the *N* side of the *P-N* junction. Those free electrons

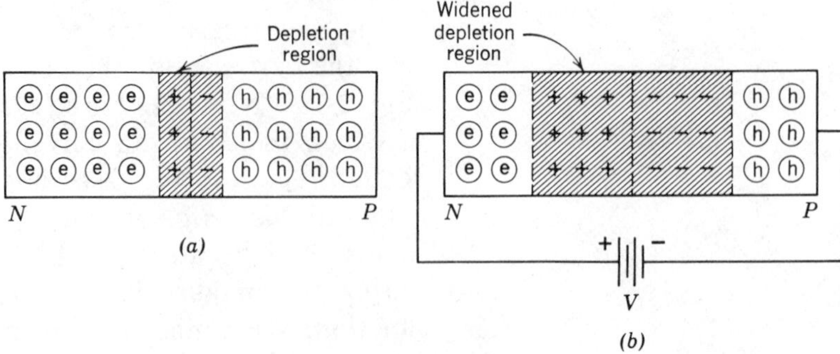

Figure 2-2 Reverse bias. (*a*) Diode without bias. (*b*) Diode with reverse bias.

that remain shift to the left. The negative terminal of the supply injects electrons into the P side of the P-N junction. These injected electrons move as far from the negative terminal as possible. They fill in the holes closest to the barrier. The result of this action is to *widen* the depletion region. In any event, there is no continuous current in the diode or in the external circuit. When the N-type material is more positive with respect to the P-type material, the diode is said to be *reverse biased.*

When the reverse bias is increased, the depletion region is widened. At some point the electrical stress of the reverse voltage is sufficient to break a covalent bond. The current carrier produced from the broken bond is accelerated through the depletion region. When it collides with another covalent bond, the energy of impact breaks that bond also. Now, two current carriers are available. This process is cumulative and the resulting flood of current carriers is called an *avalanche current.* The value of the voltage that produces the avalanche current is called the *breakdown voltage* BV_R of the diode. The numerical value of BV_R is a function of the design of the diode. The values of BV_R range from tens of volts to several thousand volts for what are called *general-purpose diodes.*

Assume that we purchase 1000 general-purpose diodes that have a rating for BV_R of 500 V. This specification means that, when they were tested by the manufacturer, all those that failed to sustain 500 V in the reverse direction were classified as rejects.

Now let us connect the supply to the diode as shown in Fig. 2-3*b*. The positive terminal of the supply is connected to the P side of the diode and the negative terminal of the supply is

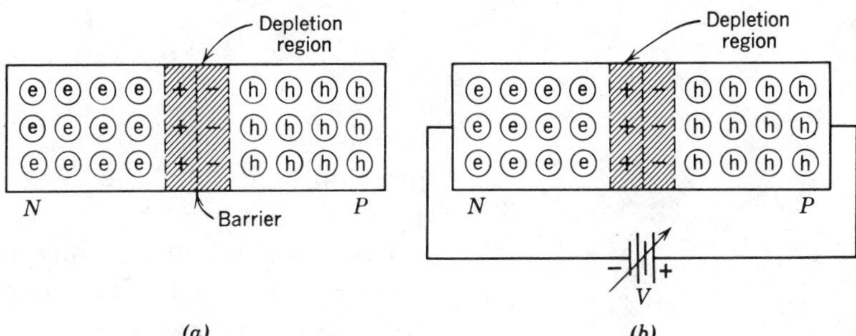

Figure 2-3 Forward bias. (*a*) The unbiased junction. (*b*) External connection for forward bias.

connected to the N side of the diode. When the supply voltage is set to zero, we have the depletion region thickness of the *unbiased state,* Fig. 2-3*a*, that resulted when the P-N junction was formed.

When the supply voltage, V, is increased from zero, electrons are injected into the N-type material from the negative terminal of the supply and electrons are extracted from the P-type material to the positive terminal of the supply. The number of positive ions in the N-type material and the number of negative ions in the P-type material are reduced. As a result, the depletion region becomes narrower.

At some value of supply voltage V, the width of the depletion region is reduced to an amount that permits electrons and holes to cross the P-N junction. Then *forward current*, I_F, will flow freely in the circuit. When this happens, we note:

1. The majority current carriers in the N-type material are electrons moving from the left terminal to the barrier.
2. The majority current carriers in the P-type material are holes moving from the right terminal to the barrier.
3. The electrons and the holes "recombine" at the barrier.

In order to have current in a diode, the supply voltage must be at least that voltage required to reduce the depletion region to zero. This voltage is called the *junction* or *barrier potential* V_j (or V_B). We find that V_J is approximately 0.3 V for germanium diodes and 0.7 V for silicon diodes at room temperature. These two numerical values are useful to memorize.

When the P material is positive with respect to the N material, Fig. 2-3*b*, we define *forward bias.* We can remember how to connect a supply to a semiconductor for forward bias by connecting

$$P \text{ to } P \qquad \text{and} \qquad N \text{ to } N$$

that is, P of the Positive of the supply to P of the P-N junction and N of the Negative of the supply to N of the P-N junction.

Conventional current flow in the diode with forward bias is shown in Fig. 2-4*a*. The standard circuit symbols for semiconductors are based on conventional current flow. The cur-

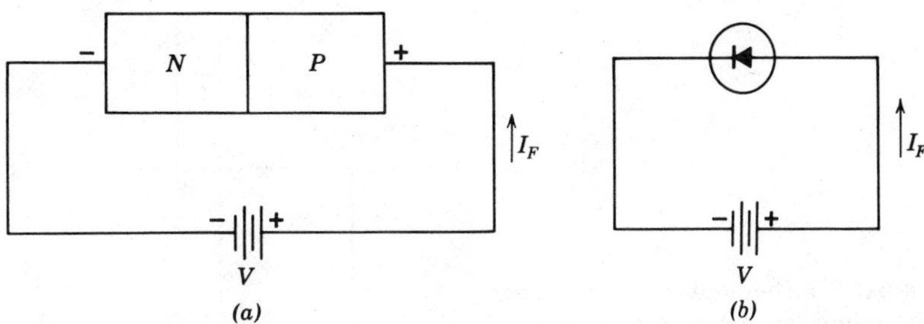

Figure 2-4 Current in a diode. (*a*) Physical layout. (*b*) Circuit symbol.

rent enters the diode into the *P* side. Therefore the circuit symbol, Fig. 2-4*b*, shows an arrowhead designating the *P* side of the diode. The thin vertical line designates the *N* side of the diode.

Commercially available diodes usually have some means to show which lead is *P* and which lead is *N*. Typical diode markings are shown in Fig. 2-5. In standard notation for diodes, a manufacturer assigns type numbers preceded by "1N" such as 1N230 or 1N1424. The color bands correspond to the final digits such as 230 for the 1N230 or 1424 for the 1N1424. Some diodes have physical shapes to show which terminal is the anode.

Now we can graph the volt-ampere characteristic for the diode. The forward region shows that the current is zero at voltages up to V_j and at V_j the current in the diode can become very large. In the analysis of many applications of diodes, we can neglect this small forward voltage value without causing serious error. The ideal reverse characteristic requires that there is no diode current. This is true as long as the reverse voltage is less than the breakdown voltage, BV_R.

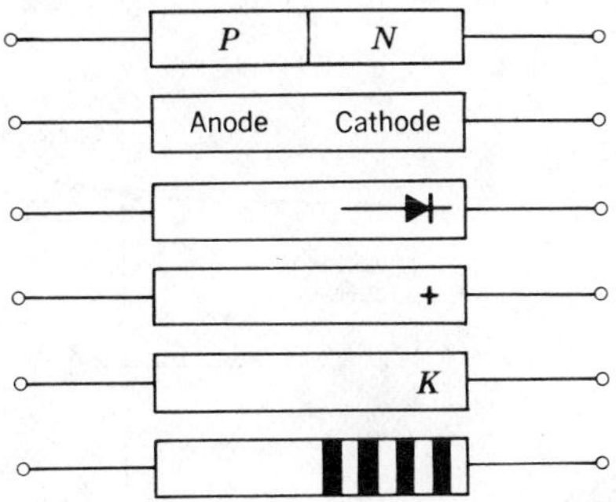

Figure 2-5 Diode markings.

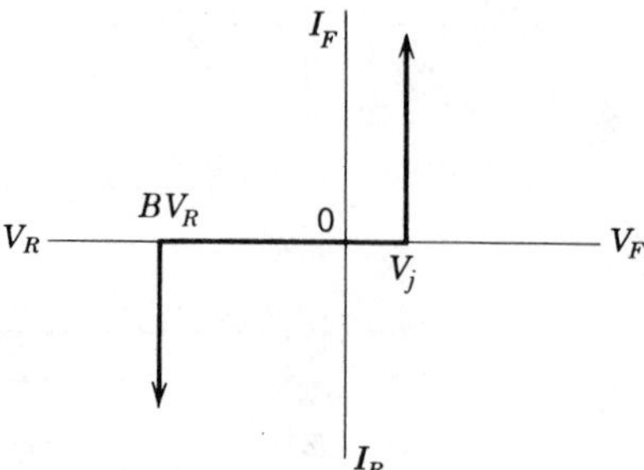

Figure 2-6 The forward and reverse characteristics of a near-ideal diode.

The forward and reverse characteristics of the near-ideal diode are given in Fig. 2-6.

General-purpose diodes are rated in terms of a current and a voltage. For example, the ratings of 2 A and 200 V mean that the diode can handle 2 A in the forward direction and that the diode will not break down in reverse as long as 200 V is not exceeded in the reverse direction.

An actual I-V characteristic curve that is obtained from a diode is shown in Fig. 2-7. It is important to note that the scales are different in the forward and reverse directions. In the forward direction, the current rises in a slight curve and then increases in a direction very close to a straight line. If this straight line is projected back to the horizontal axis, the intersection is the junction potential V_j. The slope of the straight line defines the *bulk conductance* g_B as

$$g_B \equiv \frac{\Delta I_F}{\Delta V_F} \text{ siemens} \tag{2-1a}$$

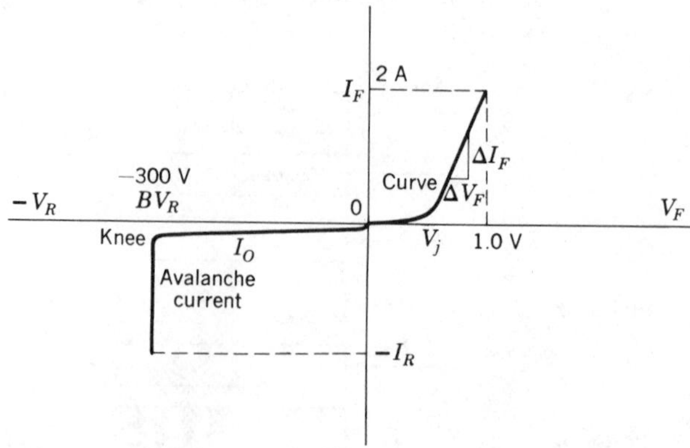

Figure 2-7 Forward and reverse characteristics of an actual diode.

The *bulk resistance* r_B is usually very small and is the sum of the resistance values of the N material and of the P material. r_B is the reciprocal of g_B.

$$r_B = \frac{1}{g_B} = \frac{1}{\text{slope}} = \frac{\Delta V_F}{\Delta I_F} \text{ ohms} \qquad (2\text{-}1b)$$

Example 2-1

A silicon diode dissipates 3 W at 2 A. Determine the forward voltage drop, V_F, across the diode and the value of r_B.

Solution

The voltage drop V_F across the diode is

$$V_F = \frac{P}{I_F} = \frac{3\text{ W}}{2\text{ A}} = \mathbf{1.5\ V}$$

Since the junction voltage V_j is 0.7 V for the silicon diode, the voltage drop $r_b I_F$ across the bulk resistance is

$$r_B I_F = V_F - V_j$$

$$r_B \times 2\text{ A} = 1.5 - 0.7 = 0.8\text{ V}$$

Therefore

$$r_B = \frac{0.8\text{ V}}{2\text{ A}} = \mathbf{0.4\ \Omega} \qquad (2\text{-}1b)$$

When reverse voltage is applied to the diode, this reverse voltage acts as a forward voltage for the minority current carriers only (electrons in the P-type material and holes in the N-type material). This minority current is labeled I_O in Fig. 2-7. This reverse current, I_O, is called *leakage current.* These minority current carriers arise from the breaking of covalent bonds at any temperature above absolute zero as we explained in the previous chapter. Doping impurities and surface discontinuities also cause leakage current. If the ambient temperature (the surrounding or room temperature) is increased, I_O increases. We will consider temperature effects on leakage currents in detail in Section 10-3. Normally I_O is a very small value when compared to rated forward current.

At some value of reverse voltage, BV_R, as shown in Fig.

Table 2-1
Commercial Diode Ratings

Rating	1N270 Germanium	1N1095 Silicon	1N1190 Silicon
Peak inverse voltage BV_R	100 V	500 V	600 V
Forward dc current I_F	200 mA	750 mA	35 A
Forward dc voltage drop, V_F	1.0 V	1.2 V	1.7 V
Maximum reverse current, I_O	100 μA at 50 V	5 μA at 500 V	10 mA at 600 V

2-7, the diode breaks down and reverse current becomes very large. The turning point of this curve is called the *knee.*

Ratings for typical general-purpose diodes are given in Table 2-1.

A measure of the effect of leakage current is the value of the *reverse resistance* r_R defined as

$$r_R \equiv \frac{BV_R}{I_O} \text{ ohms} \tag{2-1c}$$

Problems

2-1.1 Determine the value of r_B for each of the diodes listed in Table 2-1.

2-1.2 Determine the value of r_R for each of the diodes listed in Table 2-1.

Section 2-2 The Zener Diode

In a general-purpose diode, the P-type material and the N-type material are each lightly doped with the result that the reverse breakdown voltage (BV_R) is a high value. Also, when the general-purpose diode breaks down, the breakdown results in an avalanche current as shown in Fig. 2-7.

In a *Zener diode* the P-type material and the N-type material are heavily doped. This heavy doping results in a low value of the reverse breakdown voltage, BV_R. The exact value of reverse breakdown voltage can be very closely controlled during manufacture and this value of BV_R is now called the *Zener voltage* V_Z. The current that results in the reverse direction after breakdown is the *Zener current* I_Z.

Assume we purchase a lot of 1000 general-purpose diodes

that have ratings of 2 A for I_F, 1 V for V_F, and 100 V for BV_R. Each of these diodes must be able to carry 2 A continuously in the forward direction. They must be able to dissipate $V_F I_F$ or 2 W continuously. Each of these diodes must be able to take 100 V in the reverse direction to be acceptable. If a diode has an actual BV_R of 137 V, we do not care. It does meet the specification and it has a safety margin of 37 V.

Now let us say we purchase a lot of Zener diodes that have a rating of 2 A and 10 ± 0.1 V. Each diode must have a breakdown voltage, now called the *Zener voltage,* that lies between 9.9 V and 10.1 V. Any Zener diode that has a breakdown voltage outside these limits is rejected. Also the heat dissipation of the Zener diode is determined in the reverse direction whereas general-purpose diodes have the heat dissipation determined in the forward direction. In our example, the Zener diode must be capable of dissipating 10.1 V × 2 A or 20.2 W continuously.

The circuit symbols used for the Zener diode are shown in Fig. 2-8*a* and the typical electrical characteristic is shown in Fig. 2-8*b*. The reverse current I_O that exists between the

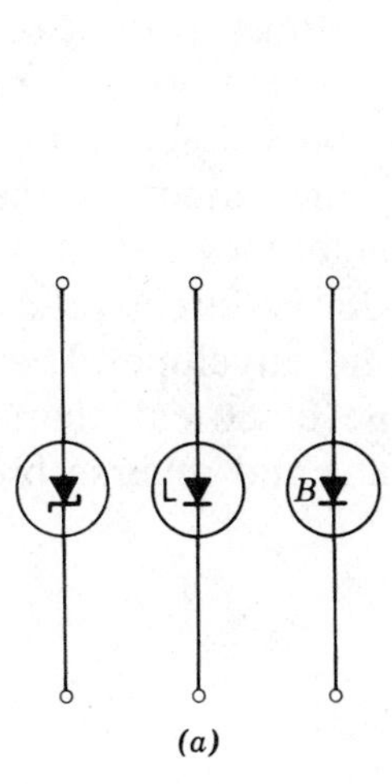

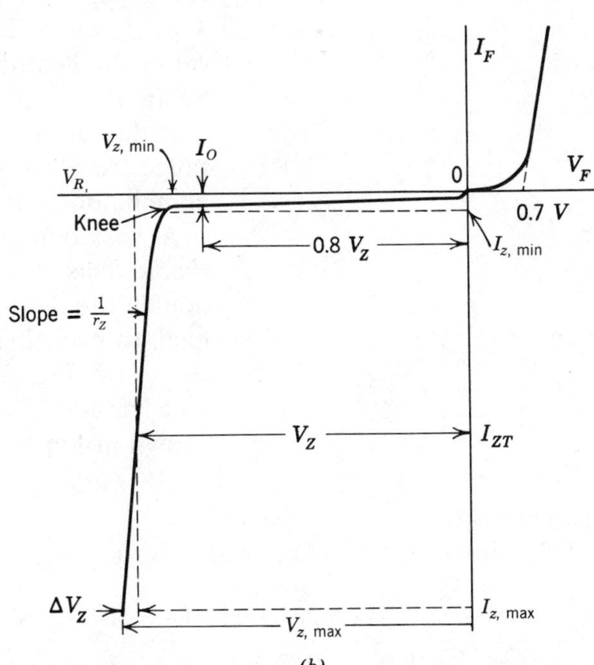

Figure 2-8 The Zener diode. (*a*) Circuit symbols. (*b*) Characteristic curve.

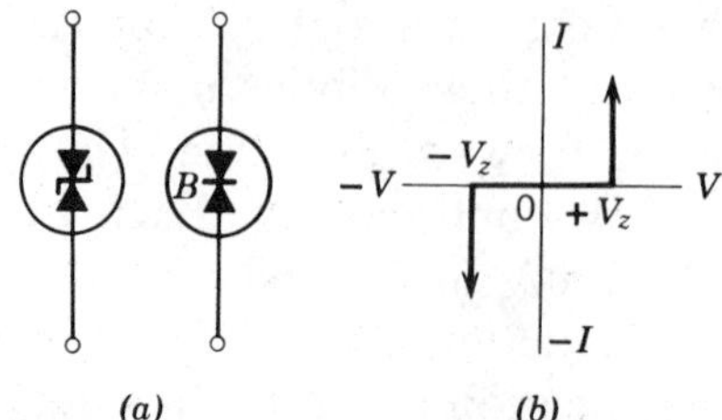

Figure 2-9 The double-breakdown diode. (*a*) Symbols. (*b*) Characteristic curve.

origin and the *knee* of the curve is the reverse leakage current of a junction caused by the minority current carriers. This current is specified by giving its value at 80% of the Zener voltage V_Z. When the reverse voltage on the diode is increased from zero, an avalanche takes place at the knee of the curve, and the current rises rapidly with only a very small change in voltage. An external resistance is required to limit this current to the maximum permissible value denoted by $I_{Z,\max}$. The least usable current in the Zener is $I_{Z,\min}$. The Zener voltage V_Z is that voltage which exists across the terminals of the diode, for a current I_{ZT} that is the approximate midpoint of its linear range. The structure of the diode and its heat sink can dissipate a maximum value of heat equivalent to $P_{Z,\max}$ or $V_Z I_{Z,\max}$ watts. The heat sink, a mechanical device used to get rid of the heat developed in a semiconductor, is considered in Section 13-2. Commercial Zener diodes are presently manufactured in power ratings from $\frac{1}{4}$ to 50 W. Voltage ratings range from 2.4 to 200 V. There are several thousand different combinations of power and voltage ratings available.

A variation of the Zener diode is the *double-breakdown diode.* This diode is sometimes called a *varistor diode.* The double-breakdown diode is effectively two Zeners placed back-to-back in the same envelope. The circuit symbols are shown in Fig. 2-9*a*. The electrical characteristics for a unit that has identical forward and reverse breakdown voltages is shown in Fig. 2-9*b*.

Example 2-2

A Zener diode has the following specifications.

$$V_Z = 20 \text{ V} \quad \text{at} \quad I_{ZT} = 10 \text{ mA}$$

$$r_Z = 8\ \Omega \qquad P_{Z,\max} = 0.5 \text{ W} \qquad I_O = 0 = I_{Z,\min}$$

Find $V_{Z,\max}$ and $V_{Z,\min}$. What is the percentage change in V_Z over this range?

Solution
The maximum allowable current, $I_{Z,\max}$, is

$$I_{Z,\max} = \frac{P_Z}{V_Z} = \frac{0.5\ \text{W}}{20\ \text{V}} = 0.025\ \text{A} = 25\ \text{mA}$$

Assume that r_Z is measured at I_{ZT}. Then

$$\begin{aligned} V_{Z,\max} &= V_Z + (I_{ZM} - I_{ZT})r_Z \\ &= 20\ \text{V} + (0.025\ \text{A} - 0.010\ \text{A}) \times 8\ \Omega = \mathbf{20.12\ V} \end{aligned}$$

and

$$\begin{aligned} V_{Z,\min} &= V_Z - (I_{ZT} - I_{Z,\min})r_Z \\ &= 20\ \text{V} - (0.010\ \text{A} - 0) \times 8\ \Omega = \mathbf{19.92\ V} \end{aligned}$$

The percentage change in V_Z is

$$\% = \frac{V_{Z,\max} - V_{Z,\min}}{V_Z} \times 100 = \frac{20.12 - 19.92}{20} \times 100 = \mathbf{1.0\%}$$

Because of this small variation in values of V_Z, Zener diodes are often referred to as *reference diodes.*

The principal application of the Zener diode is its use as a *voltage regulator.* In a voltage regulator, the voltage across the load is kept at a constant value over a range of varying load currents. In Fig. 2-10, the voltage across the load is the fixed voltage, V_Z, because the Zener diode is placed in parallel with the load. The input current is the sum of I_Z and I_L

$$I_1 = I_Z + I_L \tag{2-2}$$

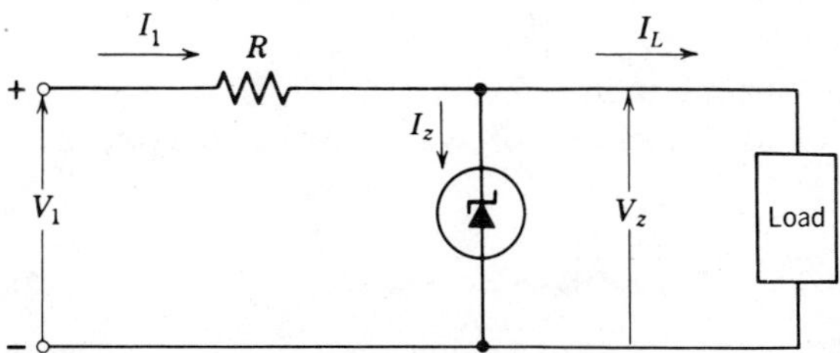

Figure 2-10 Voltage regulator circuit using a Zener diode.

and the input voltage is related to the load voltage, V_Z, by

$$V_1 = I_1 R + V_Z \tag{2-3}$$

Case I The input voltage, V_1, is fixed and the current in the load varies.

The load current can have a change of $(I_{Z,\max} - I_{Z,\min})$ without changing I_1 and without causing the load voltage, V_Z, to change. The Zener diode is used to absorb or to supply a change in load current.

Case II The load current, I_L, is fixed and the input voltage, V_1, varies.

An increase in V_1 causes I_1 to rise. The Zener diode absorbs the rise in I_1. Likewise, a decrease in V_1 causes I_1 to decrease. The Zener diode current decreases in the amount that I_1 decreases to maintain the load current at a constant value.

Example 2-3

In the circuit of Fig. 2-10, V_1 is 12.5 V and R is 50 Ω. The $\frac{1}{2}$-W Zener diode has a voltage rating of 10 V. Over what range of I_L do we have regulation?

Solution

The input current can be determined from

$$V_1 = I_1 R + V_Z \tag{2-3}$$

$$12.5\ \text{V} = I_1 \times 50\ \Omega + 10\ \text{V}$$

Solving for I_1 we have

$$I_1 = \frac{12.5 - 10}{50} = \frac{2.5\ \text{V}}{50\ \Omega} = 0.05\ \text{A} = 50\ \text{mA}$$

The maximum permissible current in the Zener diode is

$$I_{Z,\max} = \frac{P_Z}{V_Z} = \frac{0.5\ \text{W}}{10\ \text{V}} = 0.05\ \text{A} = 50\ \text{mA}$$

The least permissible current in the Zener diode, $I_{Z,\min}$, is 0 mA. Therefore, the range of load current for which the load voltage is regulated at 10 V is

from $$I_L = I_1 - I_{Z,\max} = 50 - 50 = \mathbf{0\ mA} \tag{2-2}$$

to $$I_L = I_1 - I_{Z,\min} = 50 - 0 = \mathbf{50\ mA} \tag{2-2}$$

Example 2-4
In the circuit of Fig. 2-10, R is 50 Ω and I_L is 25 mA. The $\frac{1}{2}$-W Zener diode has a voltage rating of 10 V. Over what range of V_1 do we have regulation?

Solution
The maximum allowable Zener diode current is

$$I_{Z,\text{max}} = \frac{P_Z}{V_Z} = \frac{0.5\text{ W}}{10\text{ V}} = 0.05\text{ A} = 50\text{ mA}$$

The maximum allowable line current is

$$I_{1,\text{max}} = I_L + I_{Z,\text{max}} = 25 + 50 = 75\text{ mA} \qquad (2\text{-}2)$$

and the minimum allowable line current is

$$I_{1,\text{min}} = I_L + I_{Z,\text{min}} = 25 + 0 = 25\text{ mA} \qquad (2\text{-}2)$$

The maximum value of V_1 maintaining regulation is

$$V_{1,\text{max}} = I_{1,\text{max}}R + V_Z = 0.075\text{ A} \times 50\ \Omega + 10\text{ V} = \mathbf{13.75\text{ V}} \qquad (2\text{-}3)$$

and the least value of V_1 maintaining regulation is

$$V_{1,\text{min}} = I_{1,\text{min}}R + V_Z = 0.025\text{ A} \times 50\ \Omega + 10\text{ V} = \mathbf{11.25\text{ V}} \qquad (2\text{-}3)$$

Example 2-5

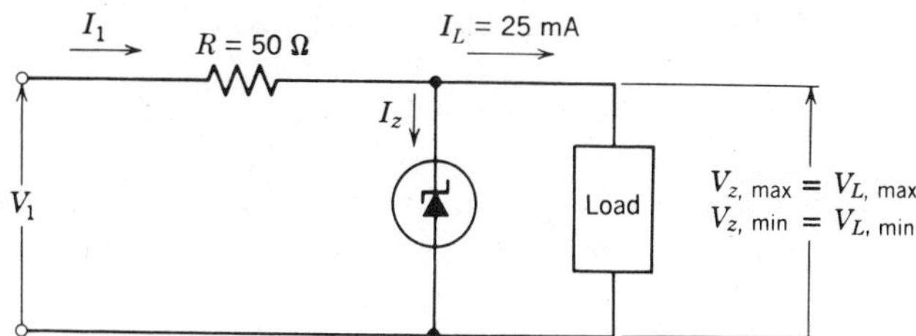

Circuit diagram for Example 2-5.

The Zener diode has the following rated values.

$$V_{ZT} = 10.1\text{ V} \quad \text{at} \quad I_{ZT} = 25\text{ mA}$$

$$I_{Z,\text{max}} = 50\text{ mA} \qquad r_Z = 4\ \Omega$$

I_L is a constant load current of 25 mA. What is the change in load voltage over the range of regulation? Over what values of input voltage, V_1, does regulation occur? What is the change in input voltage? Assume $I_{Z,\text{min}}$ is zero.

Solution

The maximum load voltage occurs when the current through the Zener diode is the maximum value

$$V_{L,\text{max}} = V_{Z,\text{max}} = V_{ZT} + (I_{Z,\text{max}} - I_{ZT})r_Z$$
$$= 10.1\text{ V} + (0.050\text{ A} - 0.025\text{ A}) \times 4\ \Omega = \mathbf{10.2\ V}$$

and the minimum load voltage occurs when the current through the Zener diode is the least value (in this example, zero).

$$V_{L,\text{min}} = V_{Z,\text{min}} = V_{ZT} - (I_{ZT} - I_{Z,\text{min}})r_Z$$
$$= 10.1\text{ V} - (0.025\text{ A} - 0) \times 4\ \Omega = \mathbf{10.0\ V}$$

When the maximum value of load voltage (and Zener diode voltage) occurs, the input current, I_1, is the maximum value

$$I_{1,\text{max}} = I_L + I_{Z,\text{max}} = 25 + 50 = 75\text{ mA} \qquad (2\text{-}2)$$

and the input voltage is the maximum value.

$$V_{1,\text{max}} = I_{1,\text{max}}R + V_{Z,\text{max}} \qquad (2\text{-}3)$$
$$= 0.075\text{ A} \times 50\ \Omega + 10.2\text{ V} = \mathbf{13.95\ V}$$

When the minimum value of load voltage (and Zener diode voltage) occurs, the input current, I_1, is the minimum value

$$I_{1,\text{min}} = I_L + I_{Z,\text{min}} = 25 + 0 = 25\text{ mA} \qquad (2\text{-}2)$$

and the input voltage is the minimum value.

$$V_{1,\text{min}} = I_{1,\text{min}}R + V_{Z,\text{min}} \qquad (2\text{-}3)$$
$$= 0.025\text{ A} \times 50\ \Omega + 10.0\text{ V} = \mathbf{11.25\ V}$$

The change in the load voltage is

$$\Delta V_L = V_{L,\text{max}} - V_{L,\text{min}} = 10.2 - 10.0 = \mathbf{0.2\ V}$$

That corresponds to a change in the input voltage of

$$\Delta V_1 = V_{L,\text{max}} - V_{L,\text{min}} = 13.95 - 11.25 = \mathbf{2.70\ V}$$

This example shows that the use of a Zener can, in this case, reduce a voltage variation from 2.70 V to 0.2 V, which is an improvement of a factor of 2.70/0.2 or of 13.5.

It should be noted that, if the input voltage exceeds 13.95 V, the Zener diode current will become excessive and cause the diode to become overheated. If the input voltage falls below 11.25 V, there is no longer a regulation and the load voltage will fall proportionally to the decrease in V_1.

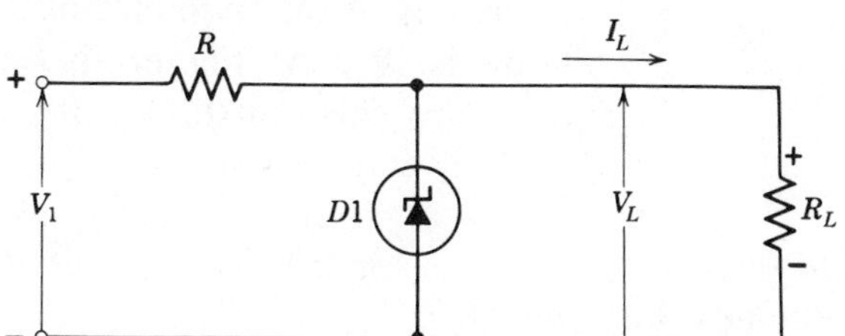

Circuit for Problems 2-2.1 through 2-2.8.

Problems Assume r_Z is zero for Problems 2-2.1 through 2-2.6.

2-2.1 V_1 is 40 V and R is 50 Ω. The Zener diode is rated at 3 W and has a value for V_Z of 20 V. Find the range over which R_L can be varied with a constant load voltage.

2-2.2 R is 4 kΩ, R_L is 10 kΩ and V_Z is 30 V. The input voltage V_1 varies between 70 V and 100 V. What are the maximum and minimum currents through the Zener diode?

2-2.3 R is 3 Ω, R_L is 10 Ω and V_Z is 10 V. The maximum power dissipation of the Zener diode is 20 W. What is the allowable range of V_1?

2-2.4 It is desired to regulate a 15-V load using a Zener diode. The source voltage is 20 V. The load current can range from 0 to 100 mA. What is the power rating for the required Zener diode and what is the value of R?

2-2.5 The maximum variation of V_1 is from 12.0 V to 14.8 V. A circuit using 15 mA must be maintained at 8.2 V. What is the value of R and what is the rating of the Zener diode that should be used?

2-2.6 A 4-W, 5-V Zener diode is placed in parallel with a 3-Ω load. R is 2 Ω. Over what range of V_1 is there voltage regulation?

2-2.7 The value of r_Z is 1 Ω for the Zener diode in Problem 2-2.1. V_Z is measured at $I_{Z,\max}$. What is the variation in load voltage caused by r_Z?

2-2.8 The value of r_Z is 8 Ω for the Zener diode in Problem 2-2.2. V_Z is measured at $I_{Z,\max}$. What is the variation in load voltage caused by r_Z?

Section 2-3 Wave-Shaping Circuits

General-purpose diodes and Zener diodes are often used to modify an incoming waveform. The objective of the discussion in this section is to show how they accomplish waveform modification. In determining the output voltage waveform, we assume that a general-purpose diode acts as a short circuit in the forward direction and acts as an open circuit in the reverse direction. A Zener diode acts as a short circuit in the forward direction. In the reverse direction, the Zener diode acts as an open circuit until the voltage across the Zener diode is V_Z. At higher incoming voltages, the Zener diode maintains this constant value of V_Z.

Example 2-6
The voltage waveform given in Fig. 2-11*a* is applied to the circuit of Fig. 2-11*b*. Find the output voltage waveform.

Solution
In the forward direction, the diode acts as a short circuit and then R_A and R_B serve as a simple voltage divider. The circuit values show that the output voltage has exactly half the amplitude of the input. When the input signal becomes negative, the input signal is a reverse voltage on the diode. The diode cannot conduct and acts as an open circuit. The current in R_B is zero and the voltage drop across R_B is zero. Therefore, the output voltage is zero for the full time that the input voltage is negative.

The output waveform is shown in Fig. 2-11*c*.

Example 2-7
The voltage waveform given in Fig. 2-11*a* is applied to the circuit of Fig. 2-11*b*. Now, however, the diode is a Zener diode with a value of 6 V for V_Z. Find the output voltage waveform.

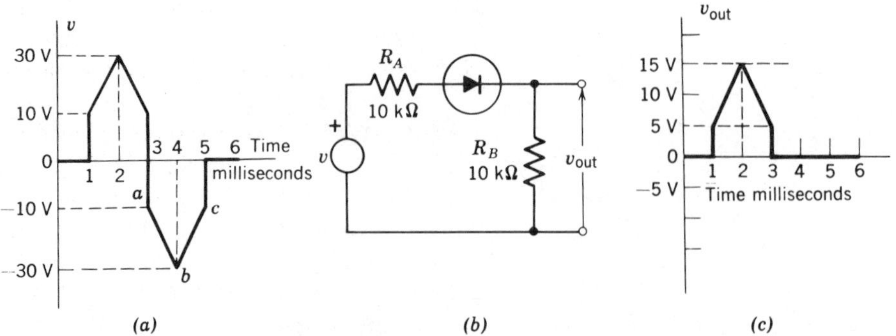

Figure 2-11 Irregular waveform applied to a diode network. (*a*) Input voltage waveform. (*b*) The circuit diagram. (*c*) The output voltage waveform.

Solution

In the forward direction, the analysis is the same as discussed for Example 2-6.

In the reverse direction the Zener diode acts as an open circuit until the incoming voltage reaches the Zener voltage, 6 V. At the 3-ms time on the incoming waveform, the reverse voltage rises from 0 to 10 V, point *a* on Fig. 2-11*a*. Of this 10 V, 6 V is required to break down the Zener diode and 4 V is left for the voltage drops across R_A and R_B. Since R_A and R_B are equal, the voltage drop across R_B is -2 V shown as point *d* on Fig. 2-12.

At the peak negative value of the incoming waveform, point *b* on Fig. 2-11, the reverse voltage applied to the circuit is -30 V. Since the Zener breakdown voltage is 6 V, the remaining -24 V is left for the voltage drops across R_A and R_B. Since R_A and R_B are equal, the output voltage is -12 V, point *e* on Fig. 2-12.

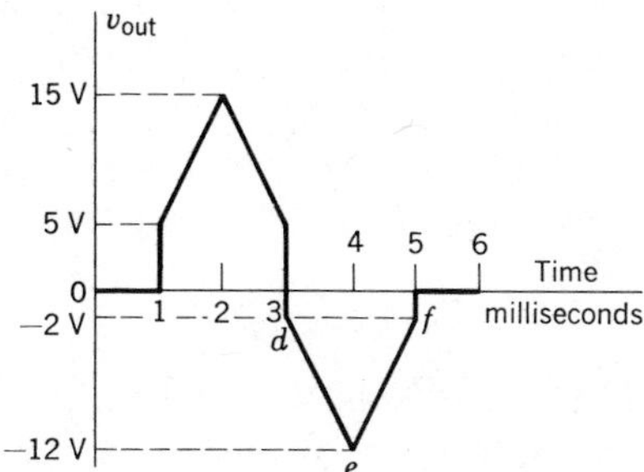

Figure 2-12 Output voltage waveform for Example 2-7.

Problems

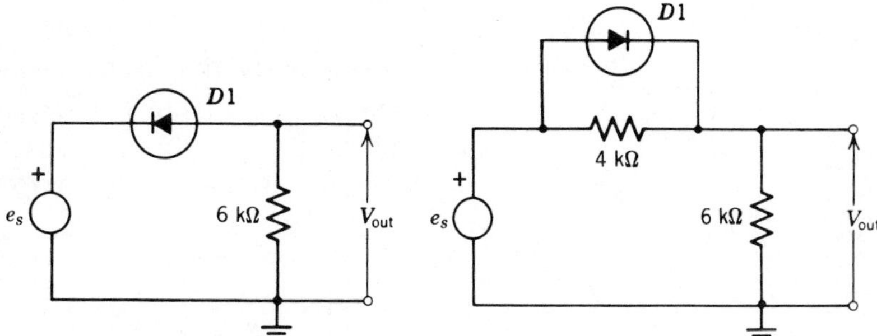

Circuit for Problems 2-3.1 and 2-3.2.

Circuit for Problems 2-3.3 and 2-3.4.

The input voltage e_s is a triangular voltage that has a peak-to-peak value of 60 V (± 30 V). For each problem, determine the output voltage waveform. Assume that the diodes are ideal. Note that Problems 2-3.7 and 2-3.8 require two output voltage waveforms.

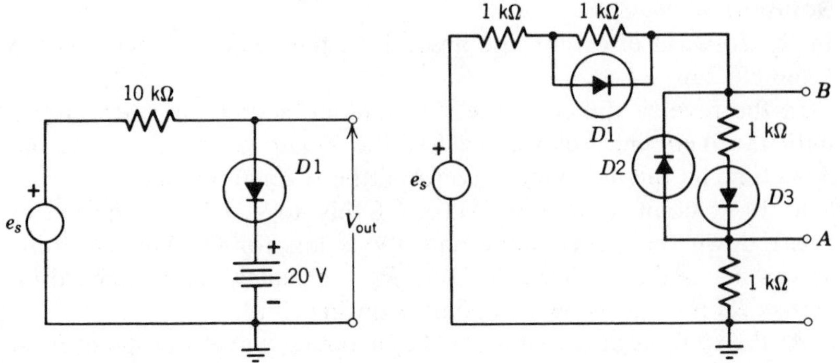

Circuit for Problems 2-3.5 and 2-3.6. Circuit for Problems 2-3.7 and 2-3.8.

2-3.1 Diode $D1$ is a general-purpose diode.
2-3.2 Diode $D1$ is a 10-V Zener diode.
2-3.3 Diode $D1$ is a general-purpose diode.
2-3.4 Diode $D1$ is a 6-V Zener diode.
2-3.5 Diode $D1$ is a general-purpose diode.
2-3.6 Diode $D1$ is a 6-V Zener diode.
2-3.7 Diodes $D1$, $D2$, and $D3$ are general-purpose diodes.
2-3.8 Diodes $D1$, $D2$, and $D3$ are 6-V Zener diodes.

Section 2-4 V-I Characteristics

The circuit shown in Fig. 2-13 can be used to show the V-I characteristic of either a device or a particular circuit. By using a sinusoidal source v we can obtain a range of both negative and positive voltages applied to the device or circuit under test. The voltage connected to the vertical amplifier of an oscilloscope is a voltage equal to iR by Ohm's law. Consequently the vertical-amplifier signal is directly proportional to the current i. Then the oscilloscope deflection can be calibrated directly in terms of current. The horizontal deflection is the voltage across the device or circuit directly.

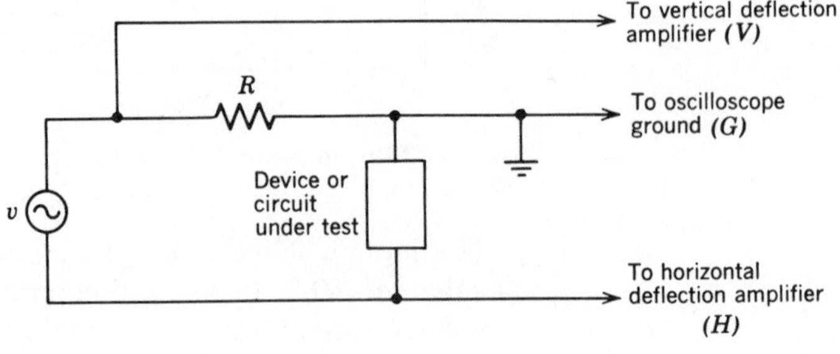

Figure 2-13 Circuit used to show V–I characteristics.

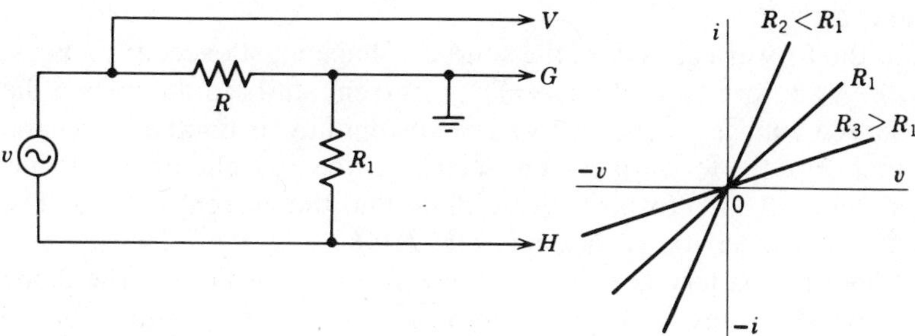

Figure 2-14 The *V–I* characteristic of a resistor.

The volt–ampere characteristic of a resistor R_1 is shown in Fig. 2-14. If the resistance value of R_1 changes, the slope of the line on the oscilloscope will change. When R_1 decreases, the current through R_1 increases and the slope increases (becomes steeper). When R_1 increases, the current through R_1 decreases and the slope decreases (becomes less steep).

Example 2-8

The circuit shown in Fig. 2-15*a* uses a silicon diode. Determine the *V-I* characteristic for the circuit. Determine the *V-I* characteristic if the 2-V battery is reversed.

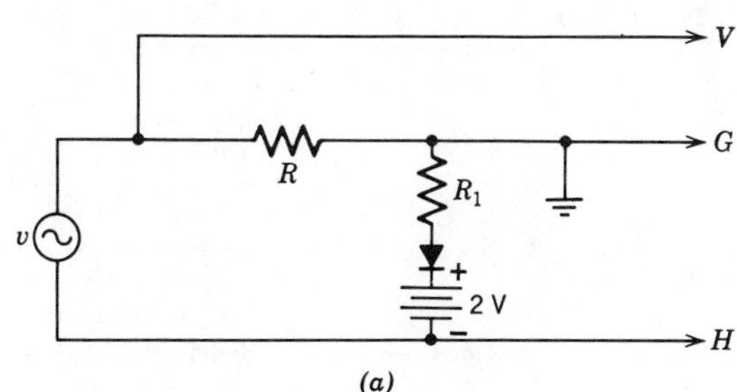

Figure 2-15 Circuit used to show the *V–I* characteristic of a reverse-biased diode. (*a*) Circuit. (*b*) *V–I* characteristic. (*c*) *V–I* characteristic obtained when the 2-V source is reversed. (*d*) Superimposed characteristics.

Solution

1. In the forward direction, the source voltage must exceed 2 V by V_j (0.7 V) before current flows. Thus, current starts to flow when the source voltage reaches 2.7 V. The current flow is limited by R_1 and therefore the forward current section of the V-I characteristic has a slope of $1/R_1$. In the reverse direction, the current is zero. The V-I characteristic is shown in **Fig. 2-15*b*.**
2. When the battery is reversed, there is a current flow in the diode when the source voltage is zero. Therefore, the source voltage must be $(2.0 - 0.7)$ or 1.3 V in the reverse direction to reduce the diode current to zero. The V-I characteristic is shown in **Fig. 2-15*c*.**
3. If both V-I characteristics are drawn together on one set of axes, Fig. 2-15*d*, the separation between the two curves is twice the battery voltage or 4 V.

Problems

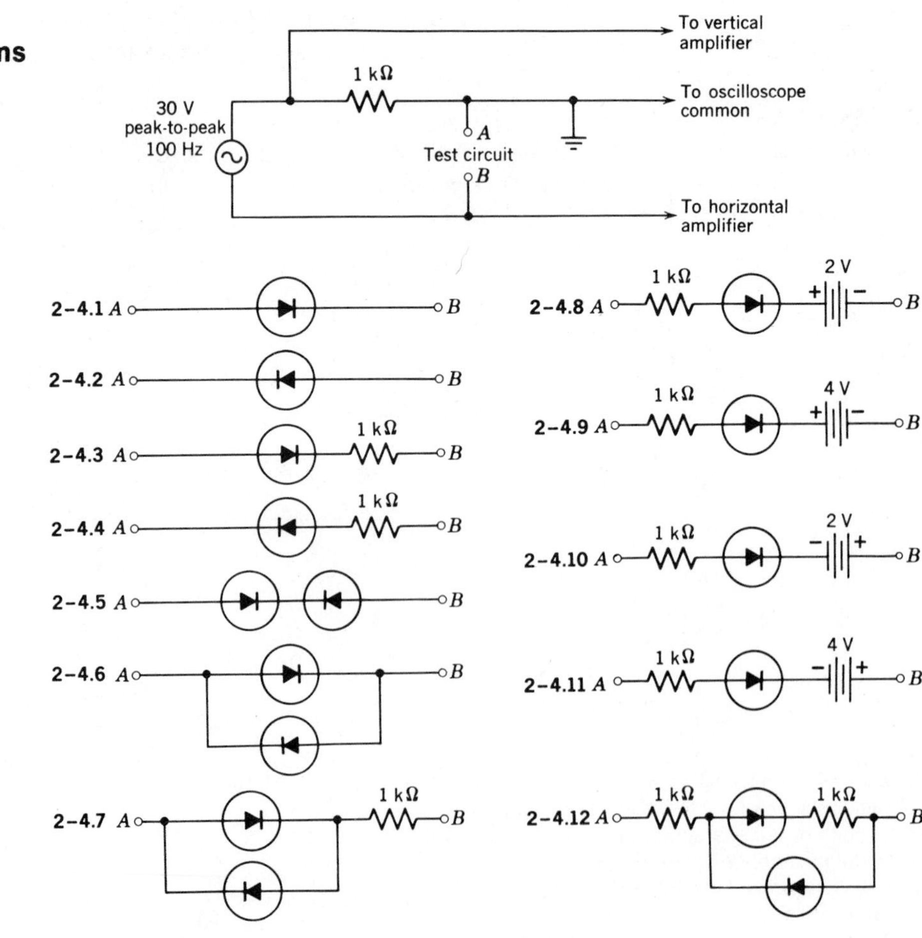

Problems 2-4.1 through 2-4.12 Assume all diodes have 0.7 V for V_j and 100 V for BV_R. For each of the 12 test circuits sketch and label the V-I characteristic.

Problems 2-4.13 through 2-4.24 Assume that all diodes have 0.7 V for V_j and that all diodes are Zener diodes with a breakdown voltage (V_Z) of 6.0 V. For each of the 12 test circuits sketch and label the V-I characteristic.

Section 2-5 The AC Model

In Section 2-1 we stated that we would defer the discussion of that part of the characteristic of a diode marked as "curve" in Fig. 2-7 to this section. This region is redrawn in Fig. 2-16. We define the *ac resistance* r_j of the diode at point G as

$$r_j \equiv \frac{\Delta V}{\Delta I} \text{ ohms} \tag{2-4}$$

This resistance value is the reciprocal of the slope of the curve at point G. As we proceed from point 0 to point H through point F and point G, we see that the curve becomes steeper. Therefore the slope increases and results in a decreasing value for r_j as we proceed from point 0 toward point H. The least value of the ac resistance is the bulk resistance value r_B at point H.

We find from the theory of conduction in metals that the value of r_j lies between the two values:

$$\boxed{\frac{25 \text{ mV}}{I_F} \leq r_j \leq \frac{50 \text{ mV}}{I_F}} \tag{2-5}$$

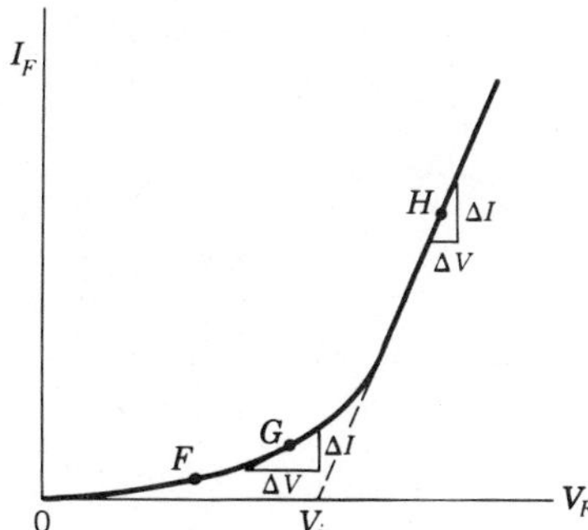

Figure 2-16 The expanded forward diode characteristic.

in which I_F is the *dc value* of the current at the point at which r_j is being evaluated.

In the problems in this text, a specific value will be given for the number of millivolts to be used for Eq. 2-5, such as 25 mV, 35 mV, or 50 mV. In checking laboratory results, we often find that closer results can be obtained if an attempt is made to determine the specific number that gives the best check for the particular diode used.

The circuit shown in Fig. 2-17*a* is used to control the output signal level v_{out} by changing the dc voltage $+V$ applied to the circuit. The capacitors C_1 and C_2 are used to keep dc currents out of the signal source and out of any circuit or circuit component connected to the output. Consequently, the term *blocking capacitor* or *coupling capacitor* is applied to this use of a capacitor. The first step in analyzing the circuit is to construct the *ac model*, Fig. 2-17*b*. In the ac model we retain only those circuit components that contribute to the ac signal circuit. The dc values are used to make the circuit work; they are not a necessary part of the ac model. Each component must be carefully examined in turn.

The supply voltage ($+V$) is an ideal dc source. By this, we mean that the voltage is a fixed value even though the dc current can vary. Since there is no change in voltage when there is a change in current, the internal resistance of the source must be zero. Therefore:

In all ac models, the dc supply voltages are shorted to the ground return.

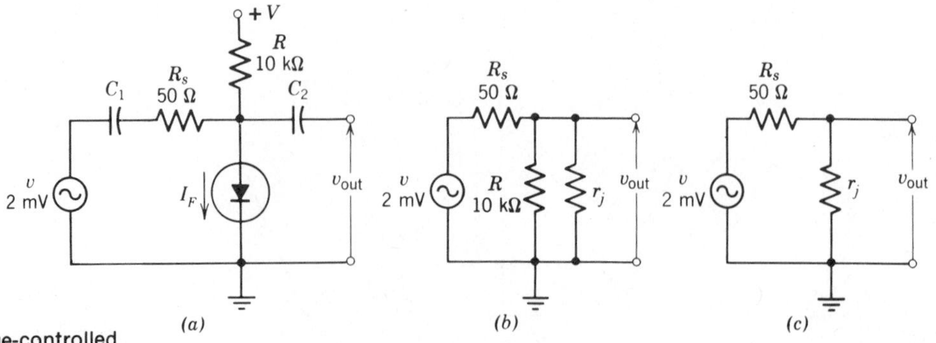

Figure 2-17 The voltage-controlled attenuator. (*a*) Circuit. (*b*) Model. (*c*) Simplified model.

Accordingly, in Fig. 2-17*b*, one end of the 10,000-Ω resistor is drawn to the ground return. Now this resistor is in parallel with r_j, the ac resistance value of the diode.

If at all times the value of r_j is much less than $\frac{1}{10}R$, we can simplify the ac model by neglecting R to obtain the model shown in Fig. 2-17*c*.

Example 2-9

Determine the ac signal output of the circuit of Fig. 2-17*a* when the dc supply V is +50 V and when the dc supply is +2 V. The diode is germanium and r_j is 25 mV/I_F.

Solution

The Kirchhoff's voltage loop equation through the diode is

$$V = RI_F + V_j$$

For the germanium diode, we use 0.3 V for V_j. For a 50-V supply voltage

$$50\text{ V} = 10{,}000\ \Omega \times I_F + 0.3\text{ V}$$

$$I_F \approx 0.005\text{ A} = 5\text{ mA}$$

For a 2-V supply

$$2\text{ V} = 10{,}000\ \Omega \times I_F + 0.3\text{ V}$$

$$I_F \approx 0.00017\text{ A} = 0.17\text{ mA}$$

When I_F is 5 mA,

$$r_j = \frac{25\text{ mV}}{I_F} = \frac{25\text{ mV}}{5\text{ mA}} = 5\ \Omega \tag{2-5}$$

When I_F is 0.17 mA,

$$r_j = \frac{25\text{ mV}}{I_F} = \frac{25\text{ mV}}{0.17\text{ mA}} = 147\ \Omega \tag{2-5}$$

The simplified model, Fig. 2-17*c*, is a voltage divider. For the 50-V supply the output signal level v_{out} is

$$v_{\text{out}} = \frac{r_j}{R_s + r_j}\, v = \frac{5\ \Omega}{50\ \Omega + 5\ \Omega}\, 2\text{ mV} = \mathbf{0.18\ mV}$$

and for the 2-V supply

$$v_{\text{out}} = \frac{r_j}{R_s + r_j} v = \frac{147\ \Omega}{50\ \Omega + 147\ \Omega} 2\ \text{mV} = \mathbf{1.49\ mV}$$

From this example we see that we can control the output voltage over a ratio of 1.49/0.18 or 8.3 to 1 by varying the dc supply voltage to the circuit. The varying dc control voltage usually comes from a source that is remote to the point at which the signal level is being controlled.

Consider two diodes *D*1 and *D*2 in series, Fig. 2-18. Diode *D*1 is a germanium diode and diode *D*2 is a silicon diode. When current flows through this circuit, the voltage drop across *A* and *B* is the sum of 0.3 V and 0.7 V or 1.0 V. Now these two diodes are placed in parallel. The forward voltage drop across *D*1 is 0.3 V. The voltage drop from *C* to *D* must be 0.3 V at all times. Current cannot flow in *D*2 unless the voltage from *C* to *D* reaches 0.7 V, but the voltage drop across *D*1 is only 0.3 V. Therefore the circuit current must flow through the germanium diode *D*1 alone.

The technique we have illustrated in Example 2-9 is the technique that will be used in analyzing amplifier circuits throughout the text:

1. A circuit (e.g., Fig. 2-17*a*) is given showing the various dc supply values and the various resistance values.
2. The *first* step in solving a problem is to solve the original circuit for the dc values of the currents in the circuit.
3. From these dc current values, we find the ac resistance values of the semiconductors by means of Eq. 2-5 (or its equivalent).
4. Having the ac resistance values, we form the *ac circuit model* (e.g., Fig. 2-17*b*).

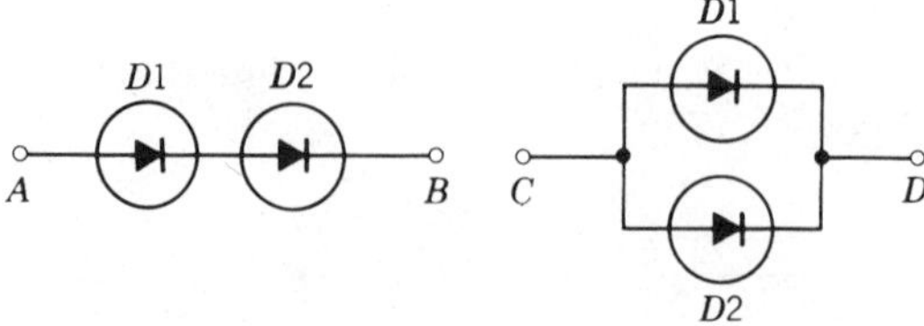

Figure 2-18 Diodes in series and in parallel.

5. An analysis of the ac circuit model yields the ac signal levels in the model.
6. These ac signal levels are the ac signal levels that we do find in the original circuit.

This technique must be thoroughly mastered.

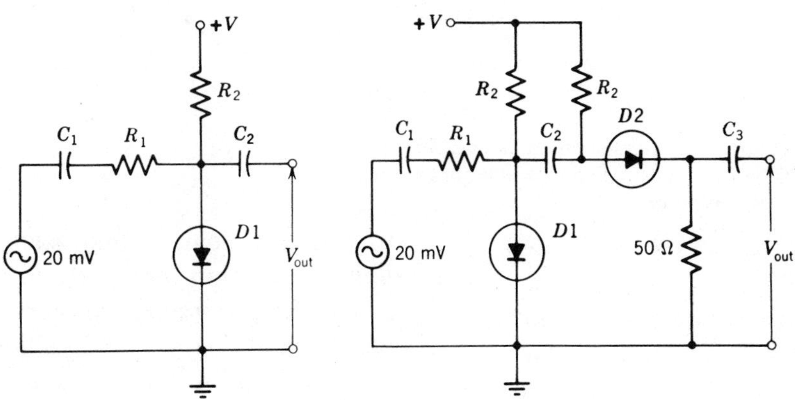

Circuit for Problems 2-5.1 to 2-5.3.   Circuit for Problems 2-5.4 to 2-5.6.

Problems Use 0.7 V for V_j and 50 mV/I_F for r_j.

2-5.1 R_1 is 50 Ω, R_2 is 20 kΩ and V is 30 V. Determine V_{out}.

2-5.2 R_1 is 5 kΩ, R_2 is 5 kΩ and V is 2 V. Determine V_{out}.

2-5.3 R_1 is 50 Ω, R_2 is 50 Ω and V is 2 V. Determine V_{out}.

2-5.4 Determine V_{out} using the component values given in Problem 2-5.1.

2-5.5 Determine V_{out} using the component values given in Problem 2-5.2.

2-5.6 Determine V_{out} using the component values given in Problem 2-5.3.

Section 2-6 Light and Diodes

When electrons recombine with holes across a P-N junction, the actual movement of the electrons in relation to the conduction band and valence band shows that there are many possible paths for the transition of the electron. When the electron falls from a high to a lower energy level, usually the energy appears as heat, although some energies are converted into light photons. This optical property has been developed

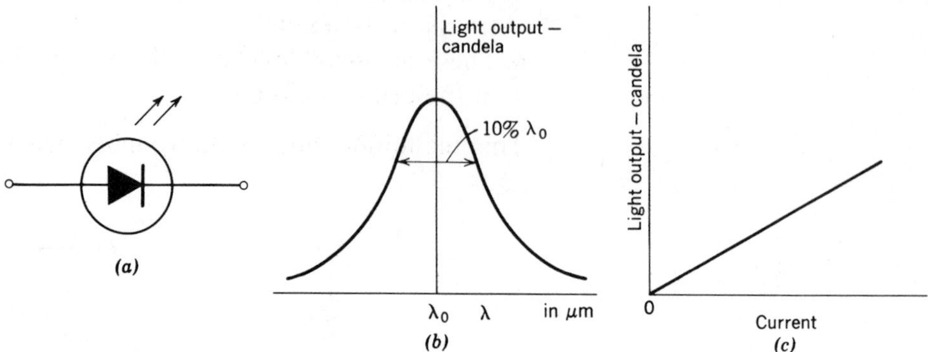

Figure 2-19 Light-emitting diodes. (*a*) Circuit symbol. (*b*) Light wavelength. (*c*) Light output.

to produce devices that have a relatively high efficiency of light conversion (presently to about 20%).

The first commercial version of the *light-emitting diode* (*LED*), Fig. 2-19*a*, used gallium arsenide with an energy gap of 1.37 eV. This diode produces a light having a wavelength of 9100 Å, which is dark red. A gallium phosphide diode has an energy gap of 2.25 eV that corresponds to 5600 Å, a green light. A number of other materials have been developed to produce other colors.

Since the recombination paths are all not exactly the same, the light is not completely monochromatic but shows a narrow band characteristic, Fig. 2-19*b*. A major advantage of the device, however, is that the light output is very nearly linear with diode current, Fig. 2-19*c*. Covalent bonds broken by an increase in ambient temperature materially reduce the light output.

The LED is operated in a forward-biased direction. There is no light output when the diode is operated in the reverse direction. The LED has a very low value of reverse voltage breakdown. The LED has a relatively high forward voltage drop (up to 2 V for V_j) in comparison to other diodes.

The advantage of the LED is that it can produce light output with very low input powers. The power taken by a conventional incandescent panel lamp is high; for example, 6.3 V at 150 mA is 945 mW. A test on a typical LED shows that a strong light is produced with a power of 30.4 mW (20 mA at 1.52 V). There is still visible light from this LED when the current is reduced to 2 mA.

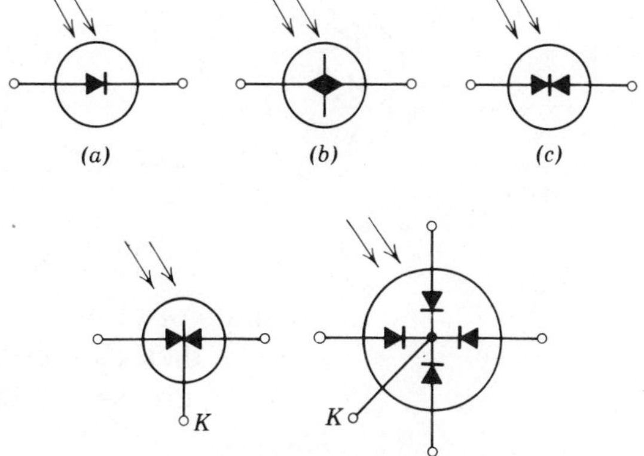

Figure 2-20 Photodiodes. (*a*) Photodiode. (*b*) NPN photoduo-diode. (*c*) PNP photoduo-diode. (*d*) PNP two-segment photodiode with cathode. (*e*) PNP four-quad photodiode with common cathode.

The converse effect is used in the *photodiode*, Fig. 2-20. The diode is reverse biased and incident light breaks covalent bonds to provide current carriers that can cross the junction barrier. The current that flows in the circuit is proportional to the incident light. There are a number of variations on this device, shown in Fig. 2-20, that are used for special applications.

Section 2-7 The Varactor Diode

In Section 2-1 we showed that, in the depletion region, there are ions or uncovered charges. Additionally, there are no free current carriers within the depletion region. A barrier voltage, V_j, appears across the depletion region because of the uncovered charges within the depletion region. The study of dc and ac circuit analysis shows that a capacitance is defined by

$$C \equiv \frac{Q}{V}$$

Thus there must be a value of *junction capacitance* C_j associated with the depletion region.

The equation for capacitance in terms of the circuit geometry is

$$C = \frac{K\kappa A}{s}\ \text{pF} \tag{2-6}$$

in which K is a constant required to evaluate C in terms of

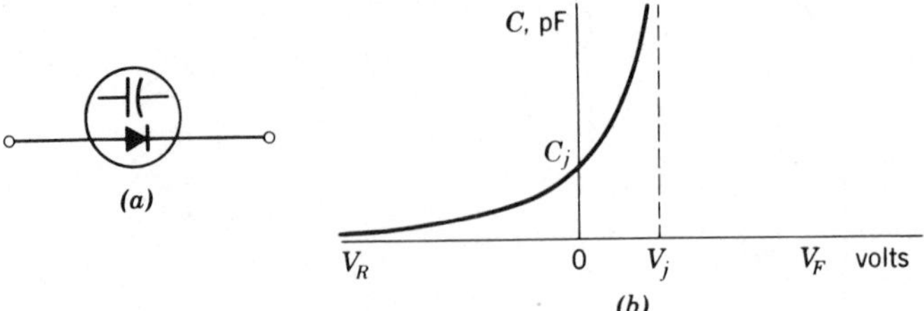

Figure 2-21 The varactor diode. (*a*) Circuit symbol. (*b*) Characteristic.

the units of A and s
κ is the dielectric constant of the depletion region
A is the cross-sectional area of the depletion region
and
s is the thickness of the depletion region

The capacitance of the diode with zero voltage across the diode is the *junction capacitance* C_j (Fig. 2-21*b*). When a reverse voltage is placed across the diode, the thickness of the depletion region increases. Consequently, s in Eq. 2-6 becomes larger and the equivalent capacitance of the diode decreases. When a forward bias is applied to the diode, the thickness of the depletion region becomes smaller and the equivalent capacitance of the diode increases. When the forward voltage causes current flow, the diode loses its capacitance properties.

Diodes designed for this purpose are called *variable-voltage capacitors* or *varactor diodes.* The circuit symbol is given in Fig. 2-21*a*. The varactor diodes are widely used in place of mechanically variable tuning capacitors to provide *solid-state tuning.* The capacitance value is controlled by the amount of the dc voltage applied in reverse to the diode.

Supplementary Problems

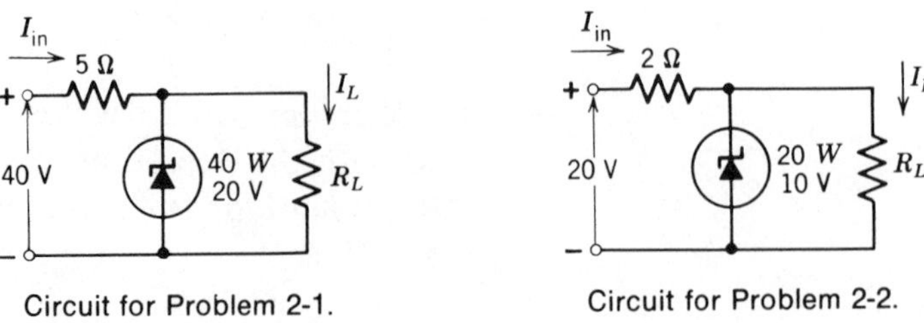

Circuit for Problem 2-1.

Circuit for Problem 2-2.

2-1 Over what range of load current do we have regulation?
2-2 Over what range of load current do we have regulation?

2-3 The ratings of a Zener diode are

$$V_Z = 6.2 \text{ V at } I_{ZT} = 20 \text{ mA and } P_{Z,\max} = 200 \text{ mW}$$

If the value for r_Z is 3 Ω, what is the maximum variation of V_Z over the allowable current range?

2-4 A cheaper Zener diode can be substituted for the Zener diode used for Problem 2-3 but the value for r_Z for this substitution is 15 Ω. Now what is the expected variation of V_Z over the allowable current range?

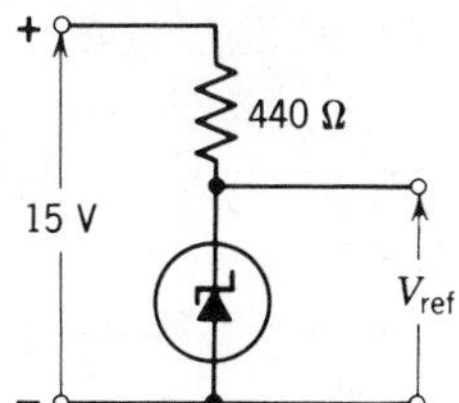

Circuit for Problems 2-5 through 2-7.

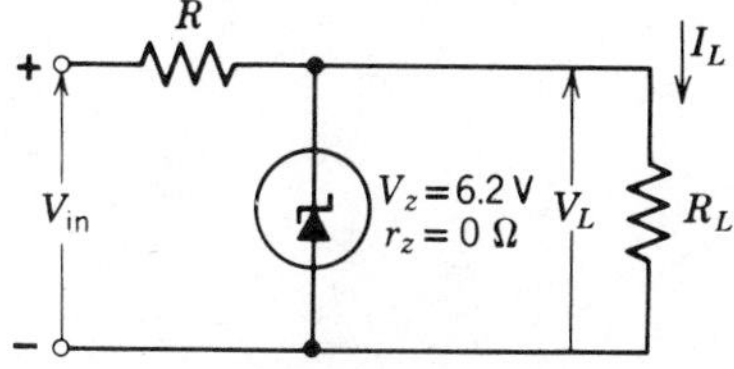

Circuit for Problems 2-8 and 2-9.

2-5 V_{ref} must be held to within ±0.2 of 1% of 6.2 V. When the Zener diode specified in Problem 2-3 is used, what is the maximum allowable variation in the 15-V source both in voltage and in percent?

2-6 Repeat Problem 2-5 if the Zener diode specified in Problem 2-4 were used.

2-7 If the 15-V source is replaced by an 80-V peak-to-peak square wave, what is the waveform across the Zener diode? Is the Zener diode being overheated? Assume V_F is 0.7 V.

2-8 R_L and R are each 100 Ω. What is the range of V_{in} over which the load voltage is maintained at 6.2 V? P_Z is 0.5 W.

2-9 V_{in} is 10 V and R is 25 Ω. What is the range of I_L and R_L over which the load voltage is maintained at 6.2 V? What is $P_{Z,\max}$?

2-10 Determine the output voltage waveform.

2-11 Determine the output voltage waveform.

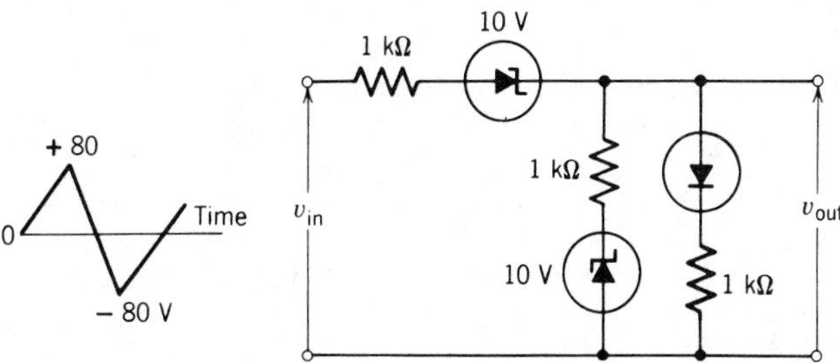

Input waveform and circuit for Problem 2-10.

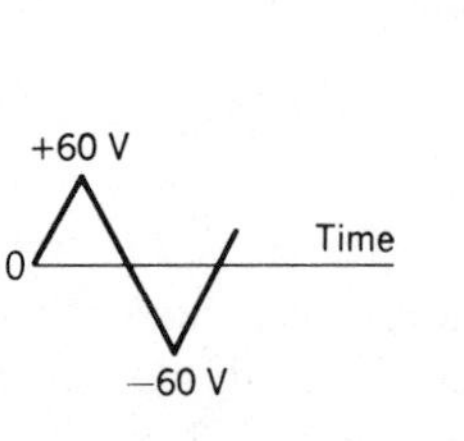

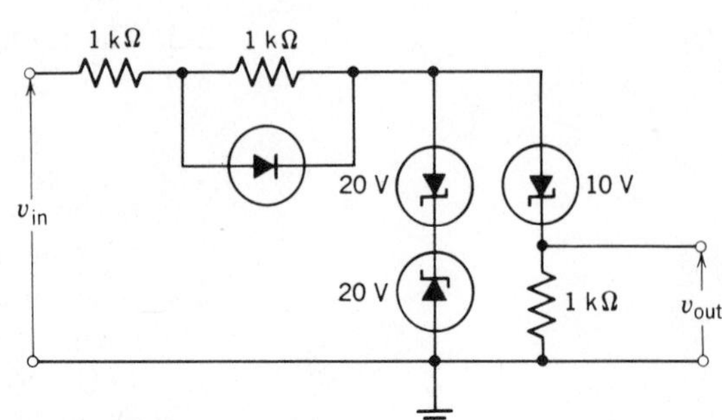

Input waveform and circuit for Problem 2-11.

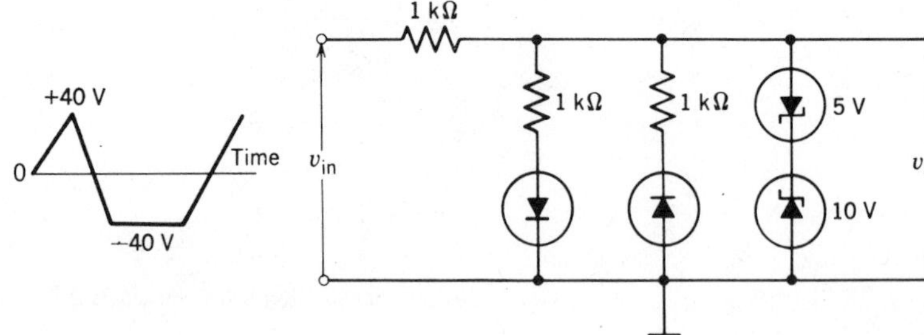

Input waveform and circuit for Problem 2-12.

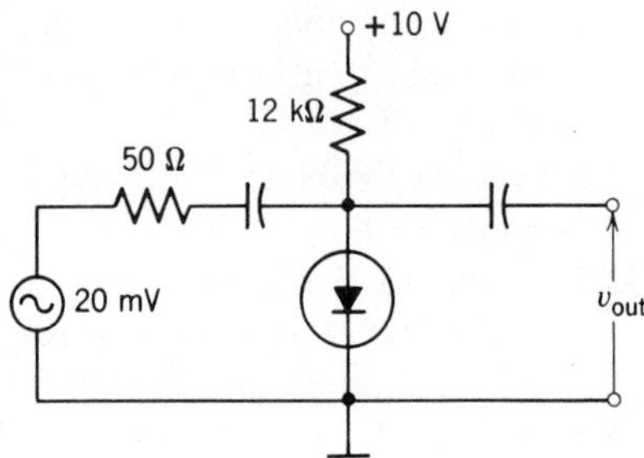

Circuit for Problem 2-13.

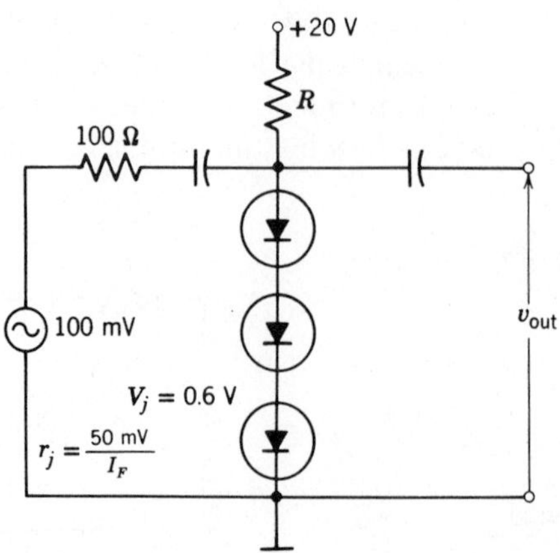

Circuit for Problems 2-14 and 2-15.

2-12 Determine the output voltage waveform.
2-13 If r_j is 50 mV/I_F and V_j is 0.7 V, determine V_{out}.
2-14 R is 510 Ω. Determine V_{out}.
2-15 R is 10 kΩ. Determine V_{out}.
2-16 The maximum ratings for a silicon diode ($V_j = 0.7$ V) are

$$V_F = 1.2\text{ V} \quad \text{and} \quad I_F = 4\text{ A}$$

Determine r_B.

Chapter 3 Rectifiers

The diode used as a rectifier converts energy from an ac source into the dc energy that is required for the operation of electronic circuits. The usual rectifier circuits are the half-wave rectifier (Section 3-1), the full-wave rectifier (Section 3-2), and the bridge rectifier (Section 3-3). Most rectifier circuits rely on a capacitor filter to smooth out the rectified waveform (Section 3-4) but some high-power applications use more complex filtering arrangements (Section 3-5). A voltage multiplier (Section 3-6) can be used to obtain a high dc output voltage. The shunt rectifier (Section 3-7), often called a clamper, is widely used as a wave-shaping circuit.

Section 3-1 The Half-Wave Rectifier

An ideal general-purpose diode and a load resistor R_L are connected in series to an ac power source (Fig. 3-1*a*). This circuit is a *half-wave rectifier.* The sinusoidal waveform of the source v has the peak value V_m volts and the effective or rms value V volts. When m is positive with respect to n, the P-type material (the *anode*) of the diode is positive with respect to the N-type material (the *cathode*). This is forward bias, and current flows through the entire series circuit. The current develops a voltage drop across the load resistor with the polarity shown on the circuit diagram. When n is positive and m is negative, the N-type material (the cathode) of the diode is positive with respect to the P-type material (the anode). This is a reverse-bias condition, and no current can flow in the circuit. Now the full voltage appears across the diode. Thus, the only voltage that can exist across the load resistor exists when m is positive. The diode is the *rectifier.* A rectifier is a device or circuit arrangement that makes one-directional (unidirectional) current or voltage from an ac source.

Since the half-wave rectifier is a series circuit, the diode current and the load current are the same (Fig. 3-1*c*). The waveform of the load voltage v_L is the positive half of the sinusoidal waveform of the source voltage, Fig. 3-1*d*. These two waveforms are related by Ohm's law at each instantaneous time.

$$v_L = R_L i_L \tag{3-1}$$

By Kirchhoff's voltage law, the voltage drop across the load plus the voltage drop across the diode must add up to equal the source voltage. Thus, if we subtract the load-voltage waveform from the source-voltage waveform, we have the voltage waveform across the diode (Fig. 3-1*e*). The maximum voltage across the diode in the reverse direction is known as the *peak inverse voltage* (PIV). The PIV is also referred to as the *peak reverse voltage* (PRV). The reverse breakdown voltage BV_R of the diode must be greater than the peak inverse voltage or the diode will fail in the circuit.

The peak voltage across the load is the peak voltage of the source, V_m. Then, by Ohm's law, the peak load current (and the peak diode current) $I_{L,\max}$ is V_m/R_L. By means of calculus, we can show that the average value of a *half* of a sine wave over a *full* ac cycle is the peak value divided by π. Then, the

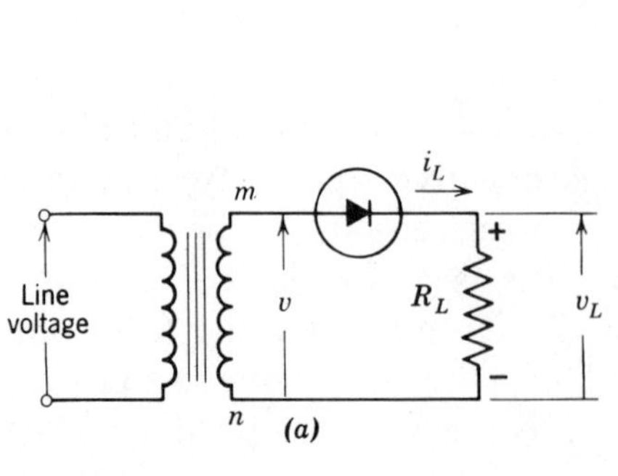

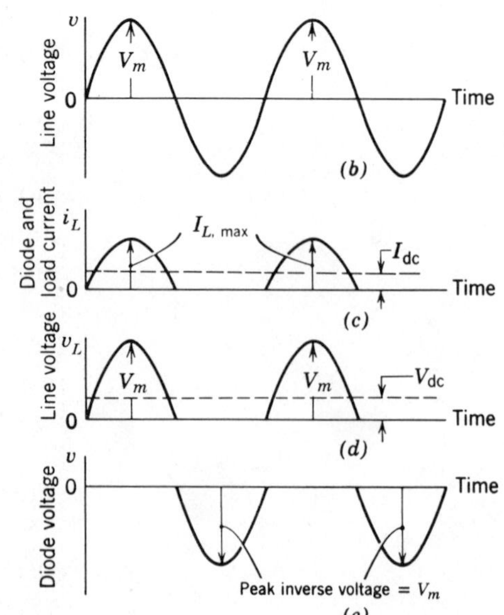

Figure 3-1 The half-wave rectifier. (*a*) Circuit. (*b*) Input voltage. (*c*) Diode and load current. (*d*) Load voltage. (*e*) Diode voltage.

average load voltage, which is the dc load voltage, is the peak value of the line voltage divided by π.

$$\boxed{V_{\text{dc}} = \frac{V_m}{\pi}} \tag{3-2a}$$

Remembering that the peak values and the effective values are related by $\sqrt{2}$ for sinusoidal waveforms

$$V_m = \sqrt{2}V$$

we have

$$V_{\text{dc}} = \frac{V_m}{\pi} = \frac{\sqrt{2}V}{\pi} = 0.318V_m = 0.450V \tag{3-2b}$$

and

$$I_{\text{dc}} = \frac{I_{L,\max}}{\pi} = \frac{V_m}{\pi R_L} = \frac{\sqrt{2}V}{\pi R_L} = \frac{V_{\text{dc}}}{R_L} = 0.318I_m = 0.450\frac{V}{R_L} \tag{3-2c}$$

If the value of the load resistor R_L is changed, the only change in the waveforms is a change in the amplitude of the current.

In the discussion of the half-wave rectifier, we assumed that the rectifying element is an ideal general-purpose diode. For an actual diode, we find there is a specific value for V_F given by the manufacturer. If we consider V_F, we must modify the diode voltage waveform, Fig. 3-1*e*, by showing a small forward voltage that is not zero. Then the load voltage waveform is reduced by this amount. The diode and load current waveform is also decreased slightly. In most practical rectifier circuits, we neglect this correction completely because V_m is much greater than V_F.

A comparison of the load voltage waveform (Fig. 3-1*d*) with the applied voltage waveform (Fig. 3-1*b*) shows that there is one pulse of load voltage for each full ac cycle of applied voltage. Therefore the fundamental frequency in the output is the frequency of the supply.

When a diode is used in a half-wave rectifier with a resistive load, three ratings are considered for the diode:

1. The peak diode current is the peak load current.
2. The average diode current is the average current in the load.
3. The peak inverse voltage is the peak voltage of the ac source.

The current rating of the transformer secondary is its maximum allowable dc load current.

Problems

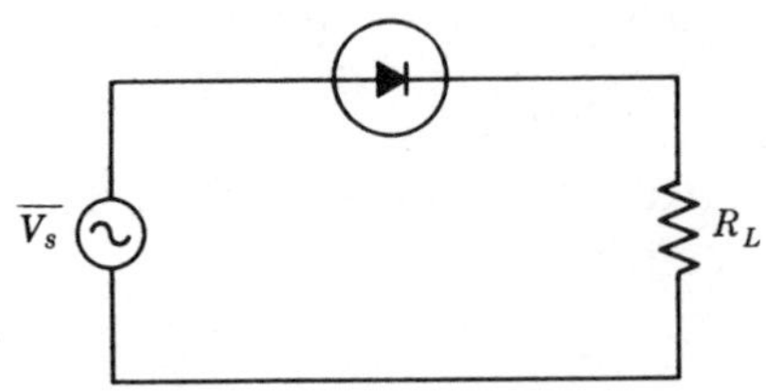

Circuit for Problems 3-1.1 and 3-1.2.

3-1.1 The supply voltage V_s is 12 V rms and the peak current rating of the diode is 1.2 A. Assuming peak rated current flows in the diode, what is the value of R_L and what is the dc current in R_L?

3-1.2 The supply voltage V_s for the rectifier is 1200 V rms and the desired average dc power in R_L is 100 W. Determine the value of R_L and the dc current in R_L.

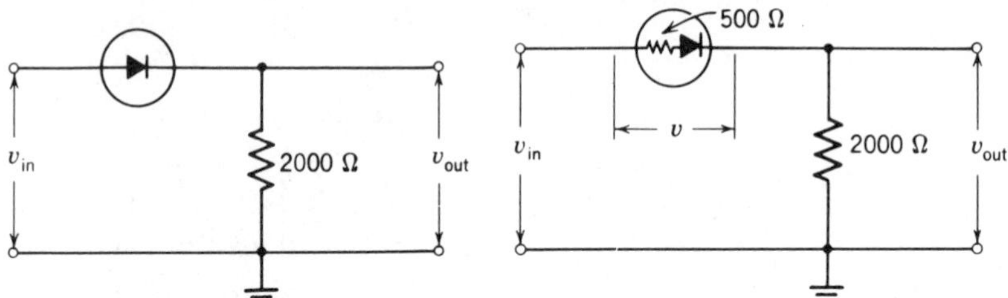

Circuit for Problem 3-1.5. Circuit for Problem 3-1.6.

3-1.3 A 60-Hz source is the input to a half-wave rectifier that supplies 100 W to a resistive load at 20 V. Determine the source voltage, the load current, and the load resistance.

3-1.4 The load on a half-wave rectifier is 5 W at 5000 V. Determine the peak voltage, the rms voltage, and the peak current of the source.

3-1.5 The input voltage is

$$v_{in} = 100 \cos 377t$$

Sketch and dimension the input–output characteristic that shows v_{out} plotted against v_{in}. The diode is ideal.

3-1.6 The input voltage is

$$v_{in} = 100 \cos 377t$$

Sketch and dimension the input–output characteristic that shows v_{out} plotted against v_{in}. Sketch and dimension two full cycles of the waveform of the voltage v.

3-1.7 If the diode in a half-wave rectifier opens, what is the effect on the operation of the circuit?

3-1.8 If the diode in a half-wave rectifier shorts, what is the effect on the operation of the circuit?

3-1.9 When a half-wave rectifier is wired up, the diode is placed in reverse in the circuit. What is the effect on the operation of the circuit?

Section 3-2 The Full-Wave Rectifier

In a *full-wave rectifier circuit* (Fig. 3-2*a*) the transformer secondary has a center tap *b*, which is the common return point of the rectifier circuit. The secondary voltage is measured from *b* to *c* and from *b* to *a* and not from *c* to *a*. The voltage of a secondary winding intended for use in this

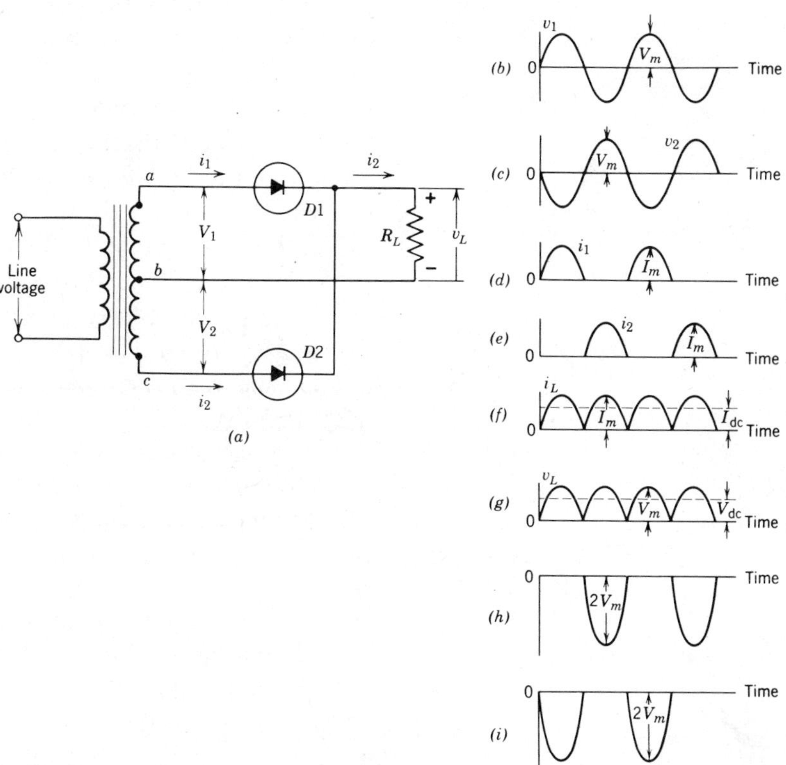

Figure 3-2 The full-wave rectifier. (*a*) Circuit. (*b*) Voltage applied to *D*1. (*c*) Voltage applied to *D*2. (*d*) Current in *D*1. (*e*) Current in *D*2. (*f*) Load current. (*g*) Load voltage. (*h*) Voltage across *D*1. (*i*) Voltage across *D*2.

circuit is specified as, for example, 35-0-35 V. This means that from b to a we have 35 V rms, and from b to c the voltage reading is also 35 V rms. Between c and a the voltage is 70 V rms. This transformer may also be rated "70 volts center-tapped."

When the voltage at point a is positive with respect to point b, the voltage at point c is negative with respect to point b. When the voltage at point c is positive with respect to point b, the voltage at point a is negative with respect to point b. Consequently, the voltage applied to diode $D2$ is 180° out of phase with respect to the voltage applied to diode $D1$. The waveform of the voltage applied to diode $D1$ is shown in Fig. 3-2b and the voltage applied to diode $D2$ is shown in Fig. 3-2c. The current in diode $D1$ (Fig. 3-2d) is the same as the waveform for the half-wave rectifier. We also have this same waveform for the current in diode $D2$ (Fig. 3-2e) but there is a 180° phase difference. The waveform for the load current is the sum of these two waveforms (Fig. 3-2f). We say that diode $D1$ rectifies the top half of the ac waveform and that diode $D2$ rectifies the bottom half of the ac waveform. The load voltage waveform (Fig. 3-2g) has the same shape as the waveform for the load current.

At the instant that peak positive voltage $+V_m$ is applied to diode $D1$, the instantaneous load voltage is $+V_m$. At this same instant, the voltage on the anode of diode $D2$ is $-V_m$. Since the cathode of diode $D2$ is connected to the load, there is a peak inverse voltage across diode $D2$ equal to $[+V_m-(-V_m)]$ or $2V_m$ volts. Similarly, the peak inverse voltage across diode $D1$ is $2V_m$ volts. The waveforms for the voltages across the diodes showing the peak inverse voltages are given in Fig. 3-2h and Fig. 3-2i.

A comparison of the load voltage waveform (Fig. 3-2g) with the applied voltage waveform (Fig. 3-2b) shows that there are two pulses of load voltage for each ac cycle. Therefore the fundamental frequency in the output is twice the frequency of the supply.

The values of the dc current in the load and the dc voltage across the load can be obtained from the results of the half-wave rectifier by using the superposition theorem. The values for diode $D1$ alone are given by Eq. 3-2a and Eq. 3-2b. The values for diode $D2$ alone are also given by Eq. 3-2a and

Eq. 3-2*b*. The addition, then, introduces a factor of 2 into the equations obtained for the half-wave rectifier.

$$\boxed{V_{\mathrm{dc}} = \frac{2}{\pi} V_m} \tag{3-3a}$$

$$V_{\mathrm{dc}} = \frac{2}{\pi} V_m = \frac{2\sqrt{2}}{\pi} V = 0.636\, V_m = 0.900\, V \tag{3-3b}$$

$$I_{\mathrm{dc}} = \frac{2}{\pi} I_m = \frac{2}{\pi}\frac{V_m}{R_L} = \frac{2\sqrt{2}}{\pi}\frac{V}{R_L} = \frac{V_{\mathrm{dc}}}{R_L} = 0.636\frac{V_m}{R_L}$$
$$= 0.900\frac{V}{R_L} \tag{3-3c}$$

where

$$V_m = \sqrt{2}V$$

When diodes are used in the full-wave rectifier using a resistive load, there are three ratings to be considered for *each* diode.

1. The peak diode current is the peak load current.
2. The average diode current is one half the average load current.
3. The peak inverse voltage is twice the peak of the transformer voltage measured between the center tap and either end.

The current rating of the transformer secondary is the maximum allowable dc (average) load current.

Problems

3-2.1 A transformer with a 117-V primary and a 275-0-275 V secondary supplies a 10-kΩ load by a full-wave rectifier. Determine the load voltage and the load current.

3-2.2 Solve Problem 3-2.1 for a 350-0-350 V secondary and a 2000-Ω load.

3-2.3 A full-wave rectifier is used to supply a load in which the dc load current is 5 A and the load voltage is 20 V. What are the

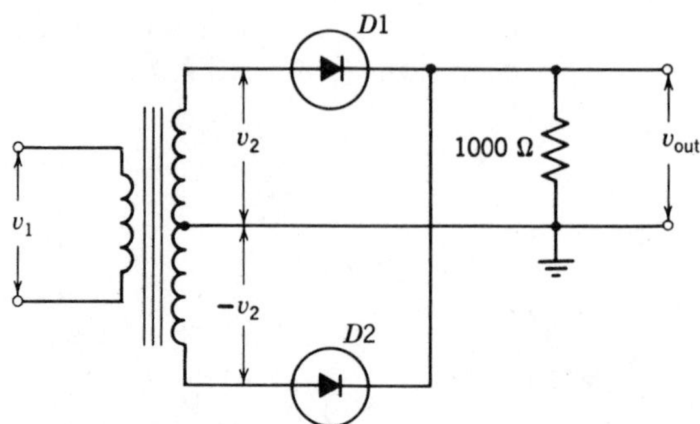

Circuit for Problems.

ratings of the power transformer if the source voltage is 117 V rms?

3-2.4 Solve Problem 3-2.3 for a 250-mA, 30-V load.

3-2.5 The voltages v_1 and v_2 are each $60 \cos 377t$. Sketch and dimension the input–output characteristic that shows v_{out} plotted against v_1.

3-2.6 If diode *D*1 opens, what is the effect on the operation of the circuit?

3-2.7 If diode *D*1 shorts, what is the effect on the operation of the circuit?

3-2.8 When the circuit is wired, *D*1 is placed in reverse in the circuit. What is the effect on the operation of the circuit?

Section 3-3 The Bridge Rectifier

The full-wave rectifier circuit requires a center tap on the source of the alternating voltage that is to be rectified. In many applications, the advantages of the higher output of the full-wave circuit are required, but the source has no center tap. The *bridge rectifier* (Fig. 3-3*a*) is used to solve this problem.

The action of the bridge rectifier can be understood by considering the two circuits shown in Fig. 3-3*b* and Fig. 3-3*c*. When terminal *b* is positive and terminal *c* is negative, there is a forward current path from *b* to *a* through diode *D*1, from *a* to *d* through the load resistance R_L, and from *d* to *c* through diode *D*2. Diodes *D*4 and *D*3 are opposed to this current direction. When terminal *c* is positive and terminal *b* is negative on the next half cycle, there is a forward current path from *c* to *a* through diode *D*3, from *a* to *d* through the load resistance R_L, and from *d* to *b* through diode *D*4. Diodes *D*1 and *D*2 are opposed to this current direction. In each of

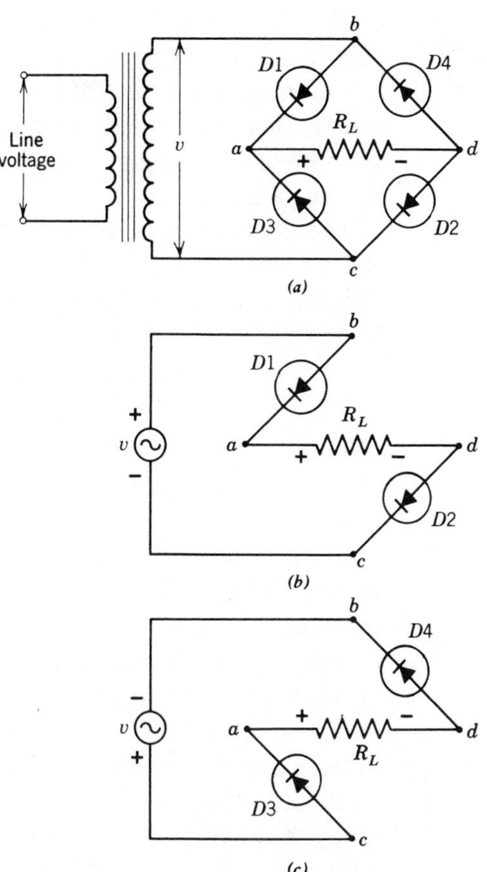

Figure 3-3 The bridge rectifier. (*a*) Circuit. (*b*) Current path when *b* is positive. (*c*) Current path when *c* is positive.

these two cases, the current flows from a to d through the load resistance R_L. Thus, there are two pulses of dc current through R_L for each full cycle of the ac supply. Full-wave rectification is accomplished.

Refer to Fig. 3-3a. When terminal b is positive and terminal c is negative, diodes $D1$ and $D2$ conduct. If the diodes are ideal, the forward voltage drop is zero. Then we can consider terminal b shorted to terminal a and terminal d shorted to terminal c. Now both diode $D4$ and diode $D3$ are in parallel with the load. Each diode ($D3$ and $D4$) sees the voltage across the load as a reverse voltage. Consequently, the peak inverse voltage across diode $D4$ and across diode $D3$ is the peak voltage of the source, V_m. By the same logic, we can show that the peak inverse voltage across diode $D1$ and across diode $D2$ is also V_m occurring when terminal c is positive and terminal b is negative.

This rectifier is a full-wave rectifier circuit in which there are two pulses of load current for each ac cycle. Therefore the fundamental frequency in the load is twice the frequency of the ac source.

The peak value of the ac source voltage is the peak value of the dc load voltage. Therefore the equations for the load voltage and for the load current are

$$V_{dc} = \frac{2}{\pi} V_m = \frac{2\sqrt{2}}{\pi} V = 0.636\,V_m = 0.900\,V \qquad \text{(3-4}a\text{)}$$

$$I_{dc} = \frac{2}{\pi} I_m = \frac{2}{\pi}\frac{V_m}{R_L} = \frac{2\sqrt{2}V}{\pi R_L} = \frac{V_{dc}}{R_L} = 0.636 I_m = 0.900 I \qquad \text{(3-4}b\text{)}$$

In both the half-wave rectifier (Fig. 3-1a) and the full-wave rectifier (Fig. 3-2a), there is a common connection between the transformer (the ac source) and the dc load. In the bridge rectifier (Fig. 3-3a), there is no common lead connection between the transformer (the ac source) and the dc load. In many applications, a common lead connection is required between the ac source and the dc load. This requirement eliminates the use of the bridge rectifier. On the other hand, in many applications, a center tap is not available at the ac source. Under this condition, the bridge rectifier must be used if full-wave rectification is required.

When a bridge rectifier is used with a resistive load, the ratings to be considered for each of the four diodes are:

1. The peak diode current is the peak load current.
2. The average diode current is one-half the average current in the load.
3. The peak inverse voltage is the peak voltage value of the ac source.

Problems 3-3.1 A transformer with a 117-V primary winding and a 250-V secondary winding is used with a bridge rectifier to supply a 10-kΩ load resistor. Determine the load voltage, the load current, and the input current and input power to the transformer.

3-3.2 Solve Problem 3-3.1 for a 5000-V secondary winding and a 200,000-Ω load resistor.

3-3.3 A bridge rectifier is used to supply a dc load with 20 A at 20 V from a 117-V source. What are the ratings of the required power transformer?

3-3.4 Solve Problem 3-3.3 if the dc load is 100 W at 117 V.

3-3.5 The bridge rectifier, Fig. 3-3*a*, is wired but diode *D*1 is accidentally wired in reverse into the bridge. What happens?

3-3.6 Repeat Problem 3-3.5 if the reversed diode is *D*4 instead of *D*1.

3-3.7 The bridge rectifier, Fig. 3-3*a*, is in operation. Diode *D*1 fails and opens. What is the effect on the operation of the circuit?

3-3.8 Repeat Problem 3-3.7 if diode *D*1 fails and shorts.

Section 3-4 The Capacitor Filter

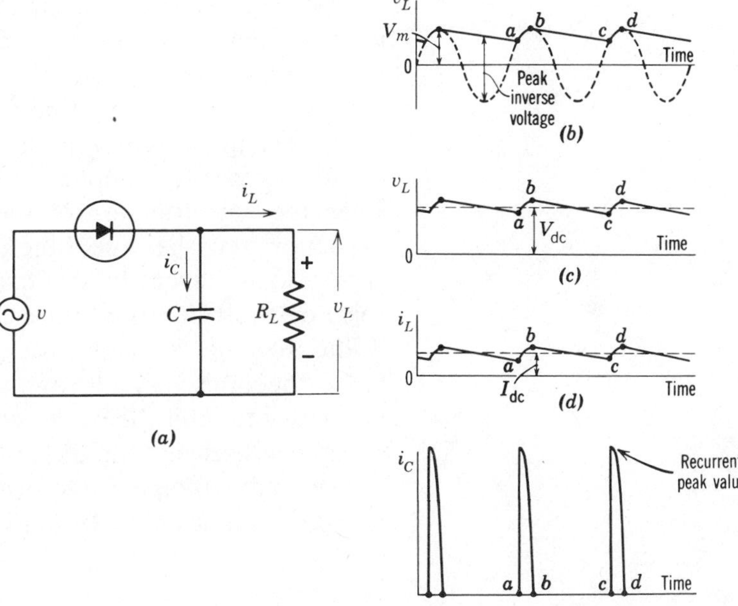

Figure 3-4 Half-wave rectifier with capacitor filter. (*a*) Circuit. (*b*) Action of the capacitor. (*c*) Load voltage. (*d*) Load current. (*e*) Diode current.

A single capacitor filter used with a half-wave rectifier circuit is shown in Fig. 3-4*a*. During the positive half of the supply cycle, the capacitor is charged in the time interval between *a* and *b*, Fig. 3-4*b*. When the applied ac wave falls below the value of the direct voltage on the capacitor, point *b*, the charging current from the diode ceases, and the load current continues to flow by the discharging action of the filter capacitor in the interval from b to c. Just after point c, the increasing supply voltage again exceeds the voltage on the capacitor and the filter capacitor recharges. The load-voltage waveform (Fig. 3-4c) is also the capacitor-voltage waveform. The peak inverse voltage is shown in Fig. 3-4b. When a filter capacitor is used, the peak inverse voltage is effectively twice the peak of the source voltage or $2V_m$. The load current (Fig. 3-4d) has the same shape as the load-voltage waveform, since the load is resistive.

The diode can only pass current during the recharging time of the capacitor, from a to b and from c to d. Thus the diode current is in the form of short pulses (Fig. 3-4e). The area under the load-current curve (Fig. 3-4d) must equal the area under the diode-current curve (Fig. 3-4e), since the total charge delivered to the capacitor is delivered to the load as load current in the form of discharge. This statement has a slight error since the diode, when it is recharging the capacitor, also at the same time supplies current into the load. However, the discussion of the operation of most rectifier circuits is greatly simplified by separating the two concepts. Then the sole function of the diode is to recharge the filter capacitor and the sole function of the filter capacitor is to supply load current by discharge. The diode current takes the form of short-duration pulses. If the load current is fixed and if the size of the capacitor is increased, the diode current pulses become very narrow with a very high amplitude. It is necessary to limit the peak current to a safe value by placing a resistance between the diode and the line-voltage source.

The load voltage waveform (the capacitor voltage waveform) shown in Fig. 3-4b (and Fig. 3-4c) is detailed in Fig. 3-5.

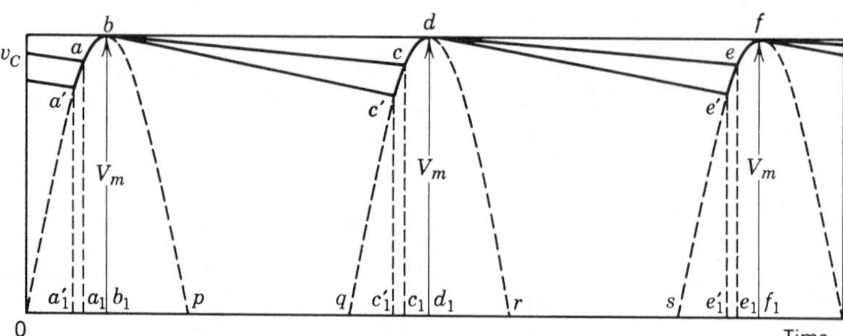

Figure 3-5 Voltage waveforms under different load conditions.

The load voltage falls from b to c and from d to e at a rate determined by the time constant of the filter capacitor and the load resistance R_LC. The capacitor is recharged by the diode from point a to point b, from point c to point d, and from point e to point f. Current flows in the diode from time a_1 to time b_1, from time c_1 to time d_1, and from time e_1 to time f_1. The load voltage waveform is the solid line

$$a\text{-}b\text{-}c\text{-}d\text{-}e\text{-}f$$

The average value or the dc value of the load voltage is slightly less than the peak value V_m. The vertical distance between a and b (or between c and d or between e and f) is the *peak-to-peak ripple voltage* in the rectifier circuit output.

When the current demand from the rectifier circuit increases, the value of R_L is reduced. Then the value of the time constant, R_LC, is reduced. The capacitor discharges more rapidly. Now the load voltage is given by the waveform

$$a'\text{-}b\text{-}c'\text{-}d\text{-}e'\text{-}f$$

The average value of the load voltage is somewhat less than in the first case. The peak-to-peak value of the ripple is increased. The increased load current demand causes an increase in the time interval that the diode passes current—from a_1-b_1 to a_1'-b_1, from c_1-d_1 to c_1'-d_1, and from e_1-f_1 to e_1'-f_1.

Consider an initial condition for which the load voltage waveform is

$$a'\text{-}b\text{-}c'\text{-}d\text{-}e'\text{-}f$$

The size of the filter capacitor is now increased. The load voltage waveform becomes

$$a\text{-}b\text{-}c\text{-}d\text{-}e\text{-}f$$

This increase in the size of the capacitor:

1. Increases the dc load voltage toward the limiting value V_m.
2. Reduces the peak-to-peak value of the ripple voltage.
3. Reduces the time that current pulses flow through the diode.
4. Increases the peak current in the diode.

Figure 3-5 is drawn for a half-wave rectifier. If a full-wave rectifier were used, Fig. 3-5 would be modified by drawing positive half waves between point p and point q and between point r and point s. This circuit change would:

1. Increase the dc load voltage only slightly toward V_m.
2. Reduce the ripple voltage by a factor of 2.
3. Double the frequency of the ripple voltage.
4. Reduce the individual diode currents by 2.
5. Make no change in the peak inverse voltage on the diodes ($2V_m$ for the half-wave and full-wave circuits).

It is important to remember that the equations developed for the half-wave rectifier circuit in Section 3-1, for the full-rectifier circuit in Section 3-2, and for the bridge rectifier circuit in Section 3-3 do not apply when a capacitor filter is used.

The 1N1764 silicon rectifier, for example, has the following typical ratings for use as a half-wave rectifier with a capacitor filter:

rms supply voltage	150 V
dc load current	0.5 A
Recurrent peak current	5.0 A
Surge current limit	35.0 A
Maximum input capacitor	250 μF

Assume the input ac power to the half-wave rectifier is turned off and the capacitor is completely discharged. Now the circuit is turned on at the instant the incoming source voltage is at its peak positive value. The discharged capacitor acts as a short and the *surge current* is limited only by the dc resistance of the circuit. The diode used in Fig. 3-6 has a *surge current limit* of 35 A. If the peak incoming voltage is $150\sqrt{2}$ or 212 V, the dc resistance required to limit the surge current to 35 A is 212/35 or 6.1 Ω. The usual procedure is to place a resistor between the source and the diode. In Fig. 3-6 the value used is the nearest commercial size, 6.2 Ω. The maximum dc current in the circuit is 0.5 A. The power rating of the resistor without considering a derating factor is

$$P = I^2R = 0.5^2 \times 6.2 = 1.6 \text{ W}$$

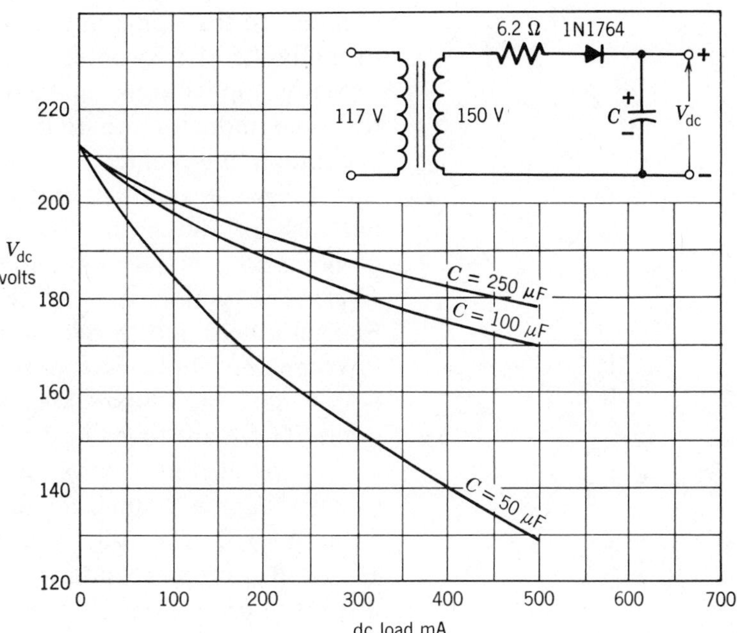

Figure 3-6 Load characteristics of a typical rectifier with capacitor filter. (*Courtesy RCA.*)

If a short circuit develops in the capacitor or in the load, this resistor will serve as a protective device for the circuit and will burn out.

The characteristics given in Fig. 3-6 for the 1N1764 diode are for actual operating conditions. The largest value that can be used for a filter capacitor is 250 μF in order to keep the recurrent peak value of diode current within the 5-A rating. Typical load curves for this diode as a half-wave rectifier are shown in Fig. 3-6.

A half-wave rectifier used with a capacitor filter provides a power supply that is used primarily where the load-current requirements are small. It provides a low-cost and lightweight solution for a filtering problem. It has the disadvantage that the direct output voltage decreases with an increase in load and that the percent ripple increases sharply with an increase in load.

Section 3-5 Complex Filters

In most applications a full-wave rectifier using a single capacitor as a filter drives a voltage-regulator circuit. Voltage regulators are examined in Chapter 20. A voltage regulator can reduce the ripple in the output voltage to a level measured in millivolts or microvolts.

Many applications, however, rely on a network of capacitors and inductors to reduce the ripple in the output to an acceptable low value.

A full-wave rectifier that uses a *choke* as part of the filter network is shown in Fig. 3-7*a*. The actual filter, the *LC* combination, is termed either an *L filter* or a *choke-input filter.* The action of the choke is to store energy in the magnetic field and to release it to the load evenly.

When the choke is too small or when the load current is very small, the choke does not deliver current over the full cycle. There are times in the cycle *ab* and *cd* (Fig. 3-7*b*) when the choke current is zero. At these times, the overall filter acts as if it were a simple capacitor filter. The load voltage falls from *A* to *B* (Fig. 3-8) with an increase of current from 0 to *B′*. At *B* a critical value is reached. At this critical value, the distances *ab* and *cd* (Fig. 3-7*b*) are just zero. Now current is flowing at all times in the choke. This flow of current in the coil prevents the capacitor from discharging, and the load voltage is maintained at a constant value from *B* to *C*. The voltage at *B* is ideally $0.63\,V_m$. The waveforms for this condition are shown in Fig. 3-7*c* and Fig. 3-7*d*. In an actual circuit, the dc resistance of the choke and the diode drop cause the voltage to fall from *B* to *D* (Fig. 3-8).

Voltage regulation is a measure of the change of load voltage with load current and is defined as

$$\text{Percent of voltage regulation} \equiv \frac{\text{no load} - \text{full load}}{\text{full load}} \times 100 \qquad (3\text{-}5)$$

A *bleeder resistor* is a resistor that is connected in parallel with the load. A bleeder has a twofold purpose in a rectifier circuit. It discharges the capacitors when the power supply is turned off so that no dangerous residual charge is left on the filter capacitors. Also, the no-load voltage is not point *A* but point *B* on Fig. 3-8. In the ideal choke-input rectifier circuit, we can see that the regulation with a bleeder is zero between *B* and *C* whereas it is $(V_m - 0.63\,V_m)/0.63\,V_m$ or 58.7% without the bleeder.

Different filtering arrangements used with rectifier circuits

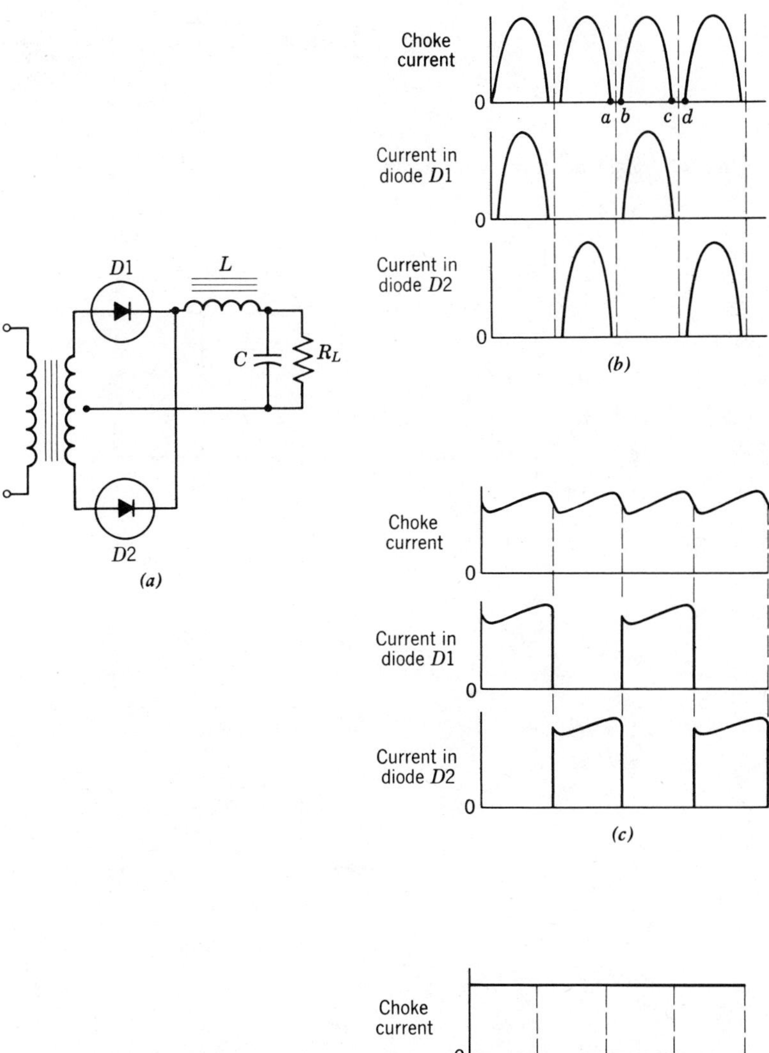

Figure 3-7 Waveforms for the choke filter. (*a*) Circuit. (*b*) Small inductance. (*c*) Normal inductance. (*d*) Infinite inductance.

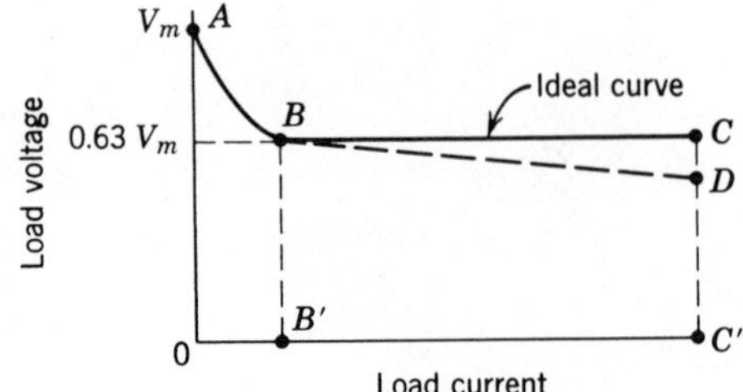

Figure 3-8 Ideal load curve for a choke filter.

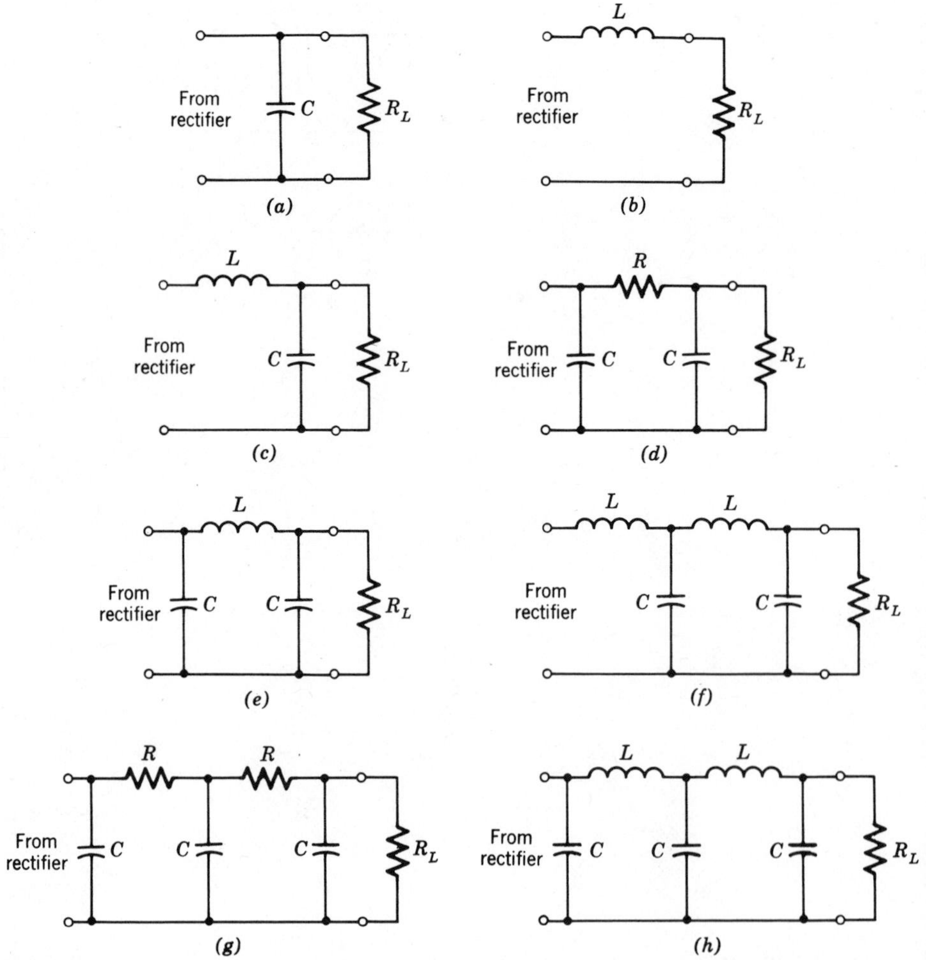

Figure 3-9 Power supply filter networks. (*a*) Capacitor filter. (*b*) Choke filter. (*c*) *L* filter. (*d*) π filter. (*e*) π filter. (*f*) π Choke-input π filter. (*g*) Double-π filter. (*h*) Double-π filter.

are shown in Fig. 3-9. The complexity of the filter is determined by the allowable ripple and the allowable voltage regulation. It should be pointed out, again, that electronic voltage regulators are used almost exclusively in new equipment designs to secure a pure dc with negligible ripple.

Section 3-6 Voltage Multipliers

A *full-wave voltage doubler* is obtained by replacing two diodes in the bridge rectifier with capacitors (Fig. 3-10*a*). Usually the circuit diagram is given in the form of Fig. 3-10*b*. Diode $D1$ charges C_A when m is positive and n is negative. When n is positive and m is negative, diode $D2$ charges C_B. If point a is the reference, the voltage across C_A, V_{C_A}, is positive and the voltage across C_B, V_{C_B}, is negative. These voltage waveforms are shown in Fig. 3-10*c*. However, the load is placed from point b to point c. Point b is the reference for the load. The voltage at point c, then, shows the total ripple variation across the load. The voltages across the two capacitors, C_A and C_B, are series aiding as far as the load is concerned. Hence, the voltage across the load is twice the voltage on each or is *doubled*. The load voltage waveform is shown in Fig. 3-10*d*.

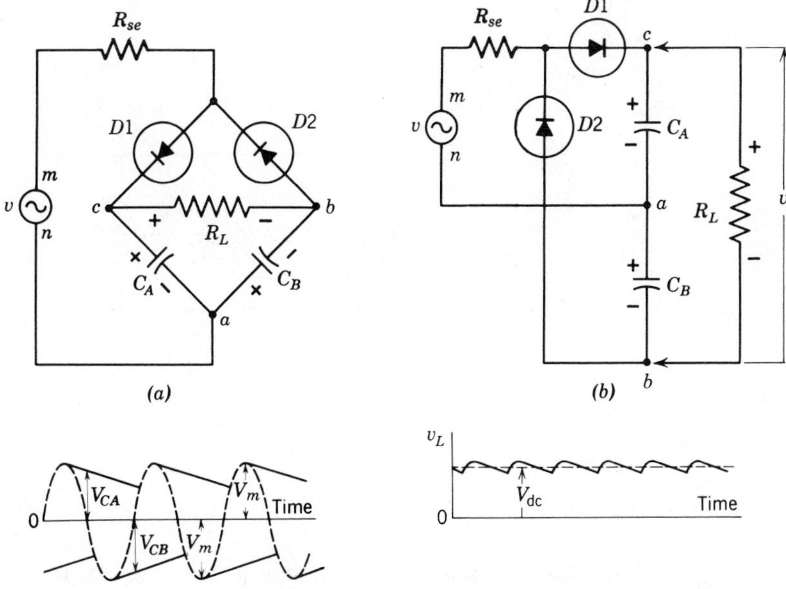

Figure 3-10 The full-wave voltage doubler. (*a*) Circuit. (*b*) Alternate form of circuit layout. (*c*) Voltage waveform across C_A and C_B. (*d*) Output voltage waveform.

When the load current is very small, the load voltage is twice the peak of the line $2V_m$. There are two impulses of charging current into the capacitors per cycle; therefore, the ripple frequency is twice the frequency of the line. The action of the two diodes in the full-wave rectifier charges the whole filter twice each cycle, whereas the charging action of this circuit charges each capacitor once per cycle, but at different times. It is in this sense a full-wave rectifier and not a half-wave rectifier. The ripple in this circuit is greater and the regulation poorer than in the equivalent full-wave rectifier. The peak inverse-voltage ratings of the diodes are twice the peak of the line voltage, $2V_m$. Since this circuit is often used on an ac line without either an isolating transformer or a step-up or step-down transformer, it is important to notice that there is no common connection between the line and the load. When the expense of a line transformer is justified, it is preferable to use the superior circuit of the conventional full-wave rectifier.

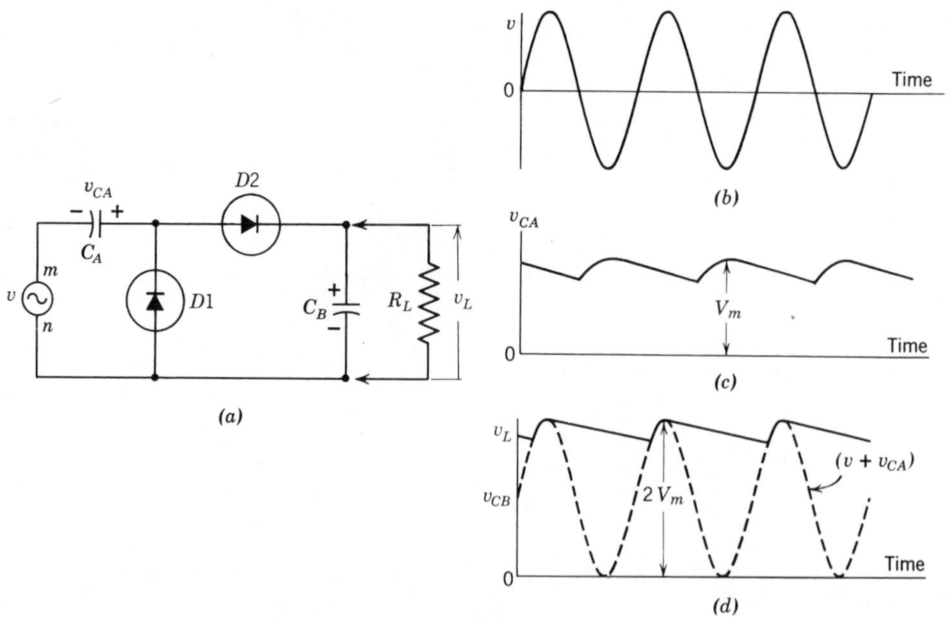

Figure 3-11 The half-wave voltage doubler. (*a*) Circuit. (*b*) Line voltage. (*c*) Waveform across C_A. (*d*) Waveform across C_B.

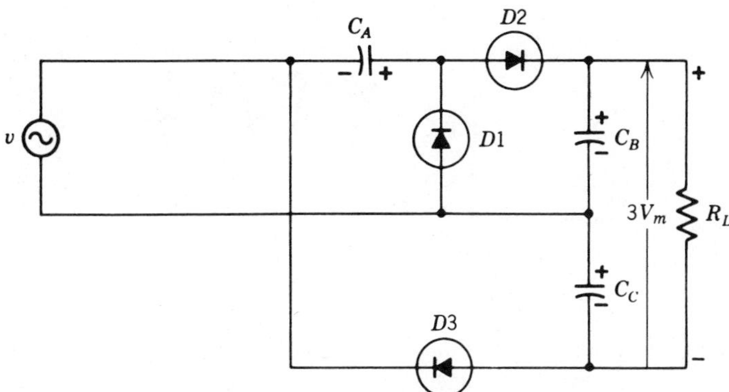

Figure 3-12 Voltage tripler.

The *half-wave voltage doubler* or the *cascade voltage-doubler* circuit is shown in Fig. 3-11*a*. When *n* is positive and *m* is negative, C_A charges through diode *D*1 to V_m, the peak of the line voltage. This action is shown in the waveform of Fig. 3-11*c*. When the cycle reverses, *n* is negative and *m* is positive. Now, the line voltage *e* and the voltage across C_A are in series aiding. The maximum value this condition can have is $2V_m$, and C_B charges to $2V_m$ through diode *D*2 (Fig. 3-11*d*). The load is connected across C_B. The load receives only one charging pulse per cycle. The ripple frequency is the line frequency giving a basis for the use of the term "half wave." The regulation of this circuit is very poor, and the ripple is very high, even with medium values of load current. The peak inverse voltage on either diode is $2V_m$. This circuit does have a common connection between the line and the load.

When a half-wave rectifier is added to the half-wave voltage doubler (Fig. 3-11*a*), the new circuit becomes a voltage tripler, Fig. 3-12. The capacitor C_B is charged to twice the peak of the supply voltage, $2V_m$. The half-wave circuit charges C_C to the peak of the line voltage V_m. The series combination of C_B and C_C results in a voltage across R_L of $3V_m$.

Section 3-7 Clampers

The basic half-wave rectifier using a capacitor filter is shown in Fig. 3-13*a*. The capacitor charges to the peak value of the supply, V_m, Fig. 3-13*c*. The peak inverse voltage across the diode is $2V_m$. The ac waveform across the diode is given in Fig. 3-13*d*.

The *clamper circuit* or *shunt-rectifier circuit*, Fig. 3-14*a*, is

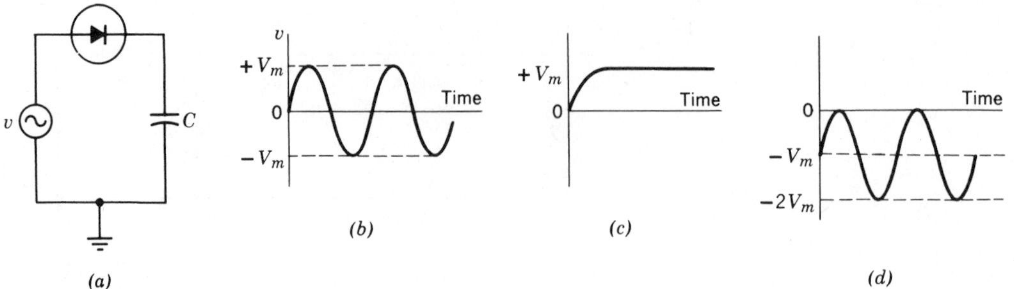

Figure 3-13 The basic half-wave rectifier. (*a*) Circuit. (*b*) Supply voltage. (*c*) Capacitor voltage. (*d*) Inverse voltage across the diode.

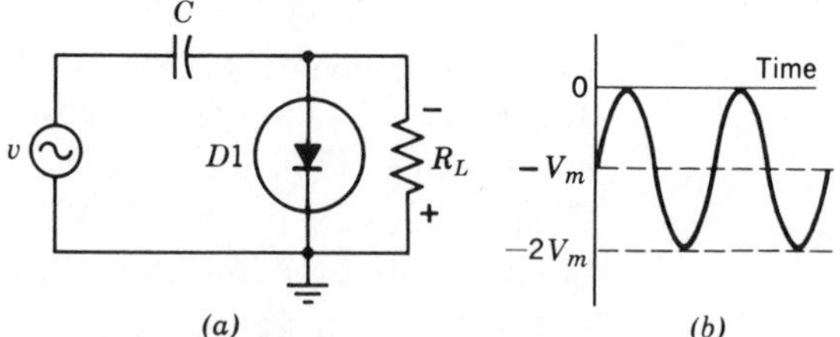

Figure 3-14 The basic shunt rectifier. (*a*) Circuit. (*b*) Output voltage waveform.

formed by interchanging the diode and the capacitor in Fig. 3-13*a*. Now a load resistor is placed in parallel with the diode. The voltage across the load is given by the waveform of the voltage across the diode, Fig. 3-14*b*. The average (dc) voltage across the load and the diode is $-V_m$.

The circuit and the waveform shown in Fig. 3-14*a* have an application of particular importance in electronics. The dc voltage obtained from this circuit as shown in Fig. 3-14*b* is $-V_m$ volts. This dc voltage is used to bias a transistor or an FET. The bias voltage is exactly proportional to an incoming signal. When this circuit is used to develop a bias voltage, it is called a *bias clamp*.

In Fig. 3-15 an *RC* filter is added to the basic shunt-rectifier circuit. The filter R_2C_2 establishes a pure direct voltage across

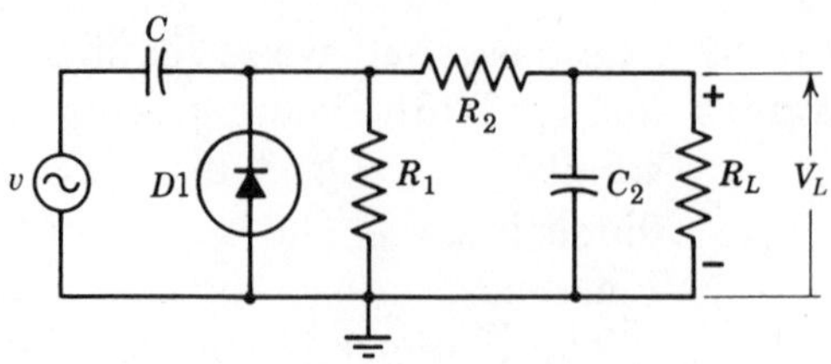

Figure 3-15 Shunt rectifier with output filter.

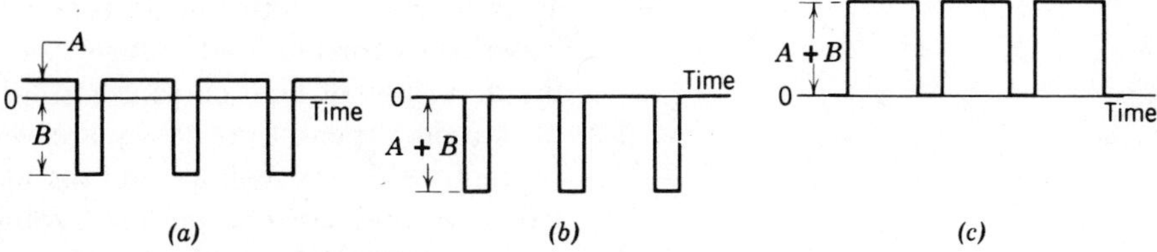

Figure 3-16 Waveforms of a clamp. (*a*) Input. (*b*) Output waveform for circuit of Figure 3-14*a*. (*c*) Output waveform with diode reversed.

R_L that is equal to twice the peak of the source voltage. The dc voltage across the load, V_L, is directly proportional to the peak value (or the rms value) of the incoming signal v. This circuit is often used in the probes of voltmeters that are designed to measure audio- and radio-frequency voltages without placing a severe shunting load impedance on the circuit where the measurement is taken.

The clamper is widely used in digital and in video signal-processing systems. The waveform shown in Fig. 3-16*a* has both positive (A) and negative values (B). When this waveform is used as the input signal to the circuit of Fig. 3-14*a*, the output of the clamper circuit causes all values of the output signal to be negative values (Fig. 3-16*b*). When the clamper diode is reversed, the opposite polarity is obtained. In this case all values of the output (Fig. 3-16*c*) are positive values. Two clamper circuits could be used across the same input signal: one would provide the positive output signal of Fig. 3-16*c* and the other would provide the negative output signal of Fig. 3-16*b*.

Supplementary Problems

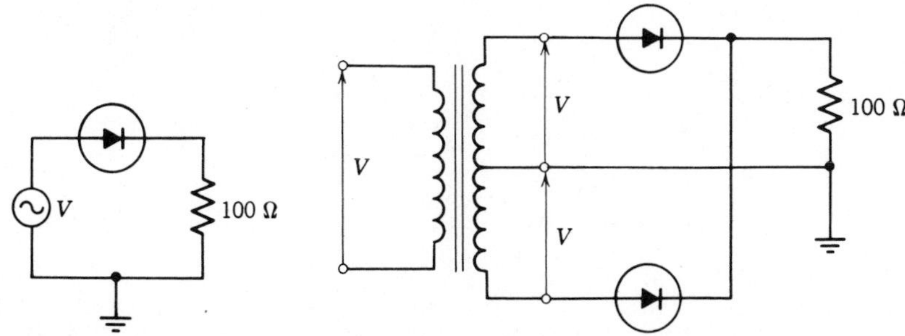

Circuit for Problems 3-1 through 3-3.

Circuit for Problems 3-4 through 3-6.

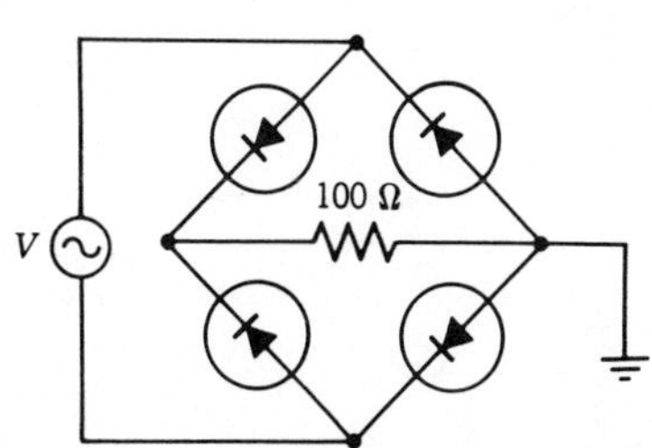

Circuit for Problems 3-7 through 3-9.

3-1 V is a 50-V peak-to-peak triangular waveform. Sketch the waveforms for the load voltage and the load current. What are the dc values of load current and load voltage?

3-2 V is a 50-V peak-to-peak square waveform. Sketch the waveforms for the load voltage and the load current. What are the dc values of load current and load voltage?

3-3 V is a sinusoidal waveform. The maximum allowable average dc current in the diode is 1 A. What is the maximum allowable peak-to-peak value for V?

3-4 Solve Problem 3-1 for the full-wave rectifier.

3-5 Solve Problem 3-2 for the full-wave rectifier.

3-6 Solve Problem 3-3 for the full-wave rectifier.

3-7 Solve Problem 3-1 for the bridge rectifier.

3-8 Solve Problem 3-2 for the bridge rectifier.

3-9 Solve Problem 3-3 for the bridge rectifier.

3-10 Sketch the output waveforms for v_A and v_B.

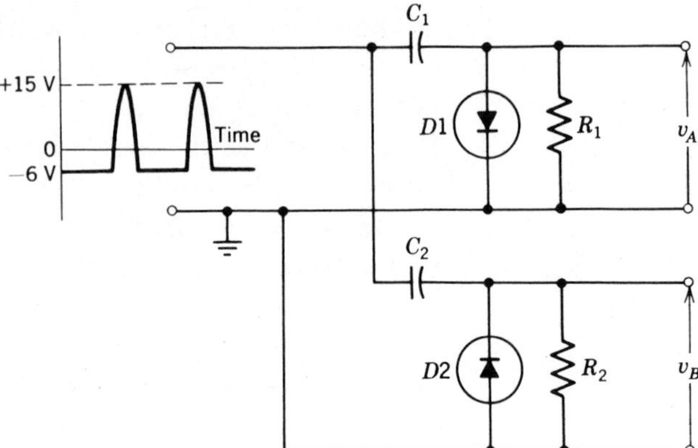

Input waveform and circuit for Problem 3-10.

Chapter 4 Transistors

The junction transistor is used as the physical model to explain how a transistor works (Section 4-1). The idealized collector characteristic for the common-emitter circuit configuration is used (Section 4-2) to form definitions of α and β. A simple common-emitter amplifier circuit using numerical values (Section 4-3) is used with waveforms to show how the circuit amplifies. This is also done for the common-collector amplifier (Section 4-4) and for the common-base amplifier (Section 4-5). The numerical results obtained from these three circuits are summarized in Table 4-1 in order to compare the properties of the three basic circuits. The basic equations to convert between any two of α, β, I_B, I_C, and I_E are developed (Section 4-6) and the results are summarized in Table 4-2.

Section 4-1 Construction and Operation

The concepts of the operation of a transistor can be understood best by considering one of the early methods of manufacturing transistors. The crystal is made in the form of a "sandwich": a thin section of P-type material between two thick slabs of N-type material (called NPN) or a thin section of N-type material between two thick slabs of P-type material (called PNP). The resulting "sandwich" is sawed up into small pieces about 0.01 by 0.01 by 0.10 in. for transistor construction (Fig. 4-1).

Supporting tabs and leads are added. One end is called the *emitter.* The other end is called the *collector.* The center section is called the *base.* In an NPN transistor, the emitter and the collector are N-type material and the base is P-type material. In a PNP transistor, the emitter and the collector are P-type material and the base is N-type material. It should be stressed that the NPN or the PNP structure is made from a single continuous crystal the same as for the PN diode. The transistor is often called the *bipolar junction transistor, BJT.*

The two junctions of a transistor (Fig. 4-2*a*) have depletion regions indicated by the shaded area which are produced

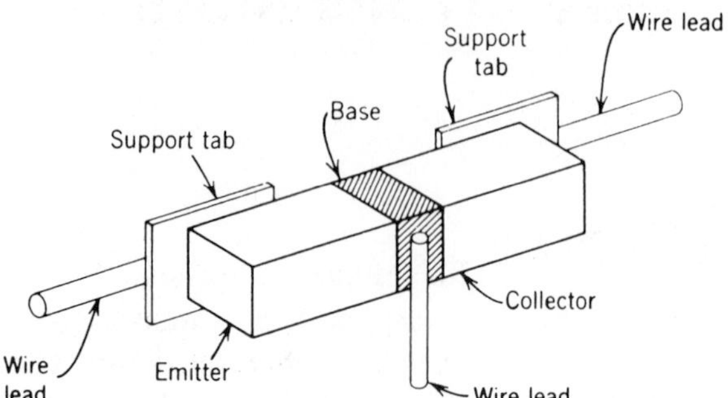

Figure 4-1 The junction transistor.

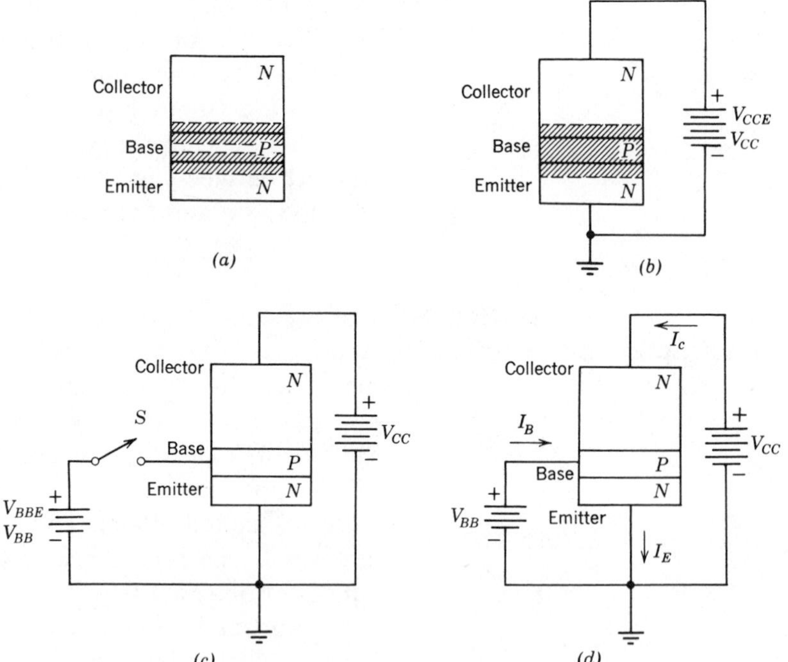

Figure 4-2 Currents in a transistor. (*a*) Depletion regions existing at formation. (*b*) Increased depletion region caused by collector bias. (*c*) Base bias. (*d*) Transistor currents.

during manufacture. Now a dc power source is connected between the collector and the emitter, Fig. 4-2*b*. In this discussion, the emitter is the reference point and the emitter connection is the common return of the circuit (the ground). The supply voltage is labeled V_{CCE}. Standard terminology uses an upper-case letter V to denote that a dc voltage is being considered *provided* the subscripts are also upper-case letters. The two-letter upper-case subscript CC states that this dc supply voltage is applied to the collector circuit. The third subscript, the upper-case *E*, shows the point to which the other side of the supply is connected, in this case the emitter. Usually, the third subscript is omitted. Thus the collector supply voltage is simply labeled V_{CC}.

The collector supply V_{CC} places a *reverse bias* on the collector-to-base junction (Fig. 4-2*b*). The effect of V_{CC} on the transistor is to widen the depletion region formed at the collector-to-base junction. The dc current in the collector, I_C (upper-case *I* with upper-case subscript *C*), is zero since this junction is reverse biased.

Now, let us add a second dc supply, V_{BBE}, connected between the base and emitter (Fig. 4-2*c*). Usually, this supply is labeled simply V_{BB}. The polarity of this supply is such that the base-to-emitter junction has a *forward bias.*

When the switch *S* of Fig. 4-2*c* is closed, current flows in the circuit because the *PN* junction of the base-to-emitter circuit has a forward bias. Current carriers are injected into the base. The depletion region in the base is materially reduced. If the base is very thin, the reduction of the depletion region is complete.

A large current flows from the collector through the base and into the emitter (Fig. 4-2*d*). The current in the emitter I_E is the sum of the base current I_B and the collector current I_C.

$$\boxed{I_E = I_B + I_C} \tag{4-1}$$

Equation 4-1 *holds true regardless of the circuit configuration employed or the transistor type that is used.*

The *dc beta** of a transistor β_{dc} is defined as the ratio of I_C to I_B at a given operating point.

$$\boxed{\beta_{\text{dc}} \equiv \frac{I_C}{I_B}} \tag{4-2}$$

* Some manufacturers use the unit symbol h_{FE} for β_{dc}.

The subscript *dc* on β signifies that this ratio is defined from the dc values I_C and I_B.

Since a small base current can control a large collector current, β_{dc} is a number much larger than 1. For this reason, a transistor is a current-controlled device. The values of β_{dc} for typical transistors can range from about 20 or 30 for low beta transistors to about 200 to 300 for high beta transistors.

The *dc alpha** of a transistor α_{dc} is defined as the ratio of I_C to I_E at a given operating point.

$$\boxed{\alpha_{dc} \equiv \frac{I_C}{I_E}} \tag{4-3}$$

The subscript *dc* on α signifies that this ratio is defined from the dc values I_C and I_E.

The value of α_{dc} is close to 1 but slightly less than 1; for example, 0.96, 0.97, 0.995, or 0.997 are typical values of α_{dc}.

The transistor and supply arrangement we have been considering is shown in Fig. 4-3*a*. This circuit using the schematic circuit symbol for the *NPN* transistor is shown in Fig. 4-3*b*. The arrowhead is assigned to the emitter side of the transistor, *not* to the collector side and relates to the transistor type.

If we consider the emitter and the base as a diode *PN* junction, we place the arrowhead to show the direction of forward bias. In this case (Fig. 4-3*b*) the base is *P* and the emitter is *N*. The current resulting from forward bias on the base–emitter junction is flowing into the base (*P*) and out of the emitter (*N*). Therefore, the arrowhead shows current flowing *out of* the emitter in the *NPN* transistor. The fact that the arrowhead points away from the base shows that the emitter must be *N*-type material. Then the base must be *P*-type material and the collector must be *N*-type material. If we use a *PNP* transistor, the dc supplies must be reversed to obtain forward bias on the emitter and reverse bias on the collector (Fig. 4-3*c*). The circuit using the circuit symbol for the *PNP* transistor is shown in Fig. 4-3*d*. Now the arrowhead points into the transistor to show that the emitter is *P*-type material.

*Some manufacturers use the unit symbol $-h_{FB}$ for α_{dc}.

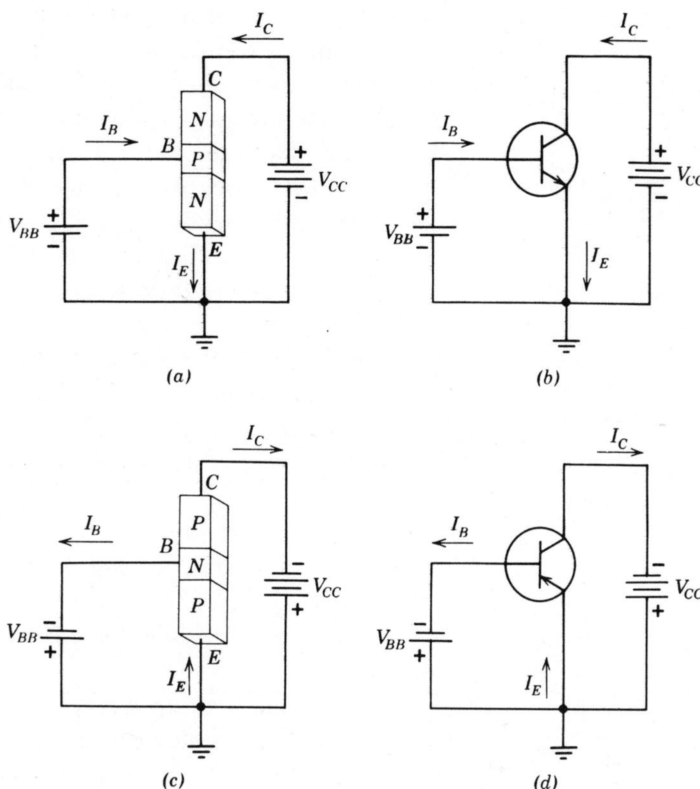

Figure 4-3 Transistor bias connections. (*a*) and (*b*) *NPN* transistor. (*c*) and (*d*) *PNP* transistor.

Section 4-2 The Common-Emitter Circuit

An *NPN* transistor is connected in a common-emitter circuit configuration to a variable base voltage supply and to a variable collector voltage supply (Fig. 4-4). Milliammeters and voltmeters are connected in the base lead and in the collector lead. In the base circuit the meters read I_B and the base-to-

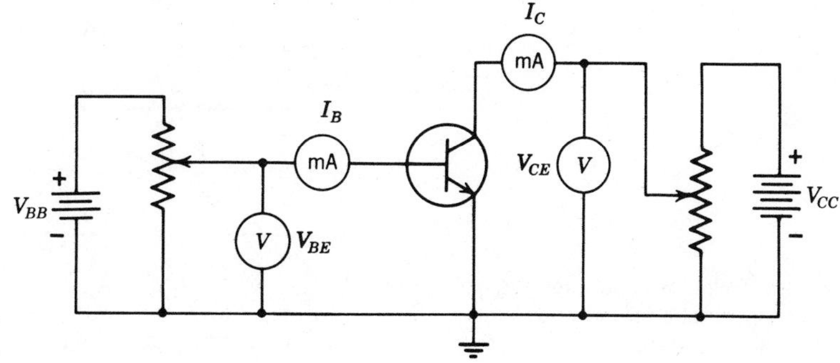

Figure 4-4 Test circuit for common-emitter *NPN* transistor circuit.

emitter voltage V_{BE}. In the collector circuit the meters read I_C and the base-to-collector voltage V_{CE}.

The base voltage supply V_{BB} in Fig. 4-4 is adjusted to set I_B at a fixed value. Then the collector-to-emitter voltage V_{CE} is raised. The collector current rises very rapidly to a particular value that does not change with a further increase in V_{CE}. A family of curves using I_B values as the step independent variable is plotted in Fig. 4-5 to show the common-emitter collector characteristic. I_C is the dependent variable and V_{CE} is the independent variable.

At a particular value of I_C, the least possible value of V_{CE} that can maintain that specific value of I_C is called the *saturation voltage,* $V_{CE,sat}$. These values are of the order of a fraction of a volt. A particular $V_{CE,sat}$ is indicated on Fig. 4-5.

In order to simplify our calculations in this chapter and in the next chapter, we use the ideal curves such as shown in Fig. 4-5. The actual collector characteristics for a transistor show a slight rise in collector current as V_{CE} *is increased.*

A *transistor curve tracer* is a form of an oscilloscope that presents transistor characteristics as a calibrated display on the screen. The power supplies are self-contained and both *NPN* and *PNP* transistors can be checked. A control establishes the *steps* of base current. The display corresponding to Fig. 4-5 would require six steps of base current at intervals of 20 μA from 0 to 100 μA.

The numerical values for three specific points on the col-

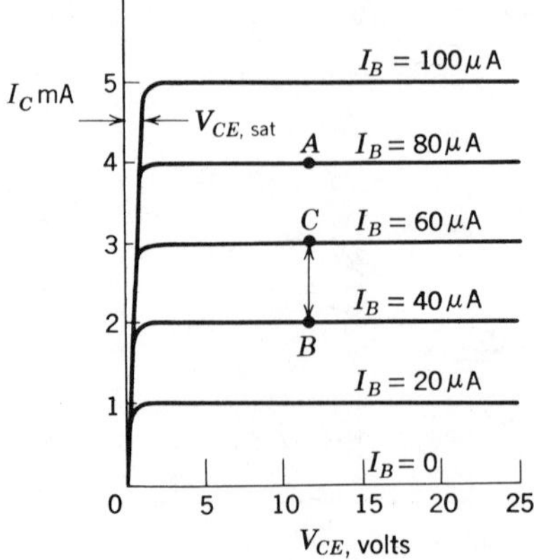

Figure 4-5 Collector characteristic or output characteristic for common-emitter circuit arrangement.

lector characteristic, Fig. 4-5, are

Point A	$I_B = 80\ \mu A$	$I_C = 4$ mA
Point B	$I_B = 40\ \mu A$	$I_C = 2$ mA
Point C	$I_B = 60\ \mu A$	$I_C = 3$ mA

The emitter current is the sum of the base current and the collector current as given by Eq. 4-1.

$$I_E = I_B + I_C$$

Then at each of these three points the emitter current is

Point A $\quad I_E = I_B + I_C = 80\ \mu A + 4\ \text{mA} = 4.08\ \text{mA} = 4080\ \mu A$

Point B $\quad I_E = I_B + I_C = 40\ \mu A + 2\ \text{mA} = 2.04\ \text{mA} = 2040\ \mu A$

Point C $\quad I_E = I_B + I_C = 60\ \mu A + 3\ \text{mA} = 3.06\ \text{mA} = 3060\ \mu A$

Now, using the definition of beta as given by Eq. 4-2, we have

$$\text{Point } A \qquad \beta_{dc} = \frac{I_C}{I_B} = \frac{4000\ \mu A}{80\ \mu A} = 50 \tag{4-2}$$

$$\text{Point } B \qquad \beta_{dc} = \frac{I_C}{I_B} = \frac{2000\ \mu A}{40\ \mu A} = 50 \tag{4-2}$$

$$\text{Point } C \qquad \beta_{dc} = \frac{I_C}{I_B} = \frac{3000\ \mu A}{60\ \mu A} = 50 \tag{4-2}$$

Using the definition of alpha as given by Eq. 4-3, we have

$$\text{Point } A \qquad \alpha_{dc} = \frac{I_C}{I_E} = \frac{4\ \text{mA}}{4.08\ \text{mA}} = 0.98 \tag{4-3}$$

$$\text{Point } B \qquad \alpha_{dc} = \frac{I_C}{I_E} = \frac{2.0\ \text{mA}}{2.04\ \text{mA}} = 0.98 \tag{4-3}$$

$$\text{Point } C \qquad \alpha_{dc} = \frac{I_C}{I_E} = \frac{3.0\ \text{mA}}{3.06\ \text{mA}} = 0.98 \tag{4-3}$$

These calculations show that the values of β_{dc} and α_{dc} are constant values over the region of the collector characteristic where the collector current curves are ideally horizontal.

The ac beta,* β_{ac}, is defined as the ratio of a *change* in collector current ΔI_C to a *change* in base current ΔI_B at a given operating point for a constant V_{CE}.

$$\boxed{\beta_{ac} \equiv \frac{\Delta I_C}{\Delta I_B}} \tag{4-4}$$

Note that the "change" in total current is equal to the ac current.

Similarly the ac alpha α_{ac} is defined as the ratio of a *change* in collector current ΔI_C to a *change* in emitter current ΔI_C to a *change* in emitter current ΔI_E at a given operating point for a constant V_{CE}.

$$\boxed{\alpha_{ac} \equiv \frac{\Delta I_C}{\Delta I_E}} \tag{4-5}$$

Now let us consider that the change is from point B to point C on the collector characteristic, Fig. 4-5. Using numerical values we have

$$\Delta I_C = 3.0 - 2.0 = 1.0 \text{ mA} = 1000\ \mu\text{A}$$
$$\Delta I_B = 60 - 40 = 20 \mu\text{A}$$

and

$$\Delta I_E = \Delta I_C + \Delta I_B = 1.0 \text{ mA} + 20\ \mu\text{A}$$
$$= 1.020 \text{ mA} = 1020\ \mu\text{A}$$

Substituting the numerical values into Eq. 4-4 we have

$$\beta_{ac} = \frac{\Delta I_C}{\Delta I_B} = \frac{1000\ \mu\text{A}}{20\ \mu\text{A}} = 50$$

Substituting the numerical values into Eq. 4-5 we find that

$$\alpha_{ac} = \frac{\Delta I_C}{\Delta I_E} = \frac{1000\ \mu\text{A}}{1020\ \mu\text{A}} = 0.98$$

* Some manufacturers use the unit symbol h_{fe} for β_{ac} and the unit symbol $-h_{fb}$ for α_{ac}.

Section 4-3 The Common-Emitter Amplifier

Now we will show how a transistor is used to amplify a signal. The basic *common-emitter amplifier* circuit is shown in Fig. 4-6*a*. Reference to Fig. 4-6*a* shows that the voltage loop that contains the input signal is from the signal source (e_s), then from the base to the emitter through the transistor to ground (the common reference point) and, then through V_{BB} back to the signal source. The output voltage loop is from the ground (the reference point), then from the emitter to the collector through the transistor, then through R_C, and then through V_{CC} back to the ground. The emitter is connected to the ground

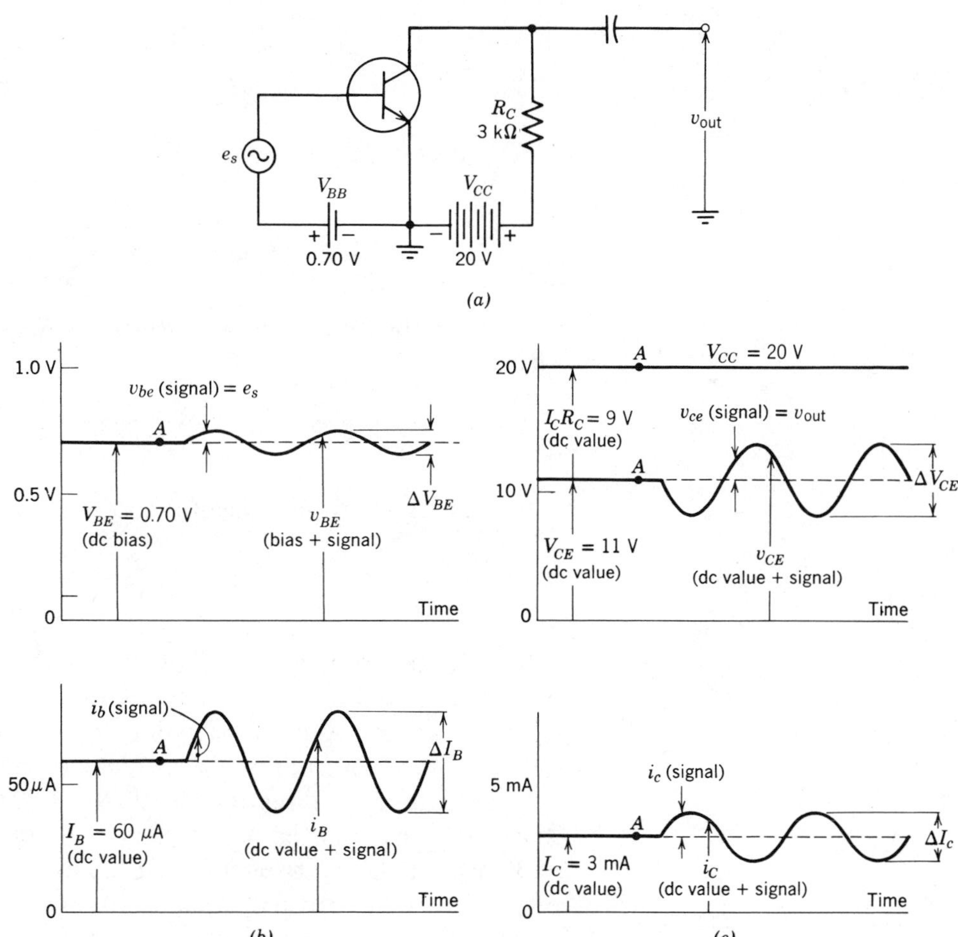

Figure 4-6 The common-emitter amplifier. (*a*) Circuit. (*b*) Base waveforms. (*c*) Collector waveforms.

(the reference point). The emitter is both in the input voltage loop and in the output voltage loop—hence the term *common-emitter.*

In the example discussed below we will assume values for the base current and for the collector current. In the following chapters on transistors we will show how these values are established and determined. At this point we are interested only in how a transistor amplifies and what the general characteristics of a transistor amplifier are.

Let us assume the following values for the circuit without a signal input ($e_s = 0$).

$$V_{BE} = V_{BB} = 0.70\ \text{V} \qquad V_{CC} = 20\ \text{V}$$
$$I_B = 60\ \mu\text{A} \qquad I_C = 3\ \text{mA}$$

The value of β_{dc} for this silicon transistor is

$$\beta_{\text{dc}} = \frac{I_C}{I_B} = \frac{3\ \text{mA}}{60\ \mu\text{A}} = 50 \tag{4-2}$$

The collector current flowing through R_C results in a voltage drop of

$$I_C R_C = 0.003\ \text{A} \times 3000\ \Omega = 9.0\ \text{V}$$

The voltage from the collector to the emitter V_{CE} is

$$V_{CE} = V_{CC} - I_C R_C = 20\ \text{V} - 9\ \text{V} = 11.0\ \text{V}$$

These values are indicated by the points marked A on the waveforms of Fig. 4-6b and Fig. 4-6c.

Now we introduce a signal e_s that has a peak value of 50 mV. Since the supply V_{BB} is ideal, its ac resistance is zero. Then e_s appears directly across the transistor between the base and the emitter as v_{be}. Also, assume that, when e_s is +50 mV peak, I_B rises to 80 μA and that, when e_s is the negative peak, −50 mV, I_B falls to 40 μA. Then the peak-to-peak value of the ac signal voltage is 100 mV. The peak-to-peak value of the signal component of I_B is (80 − 40) or 40 μA. The peak value is 40/2 or 20 μA. The waveforms for the input signal voltage and the waveform for the base current are shown in Fig. 4-6*b*.

The numerical value of β_{ac} is 50. When the peak signal current in the base is 20 μA, the peak value of the signal current in the collector is

$$\Delta I_C = \beta_{ac}\,\Delta I_B = 50 \times 20 = 1000\ \mu\text{A} = 1\ \text{mA} \tag{4-4}$$

Thus, when the base current rises 20 μA from 50 μA to 70 μA, the collector current rises 1 mA from 3 mA to 4 mA. Similarly, when the base current falls 20 μA from 50 μA to 30 μA, the collector current falls 1 mA from 3 mA to 2 mA. The signal current in the collector has a peak value of 1 mA and a peak-to-peak value of 2 mA. This action is shown on the current waveforms of Fig. 4-6.

The voltage from the collector to ground (the emitter) determines the output signal. When I_C increases, V_{CE} decreases. This decrease is caused by an increased $I_C R_C$ voltage drop across R_C. When I_C changes 1 mA, the change in the voltage drop across R_C is 0.001 A $\times$ 3000 Ω or 3.0 V. Thus, when I_C rises from 3 mA to 4 mA, V_{CE} falls from 11 V to 8 V. When I_C falls from 3 mA to 2 mA, V_{CE} rises from 11 V to 14 V. The peak value of the ac output signal voltage is 3 V and its peak-to-peak value is 6 V. These voltage relationships are shown in Fig. 4-6. Examination of these waveforms leads to the very important conclusion that

> *The output voltage is* 180° *out of phase with the input signal in the common-emitter amplifier.*

We define the *current gain* A_i as

$$\boxed{A_i \equiv \frac{\text{the change in load current}}{\text{the change in input current}}} \tag{4-6}$$

Therefore, for this common-emitter amplifier, the current gain is:

$$A_i = \frac{\Delta I_C}{\Delta I_B} = \frac{2\ \text{mA}}{40\ \mu\text{A}} = \frac{2000\ \mu\text{A}}{40\ \mu\text{A}} = 50$$

We define the *voltage gain* A_v as

$$A_v = \frac{\text{the change in load voltage}}{\text{the change in input voltage}} \quad (4\text{-}7)$$

Then, for this common-emitter amplifier, the voltage gain is:

$$A_v = \frac{\Delta V_{CE}}{\Delta V_{BE}} = \frac{6\text{ V}}{100\text{ mV}} = \frac{6000\text{ mV}}{100\text{ mV}} = 60$$

We define the *power gain* A_p as the product of the current gain and the voltage gain.

$$A_p \equiv A_i \times A_v \quad (4\text{-}8)$$

Then, for this common-emitter amplifier, the power gain is

$$A_p = A_i \times A_v = 50 \times 60 = 3000$$

The ac load on the signal source e_s is the *input resistance* r_{in}, and it is defined as

$$r_{\text{in}} \equiv \frac{\text{the change in the input voltage}}{\text{the change in the input current}} \quad (4\text{-}9)$$

Then, for this common-emitter amplifier, the input resistance is

$$r_{\text{in}} = \frac{\Delta V_{BE}}{\Delta I_B} = \frac{100\text{ mV}}{40\ \mu\text{A}} = \frac{100\text{ mV}}{0.040\text{ mA}} = 2500\ \Omega$$

Section 4-4 The Common-Collector Amplifier—The Emitter Follower

In the *common-collector amplifier* or *emitter follower*, the load resistor is moved from the collector branch into the emitter branch, Fig. 4-7*a*. The output signal is taken from the emitter instead of from the collector. Thus, the output signal v_{out} is the ac voltage drop across R_E.

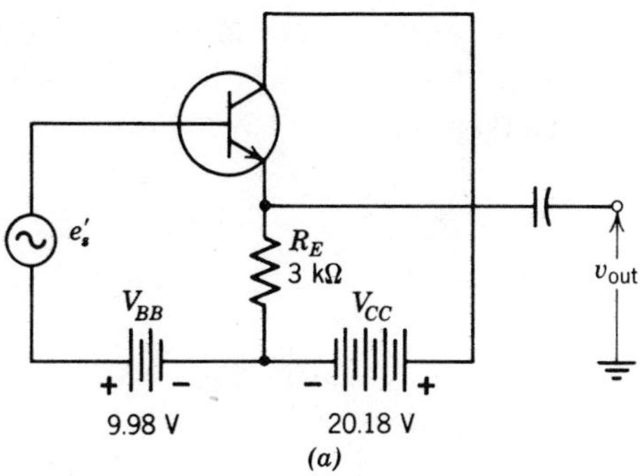

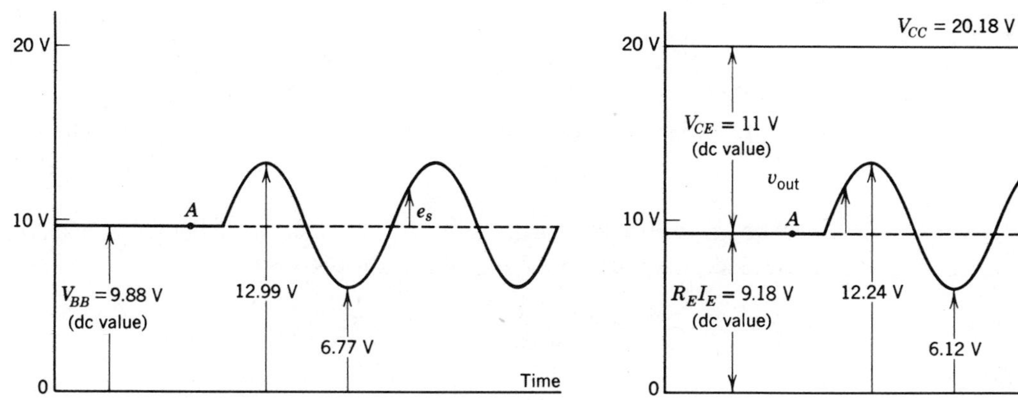

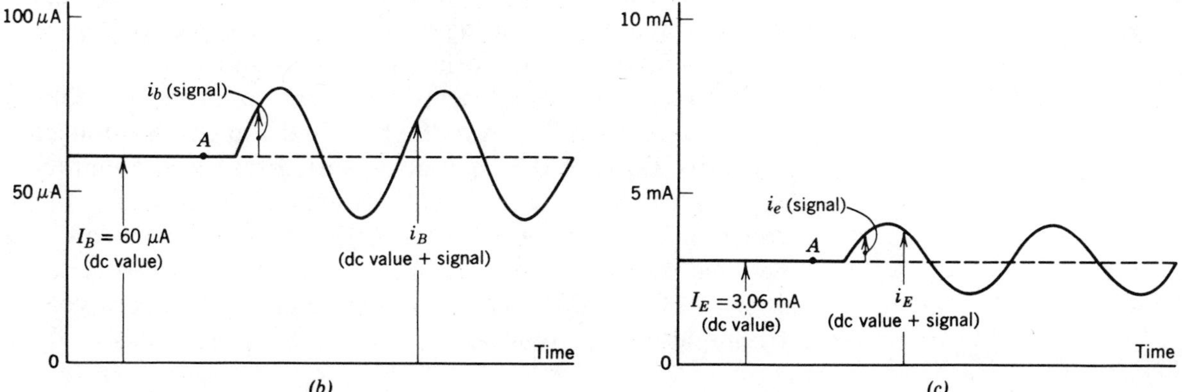

Figure 4-7 The common-collector amplifier. (*a*) Circuit. (*b*) Input waveforms. (*c*) Output waveforms.

In order to compare this circuit with the common-emitter amplifier, Fig. 4-6, we will use the same dc values for the transistor.

$$V_{BE} = 0.70\text{ V} \qquad \beta_{\text{dc}} = 50 \qquad V_{CE} = 11\text{ V}$$
$$I_B = 60\ \mu\text{A} \quad \text{and} \quad I_C = 3\text{ mA}$$

We will also use the same peak-to-peak signal values for the transistor.

$$\Delta V_{BE} = 100\text{ mV} \qquad \Delta I_B = 40\ \mu\text{A} \qquad \Delta I_C = 2\text{ mA} \quad \text{and} \quad \beta_{\text{ac}} = 50$$

The dc current in the emitter I_E is

$$I_E = I_B + I_C = 60\ \mu\text{A} + 3\text{ mA} = 0.06\text{ mA} + 3\text{ mA} = 3.06\text{ mA} \tag{4-1}$$

The dc voltage drop across R_E is $I_E R_E$ or $3.06\text{ mA} \times 3\text{ k}\Omega$ or 9.18 V. Inspection of Fig. 4-7*a* shows that

$$V_{CC} = I_E R_E + V_{CE} = 9.18 + 11 = 20.18\text{ V}$$

From this circuit, we also see that

$$V_{BB} = I_E R_E + V_{BE} = 9.18 + 0.70 = 9.88\text{ V}$$

These supply voltages are shown on the circuit. The supply voltages and the operating point values, labeled as points *A*, are shown on the waveforms, Figs. 4-7*b* and 4-7*c*.

The incoming signal e_s is sufficient to produce peak ac signal currents of 20 μA in the base and 1 mA in the collector. Then, by Eq. 4-1, the peak ac signal current in the emitter is the sum of the peak ac signal currents in the base and in the collector or $20\ \mu\text{A} + 1\text{ mA}$ or 1.02 mA. By Ohm's law, the peak ac output signal voltage is the ac voltage drop in R_E or $1.02 \times 3\text{ k}\Omega$ or 3.06 V. The maximum total instantaneous voltage from emitter to ground is

$$I_E R_E + 3.06 = 9.18 + 3.06 = 12.24\ V.$$

The minimum total instantaneous voltage from emitter to ground is

$$I_E R_E - 3.06 = 9.18 - 3.06 = 6.12 \text{ V}$$

The peak-to-peak output signal voltage is 2×3.06 or 6.12 V.

The incoming signal e_s is applied across the series circuit of the base to emitter of the transistor and R_E. The peak value of e_s is the peak value of 50 mV required from base to emitter plus the peak signal voltage across R_E, 3.06 V. The peak value of e_s is 50 mV + 3.06 V or 3.11 V. The maximum instantaneous voltage from base to ground is the sum of V_{BB} plus the peak signal voltage or $9.88 + 3.11$ or 12.99 V. The minimum instantaneous voltage from base to ground is V_{BB} minus the peak signal voltage or $9.88 - 3.11$ or 6.77 V. The peak-to-peak input ac signal voltage is 2×3.11 or 6.22 V. These values are shown on the waveforms of Fig. 4-7.

Examination of the waveforms shown in Fig. 4-7 shows that, when e_s increases in the positive direction, v_{out} also increases in the positive direction. The important conclusion we draw is:

The output voltage of the emitter-follower circuit is in phase with the input signal.

The current gain is the ratio of the signal current in R_E to the signal current in the base.

$$A_i = \frac{1.02 \text{ mA}}{20 \ \mu\text{A}} = \frac{1020 \ \mu\text{A}}{20 \ \mu\text{A}} = 51$$

The voltage gain is the ratio of the output signal voltage across R_E to the input signal voltage.

$$A_v = \frac{3.06 \text{ V}}{3.11 \text{ V}} = \frac{6.12 \text{ V}}{6.22 \text{ V}} = 0.984 \approx 1.000$$

The power gain is the product of the current gain and the voltage gain.

$$A_p = A_i \times A_v = 51 \times 1 = 51$$

The input resistance is the peak input signal voltage divided by the peak signal base current.

$$r_{in} = \frac{3.11 \text{ V}}{20 \ \mu\text{A}} = 155\ 500 \ \Omega = 155.5 \text{ k}\Omega$$

A comparison with the common-emitter amplifier will be made at the end of the next section.

Section 4-5 The Common-Base Amplifier

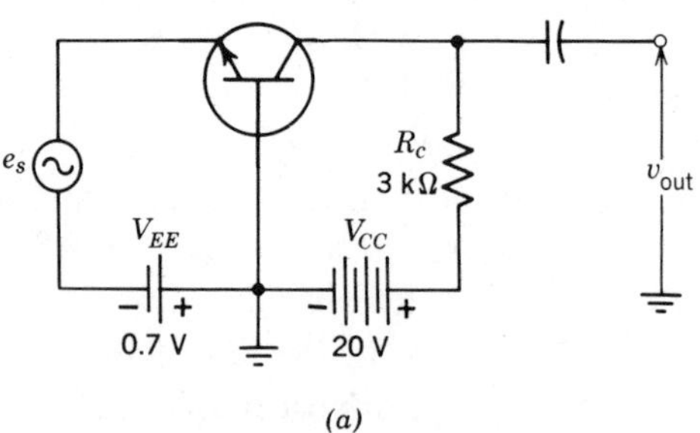

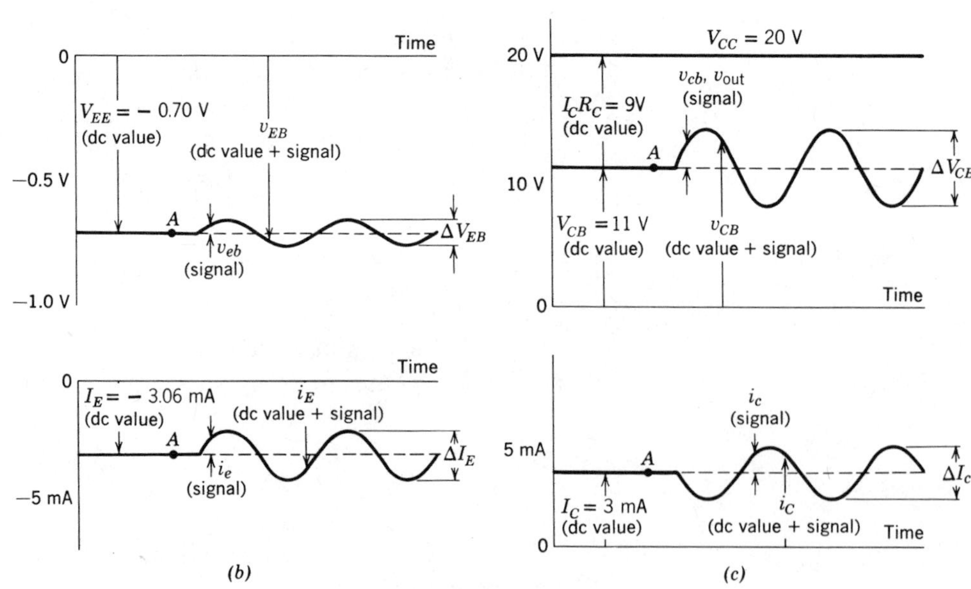

Figure 4-8 The common-base amplifier. (*a*) Circuit. (*b*) Emitter waveforms. (*c*) Collector waveforms.

In the *common-base amplifier* the signal is fed into the emitter and taken from the collector (Fig. 4-8*a*). The common-base amplifier is used primarily in radio-frequency amplifiers, which are beyond the scope of this text. The circuit is used only occasionally at low frequencies.

The bias supply V_{EE} provides the required forward bias on the emitter-to-base junction of the *NPN* silicon transistor. In this circuit a negative bias supply is required for the emitter whereas V_{CC} must be positive to place a reverse voltage on the collector. In order to make a comparison with the common-emitter amplifier (Section 4-3) and the emitter-follower amplifier (Section 4-4), we will use the values:

$$V_{EE} = -0.7 \text{ V} \qquad V_{CC} = +20 \text{ V} \qquad \beta = 50$$
$$I_B = 60\ \mu\text{A} \qquad I_C = 3 \text{ mA} \qquad I_E = 3.06 \text{ mA}$$
$$\Delta I_B = 40\ \mu\text{A} \qquad \Delta I_C = 2 \text{ mA} \qquad \Delta I_E = 2.04 \text{ mA}$$

The signal voltage e_s required is the same as used for the common-emitter amplifier.

$$\Delta V_{EB} = 100 \text{ mV}$$

and the corresponding change in the voltage drop across R_C is the peak-to-peak value of the output voltage.

$$\Delta I_C R_C = \Delta V_{CB} = 6.0 \text{ V}$$

The operating point values (the dc values) are shown on the waveforms, Fig. 4-8*b* and Fig. 4-8*c*, as the points marked *A*. When the applied signal e_s increases in a positive direction, the increase opposes the polarity of V_{EE}. The new forward bias on the emitter-to-base junction is reduced and, consequently, both the emitter current and the collector current decrease. This decrease in I_C reduces the voltage drop across R_C, and V_{CB}, the output voltage, increases. The waveforms that are produced are shown in Fig. 4-8*b* and in Fig. 4-8*c*.

The important conclusion to be made from these waveforms is:

The output voltage is in phase with the input voltage in the common-base amplifier.

The current gain of the amplifier is

$$A_i = \frac{\Delta I_C}{\Delta I_E} = \frac{2.00 \text{ mA}}{2.04 \text{ mA}} = 0.98 \approx 1 \tag{4-6}$$

The voltage gain of the amplifier is

$$A_v = \frac{\Delta V_{CB}}{\Delta V_{EB}} = \frac{6.0 \text{ V}}{100 \text{ mV}} = 60 \tag{4-7}$$

The power gain of the amplifier is

$$A_p = A_i A_v = 0.98 \times 60 = 58.8 \approx 60 \tag{4-8}$$

The input resistance to the circuit is

$$r_{\text{in}} = \frac{\Delta V_{EB}}{\Delta I_E} = \frac{100 \text{ mV}}{2.04 \text{ mA}} = 49\ \Omega \tag{4-9}$$

The results of these three basic amplifier circuits are summarized in Table 4-1. Most amplifiers use the common-emitter circuit configuration because the circuit offers both voltage and current gain resulting in a much higher power gain than can be obtained by either the emitter follower or the common-base amplifier. One other important consideration for the use of the common-emitter amplifier is that its input resistance r_{in} (2500 Ω in our example) is of the order of the load resistance (3000 Ω in our example).

Table 4-1
Comparison of the Basic Amplifier Circuits

	Common-Emitter	**Common-Collector (Emitter Follower)**	**Common-Base**
Current gain, A_i	50	51	$0.98 \approx 1$
Voltage gain, A_v	60 out of phase	1 in phase	60 in phase
Power gain, A_p	3000	51	$58.8 \approx 60$
Input resistance, r_{in}	250 Ω	155,500 Ω	49 Ω
Phase shift	180°	0°	0°

The emitter follower is used whenever a very high input resistance is required. The emitter follower is especially useful to match a high-resistance (impedance) source to a low-resistance (impedance) load. The common-base amplifier has the very material disadvantage that the input resistance to the circuit is very low especially when compared to the order of magnitude of the load resistor used in the circuit.

Section 4-6 Relationships between α and β

In Section 4-1 we showed that the currents in a transistor are related by

$$I_E = I_B + I_C \tag{4-1}$$

We defined the ratio of collector current to base current as beta.

$$\beta_{\text{dc}} \equiv \frac{I_C}{I_B} \tag{4-2}$$

We defined the ratio of collector current to emitter current as alpha.

$$\alpha_{\text{dc}} \equiv \frac{I_C}{I_E} \tag{4-3}$$

We now wish to show the relationships between α_{dc} and β_{dc} and to establish conversions from one current to another current in terms of α_{dc} and β_{dc}.

Solving Eq. 4-1 for I_B we have

$$I_B = I_E - I_C$$

and substituting this expression for I_B in Eq. 4-2 we find that

$$\beta_{\text{dc}} = \frac{I_C}{I_E - I_C}$$

Dividing each term by I_E, we have

$$\beta_{\text{dc}} = \frac{I_C/I_E}{I_E/I_E - I_C/I_E}$$

Replacing I_C/I_E by α_{dc} (Eq. 4-3) we have

$$\beta_{dc} = \frac{\alpha_{dc}}{1-\alpha_{dc}}$$

We could have derived this equation in terms of β_{ac} and α_{ac} by using the definition for β_{ac} (Eq. 4-4) and by using the definition for α_{ac} (Eq. 4-5). In order to simplify future equations, we will, at this point, drop the subscripts *dc* and *ac* on β and α. It should be obvious from a specific equation which subscript is implied. Now we may write the last equation as

$$\boxed{\beta = \frac{\alpha}{1-\alpha}} \tag{4-10}$$

Now let us cross multiply Eq. 4-10.

$$\beta(1-\alpha) = \alpha$$

Expanding we have

$$\beta - \alpha\beta = \alpha$$

Rearranging we have

$$\beta = \alpha + \alpha\beta$$

Factoring we have

$$\beta = \alpha(1+\beta)$$

Solving for α we find

$$\boxed{\alpha = \frac{\beta}{1+\beta}} \tag{4-11}$$

Equation 4-10 and Eq. 4-11 are very important because they give us the means to convert from α to β or from β to α. Now we require means of converting from one current to another in terms of α and β.

From Eq. 4-2 we can write

$$I_C = \beta I_B$$

and

$$I_B = \frac{1}{\beta} I_C$$

and from Eq. 4-3 we can write

$$I_C = \alpha I_E$$

and

$$I_E = \frac{1}{\alpha} I_C$$

If we substitute Eq. 4-11 into these last two equations, we have

$$I_C = \alpha I_E = \frac{\beta}{1+\beta} I_E$$

and

$$I_E = \frac{1}{\alpha} I_C = \frac{1+\beta}{\beta} I_C$$

If we take Eq. 4-1

$$I_E = I_B + I_C \tag{4-1}$$

and substitute βI_B for I_C, we have

$$I_E = I_B + \beta I_B = (1+\beta) I_B$$

and

$$I_B = \frac{1}{1+\beta} I_E$$

We derived these conversions in terms of the dc values (I_B, I_C, and I_E). Identical results are obtained if the ac signal values (i_b, i_c, and i_e) are used.

The results of this section are summarized in Table 4-2. These conversions are used continuously in the study of transistor circuits. It is most important that the student learn them at the earliest possible time in his or her study.

Table 4-2
Transistor Current Relationships
Multiplying Factors To Convert

From \ To	I_B (or i_b)	I_C (or i_c)	I_E (or i_e)
I_B (or i_b)	1	β	$1+\beta$
I_C (or i_c)	$\frac{1}{\beta}$	1	$\frac{1+\beta}{\beta}$ or $\frac{1}{\alpha}$
I_E (or i_e)	$\frac{1}{1+\beta}$	$\frac{\beta}{1+\beta}$ or α	1

$$\beta = \frac{\alpha}{1-\alpha} \qquad \alpha = \frac{\beta}{1+\beta}$$

Problems

4-6.1 Find α for each of the following values of β.
50, 100, 120, 150, and 200

4-6.2 Find α for each of the following values of β.
46, 65, 84, 125, and 165

4-6.3 Find β for each of the following values of α.
0.995, 0.990, 0.9875, and 0.9765

4-6.4 Find β for each of the following values of α.
0.991, 0.962, 0.946, and 0.983

4-6.5 If the base current in a transistor is 20 μA when the emitter current is 6.4 mA, what are the values for α and β?

4-6.6 The published value of β for a transistor states that β can vary from 40 to 90. If I_B is fixed at 16 μA, what is the expected variation in I_C?

Questions

4-1 Describe the construction of a junction transistor.

4-2 Where are depletion regions found in a transistor?

4-3 What is the polarity of V_{CC} when applied to a *PNP* transistor? To an *NPN* transistor?

4-4 Why is forward bias applied between the base and the emitter of a transistor?

4-5 What is the polarity of forward bias on the base-to-emitter junction of a *PNP* transistor? Of an *NPN* transistor?

4-6 How is I_C related to I_B and I_E?

4-7 Define β_{dc}, α_{dc}, β_{ac}, and α_{ac}.

4-8 Can two separate diodes be used to take the place of a transistor? Explain.

4-9 What are the characteristics of a common-emitter amplifier circuit (A_i, A_v, r_{in}, and phase shift)?
4-10 What are the characteristics of an emitter follower?
4-11 What are the characteristics of a common-base amplifier circuit?
4-12 If α is given, how is β obtained?
4-13 If β is given, how is α obtained?
4-14 What is I_B in terms of I_C?
4-15 What is I_B in terms of I_E?
4-16 What is I_C in terms of I_B?
4-17 What is I_C in terms of I_E?
4-18 What is I_E in terms of I_B?
4-19 What is I_E in terms of I_C?

Chapter 5 Transistor Bias

The resistors used in an amplifier circuit determine the dc operating point of the transistor used in the circuit (Section 5-1). Details of the calculations are given for the common-emitter amplifier circuit, for the common-collector amplifier circuit, and for the common-base amplifier circuit (Section 5-2). The more complex circuits for the common-emitter amplifier using emitter feedback and using collector-to-base feedback are examined (Section 5-3). Methods of treating circuits using voltage dividers to derive the desired voltages are also considered in the last section.

Section 5-1 Transistor Bias Circuits

In order to use a transistor as an amplifier, an associated network of resistors together with suitable dc supply voltages must be used. The supply voltages and the resistors establish a set of dc electrode voltages and currents for the transistor, called *quiescent values*, that determine the *operating point* or the *Q point* for the transistor. In most cases, the quiescent values are *not* changed by the application of an ac input signal to the circuit. We will show in Chapter 6 and in Chapter 7 how the operating-point values determine the gain characteristics of the amplifier.

Before we analyze an actual circuit, we should outline the general procedures and concepts that apply to all the circuits in this chapter.

The general procedure to determine the operating-point values is, in itself, simple:

1. Equations are written for the circuit based on Kirchhoff's voltage law.
2. Equations are written for the circuit based on Kirchhoff's current law.
3. Known numerical values are substituted into the equations.
4. The equations are solved for the missing numerical values.

If we have an equation that contains I_B, I_C, and I_E as unknowns, we can reduce the number of unknowns to one by making use of the conversion factors developed for Table 4-2 (Page 98). These conversion factors were derived from Kirchhoff's current law ($I_E = I_B + I_C$) and from the definitions of α and β. Whenever we use one of the conversions of Table 4-2, we are actually using Kirchhoff's current law.

Most low- to medium-power semiconductor circuits use resistor component values that are conveniently expressed in kilohms (kΩ). The semiconductors have currents that are measured in milliamperes (mA). The numerical calculations are greatly simplified if Ohm's law is revised to:

$$\begin{aligned} \text{Volts} &= \text{milliamperes} \times \text{kilohms} \\ \text{Volts} &= \text{mA} \times \text{k}\Omega \end{aligned} \tag{5-1}$$

In this chapter, we will use Eq. 5-1 for all numerical calculations.

Example

$$V = IR = 0.002 \text{ A} \times 10{,}000\ \Omega = 20 \text{ V}$$

or

$$V = 2 \text{ mA} \times 10 \text{ k}\Omega = 20 \text{ V}$$

Example

$$R = \frac{V}{I} = \frac{15 \text{ V}}{0.003 \text{ A}} = 5000\ \Omega$$

or

$$R = \frac{15 \text{ V}}{3 \text{ mA}} = 5 \text{ k}\Omega$$

Section 5-2 Biasing the Basic Transistor Circuits

The common-emitter amplifier circuit is shown in Fig. 5-1. *Usually V_{BB} and V_{CC} are obtained from the same source voltage so that only one source voltage is required for the circuit.* The Kirchhoff's voltage loop equation for the input circuit is

$$V_{BB} = R_B I_B + V_{BE} \tag{5-2a}$$

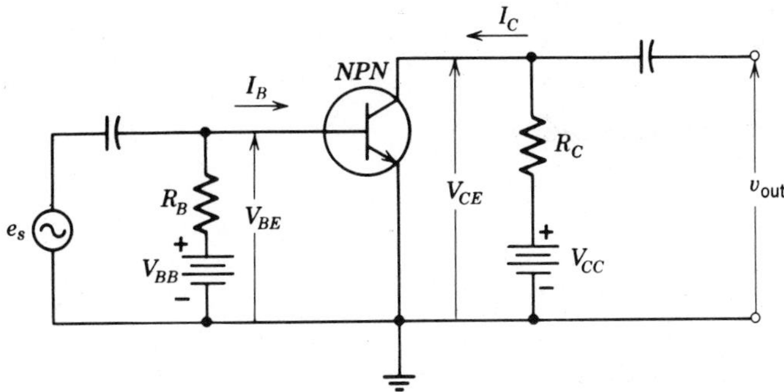

Figure 5-1 The common-emitter amplifier circuit.

where V_{BE} is the voltage measured *from* the base *to* the emitter.

The Kirchhoff's voltage loop equation for the output circuit is

$$V_{CC} = R_C I_C + V_{CE} \tag{5-2b}$$

where V_{CE} is the voltage measured *from* the collector *to* the emitter.

Using the conversion of Table 4-2, we can write

$$I_C = \beta I_B \quad \text{or} \quad I_B = \frac{I_C}{\beta} \tag{5-2c}$$

Example 5-1

Assume for the circuit of Fig. 5-1 that we have the following numerical values for a silicon transistor.

$$V_{BB} = +10\text{ V} \qquad V_{CC} = +10\text{ V} \qquad R_C = 4\text{ k}\Omega$$
$$V_{BE} = 0.7\text{ V} \quad \text{and} \quad \beta = 50$$

The value of R_B is required to set V_{CE} to +5 V.

Solution

We substitute the numerical values into the loop equations for the circuit.

$$V_{BB} = R_B I_B + V_{BE} \tag{5-2a}$$

$$10 = R_B I_B + 0.7 \tag{1}$$

$$V_{CC} = R_C I_C + V_{CE} \tag{5-2b}$$

$$10\text{ V} = 4\text{ k}\Omega \times I_C + 5\text{ V} \tag{2}$$

and

$$I_B = \frac{I_C}{\beta} = \frac{I_C}{50} \qquad (3) \qquad (5\text{-}2c)$$

Equation (2) can be solved for I_C

$$10\text{ V} = 4\text{ k}\Omega \times I_C + 5\text{ V}$$

$$I_C = \frac{5\text{ V}}{4\text{ k}\Omega} = 1.25\text{ mA}$$

and substituted into Eq. (3).

$$I_B = \frac{I_C}{50} = \frac{1.25\text{ mA}}{50} = 0.025\text{ mA} = 25\ \mu\text{A}$$

Using this value in Eq. (1) we have

$$10\text{ V} = R_B I_B + 9.7\text{ V}$$

$$10\text{ V} = R_B \times 0.025\text{ mA} + 0.7\text{ V}$$

$$R_B = \frac{9.3\text{ V}}{0.025\text{ mA}} = \mathbf{372\text{ k}\Omega}$$

The common-collector amplifier circuit is shown in Fig. 5-2. Usually V_{BB} and V_{CC} are the same voltage source so that only one source voltage is required for the circuit. The Kirchhoff's voltage loop equation through the base for the input circuit is

$$V_{BB} = R_B I_B + V_{BE} + R_E I_E \qquad (5\text{-}3a)$$

and Kirchhoff's voltage loop equation through the collector

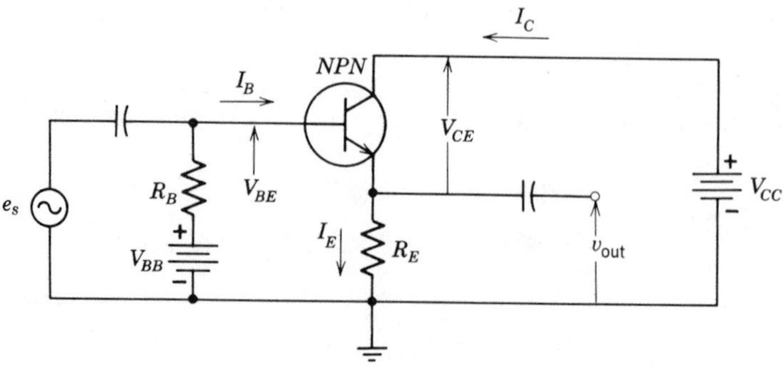

Figure 5-2 The common-collector amplifier circuit.

for the output circuit is

$$V_{CC} = V_{CE} + R_E I_E \tag{5-3b}$$

From Table 4-2, we use the conversion factors to write

$$I_E = (1+\beta)I_B \quad \text{or} \quad I_B = \frac{I_E}{1+\beta} \tag{5-3c}$$

Example 5-2

Assume for Fig. 5-2 that we have the following numerical values for a silicon transistor.

$$V_{BB} = +10\text{ V} \qquad V_{CC} = +10\text{ V} \qquad R_E = 4\text{ k}\Omega$$
$$V_{BE} = 0.7\text{ V} \quad \text{and} \quad \beta = 50$$

Find the value for R_B required to set V_{CE} to 5 V.

Solution

Placing these values into Eqs. 5-3*a*, 5-3*b*, and 5-3*c* we have

$$V_{BB} = R_B I_B + V_{BE} + R_E I_E \qquad (5\text{-}3a)$$

$$10\text{ V} = R_B I_B + 0.7\text{ V} + 4\text{ k}\Omega \times I_E \qquad (1)$$

$$V_{CC} = V_{CE} + R_E I_E \qquad (5\text{-}3b)$$

$$10\text{ V} = 5\text{ V} + 4\text{ k}\Omega \times I_E \qquad (2)$$

and

$$I_B = \frac{I_E}{1+\beta} = \frac{I_E}{51} \qquad (3) \qquad (5\text{-}3c)$$

Solving Eq. (2) for I_E we have

$$4\text{ k}\Omega \times I_E = 5\text{ V}$$

$$I_E = \frac{5\text{ V}}{4\text{ k}\Omega} = 1.25\text{ mA}$$

and substituting into Eq. (3) we have

$$I_B = \frac{1.25\text{ mA}}{51} = 0.0245\text{ mA} = 24.5\ \mu\text{A}$$

Using these values in Eq. (1) we find that

$$10\text{ V} = R_B I_B + 0.7\text{ V} + 4\text{ k}\Omega \times I_E$$

$$10\text{ V} = R_B(0.0245\text{ mA}) + 0.7\text{ V} + 4\text{ k}\Omega \times 1.25\text{ mA}$$

$$10\text{ V} = 0.0245\text{ mA} \times R_B + 0.7\text{ V} + 5\text{ V}$$

$$R_B = \frac{4.3\text{ V}}{0.0245\text{ mA}} = \mathbf{175.5\ k\Omega}$$

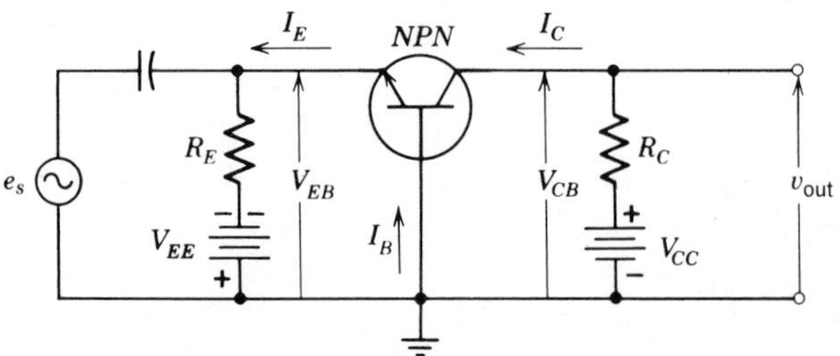

Figure 5-3 The common-base amplifier circuit.

The common-base amplifier circuit is shown in Fig. 5-3. In this circuit we must have two different dc power sources because two different polarities are required for the source voltages. The input Kirchhoff's voltage loop equation through the emitter is

$$V_{EE} = R_E I_E + V_{EB} \tag{5-4a}$$

and the output Kirchhoff's voltage loop equation through the collector is

$$V_{CC} = R_C I_C + V_{CB} \tag{5-4b}$$

Using the conversion factors of Table 4-2, we can write

$$I_C = \alpha I_E = \frac{\beta}{1+\beta} I_E \tag{5-4c}$$

Example 5-3

Determine I_C, I_E, and R_E in the circuit of Fig. 5-3 if V_{CB} is 5 V and the circuit values are:

$$V_{EE} = -10\text{ V} \qquad V_{CC} = +10\text{ V} \qquad R_C = 4\text{ k}\Omega$$
$$V_{EB} = 0.7\text{ V} \quad \text{and} \quad \alpha = 0.98$$

Solution
Placing these values into Eqs. 5-4*a*, 5-4*b*, and 5-4*c* we have

$$10\text{ V} = R_E I_E + 0.7\text{ V} \qquad (1)$$

$$10\text{ V} = 4\text{ k}\Omega \times I_C + 5\text{ V} \qquad (2)$$

and

$$I_C = 0.98 I_E \qquad (3)$$

Eq. (2) can be solved for I_C.

$$10\text{ V} = 4\text{ k}\Omega \times I_C + 5\text{ V}$$

$$4\text{ k}\Omega \times I_C = 5\text{ V}$$

$$I_C = \frac{5\text{ V}}{\text{k}\Omega} = \mathbf{1.25\text{ mA}}$$

If we use Eq. (3)

$$I_E = \frac{1}{0.98} I_C = \frac{1.25\text{ mA}}{0.98} = \mathbf{1.2755\text{ mA}}$$

and substitute into Eq. (1)

$$10\text{ V} = R_E \times 1.2755\text{ mA} + 0.7\text{ V}$$

$$R_E = \frac{10\text{ V} - 0.7\text{ V}}{1.2755\text{ mA}} = \frac{9.3\text{ V}}{1.2755\text{ mA}} = \mathbf{7.29\text{ k}\Omega}$$

Problems

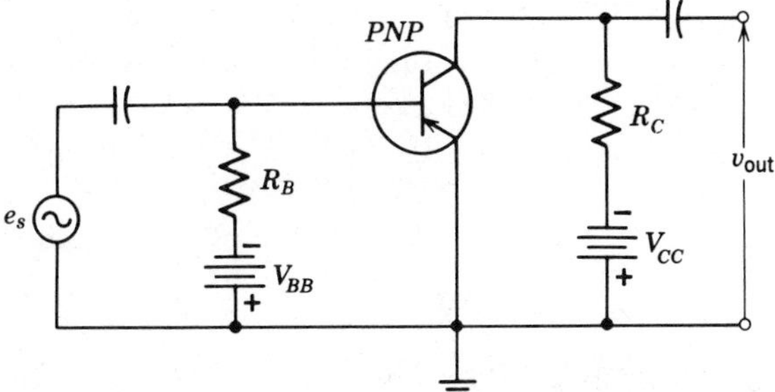

Circuit for Problems 5-2.1 through 5-2.4.

For all problems, assume the transistor is silicon.

$V_{BE} = 0.7\text{ V} \qquad \beta = 70 \qquad \alpha = 1.00$ and $V_{CE,\text{sat}} = 0.2\text{ V}$

5-2.1 V_{BB} and V_{CC} are each -6 V. R_B is 50 kΩ. What is R_C if V_{CE} is 2 V?

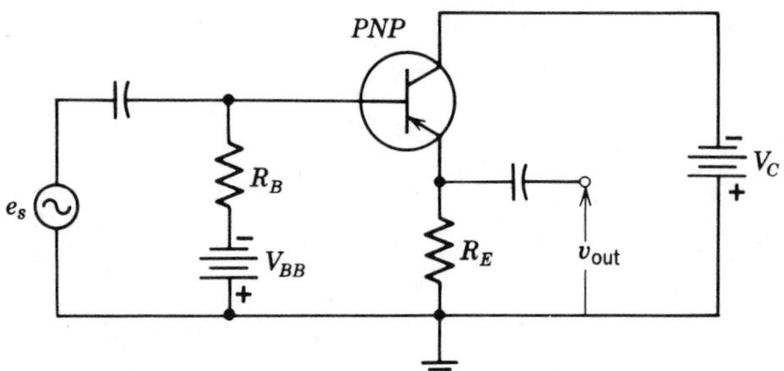

Circuit for Problems 5-2.5 through 5-2.8.

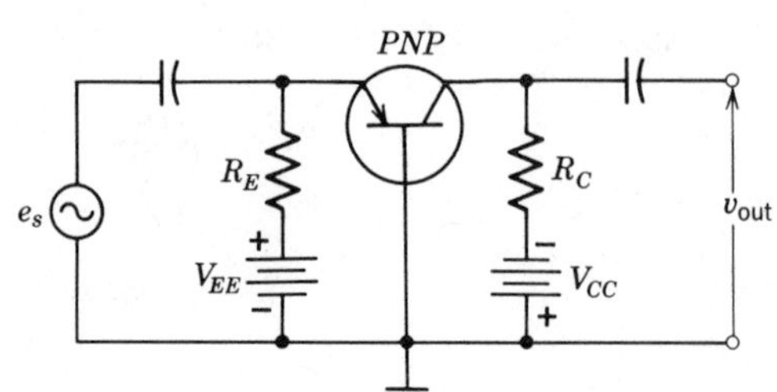

Circuit for Problems 5-2.9 through 5-2.12.

5-2.2 V_{BB} and V_{CC} are each −6 V. If V_{CE} is 3 V when I_B is 50 μA, what are the values for R_B and R_C?

5-2.3 What value of R_C causes saturation ($V_{CE,sat}$) in Problem 5-2.1?

5-2.4 V_{BB} and V_{CC} are each −8 V. R_C is 2 kΩ and V_{CE} is 4 V. When both V_{BB} and V_{CC} are raised to −10 V, what is the new value of V_{CE}?

5-2.5 V_{BB} and V_{CC} are each −9 V. R_E is 100 Ω. Determine R_B to set V_{CE} to 4.5 V.

5-2.6 V_{BB} and V_{CC} are each −4 V. R_E is 2 kΩ. What value of R_B is required to set V_{CE} to 2 V?

5-2.7 V_{BB} and V_{CC} are each 12 V. R_E is 10 kΩ. What value of R_B sets V_{CE} to 10 V?

5-2.8 If R_B in Problem 5-2.6 is halved, what is V_{CE}? If R_B in Problem 5-2.6 is doubled, what is V_{CE}?

5-2.9 $V_{EE} = +6$ V $V_{CC} = -6$ V $R_C = 2$ kΩ
Find the value of R_E required to set V_{CB} to 4 V.

5-2.10 $V_{EE} = +6$ V $V_{CC} = -20$ V $R_C = 10$ kΩ
Find the value of R_E required to set V_{CB} to 10 V.

5-2.11 $V_{EE} = +6$ V $V_{CC} = -6$ V $R_C = 2$ kΩ
If V_{CB} is 2 V, what is R_E?

5-2.12 $V_{EE} = +12$ V $V_{CC} = -12$ V $R_E = 5$ kΩ
What value of R_C establishes V_{CB} at 9 V?

Section 5-3 Complex Transistor Bias Circuits

The common-emitter amplifier circuit with emitter feedback is shown in Fig. 5-4. In the previous circuits in Section 5-2, we used the complete battery symbols for the dc supplies. From this point on, we will indicate the supply sources simply by indicating a single terminal. The terminal of each power source is returned to the common point of the circuit, in this case, indicated by the ground symbol.

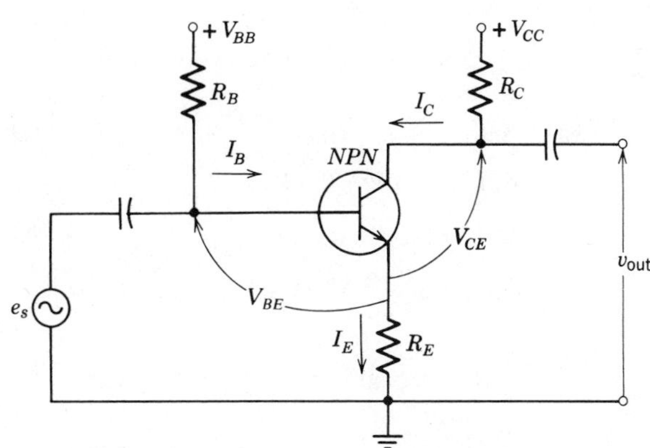

Figure 5-4 The common-emitter amplifier circuit with emitter feedback.

Usually V_{BB} and V_{CC} are the same source voltage so that only one source voltage is required for the circuit. Then we use V_{CC} for V_{BB} in all the equations.

The Kirchhoff's voltage loop equation for the input circuit through the base is

$$V_{BB} = R_B I_B + V_{BE} + R_E I_E \tag{5-5a}$$

and the Kirchhoff's voltage loop equation for the output circuit through the collector is

$$V_{CC} = R_C I_C + V_{CE} + R_E I_E \tag{5-5b}$$

Using the conversions from Table 4-2, we have

$$I_E = \frac{1+\beta}{\beta} I_C \tag{5-5c}$$

and

$$I_B = \frac{I_C}{\beta} \tag{5-5d}$$

Example 5-4

Assume for Fig. 5-4 we have the following numerical values for a germanium transistor.

$$V_{BB} = +15\text{ V} \qquad V_{CC} = +15\text{ V} \qquad V_{BE} = 0.3\text{ V}$$
$$R_C = 4\text{ k}\Omega \qquad R_E = 600\ \Omega \qquad \beta = 60$$

The value of R_B is required to set V_{CE} to 8 V.

Solution
Placing these values into Eqs. 5-5*a*, 5-5*b*, and 5-5*c* we have

$$V_{BB} = R_B I_B + V_{BE} + R_E I_E \qquad (5\text{-}5a)$$

$$15\text{ V} = R_B I_B + 0.3\text{ V} + 0.6\text{ k}\Omega \times I_E \quad (1)$$

and

$$V_{CC} = R_C I_C + V_{CE} + R_E I_E \qquad (5\text{-}5b)$$

$$15\ V = 4\text{ k}\Omega \times I_C + 8\text{ V} + 0.6\text{ k}\Omega \times I_E \quad (2)$$

and

$$I_E = \frac{1+\beta}{\beta} I_C = \frac{61}{60} I_C \qquad (3) \qquad (5\text{-}5c)$$

and

$$I_B = \frac{I_C}{\beta} = \frac{I_C}{60} \qquad (4)$$

Eq. (1) has three unknowns. Eq. (2) has two unknowns. We can reduce Eq. (2) to one unknown by using Eq. (3).

$$15\text{ V} = 4\text{ k}\Omega \times I_C + 8\text{ V} + 0.6\text{ k}\Omega \times I_E$$

$$15\text{ V} = 4\text{ k}\Omega \times I_C + 8\text{ V} + 0.6\text{ k}\Omega \times \frac{61}{60} I_C$$

$$4.61\text{ k}\Omega \times I_C = 7\text{ V}$$

$$I_C = 1.518\text{ mA}$$

Using this value for I_C in Eq. (4) we have

$$I_B = \frac{I_C}{60} = \frac{1.518\text{ mA}}{60} = 0.025\text{ mA} = 25\ \mu\text{A}$$

Recalling that

$$I_E = I_B + I_C$$

then

$$I_E = 0.025 + 1.518 = 1.543\text{ mA}$$

Now we can substitute these numbers into Eq. (1).

$$15\text{ V} = R_B I_B + 0.3\text{ V} + 0.6\text{ k}\Omega \times I_E$$

$$15\text{ V} = R_B \times 0.025\text{ mA} + 0.3\text{ V} + 0.6\text{ k}\Omega \times 1.543\text{ mA}$$

$$0.025\text{ mA} \times R_B = 13.77\text{ V}$$

$$R_B = \frac{13.77\text{ V}}{0.025\text{ mA}} = \mathbf{551\text{ k}\Omega}$$

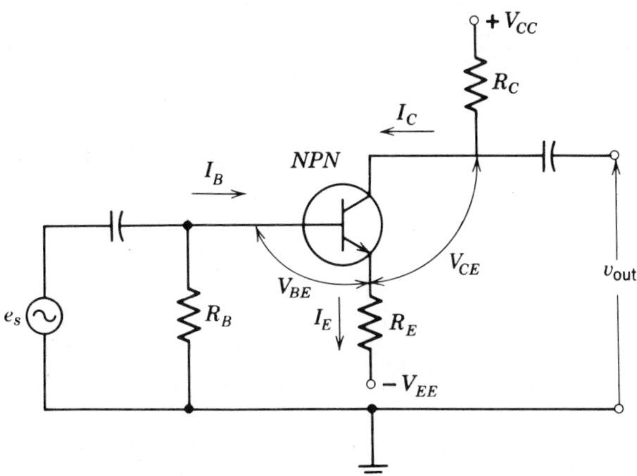

Figure 5-5 Common-emitter amplifier circuit with emitter feedback.

A different arrangement of the common-emitter amplifier circuit is shown in Fig. 5-5. When the voltage loop equation is formed through the collector, the total voltage on the circuit is the *difference of potential* between the two ends of the circuit.

$$V_{CC} - (-V_{EE}) = V_{CC} + V_{EE}$$

When the voltage loop is formed through the base, there is only one supply, V_{EE}, that is considered for the circuit. Then, the two voltage loop equations are

$$V_{CC} + V_{EE} = R_C I_C + V_{CE} + R_E I_E \qquad (5\text{-}6a)$$

and

$$V_{EE} = R_B I_B + V_{BE} + R_E I_E \qquad (5\text{-}6b)$$

From Table 4-2, the two current equations are

$$I_E = \frac{1+\beta}{\beta} I_C \qquad (5\text{-}6c)$$

and

$$I_B = \frac{I_C}{\beta} \qquad (5\text{-}6d)$$

The method of numerical solution of Eqs. 5-6*a*, 5-6*b*, 5-6*c*, and 5-6*d* does not introduce any new concepts. Therefore, we will leave the numerical method to the problem set.

The circuit shown in Fig. 5-6 uses both emitter feedback through R_E and collector-to-base feedback through R_B. Here,

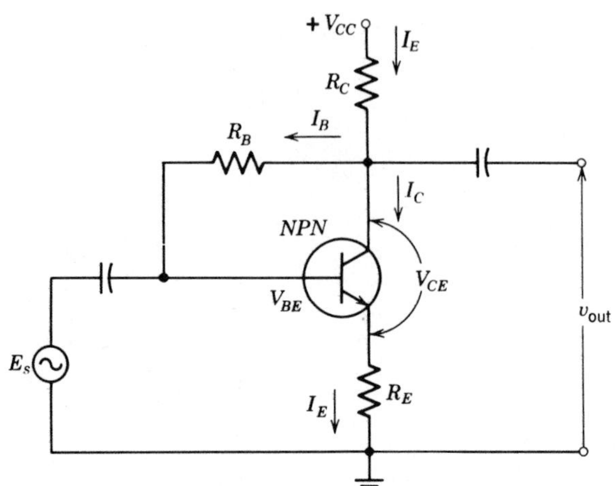

Figure 5-6 Common-emitter amplifier circuit with collector-to-base feedback and with emitter feedback.

R_B is returned to the collector instead of to a base-supply voltage. It is important to realize that the current in R_C is I_E and not I_C. The voltage loop equation through the collector is

$$V_{CC} = R_C I_E + V_{CE} + R_E I_E \tag{5-7a}$$

The voltage loop equation through the base is

$$V_{CC} = R_C I_E + R_B I_B + V_{BE} + R_E I_E \tag{5-7b}$$

The current equation from Table 4-2 relating I_B and I_E is

$$I_B = \frac{I_E}{1+\beta} \quad \text{or} \quad I_E = (1+\beta)I_B \tag{5-7c}$$

Again, the solution of these equations for numerical values is left to the problem set.

The circuit shown in Fig. 5-7*a* uses a voltage divider (R_1 and R_2) to provide the bias for the base. This arrangement is commonly used for transistors incorporated in an IC (integrated circuit). The procedure that must be used to determine the currents in the circuit requires the application of Thévenin's theorem. The base lead is opened at point A in Fig. 5-7*a*. The divider formed by R_1 and R_2 is replaced by a source in series with a resistor (Fig. 5-7*b*). The source in the equivalent circuit by Thévenin's theorem is the open circuit

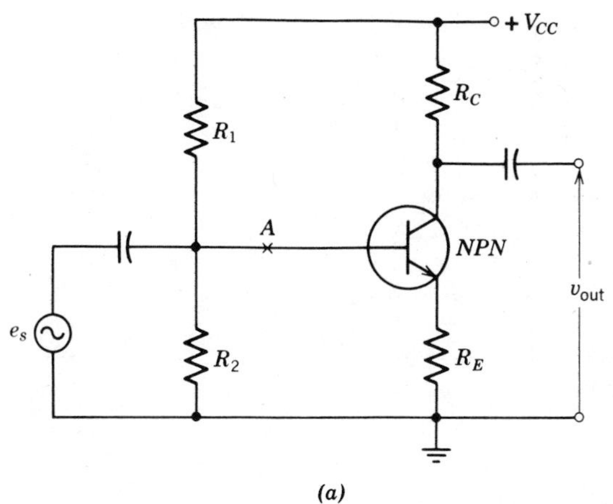

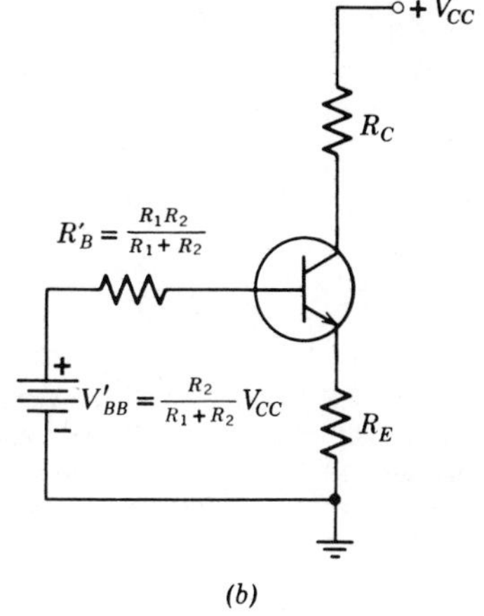

Figure 5-7 Common-emitter amplifier circuit with bias derived from a voltage divider. (*a*) Circuit. (*b*) Equivalent bias circuit.

voltage V'_{BB} measured at point A.

$$V'_{BB} = \frac{R_2}{R_1 + R_2} V_{CC} \qquad (5\text{-}8a)$$

The equivalent circuit resistance as specified by Thévenin's theorem "looks back" into the circuit at point A with the source voltage (V_{CC}) short circuited. Thus the equivalent circuit resistance R'_B is R_1 in parallel with R_2.

$$R'_B = \frac{R_1 R_2}{R_1 + R_2} \qquad (5\text{-}8b)$$

Example 5-5

Find the value of V_{CE} for the circuit shown in Fig. 5-8*a*. The transistor is germanium with a β of 50 and a value for V_{BE} of 0.3 V.

Solution

The first step is to reduce the voltage divider by Thévenin's theorem.

$$V'_{BB} = \frac{R_2}{R_1 + R_2} V_{CC} = \frac{20\ \text{k}\Omega}{20\ \text{k}\Omega + 120\ \text{k}\Omega} 10 = 1.43\ \text{V} \qquad (5\text{-}8a)$$

and

$$R'_B = \frac{R_1 R_2}{R_1 + R_2} = \frac{20\ \text{k}\Omega \times 120\ \text{k}\Omega}{20\ \text{k}\Omega + 120\ \text{k}\Omega} = 17.1\ \text{k}\Omega \qquad (5\text{-}8b)$$

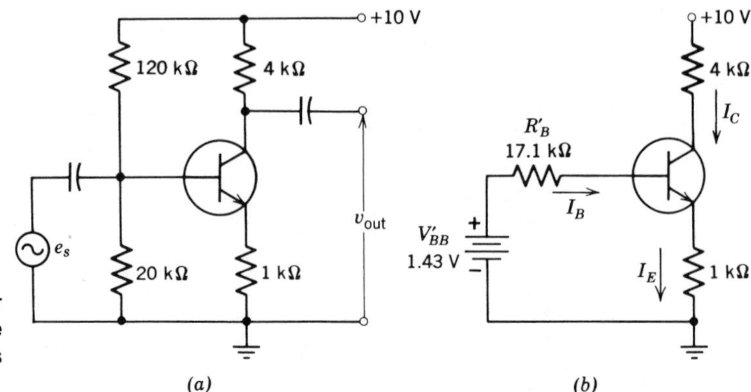

Figure 5-8 Common-emitter amplifier circuit with bias derived from a voltage divider. (*a*) Circuit. (*b*) Equivalent bias circuit.

These values are shown in the equivalent circuit of Fig. 5-8*b*. Now the voltage loop equations for the equivalent circuit are:

$$V_{CC} = R_C I_C + V_{CE} + R_E I_E \qquad (1) \quad (5\text{-}5b)$$

and

$$V'_{BB} = R'_B I_B + V_{BE} + R_E I_E \qquad (2) \quad (5\text{-}5a)$$

Substituting numerical values, we find that

$$10\text{ V} = 4\text{ k}\Omega \times I_C + V_{CE} + 1\text{ k}\Omega \times I_E \qquad (1)$$

and

$$1.43\text{ V} = 17.1\text{ k}\Omega \times I_B + 0.3\text{ V} + 1\text{ k}\Omega \times I_E \qquad (2)$$

Using Table 4-2 for the current relations, we find that

$$I_E = (1+\beta)I_B = 51 I_B \qquad (3)$$

and

$$I_C = \beta I_B = 50 I_B \qquad (4)$$

Substituting Eq. (3) into Eq. (2), we find that

$$1.43\text{ V} = 17.1\text{ k}\Omega \times I_B + 0.3\text{ V} + 1\text{ k}\Omega \times 51\ I_B$$

$$68.1\text{ k}\Omega \times I_B = 1.13\text{ V}$$

$$I_B = 0.0166\text{ mA} = 16.6\ \mu\text{A}$$

Evaluating Eq. (3) and Eq. (4) we have

$$I_C = 50 I_B = 50 \times 0.0166\text{ mA} = 0.830\text{ mA}$$

and

$$I_E = 51 I_B = 51 \times 0.0166\text{ mA} = 0.846\text{ mA}$$

Substituting these values into Eq. (1) we have

$$10\ \text{V} = 4\ \text{k}\Omega \times 0.830\ \text{mA} + V_{CE} + 0.846\ \text{V}$$

$$V_{CE} = 10\ \text{V} - 0.846\ \text{V} - 4\ \text{k}\Omega \times 0.830\ \text{mA} = \mathbf{5.83\ V}$$

Example 5-6
The value of R_2 is required in the circuit given in Fig. 5-9*a* to provide a value of 10 V for V_{CE}.

Solution
The Kirchhoff's voltage loop equation through the collector is

$$V_{CC} = R_C I_C + V_{CE} + R_E I_E$$

Using numerical values we have

$$20\ \text{V} = 6\ \text{k}\Omega \times I_C + 10\ \text{V} + 2\ \text{k}\Omega \times I_E \qquad (1)$$

But, from Table 4-2,

$$I_C = \beta I_B = 60\ I_B \qquad (2)$$

and

$$I_E = (1 + \beta) I_B = 61 I_B \qquad (3)$$

Substituting into Eq. (1) we find that

$$20\ \text{V} = 6\ \text{k}\Omega \times 60 I_B\ \text{mA} + 10\ \text{V} + 2\ \text{k}\Omega \times 61 I_B\ \text{mA}$$

$$482\ \text{k}\Omega \times I_B = 10\ \text{V}$$

$$I_B = 0.0207\ \text{mA} = 20.7\ \mu\text{A}$$

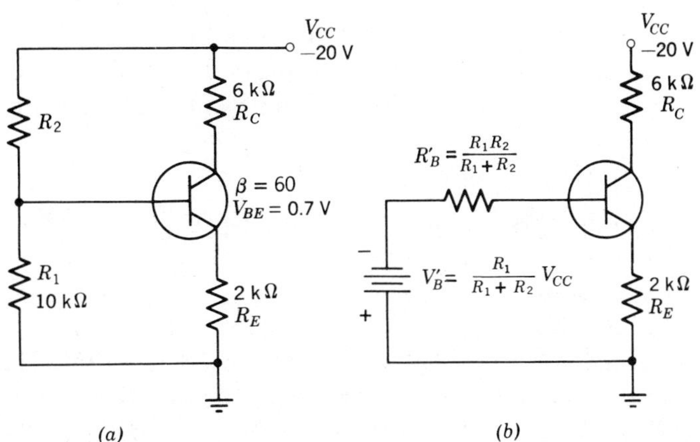

Figure 5-9 The determination of the base-bias divider circuit. (*a*) circuit (*b*) equivalent by Thévenin's theorem.

The value of the emitter current is

$$I_E = (1+\beta)I_E = (1+60)(0.0207) = 1.263 \text{ mA}$$

The voltage divider circuit in the base is converted to the circuit given in Fig. 5-9*b* by means of Thévenin's theorem. The Kirchhoff's voltage loop equation through the base is

$$\frac{R_1}{R_1+R_2}V_{CC} = \frac{R_1R_2}{R_1+R_2}I_B + V_{BE} + I_ER_E$$

Substituting numerical values we find that

$$\frac{10\text{ k}\Omega}{10\text{ k}\Omega + R_2}20\text{ V} = \frac{10\text{ k}\Omega \times R_2}{10\text{ k}\Omega + R_2}0.0207\text{ V} + 0.7\text{ V} + 1.263\text{ mA} \times 2\text{ k}\Omega$$

$$\frac{200}{10\text{ k}\Omega + R_2} = \frac{0.207R_2}{10\text{ k}\Omega + R_2} + 3.226\text{ V}$$

Clearing fractions we have

$$200 = 0.207R_2 + 3.226(10+R_2)$$

$$200 = 0.207R_2 + 32.26 + 3.226R_2$$

$$3.433R_2 = 167.74$$

and

$$R_2 = \mathbf{48.9\ k\Omega}$$

If the value of R_2 had been given, the same procedure would have yielded a numerical value for R_1.

There are a great many variations in the means that can be used to establish the operating values for a transistor. The general procedure is to reduce the circuit to one of the forms we have considered so far in this chapter.

Example 5-7

Find the value of R_B in the circuit of Fig. 5-10*a* that sets V_{CE} to 5 V.

Solution

The voltage divider circuit itself is shown isolated in Fig. 5-10*b*. The voltage at one end of the voltage divider (point C) is +10 V. The voltage at the other end of the voltage divider (point A) is −30 V. Therefore, the potential difference or the voltage across the whole

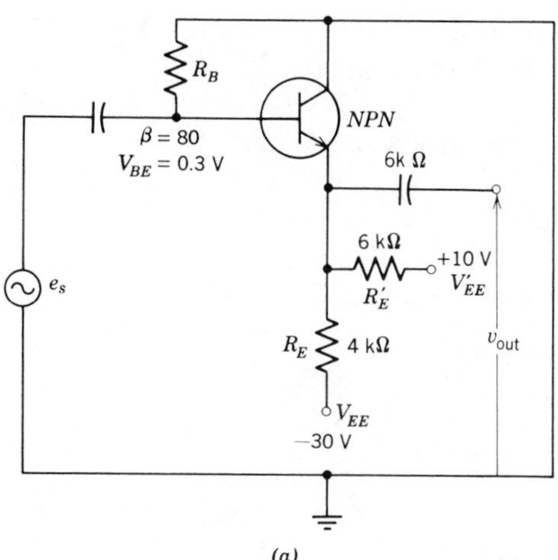

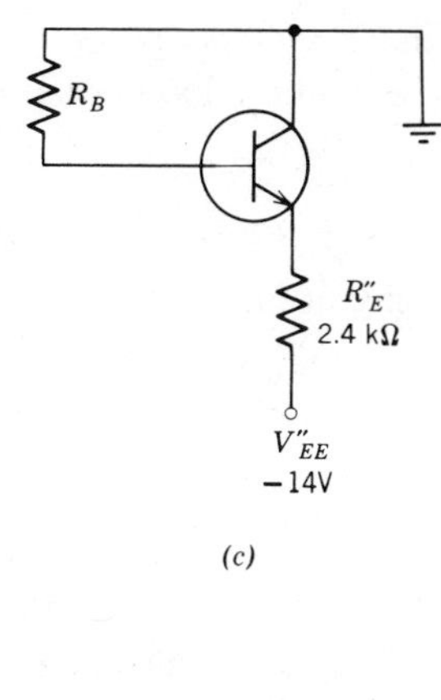

C
+ 10 V
6 kΩ
B
4 kΩ
V_AB
A
−30 V
(b)

Figure 5-10 Amplifier with voltage-divider in the emitter circuit. (*a*) Circuit. (*b*) Emitter voltage divider. (*c*) Reduced circuit.

divider $(R_E + R'_E)$ is

$$V'_{EE} - (-V_{EE}) = 10 - (-30) = 40 \text{ V}$$

Then the voltage from point A to point B, V_{AB}, by the voltage divider rule is

$$V_{AB} = 40 \text{ V} \frac{4 \text{ k}\Omega}{4 \text{ k}\Omega + 6 \text{ k}\Omega} = 16 \text{ V}$$

Then the potential of point B to ground is V''_{EE}.

$$V''_{EE} = (-30) + (+16) = -14 \text{ V}$$

The resistance of the equivalent circuit R of the emitter voltage divider is found using Thévenin's theorem as

$$R''_E = \frac{4 \text{ k}\Omega \times 6 \text{ k}\Omega}{4 \text{ k}\Omega + 6 \text{ k}\Omega} = 2.4 \text{ k}\Omega$$

These values are now placed on the equivalent circuit, Fig. 5-10*c*. The voltage loop equation through the collector is

$$V''_{EE} = R''_E I_E + V_{CE}$$

using numerical values we have

$$14\text{ V} = 2.4\text{ k}\Omega \times I_E + 5\text{ V}$$

$$2.4\text{ k}\Omega \times I_E = 9\text{ V}$$

$$I_E = 3.75\text{ mA}$$

and the base current is

$$I_B = \frac{I_E}{1+\beta} = \frac{3.75\text{ mA}}{1+80} = 0.046\text{ mA} = 46\ \mu A$$

The voltage loop equation through the base is

$$V''_{EE} = R''_E I_E + V_{BE} + R_B I_B$$

$$14\text{ V} = 2.4\text{ k}\Omega \times 3.75\text{ mA} + 0.3\text{ V} + R_B \times 0.046\text{ mA}$$

$$0.046\text{ mA} \times R_B = 4.7\text{ V}$$

$$R_B = \frac{4.7\text{ V}}{0.046\text{ mA}} = \mathbf{102\ k\Omega}$$

Problems

5-3.1 In part *a* of the given circuits, V_{CC} is +20 V, R_C is 5 kΩ, R_E is 4 kΩ, and R_B is 750 kΩ. Find V_{CE} and I_C.

5-3.2 In part *a*, V_{CC} is +10 V, R_E is 2 kΩ, R_C is 4 kΩ, and R_B is 750 kΩ. Find V_{CE} and I_C.

5-3.3 In part *a*, V_{CC} is +45 V, R_E is 5 kΩ, and R_C is 8 kΩ. Find R_B to set V_{CE} to 25 V.

5-3.4 What value of R_B will saturate the transistor in Problem 5-3.1?

5-3.5 In part *b*, V_{CC} is −18 V and V_{EE} is +4 V. R_E is 2000 Ω and R_C is 4000 Ω. What value of R_B establishes an operating current of 1.5 mA for I_C? What is V_{CE}?

5-3.6 In part *b*, V_{CC} is −12 V and V_{EE} is +5 V. R_E is 0.8 kΩ, R_B is 50 kΩ, and R_C is 2.4 kΩ. What are I_C and V_{CE}?

5-3.7 In part *c*, V_{CC} is +10 V and R_E is 2 kΩ. What values of R_C and R_B yield a value of V_{CE} equal to 6 V?

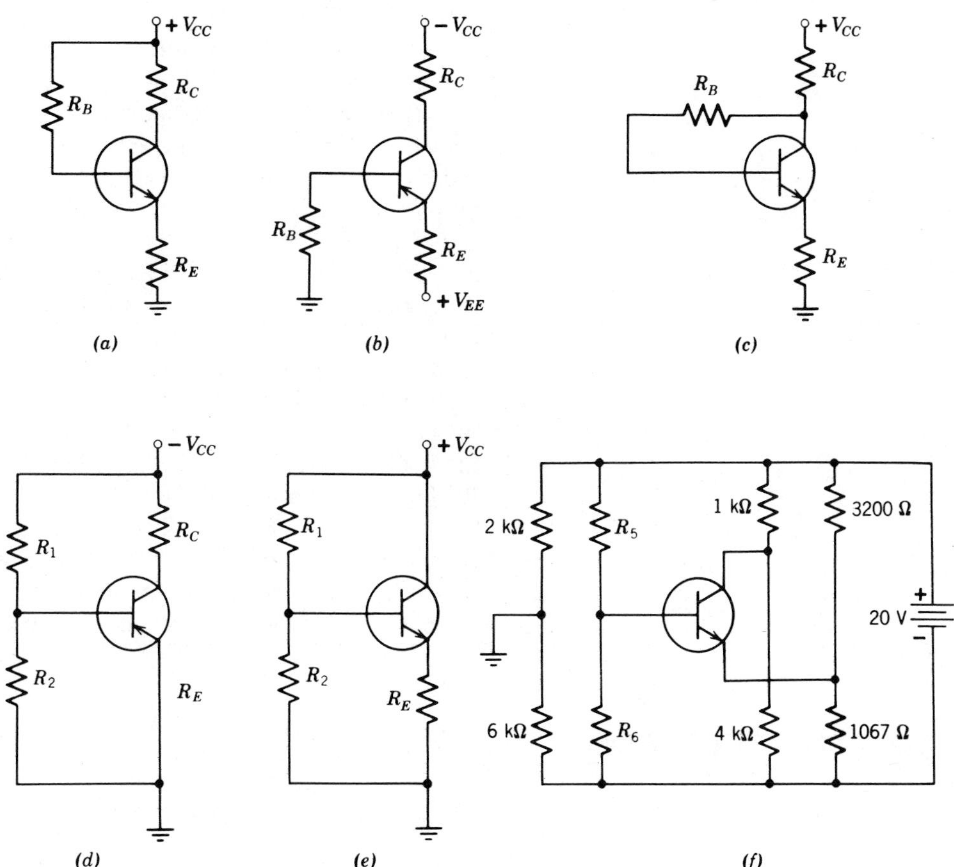

Circuits for Problems 5-3.1 through 5-3.22. All transistors are silicon ($V_{BE} = 0.7$ V) and the value of β is 60. Also, $V_{CE,\text{sat}} = 0.2$ V.

5-3.8 In part *c*, V_{CC} is +12 V and R_E is 1500 Ω. R_C is 5 kΩ. What value of R_B sets V_{CE} to 4 V?

5-3.9 In part *c*, V_{CC} is +12 V and R_E is 1800 Ω. What values of R_C and R_B make V_{CE} 3 V when I_C is 2.0 mA?

5-3.10 In part *c*, V_{CC} is +15 V. R_E and R_C are each 1500 Ω. If R_B is very large, say 10 MΩ, V_C is a large value. R_B is slowly decreased. What is the base current and what is V_{CE} when V_{CE} is decreased to its lowest possible value?

5-3.11 In part *d*, V_{CC} is −10 V, R_E is 2 kΩ, R_C is 3 kΩ, and R_1 and R_2 are each 200 kΩ. Find I_C and V_{CE}.

5-3.12 In part *d*, V_{CC} is −4 V, R_E is 1 kΩ, R_2 is 50 kΩ, R_C is 10 kΩ, and R_1 is 150 kΩ. Find V_{CE}.

5-3.13 In part *d*, V_{CC} is -20 V, R_C is 5 kΩ, R_E is 2 kΩ, R_1 is 70 kΩ, and R_2 is 30 kΩ. Find I_C and V_{CE}.
5-3.14 In part *d*, V_{CC} is -10 V, I_C is 1 mA, V_{CE} is 3 V, R_E is 1500 Ω, and R_2 is 100 kΩ. Find R_C and R_1.
5-3.15 In part *d*, V_{CC} is -10 V, R_C is 2 kΩ, R_E is 1 kΩ, R_2 is 10 kΩ, and V_{CE} is 4 V. Find R_1.
5-3.16 In part *d*, V_{CC} is -16 V, R_C is 10 kΩ, R_E is 600 Ω, R_2 is 10 kΩ, and V_{CE} is 8 V. Find R_1.
5-3.17 In part *e*, V_{CC} is $+12$ V, R_E is 1500 Ω, and R_2 is 30 kΩ. Determine R_1 to make V_{CE} 6 V.
5-3.18 In part *e*, V_{CC} is $+15$ V, R_1 is 300 kΩ, and R_2 is 120 kΩ. R_E is 1200 Ω. Determine I_C and V_{CE}.
5-3.19 What value of R_1 makes V_{CE} 4 V for the transistor in Problem 5-3.17?
5-3.20 In part *e*, V_{CC} is $+30$ V, R_E is 10 kΩ, R_1 is 100 kΩ, and R_2 is 100 kΩ. Find I_B and V_{CE}.
5-3.21 Using part *f*, if R_5 and R_6 are each 100 kΩ, what is V_{CE}? Determine the voltages measured from ground to the collector, to the base and to the emitter.
5-3.22 Repeat Problem 5-3.21 if R_5 and R_6 are each 10 kΩ.

Supplementary Problems

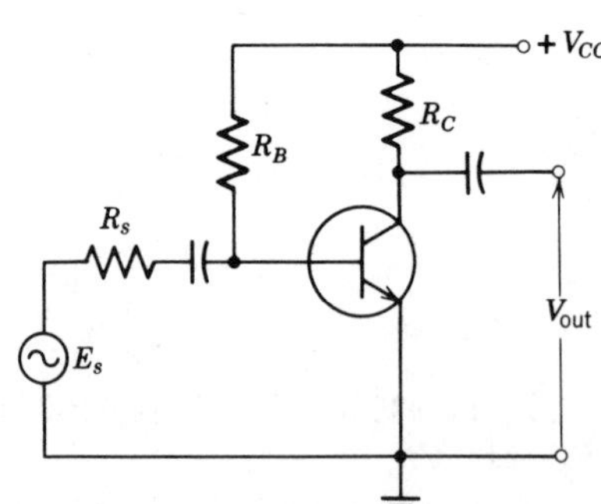

Circuit for Problems 5-1 and 5-2.

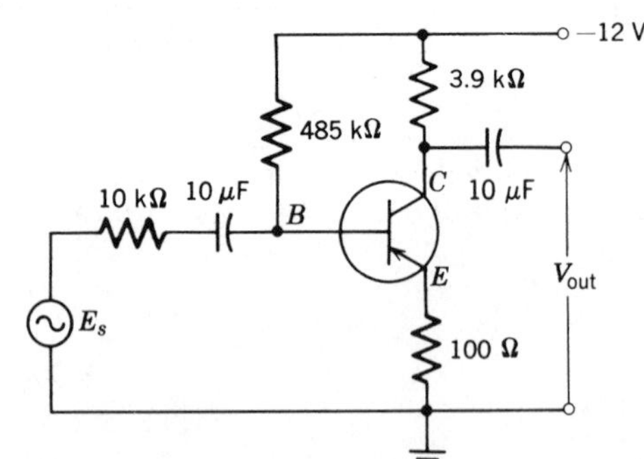

Circuit for Problems 5-3 and 5-4.

5-1 $V_{CC} = +12$ V; $\beta = 40$; $V_{BE} = 0.7$ V; $R_C = 4$ kΩ; and $V_{CE,sat} = 0.2$ V. Find the value of R_B that just causes saturation.
5-2 $V_{CC} = +15$ V; $\beta = 50$; $V_{BE} = 0.7$ V; $R_C = 4$ kΩ; and $R_B = 470$ kΩ. Find V_{CE}.
5-3 Experimental results show that I_B is 23 μA and that $I_E =$

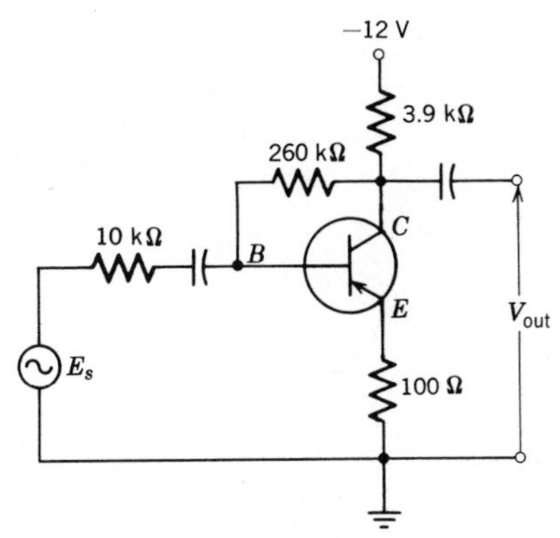

Circuit for Problem 5-5.

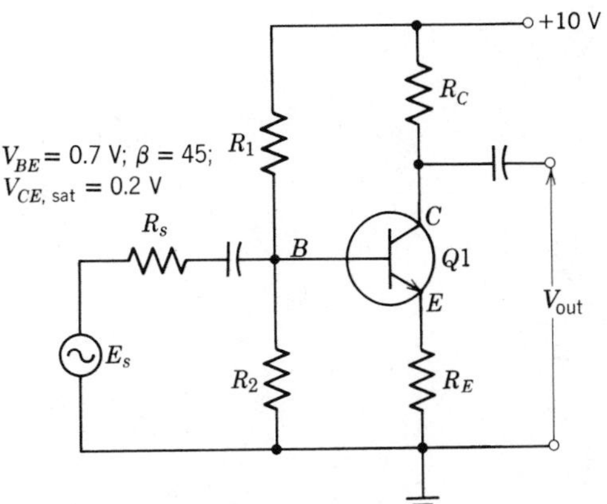

Circuit for Problems 5-6 through 5-10.

1.5 mA. Find V_{BE} and determine the voltages measured from each of *B*, *C*, and *E* to ground.

5-4 Repeat Problem 5-3 if the supply voltage is reversed to +12 V.

5-5 V_{BE} is 0.3 V and β is 35. Find V_{CE}.

5-6 $R_1 = 400\ \text{k}\Omega$; $R_2 = 160\ \text{k}\Omega$; $R_E = 4\ \text{k}\Omega$; and $R_C = 20\ \text{k}\Omega$. Find V_{CE}.

5-7 $R_2 = 50\ \text{k}\Omega$; $R_E = 1\ \text{k}\Omega$; $R_C = 3\ \text{k}\Omega$; and $V_{CE} = 6$ V. Find R_1.

5-8 $R_1 = 100\ \text{k}\Omega$; $R_2 = 35\ \text{k}\Omega$; $R_E = 1\ \text{k}\Omega$; and $R_C = 5\ \text{k}\Omega$. Determine V_{CE}.

5-9 $R_1 = 200\ \text{k}\Omega$; $R_E = 2\ \text{k}\Omega$; $R_C = 6\ \text{k}\Omega$; and $V_{CE} = 6$ V. Find R_2.

5-10 Use the data of Problem 5-6. R_C is increased until saturation occurs. What is the value of R_C?

5-11 Experimental measurements yield a value of 23 μA for I_B and 1.5 mA for I_E. Determine V_{BE}. What are the expected voltage values to be obtained from *B* to ground, from *C* to ground, and from *E* to ground?

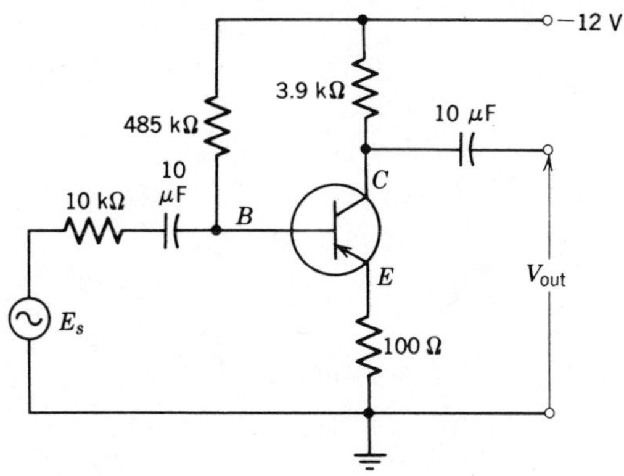

Circuit for Problem 5-11.

Chapter 6 Transistor Load Lines

The principles of a load line can be demonstrated by solving a series circuit of two resistors graphically (Section 6-1). The intersection of the two load lines yields values for the quiescent operating point as shown by the case of a diode in series with a resistor. The bias resistor R_B (or the bias network) establishes the quiescent operating point for the transistor amplifier (Section 6-2). The load line can also be used to determine maximum output signal levels. When the common-emitter amplifier circuit has different resistance values for the load for the dc circuit and for the ac circuit, we require both a dc load line and an ac load line (Section 6-3). We can set the quiescent operating point to that value that yields the maximum possible peak-to-peak output voltage without clipping.

Section 6-1 The Load Line Concept

Load line is the term used to describe the graphical relation between the voltage and current values that are possible for a particular component or for a particular circuit.

Ohm's law was determined originally from a graphical approach by showing that the load line for resistor is a straight line.

Consider a series circuit of two resistors R_1 and R_2 connected to a supply voltage V (Fig. 6-1*a*). The graph on which load lines for the resistors are to be drawn is given in Fig. 6-1*b*. The total width of the graph is V volts, the supply voltage. The load line for R_2 is drawn from A to E. It is a straight line because R_2 is a fixed resistor. When the whole voltage V is applied across R_2, the current value is V/R_2 and, thus, locates point E. This line has a positive slope of value $+1/R_2$.

The load line for R_1 is drawn by considering point B to be

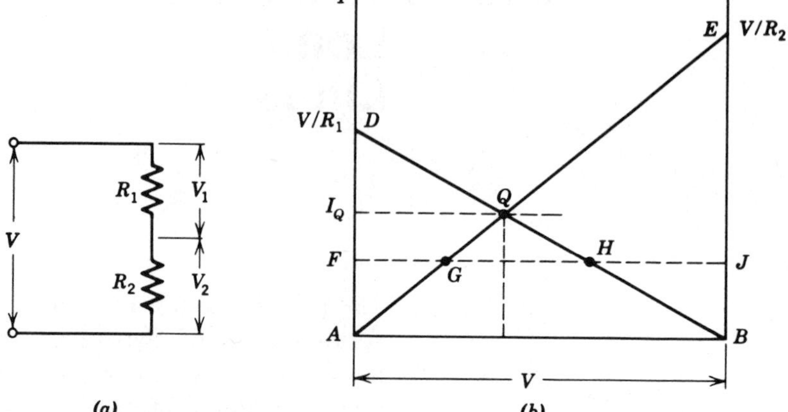

Figure 6-1 A graphical solution for a series circuit. (*a*) Circuit. (*b*) Graphical solution.

zero volts and point A to be V volts; that is, the scale of V reads from B to A instead of from A to B. When the whole voltage V is applied across R_1, the current value is V/R_1 locating point D. In terms of the coordinate axes that have the zero at point A, this line has a negative slope of value $-1/R_1$. This negative slope does not imply a negative resistance. It is only $-1/R_1$ because of the way in which the slope is being considered.

The line from A to E represents all values of current through R_2 when the voltage across R_2 is varied from 0 to V. Likewise the line from B to D represents all values of current through R_1 when the voltage across R_1 is varied from 0 to V.

Consider the original circuit again. It is a series circuit and the requirement of a series circuit is that the current in all parts of the series circuit is the same. The horizontal line FJ represents a current F that is common to the two resistors. The voltage drop across R_2 for this current is FG and across R_1 is JH. Obviously, FG plus JH does not equal V so that this value of current F cannot be the solution for the network.

The only value of current that can be the solution for the network is the value I_Q, given by the intersection of the two load lines, point Q. Point Q is called the *Q-point* or the *quiescent point* or the *operating point*. At Q, the voltage across R_2 is AC and the voltage across R_1 is BC. These two values properly add up to equal the supply voltage V.

In using this graphical approach, we notice that the slope and direction of the load line for R_2 do not change if the source voltage V changes. On the other hand, if the source voltage changes, the location of B changes and the value of V/R_1 changes. The slope of this load line does not change; the slope remains at the value $-1/R_1$. The important conclusion we observe is that *any line* parallel to B-H-Q-D has the slope $-1/R_1$ and has the resistance value R_1.

Figure 6-2*a* shows a diode in series with a load resistor R_L placed across a source voltage V. The forward diode characteristic is nonlinear as shown on Fig. 6-2*b*. A load line for R_L is now drawn on this characteristic curve. The X axis intercept of the load line is V_A and the Y axis intercept is V_A/R_L. The intersection of the diode characteristic with the load line is the operating point (the Q-point) of the circuit.

The endpoints of the load line are equivalent to:

1. The voltage that exists across the diode-socket terminals if the diode is removed from the circuit.
2. The current in the circuit if a short circuit is placed across the diode.

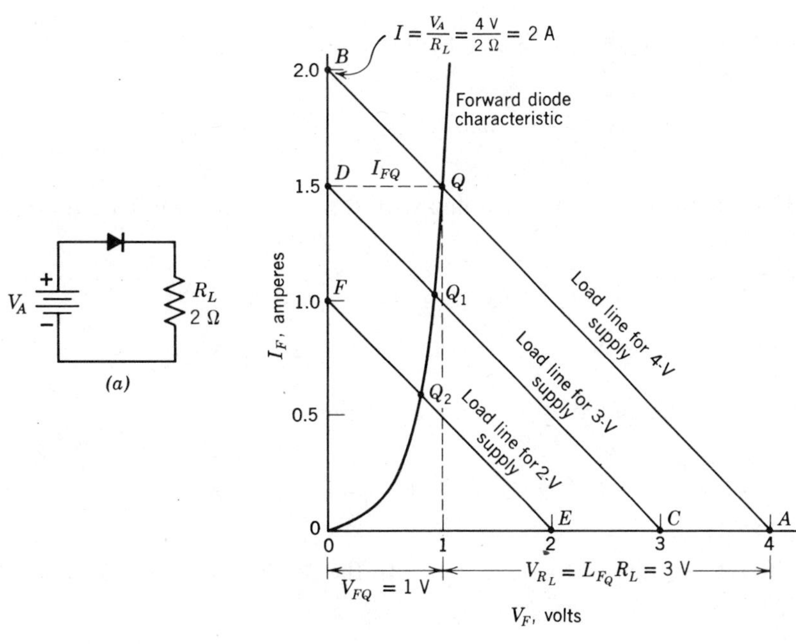

Figure 6-2 The Q-point for a diode. (*a*) Circuit. (*b*) Load lines.

Example 6-1
Using the circuit and the diode characteristic of Fig. 6-2, determine the diode current and the diode voltage for:

Case I. $V_A = 4\text{ V}$
Case II. $V_A = 3\text{ V}$
Case III. $V_A = 2\text{ V}$

Solution
Case I. The endpoints of the load line are

$$V_A = 4\text{ V} \quad \text{and} \quad I = \frac{V_A}{R_L} = \frac{4\text{ V}}{2\ \Omega} = 2\text{ A}$$

These points are located at A and B on Fig. 6-2b. The load line is drawn between A and B. The intersection of the load line with the diode characteristic (point Q) gives

$$I_{FQ} = \mathbf{1.5\ A} \quad \text{and} \quad V_{FQ} = \mathbf{1.0\ V}$$

The voltage drop across R_L is

$$V_{R_L} = V_A - V_{FQ} = 4.0 - 1.0 = \mathbf{3.0\ V}$$

Case II. The endpoints of the load line are

$$V_A = 3\text{ V} \quad \text{and} \quad I = \frac{V_A}{R_L} = \frac{3\text{ V}}{2\ \Omega} = 1.5\text{ A}$$

The intersection of the load line with the diode characteristic (point Q_1) gives

$$I_{FQ_1} = \mathbf{1.05\ A} \quad \text{and} \quad V_{FQ_1} = \mathbf{0.90\ V}$$

and $$V_{R_L} = V_A - V_{FQ_1} = 3.0 - 0.90 = \mathbf{2.1\ V}$$

Case III. The endpoints of the load line are

$$V_A = 2\text{ V} \quad \text{and} \quad I = \frac{V}{R_L} = \frac{2\text{ V}}{2\ \Omega} = 1\text{ A}$$

The intersection of the load line with the diode characteristic (point Q_2) gives

$$I_{FQ_2} = \mathbf{0.6\ A} \quad \text{and} \quad V_{FQ_2} = \mathbf{0.8\ V}$$

and $$V_{R_L} = V_A - V_{FQ_2} = 2.0 - 0.8 = \mathbf{1.2\ V}$$

It should be noted that, as the source voltage increases from 2 V to 3 V to 4 V, the change is not proportional in V_{FQ}

from 0.8 V to 0.9 V to 1.0 V

although the change in I_{FQ} is more linear.

from 0.6 A to 1.05 A to 1.5 A

Problems

6-1.1 A 30-Ω resistor and a 40-Ω resistor are connected in series across a 120-V source. By using a graphical approach, determine the current in the circuit and the voltage drop across each resistor.

6-1.2 A 10-Ω resistor and a 3-Ω resistor are connected in series across a 15-V source. By using a graphical approach, determine the current in the circuit and the voltage drop across each resistor.

Data for the diode characteristic:

V_F (volts)	0	0.4	0.6	0.8	1.0
I_F (mA)	0	10	20	60	150

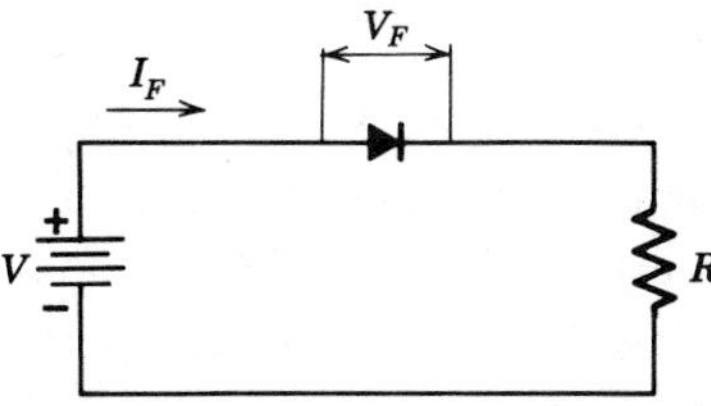

Circuit and data for Problems 6-1.3 and 6-1.4.

6-1.3 If R is 200 Ω, what is the change in V_F and in I_F if V is increased from 8 V to 12 V?

6-1.4 If R is 50 Ω and V is 5 V, what is the heating power ($V_{FQ}I_{FQ}$) in the diode?

Section 6-2 The DC Load Line for the Transistor

The Kirchhoff's voltage loop equation through the collector for the common-emitter amplifier circuit shown in Fig. 6-3*a* is

$$V_{CC} = I_C R_C + V_{CE}$$

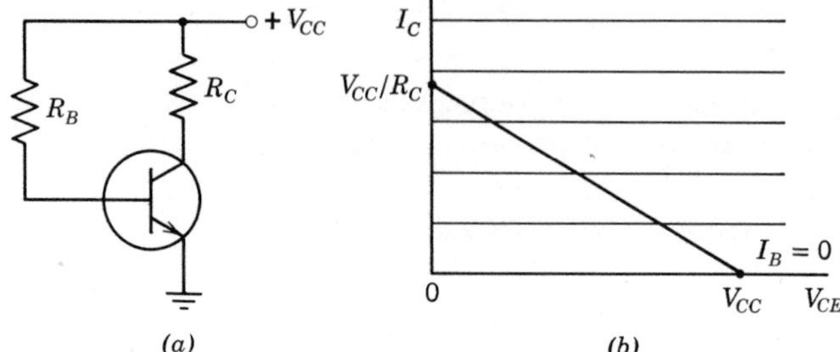

Figure 6-3 The dc load line. (*a*) Circuit. (*b*) The load line on the collector characteristic.

For a particular circuit, V_{CC} and R_C are fixed quantities and I_C and V_{CE} are variables that depend upon the value of R_B. If this equation is solved for I_C, we have

$$I_C R_C = -V_{CE} + V_{CC}$$

Dividing through by R_C we have

$$I_C = -\frac{1}{R_C} V_{CE} + \frac{V_{CC}}{R_C} \qquad (6\text{-}1)$$

This equation has the form

$$y = mx + b$$

which is one of the standard forms of the equation for a straight line. In this form, b is the Y axis intercept and m is the slope of the load line. The Y axis intercept of the load line is V_{CC}/R_C and the slope of the load line is $(-1/R_C)$. Thus, we show that the load line for a transistor is a straight line when it is drawn on the collector characteristic (Fig. 6-3*b*).

When I_C is zero in Eq. 6-1, we have

$$0 = -\frac{1}{R_C} V_{CE} + \frac{V_{CC}}{R_C}$$

Then

$$\frac{V_{CE}}{R_C} = \frac{V_{CC}}{R_C}$$

or

$$V_{CE} = V_{CC} \qquad (6\text{-}2)$$

Equation 6-2 states that one end of the load line has the coordinates

$$I_C = 0 \quad \text{and} \quad V_{CE} = V_{CC}$$

The intersection of the load line with the X axis is V_{CC}.

When V_{CE} is zero in Eq. 6-1, we have

$$I_C = \frac{V_{CC}}{R_C} \tag{6-3}$$

Equation 6-3 states that the other end of the load line has the coordinates

$$I_C = \frac{V_{CC}}{R_C} \quad \text{and} \quad V_{CE} = 0$$

The intersection of the load line with the y axis is V_{CC}/R_C.

Let us assume that the transistor shown in the circuit of Fig. 6-3*a* can be removed from a transistor socket. The *X* axis intercept is the voltage we measure across the transistor socket terminals (*C* terminal to *E* terminal). The *Y* axis intercept is the "short-circuit" current measured in a lead that forms a short from the *C* terminal to the *E* terminal.

A signal source E_s having a source resistance R_s is connected to the common-emitter amplifier (Fig. 6-4). When E_s is

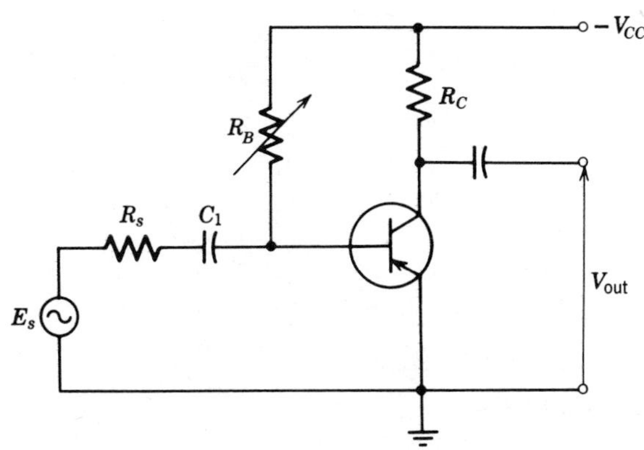

Figure 6-4 Circuit providing an ac output signal.

raised from zero, an ac signal current is produced in the base. In turn, this ac signal current in the base produces an ac signal current in the collector. The ac signal current in R_C produces the ac voltage drop across R_C that we observe as the output signal V_{out}.

Let us adjust R_B in the circuit given in Fig. 6-4 to a value of I_B that locates the operating point of the transistor at point Q on the load line, Fig. 6-5. The small sinusoidal signal current fed into the base varies the base current sinusoidally from Q to B to Q to A and to Q. This sinusoidal variation of base current is shown on Fig. 6-5 diagonally for two cycles plotted against time. The sinusoidal variation of base current causes the collector current to vary sinusoidally. Two cycles of collector current are drawn to a horizontal time scale. The corresponding variation of the collector voltage is drawn vertically for two cycles.

Consider the circuit given in Fig. 6-4. R_B is adjusted to give successively three different operating points: Q_1, Q_2, and Q_3. These operating points are shown on the load lines of Figs.

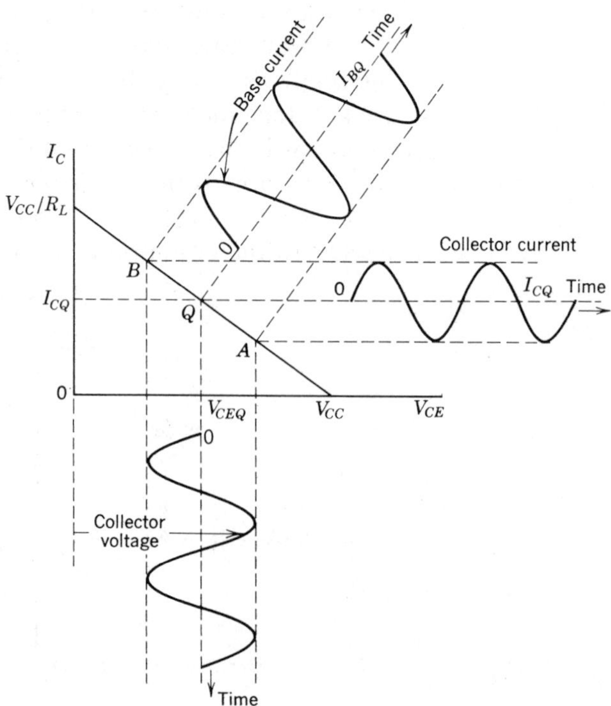

Figure 6-5 AC signals obtained from a load line.

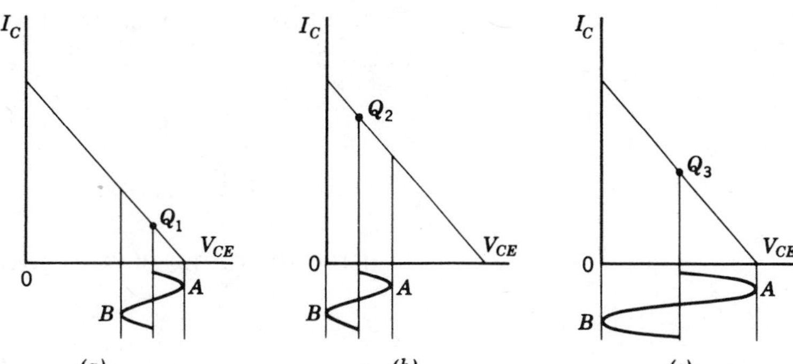

Figure 6-6 Effect of the variation of the *Q*-point on maximum undistorted output signal.

6-6*a*, 6-6*b*, and 6-6*c*. In each case, the signal source E_s is raised from zero until one or both sides of the output signal just start to "clip" or flatten. If the operating point is Q_1 (Fig. 6-6*a*) the output signal first starts to clip at *A*. This condition of clipping is called cutoff since the collector current at this point is zero, its least value.

Now consider that the operating point is located toward the other end of the load line, point Q_2 in Fig. 6-6*b*. When E_s is raised from zero, the output waveform now first starts to clip at *B*. This clipping is caused by saturation. *Saturation* is the condition at which the collector current is the maximum possible value, the *Y* axis intercept value on the load line.

In Fig. 6-6*c* the operating point Q_3 is located at the center of the load line. When E_s is raised from zero, clipping occurs simultaneously at *A* and at *B*. Now we have the condition from which we can obtain the maximum possible output signal. This operating point is called the *optimum Q-point.*

In Fig. 6-7 the distance from the *Q*-point to the origin measured along the *X* axis is *A*. The distance from the

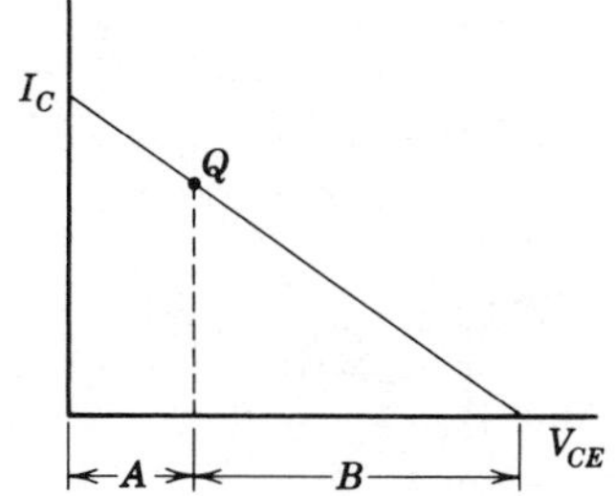

Figure 6-7 Load line showing the *Q*-point.

Q-point to V_{CC} measured along the X axis is B. In this diagram

$$A < B$$

Therefore, the maximum possible peak-to-peak output signal is $2A$ volts.

If the operating point is located so that

$$A > B$$

the maximum possible peak-to-peak output signal is $2B$ volts.

When the Q-point is located exactly in the center of the load line,

$$A = B$$

and the maximum possible peak-to-peak output signal is

$$2A = 2B = V_{CC} \text{ volts}$$

or

$$V_{out} = V_{CC} \text{ volts, peak to peak, maximum} \qquad (6\text{-}4)$$

This Q-point yields the optimum bias point for the circuit since the peak-to-peak output voltage is its largest possible value.

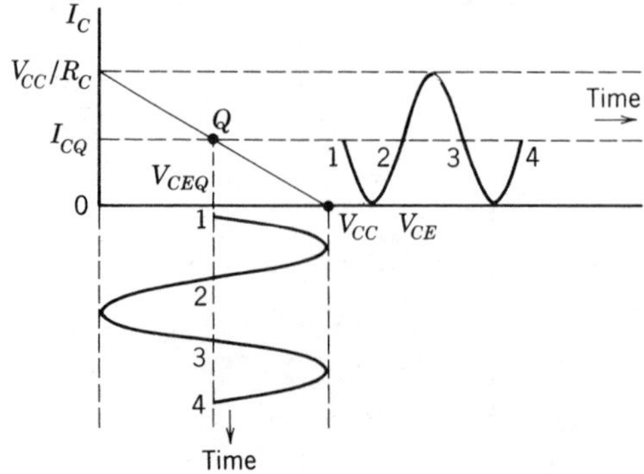

Figure 6-8 Optimum bias condition.

We show the optimum operating point in detail in Fig. 6-8. The Q-point is in the center of the load line. Evidently the value of I_{CQ} is midway between the origin and the load-line intercept on the I_C axis.

$$I_{CQ} = \frac{1}{2}\frac{V_{CC}}{R_C} = \frac{V_{CC}}{2R_C}$$

In order to have only one equation for all possible cases for a load line, this equation is rearranged to conform to a general equation to be developed in Section 6-3.

$$I_{CQ} = \frac{V_{CC}}{R_C + R_C} \tag{6-5}$$

Example 6-2
The transistor used in the circuit of Fig. 6-4 has a β of 50 and a value of 0.7 V for V_{BE}. R_C is 5.6 kΩ and V_{CC} is -15 V. Determine the value of R_B required to adjust the circuit to the optimum operating point.

Solution
The collector current I_{CQ} is

$$I_C = \frac{V_{CC}}{R_C + R_C} = \frac{15\text{ V}}{5.6\text{ k}\Omega + 5.6\text{ k}\Omega} = 1.34\text{ A} \tag{6-5}$$

The base current is

$$I_{BQ} = \frac{I_{CQ}}{\beta} = \frac{1.34\text{ mA}}{50} = 0.0268\text{ mA} = 26.8\ \mu\text{A}$$

The Kirchhoff's voltage loop equation through the base is

$$V_{CC} = I_B R_B + V_{BE}$$

$$15\text{ V} = 0.0268\text{ mA} \times R_B\text{ k}\Omega + 0.7\text{ V}$$

$$R_B = \mathbf{534\ k\Omega}$$

The common-emitter amplifier shown in Fig. 6-9a uses two power supply voltages, -6 V for V_{CC} and $+4$ V for V_{EE}. The bias resistor is varied over a wide range and the dc voltages

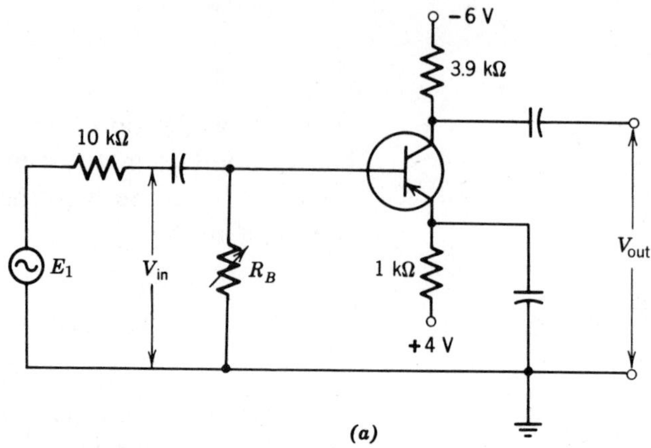

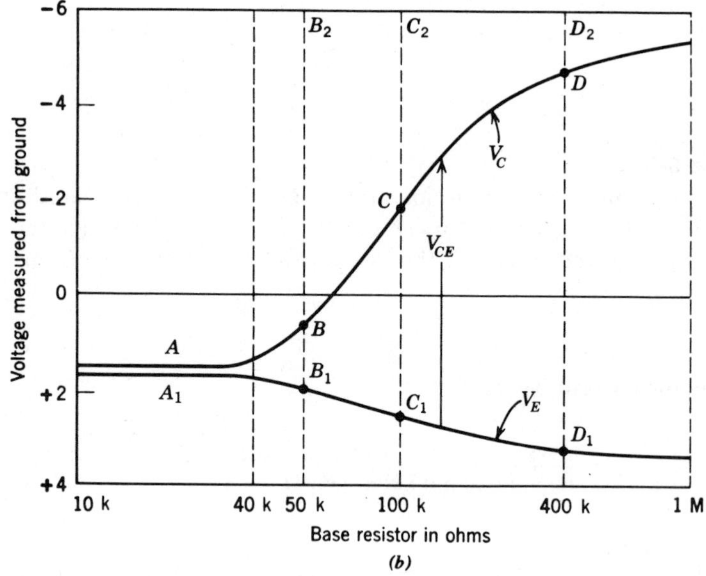

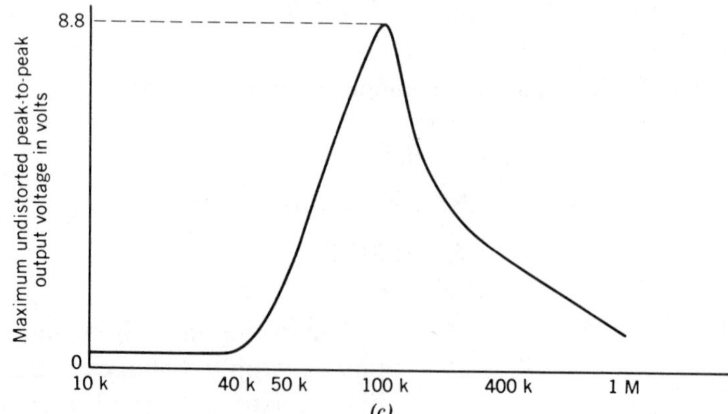

Figure 6-9 Effects of varying the bias on an amplifier stage. (*a*) Circuit. (*b*) Absolute voltage levels. (*c*) Maximum undistorted peak-to-peak output voltage.

from the collector to ground and from the emitter to ground are measured. The values are plotted on the graph shown in Fig. 6-9*b*.

When the transistor is saturated, the transistor cannot provide a useful output signal. The saturation condition is located at point *A*.

When R_B is increased so that the *Q*-point is at point *B*, the maximum possible output voltage swing across the transistor is the distance B-B_1 and the possible signal swing across R_C is the distance B-B_2. Since B-B_1 is less than B-B_2, the maximum undistorted peak-to-peak output signal is $2(B\text{-}B_1)$ volts.

When R_B is increased to set the *Q*-point to point *C*, the distances C-C_1 and C-C_2 are equal. Now we have the optimum condition when the output signal has the maximum possible peak-to-peak value

$$2(C\text{-}C_1) = 2(C\text{-}C_2) \text{ volts}$$

When R_B is increased to set the *Q*-point to point *D*, the maximum peak-to-peak output signal is reduced to $2(D\text{-}D_2)$ volts.

The maximum undistorted peak-to-peak output voltage values are shown plotted against R_B in Fig. 6-9*c*.

Problems

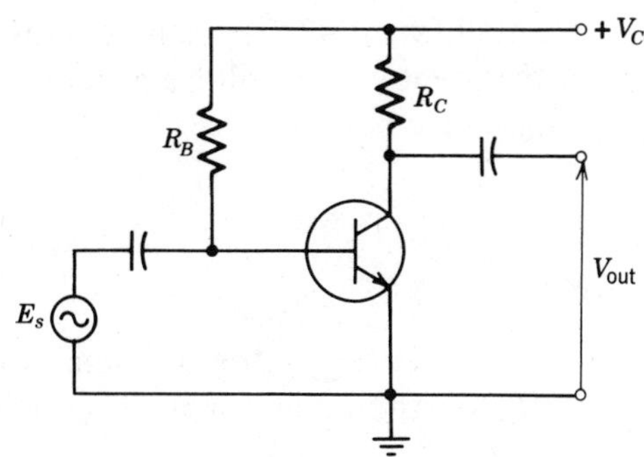

Circuit for Problems 6-2.1 to 6-2.4.

All transistors are silicon and have a β of 50. Draw the load line for each problem and label the *Q*-point on the load line.

6-2.1 The values for the circuit are: V_{CC} is +20 V, R_C is 2 kΩ, and R_B is 300 Ω. Find the operating point and the maximum value of V_{out}.

6-2.2 The value of R_B in Problem 6-2.1 is changed to 120 kΩ. Find the operating point and the maximum value of V_{out}.
6-2.3 What is the value of R_B to adjust the circuit to the optimum operating point? What is V_{out}?
6-2.4 V_{CC} is +12 V and R_C is 500 Ω. The maximum required value for V_{out} is 5 V peak to peak. What is the upper limit and what is the lower limit for the value that can be used for R_B?

Section 6-3 The AC Load Line for the Transistor

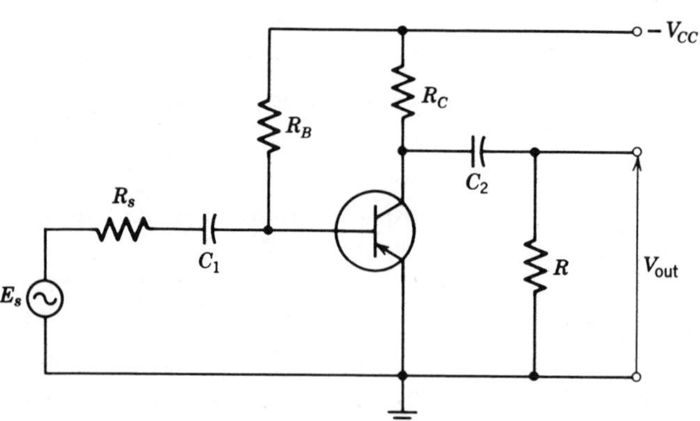

Figure 6-10 Amplifier having an R–C coupled load.

In the circuit shown in Fig. 6-10, there is a second resistor R in the collector circuit that is coupled to the collector by means of the capacitor C_2. The reactance of the capacitor is small so that its effect may be neglected at the lowest frequency that will be processed by the circuit. This condition is met when

$$X_{C2} \leq \frac{1}{10} R$$

Then, as far as the ac signal circuit is concerned, the ac load R_{ac} on the collector is R_C in parallel with R:

$$R_{ac} = \frac{R_C R}{R_C + R} \tag{6-6}$$

We can draw an ac load line for R_{ac} on the collector characteristic by using an assumed current approach. For example, assume that R_{ac} is 5 kΩ. We choose *any* convenient

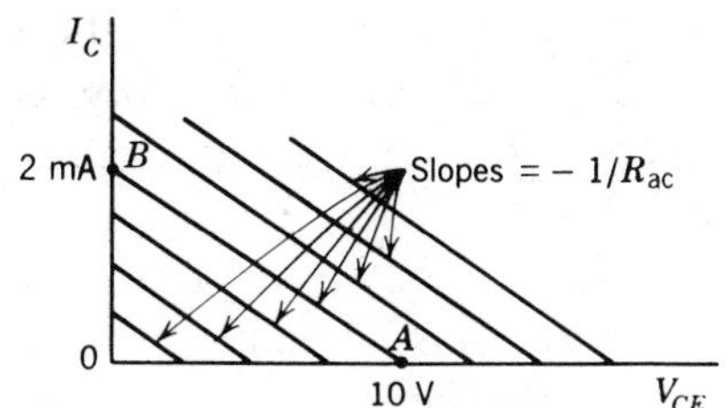

Figure 6-11 Load lines for an ac resistance shown on the axes for the collector characteristic.

value for current as long as that current value is within the current scale of the collector characteristic. Let us assume 2 mA. This current value is shown on the collector characteristic as point B, Fig. 6-11. Now, if we take the product of IR_{ac} using the assumed numerical value of current, we have $2\,\text{mA} \times 5\,\text{k}\Omega$ or 10 V. The voltage value is shown on the collector characteristic as point A. A line is drawn between point A and point B. This line has a slope of $(-1/R_{ac})$. *Any line* drawn parallel to the line between A and B has the slope $(-1/R_{ac})$. Therefore, all the parallel lines are load lines representing this particular value of R_{ac}.

Now we superimpose the dc load line on the collector characteristic showing the various ac load lines (Fig. 6-12). The endpoints of the dc load line are V_{CC} and V_{CC}/R_C. There can be only one operating point for a circuit that has a specific bias value. That operating point is the intersection point of the ac load line with the dc load line. We show typical Q-points on Fig. 6-12 as Q_1, Q_2, Q_3, Q_4, Q_5, Q_6, Q_7, and Q_8. The particular operating point is established by the value of R_B. As

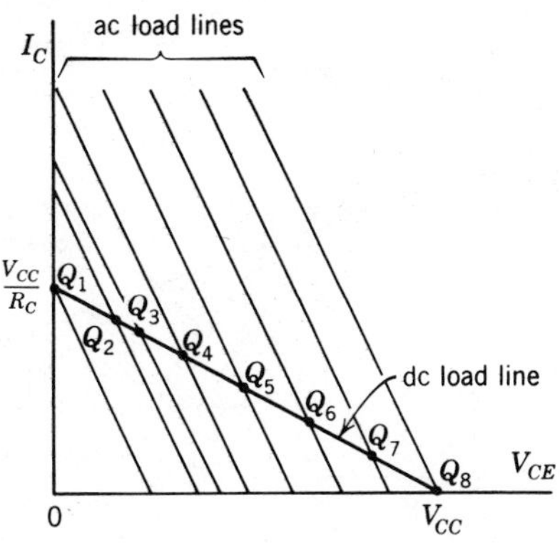

Figure 6-12 The dc load line with various ac load lines shown on the axes for the collector characteristic.

R_B is increased, we shift the operating point from saturation at Q_1 to cutoff at Q_8. Any intermediate operating point between Q_1 and Q_8 can serve as the operating point for an amplifier.

There is, however, only one operating point that provides optimum bias. When a circuit is operated at optimum bias, we obtain the largest possible value of undistorted peak-to-peak output voltage V_{out} from the amplifier. Now we must take the signal swing along the ac load line. Therefore, the optimum bias occurs when the Q-point bisects the ac load line. Inspection of Fig. 6-12 shows that the optimum bias occurs approximately at the operating point Q_3.

In Fig. 6-13 we show the dc load line and the ac load line that provides the condition of optimum bias. When optimum bias occurs, the Y axis intercept of the ac load line must be $2I_{CQ}$ since the operating point Q is at the center of the ac load line. The greatest undistorted output signal that can be obtained from the circuit is the peak-to-peak value

$$V_{\text{out}} = 2A = 2V_{CEQ} \text{ volts, peak to peak, maximum}$$

To form an equation for the optimum bias, we can write

$$V_{CC} = A + B$$

By applying Ohm's law to the shaded triangle (1) on Fig. 6-13, we have

$$A = (2I_{CQ} - I_{CQ})R_{\text{ac}} = I_{CQ}R_{\text{ac}}$$

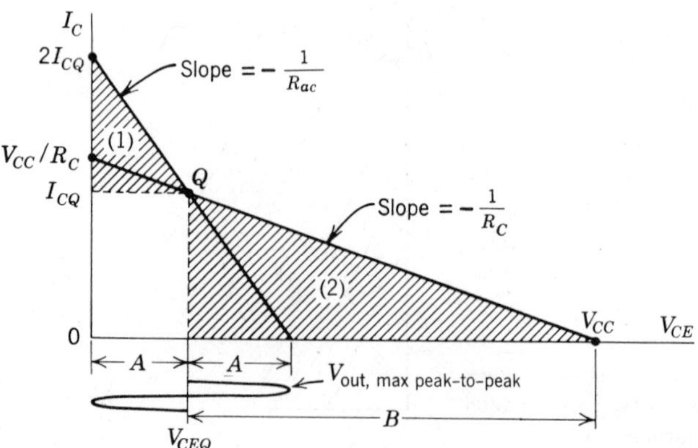

Figure 6-13 Determination of the optimum bias for an ac load line.

and, by applying Ohm's law to the shaded triangle (2), we have

$$B = I_{CQ}R_C$$

Then

$$V_{CC} = A + B = I_{CQ}R_{ac} + I_{CQ}R_C$$

Solving for I_{CQ}, we find

$$I_{CQ} = \frac{V_{CC}}{R_C + R_{ac}} \tag{6-7}$$

We may generalize this equation by using the form

$$\boxed{I_{CQ} = \frac{V_{dc}}{R_{dc} + R_{ac}}} \tag{6-8}$$

where V_{dc} is the total dc voltage applied to the collector circuit,
R_{dc} is the total resistance in which dc current flows in the collector circuit, and
R_{ac} is the ac load resistance in the collector circuit.

Inspection of the circuit of Fig. 6-10 shows that

$$V_{dc} = V_{CC}$$

$$R_{dc} = R_C$$

and

$$R_{ac} = \frac{R_C R}{R_C + R} \tag{6-6}$$

Figure 6-13 shows that the maximum peak-to-peak output voltage is $2A$. But the distance A in triangle (1) is $I_{CQ}R_{ac}$. Then the maximum possible output voltage is

$$\boxed{V_{out} = 2V_{CEQ} = 2I_{CQ}R_{ac} \text{ volts, peak to peak, maximum}} \tag{6-9}$$

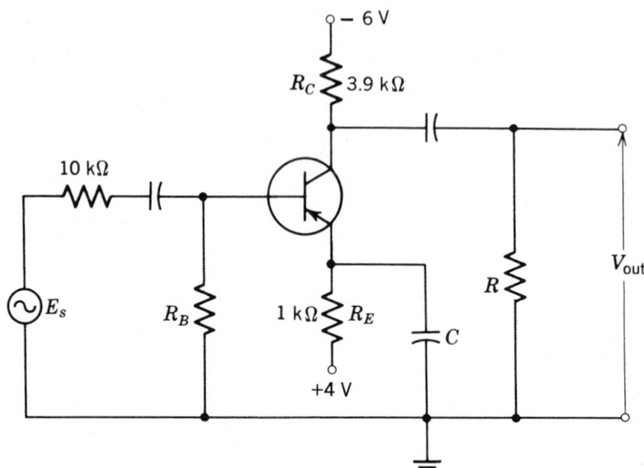

Circuit for Example 6-3 and Example 6-4.

Example 6-3

In the circuit for Example 6-3, R is not placed in the circuit. Determine the operating point and the maximum possible peak-to-peak output voltage.

Solution

Examination of the circuit shows that the value of V_{dc} is

$$V_{dc} = |V_{CC}| + |V_{EE}| = |-6| + |+4| = 10 \text{ V}$$

and the value of R_{dc} is

$$R_{dc} = R_C + R_E = 3.9 + 1.0 = 4.9 \text{ k}\Omega$$

and the value of R_{ac} is

$$R_{ac} = R_C = 3.9 \text{ k}\Omega$$

$$I_{CQ} = \frac{V_{dc}}{R_{dc} + R_{ac}} = \frac{10 \text{ V}}{4.9 \text{ k}\Omega + 3.9 \text{ k}\Omega} = \mathbf{1.14\ mA} \qquad (6\text{-}8)$$

$$V_{CQ} = I_{CQ}R_{ac} = 1.14 \text{ mA} \times 3.9 \text{ k}\Omega = \mathbf{4.4\ V}$$

The maximum peak-to-peak output voltage is

$$V_{out,max,p-p} = 2V_{CEQ} = 2I_{CQ}R_{ac} = 2 \times 4.4 \text{ V} = \mathbf{8.8\ V} \qquad (6.9)$$

Example 6-4
In the circuit for Example 6-3, R is 5.1 kΩ. Determine the operating point and the maximum possible peak-to-peak output voltage.

Solution
Examination of the circuit shows that the values of V_{dc} (10 V) and of R_{dc} (4.9 kΩ) are unchanged from Example 6-3. The value of R_{ac}, the ac load, is the parallel combination of the 3.9-kΩ resistor and the added 5.1-kΩ resistor.

$$R_{ac} = \frac{3.9\text{ k}\Omega \times 5.1\text{ k}\Omega}{3.9\text{ k}\Omega + 5.1\text{ k}\Omega} = 2.21\text{ k}\Omega \qquad (6\text{-}6)$$

Then

$$I_{CQ} = \frac{V_{dc}}{R_{dc} + R_{ac}} = \frac{10\text{ V}}{4.9\text{ k}\Omega + 2.21\text{ k}\Omega} = \mathbf{1.41\text{ mA}} \qquad (6\text{-}8)$$

Now

$$V_{CE} = I_{CQ}R_{ac} = 1.41\text{ mA} \times 2.21\text{ k}\Omega = \mathbf{3.1\text{ V}}$$

and the maximum possible undistorted output signal is reduced by the addition of the 5.1-kΩ resistor to

$$V_{\text{out,max},p-p} = 2V_{CE} = 2 \times 3.1\text{ V} = \mathbf{6.2\text{ V}} \qquad (6\text{-}9)$$

If I_{CQ} is greater than the optimum operating point value (Fig. 6-14a) the shaded triangle (1) is smaller than the shaded triangle (2). Consequently, when the output signal v_{out} is increased from zero, we see that saturation will limit the

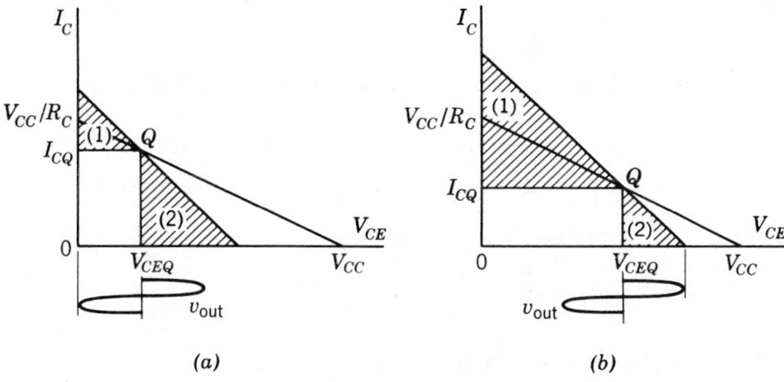

Figure 6-14 Graphical determination of output voltage. (a) I_{CQ} is greater than the optimum value. (b) I_{CQ} is less than the optimum value.

undistorted output voltage to

$$V_{out} = 2V_{CEQ} \text{ volts, peak to peak, maximum} \qquad (6\text{-}10)$$

On the other hand, if I_{CQ} is less than the optimum operating point value (Fig. 6-14*b*) the shaded triangle (2) is smaller than the shaded triangle (1). Then, when the output signal v_{out} is increased from zero, we see that cutoff will limit the undistorted output voltage to

$$V_{out} = 2I_{CQ}R_{ac} \text{ volts, peak to peak, maximum} \qquad (6\text{-}11)$$

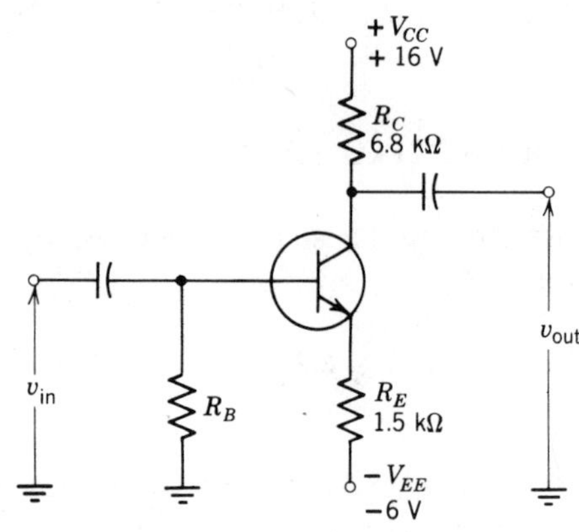

Circuit for Example 6-5.

Example 6-5

Determine the maximum peak-to-peak undistorted output voltage obtainable from the circuit.

Solution

In order to make use of Eq. 6-8, we require numerical values for V_{dc}, R_{dc}, and R_{ac}.

$$V_{dc} = |+V_{CC}| + |-V_{EE}| = |+16| + |-6| = 16 + 6 = 22 \text{ V}$$

$$R_{dc} = R_C + R_E = 6.8 + 1.5 = 8.3 \text{ k}\Omega$$

and $$R_{ac} = R_C + R_E = 6.8 + 1.5 = 8.3 \text{ k}\Omega$$

Then substituting into Eq. 6-8, we have

$$I_{CQ} = \frac{V_{dc}}{R_{dc} + R_{ac}} = \frac{22\ \text{V}}{8.3\ \text{k}\Omega + 8.3\ \text{k}\Omega} = 1.33\ \text{mA} \qquad (6\text{-}8)$$

The Kirchhoff's voltage loop equation through the collector is

$$|+V_{CC}| + |-V_{EE}| = I_C R_C + V_{CE} + I_E R_E$$

Assuming I_C and I_E are equal and substituting numerical values, we have

$$22\ \text{V} = 1.33\ \text{mA} \times 6.8\ \text{k}\Omega + V_{CE} + 1.33\ \text{mA} \times 1.5\ \text{k}\Omega$$

$$V_{CE} = 11\ \text{V}$$

Note that the sum of the voltage drops across R_C and R_E is the same value.

$$I_{CQ}(R_C + R_E) = I_{CQ}R_{dc} = 1.33\ \text{mA} \times (6.8\ \text{k}\Omega + 1.5\ \text{k}\Omega) = 11\ \text{V}$$

Then the maximum peak-to-peak ac voltage across both R_C and R_E is

$$2V_{CE} = 2I_{CQ}(R_C + R_E) = 2I_{CQ}R_{ac} = 22\ \text{V}$$

However, the output voltage is taken across R_C only. Then the peak-to-peak undistorted output voltage is

$$V_{out} = 2I_{CQ}R_C = 2 \times 1.33\ \text{mA} \times 6.9\ \text{k}\Omega$$

$$= \mathbf{18\ V\ peak\ to\ peak,\ maximum}$$

This result can also be determined by using the voltage-divider rule. The whole 22 volts appears across $R_{ac}(R_C + R_E)$ but the output voltage is only that fraction of the voltage that appears across the load R_C. Then

$$V_{out} = \frac{6.8\ \text{k}\Omega}{6.8\ \text{k}\Omega + 1.5\ \text{k}\Omega} 22\ \text{V} = \mathbf{18\ V\ peak\ to\ peak,\ maximum}$$

If this circuit is not biased at the optimum Q-point, we can modify the final results by using the approach taken for Fig. 6-14 (Eq. 6-10 or Eq. 6-11).

Example 6-6

Determine the maximum peak-to-peak undistorted output voltage obtainable from the circuit.

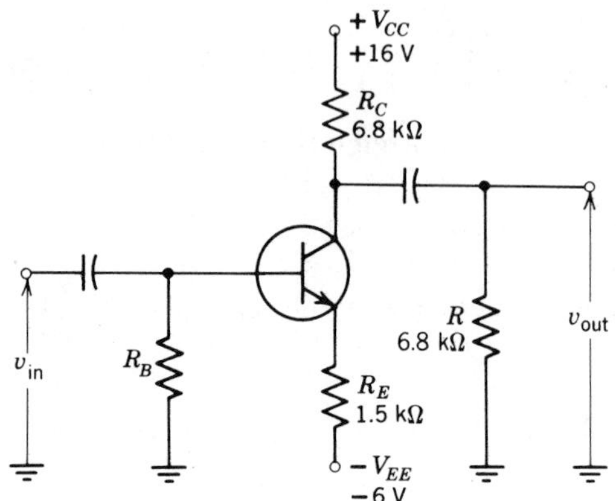

Circuit for Example 6-6.

Solution

The values for V_{dc} and for R_{dc} are the same as those for Example 6-5.

$$V_{dc} = 22\text{ V} \quad \text{and} \quad R_{dc} = 8.3\text{ k}\Omega$$

The ac load resistance on the collector R_L is the parallel combination of R_C and R.

$$R_L = \frac{R_C R_0}{R_C + R_0} = \frac{6.8\text{ k}\Omega \times 6.8\text{ k}\Omega}{6.8\text{ k}\Omega + 6.8\text{ k}\Omega} = 3.4\text{ k}\Omega$$

The whole ac resistance R_{ac} is the sum of R_L and R_E.

$$R_{ac} = 3.4 + 1.5 = 4.9\text{ k}\Omega$$

Substituting into Eq. 6-8, we have

$$I_{CQ} = \frac{V_{dc}}{R_{dc} + R_{ac}} = \frac{22\text{ V}}{8.3\text{ k}\Omega + 4.9\text{ k}\Omega} = 1.67\text{ mA} \qquad (6\text{-}8)$$

and the maximum peak-to-peak undistorted ac voltage across the load is

$$V_{out} = 2I_{CQ}R_L = 2 \times 1.67\text{ mA} \times 3.4\text{ k}\Omega$$

$$= \mathbf{11.33\ V\ peak\ to\ peak,\ maximum}$$

We can use an alternative approach to the solution. The Kirchhoff's voltage loop through the collector circuit is

$$|+V_{CC}| + |-V_{EE}| = I_{CQ}R_C + V_{CE} + I_{CQ}R_E$$

$$22 = 1.67 \text{ mA} \times 6.8 \text{ k}\Omega + V_{CE} + 1.67 \text{ mA} \times 1.5 \text{ k}\Omega$$

$$V_{CE} = 8.17 \text{ V}$$

Then the maximum peak-to-peak voltage across the whole ac resistance R_{ac} is

$$2V_{CE} = 16.34 \text{ V}$$

or

$$2I_{CQ}R_{ac} = 16.34 \text{ V}$$

Considering the voltage divider formed by R_L and R_E,

$$V_{out} = \frac{R_L}{R_L + R_E} 16.34 = \frac{R_L}{R_{ac}} 16.34$$

$$= \frac{3.4 \text{ k}\Omega}{4.9 \text{ k}\Omega} 16.34 = \mathbf{11.33 \text{ V peak to peak, maximum}}$$

Problems

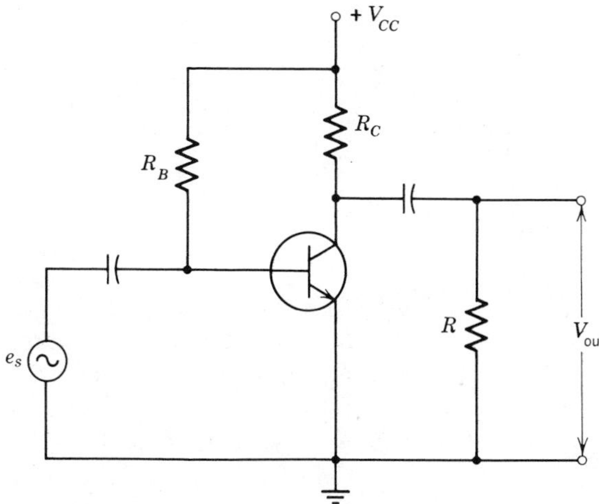

Circuit for Problems 6-3.1 through 6-3.6.

All transistors are silicon.

6-3.1 V_{CC} is 30 V, R_C is 10 kΩ, and R is 15 kΩ. β is 80. Determine the value of R_B that provides optimum bias to give the maximum available peak-to-peak output voltage. What is this value of V_{out}?

6-3.2 Repeat Problem 6-3.1 for V_{CC} equal to 8 V, R_C equal to 8 kΩ, R equal to 12 kΩ, and β equal to 60.

6-3.3 V_{CC} is 40 V, R_C is 10 kΩ, R is 10 kΩ, and β is 100. Determine R_B for optimum bias conditions. What is the maximum value of V_{out}?

6-3.4 Repeat Problem 6-3.2 if R is changed to 5 kΩ.

6-3.5 A properly bypassed 2000-Ω emitter resistor R_E is added to the circuit of Problem 6-3.1. What is the optimum value for R_B and what is the maximum value of V_{out}?

6-3.6 A properly bypassed 2000-Ω emitter resistor R_E is added to the circuit of Problem 6-3.2. What is the optimum value of R_B and what is the maximum value of V_{out}?

Supplementary Problems

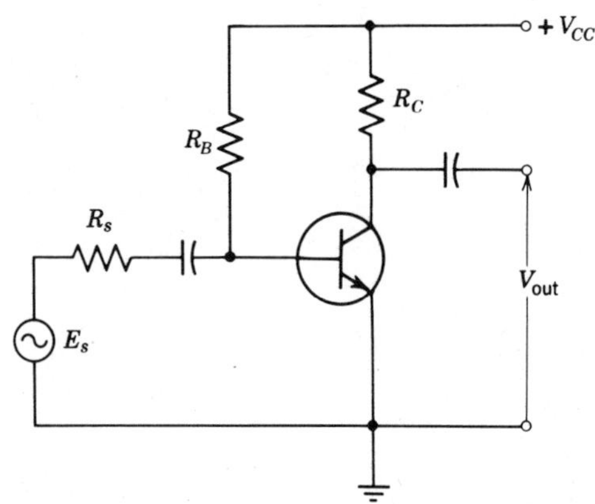

Circuit for Problems 6-1 through 6-6.

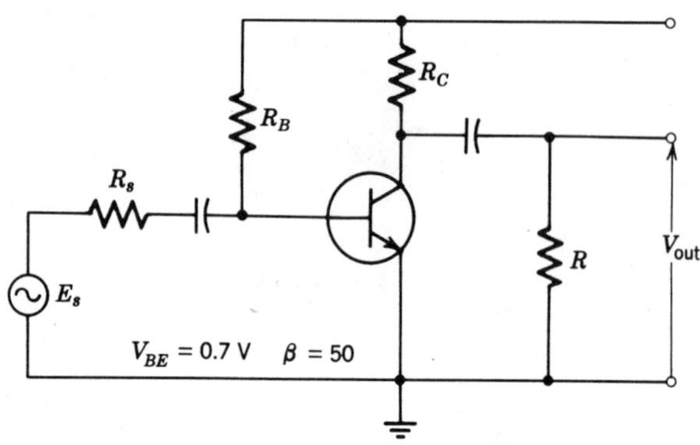

Circuit for Problems 6-7 through 6-10.

6-1 $R_B = 200$ kΩ; $R_C = 2$ kΩ; $\beta = 40$; $V_{BE} = 0.7$ V; $V_{CC} = +8$ V. Find I_{CQ}, V_{CEQ}, and the maximum value of V_{out} without clipping.

6-2 $R_B = 100$ kΩ; $R_C = 2$ kΩ; $\beta = 40$; $V_{BE} = 0.7$ V; $V_{CC} = +8$ V. Find I_{CQ}, V_{CEQ}, and the maximum value of V_{out} without clipping.

6-3 $R_B = 800$ kΩ; $R_C = 12$ kΩ; $\beta = 50$; $V_{BE} = 0.7$ V; $V_{CC} = +20$ V. Find I_{CQ}, V_{CEQ}, and the maximum value of V_{out} without clipping.

6-4 $R_B = 2$ MΩ; $R_C = 12$ kΩ; $\beta = 50$; $V_{BE} = 0.7$ V; $V_{CC} = +20$ V. Find I_{CQ}, V_{CEQ}, and the maximum value of V_{out} without clipping.

6-5 What value of R_B in Problem 6-3 yields the maximum possible peak-to-peak value of V_{out} without clipping?

6-6 What value of R_C in Problem 6-4 yields the maximum possible peak-to-peak value of V_{out} without clipping?

6-7 $R_B = 2$ MΩ; $R_C = 12$ kΩ; $R = 12$ kΩ; and $V_{CC} = +20$ V. Find the value of the maximum peak-to-peak output voltage without clipping.

6-8 $R_B = 400\ \text{k}\Omega$; $R_C = 2\ \text{k}\Omega$; $R = 2\ \text{k}\Omega$; and $V_{CC} = +10\ \text{V}$. Find the value of the maximum peak-to-peak output voltage without clipping.

6-9 What value of R_B sets the circuit in Problem 6-7 to optimum and what is this resulting output voltage?

6-10 What value of R_B sets the circuit in Problem 6-8 to optimum and what is this resulting output voltage?

6-11 $R_C = 18\ \text{k}\Omega$; $R_E = 6\ \text{k}\Omega$; $R = 30\ \text{k}\Omega$; $V_{CC} = +30\ \text{V}$; and $V_{EE} = -4\ \text{V}$. What value of R_B adjusts the circuit for optimum output voltage? What is the resulting output voltage?

6-12 Repeat Problem 6-11 for the following values: $R_C = 8\ \text{k}\Omega$; $R_E = 2\ \text{k}\Omega$; $R = 8\ \text{k}\Omega$; $V_{CC} = +20\ \text{V}$; and $V_{EE} = -6\ \text{V}$.

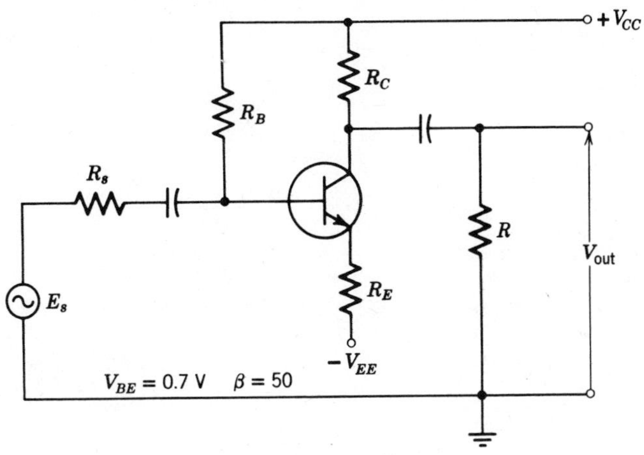

Circuit for Problems 6-11 and 6-12.

Chapter 7 Transistor Small-Signal Amplifiers

The calculation of the ac gain of a transistor amplifier circuit can be reduced to a simple procedure that can be applied to any circuit configuration. The definitions of gains and of input resistance are established. Circuit gains are expressed in terms of source resistance and input resistance (Section 7-1). The concepts of the ac emitter resistance r_e' and of the ac model for a transistor are examined. The gain and the input resistance for each of the basic amplifier circuits are developed by means of a model: the common-emitter amplifier (Section 7-3), the common-collector amplifier (Section 7-4), and the common-base amplifier (Section 7-5). These concepts are extended to the same more complex amplifier circuits that were considered for biasing methods in Chapter 5: the common-emitter amplifier using emitter feedback (Section 7-6) and the common-emitter amplifier using collector-to-base feedback (Section 7-7). These techniques are extended to show how gains are calculated in cascaded amplifiers (Section 7-8).

Section 7-1 General Considerations

In Chapters 5 and 6, we studied methods of biasing transistors. These techniques establish the Q-point. Now, *after* a semiconductor is biased for a particular operating point, an ac signal can be applied to the circuit. The ac signal E_s can be either a laboratory generator or an information source.

In Fig. 7-1, we connect a signal source to an amplifier that is represented by a "black box." The signal source has an emf of E_s volts and an internal resistance (impedance) of R_s (Z_s) ohms. The voltage at the input of the amplifier is V_{in}. The output voltage measured across the output terminals of the amplifier is V_{out}. The voltages E_s, V_{in}, and V_{out} can be rms (effective) values, peak values, peak-to-peak values, or instantaneous values.

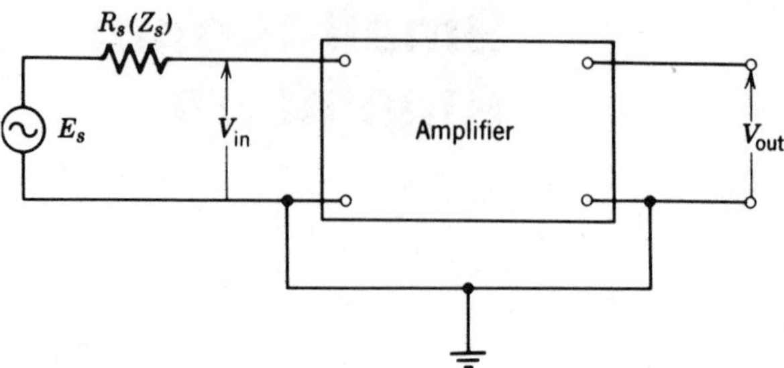

Figure 7-1 Block diagram of a signal amplifier.

Let us form two very important definitions that we will use throughout the text.

$$A_v \equiv \frac{V_{out}}{V_{in}} \tag{7-1a}$$

and

$$A_e \equiv \frac{V_{out}}{E_s} \tag{7-1b}$$

A_v is the voltage gain across the circuit, the "black box," and A_e is the voltage gain over the whole circuit from the source emf E_s to the output voltage V_{out}.

The amplifier circuit, the "black box" of Fig. 7-1, has a finite input resistance r'_{in}. The effect of r'_{in} is to form a voltage divider with R_s, Fig. 7-2. We use the voltage-divider rule to state that

$$V_{in} = \frac{r'_{in}}{r'_{in} + R_s} E_s \tag{7-2}$$

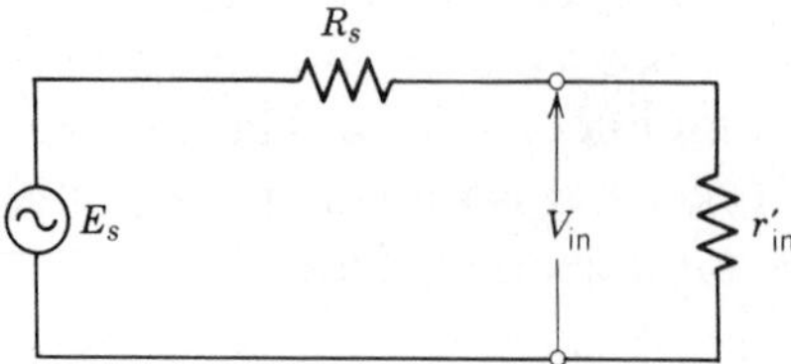

Figure 7-2 Effect of the input resistance of an amplifier.

Multiply both sides of Eq. 7-2 by $V_{out}/E_s V_{in}$ to obtain

$$V_{in} \times \frac{V_{out}}{E_s V_{in}} = \frac{r'_{in}}{r'_{in} + R_s} E_s \times \frac{V_{out}}{E_s V_{in}}$$

Cancelling common factors, we have

$$\frac{V_{out}}{E_s} = \frac{r'_{in}}{r'_{in} + R_s} \times \frac{V_{out}}{V_{in}}$$

And using the definitions given by Eq. 7-1*a* and Eq. 7-1*b*, we have

$$\boxed{A_e = \frac{r'_{in}}{r'_{in} + R_s} A_v} \tag{7-3}$$

Equation 7-3 is a fundamental equation for understanding the concepts of signal gain. Equation 7-3 shows that, when the signal source has a finite value of internal resistance, the overall voltage gain A_e must be less than the voltage gain across the semiconductor circuit A_v. If the application involves a maximum power transfer from E_s, obviously the semiconductor circuit must be designed so that r'_{in} equals R_s. If r'_{in} is much smaller than R_s, the overall voltage gain A_e is much smaller than A_v. When we wish to maximize the overall circuit voltage gain, the semiconductor circuit must be designed so that r'_{in} is much greater than R_s.

This discussion shows that the information we require to analyze an amplifier circuit consists of our ability to determine

$$r'_{in} \text{ and } A_v.$$

If we know these values, we can determine the voltage gain, the current gain, and the power gain for any semiconductor circuit.

The methods we will develop in this chapter will enable us to predetermine values of A_v and r'_{in} for the basic circuit configurations.

Example 7-1
In Fig. 7-2, E_s is 200 mV, R_s is 10 kΩ, and r'_{in} is 1200 Ω. Determine V_{in}.

Solution
V_{in} is found by the application of the voltage-divider rule.

$$V_{in} = \frac{r'_{in}}{r'_{in} + R_s} E_s = \frac{1200\ \Omega}{1200\ \Omega + 10{,}000\ \Omega} 200\ \text{mV} = \mathbf{21.4\ mV} \qquad (7\text{-}2)$$

These numerical values were selected to show that much of the signal level can be lost because of the effect of a low input resistance compared to R_s. It should be recalled that, when a circuit is matched (maximum power transfer), R_s and r'_{in} are equal. Then V_{in} is half E_s.

When a signal source is connected to an amplifier (Fig. 7-3), there is a source current I_s. The voltage drop across R_s is the difference in potential between the ends of R_s.

$$(E_s - V_{in})$$

The current in R_s is

$$I_s = \frac{E_s - V_{in}}{R_s}$$

The input resistance to the semiconductor circuit is r'_{in}. From Ohm's law, r'_{in} is

$$r'_{in} = \frac{V_{in}}{I_s}$$

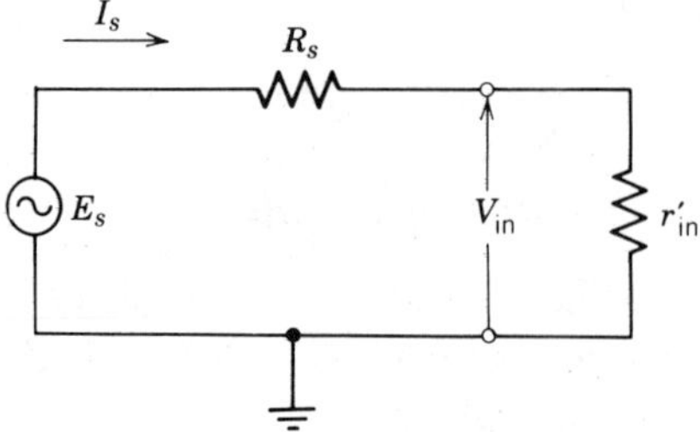

Figure 7-3 Measuring the value of r'_{in}.

Substituting the equation for I_s, we have

$$r'_{in} = \frac{V_{in}}{\left(\dfrac{E_s - V_{in}}{R_s}\right)}$$

Rearranging, we have

$$\boxed{r'_{in} = \frac{V_{in}}{E_s - V_{in}} R_s} \qquad (7\text{-}4)$$

In the laboratory we can measure E_s, V_{in}, and V_{out} to obtain the voltage gains. Equation 7-4 is very important because we can use it to determine the value of r'_{in} from the ac voltage measurements.

Example 7-2
The following values were obtained in the laboratory for the circuit of Fig. 7-3 where r'_{in} is the equivalent of the input resistance of a complete amplifier.

$$E_s = 20\ \text{V} \qquad R_s = 100\ \text{k}\Omega \quad \text{and} \quad V_{in} = 30\ \text{mV}$$

The voltages are the peak-to-peak values of sinusoidal waveforms obtained by an oscilloscope.

Solution
The current I_s of the source is

$$I_s = \frac{E_s - V_{in}}{R_s} = \frac{20\ \text{V} - 0.030\ \text{V}}{100\ \text{k}\Omega} \approx \frac{20\ \text{V}}{100\ \text{k}\Omega} = 0.2\ \text{mA} = 200\ \mu\text{A}$$

The input resistance is

$$r'_{in} = \frac{V_{in}}{I_s} = \frac{30\ \text{mV}}{0.2\ \text{mA}} = \mathbf{150\ \Omega}$$

Using Eq. 7-4, we can find r'_{in} directly.

$$r'_{in} = \frac{V_{in}}{E_s - V_{in}} R_s = \frac{0.030\ \text{V}}{20\ \text{V} - 0.030\ \text{V}} 100\ 000\ \Omega = \mathbf{150\ \Omega} \qquad (7\text{-}4)$$

Section 7-2 Emitter Resistance(r'_e)

In Section 2-5, we developed the ac model (the ac equivalent) circuit for a diode. We stated that the ac resistance of a diode is

$$\frac{25\text{ mV}}{I_F} \leq r_j \leq \frac{50\text{ mV}}{I_F} \tag{2-5}$$

In the ac model for a transistor (Fig. 7-4), we show the ac resistance for the transistor as the ac resistance r'_e in series with the emitter. The value of r'_e at room temperature is found by determining the dc current in the emitter I_E and by using

$$\boxed{\frac{25\text{ mV}}{I_E} \leq r'_e \leq \frac{50\text{ mV}}{I_E}} \tag{7-5}$$

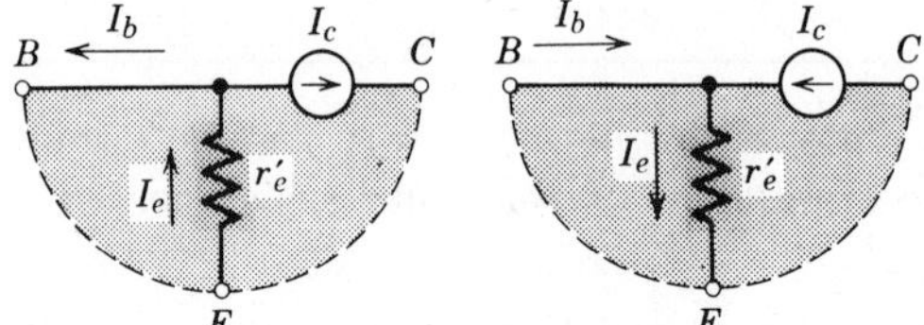

Figure 7-4 The ac model for a transistor.

Example 7-3

If the value of I_E is 0.2 mA, what is the range of the expected value of r'_e?

Solution

Placing the value of I_E into Eq. 10-5, we have

$$\frac{25\text{ mV}}{I_E} \leq r'_e \leq \frac{50\text{ mV}}{I_E} \tag{7-5}$$

$$\frac{25\text{ mV}}{0.2\text{ mA}} \leq r'_e \leq \frac{50\text{ mV}}{0.2\text{ mA}}$$

$$\mathbf{125\ \Omega \leq r'_e \leq 250\ \Omega}$$

In the ac model, Fig. 7-4, the currents are shown as effective or rms values.

$$I_c,\ I_b,\text{ and } I_e$$

Any equation or concept developed in terms of rms values is also valid when the currents and voltages are expressed as peak values ($I_{c,\max}$, $I_{b,\max}$, and $I_{e,\max}$), as peak-to-peak values, or as instantaneous values (i_c, i_b, and i_e).

In the model (Fig. 7-4) we show a current generator for I_c. This is valid because collector current exists only when base current exists. Also if the value of I_b is multiplied by a constant (β) we have I_c. If a signal source connected to a transistor circuit produces a signal current in the transistor, signal currents are produced in the other two leads of the transistor since

$$I_e = I_b + I_C \tag{4-1}$$

If we assign a direction to any one of the three currents, the directions of the other two currents are automatically determined.

The ac models given in Fig. 7-4 show the currents in different directions. The results of an ac circuit analysis do not depend on which set of current directions is used. The directions of the currents are usually determined initially by assigning an instantaneous polarity to the ac signal source. It must be emphasized that the assignment of the polarity of the signal currents is completely independent of the dc current directions in a transistor. Either diagram of Fig. 7-4 is valid *either* for an *NPN* transistor *or* for a *PNP* transistor.

Section 7-3 The Common-Emitter Amplifier Model

A very important concept in electronics is the function of the "ground" in Fig. 7-5. The "ground" is simply the zero-voltage reference point from which all other voltages are measured. The voltage measured between "ground" and the terminal marked $-V_{BB}$ is the voltage of the emitter supply, which is a pure dc voltage. Likewise, the voltage measured between "ground" and the terminal marked $-V_{CC}$ is the voltage of the collector supply, which is again an ideal dc voltage.

If the supplies ($-V_{BB}$ and $-V_{CC}$) are ideal, when there is a change in current (ΔI_B or ΔI_C), the change in voltage (ΔV_{BB} or ΔV_{CC}) is zero. Consequently, the ratios, $\Delta V_{BB}/\Delta I_B$ and $\Delta V_{CC}/\Delta I_C$, are zero and the impedances of the supplies are each zero ohms.

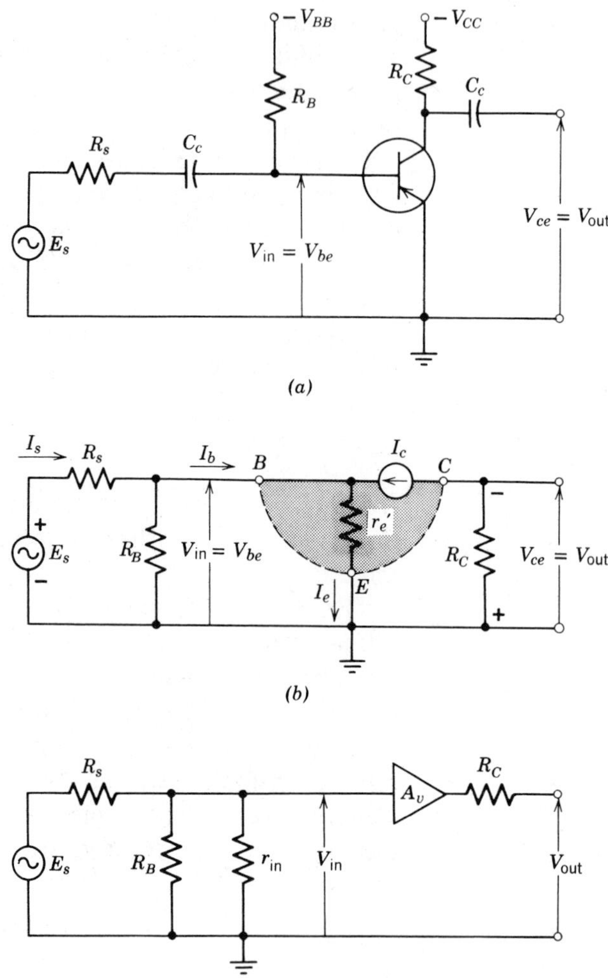

Figure 7-5 The common-emitter amplifier. (*a*) Complete circuit. (*b*) The formal model. (*c*) The simplified model.

We can arrive at this same conclusion from another viewpoint. In order to reduce the ripple toward zero, we place very large capacitors across the output of the rectifiers supplying $-V_{BB}$ and $-V_{CC}$. The reactances of these capacitors are very low at the frequencies of the signal E_s.

The actual amplifier circuit (Fig. 7-5*a*) shows the connections to the dc supplies $-V_{BB}$ and $-V_{CC}$. An *equivalent circuit* for the ac signal flow is called the *formal model* (Fig. 7-5*b*). The model contains only those elements that are required to analyze the circuit from the viewpoint of the ac signal. Consequently all blocking and coupling capacitors are assumed to have a reactance of zero ohms. Also, since the ac impedance of the dc supplies is zero, the leads that go to the supplies are returned to the common reference line or "ground." The equivalent circuit for the transistor is the ac model used for Fig. 7-4.

Our method of analysis for the different amplifier circuits will all follow this pattern:

1. A formal model is made from the actual circuit.
2. An algebraic analysis of the formal model gives equations for r_{in} and A_v.
3. If the values of r_{in} and A_v are transferred to the simplified model, it is apparent that V_{in} is the result of a voltage divider placed across E_s and that V_{out} is V_{in} multiplied by A_v.

The objective of this procedure is to form a simplified approach to the amplifier. We will show that all amplifier types can be treated in this same manner. We will show that r_{in} can be determined from a simple equation and that A_v can be determined from a simple equation. Also we will show that the equations for r_{in} and for A_v follow the same pattern for the different amplifier type and circuit configurations.

In this way, the student should develop a "feeling" for semiconductor amplifiers so that he or she can look at a circuit, can readily reduce the circuit to the simplified model, and then, without having to refer to detailed derivations, can write down the gain equations for the circuit.

Now, let us examine the formal model for the common-emitter amplifier (Fig. 7-5*b*). Let us start by arbitrarily assigning a positive and a negative instantaneous polarity to E_s. This polarity determines the direction for the source current I_s. The direction of I_b is determined by the direction of I_s. Having the direction of I_b given, the directions of I_e and I_c are mandated. I_c flows through R_C. Knowing the direction of I_c through R_C, we place the polarity markings on R_C. These polarity markings show that V_{out} (V_{ce}) is 180° out of phase with E_s.

The dc circuit analysis of the previous chapter is used to determine the Q-point values. We need the value of the dc current in the emitter I_E to determine r'_e from

$$\frac{25 \text{ mV}}{I_E} \le r'_e \le \frac{50 \text{ mV}}{I_E} \qquad (7\text{-}5)$$

The input voltage to the transistor V_{in} is given by Ohm's law as

$$V_{in} = V_{be} = I_e r'_e$$

Let us replace I_e by I_b, using the conversion from Table 4-2 (Page 98)

$$I_e = (1 + \beta)I_b$$

to give

$$V_{in} = V_{be} = (1 + \beta)r'_e I_b \tag{7-6}$$

The input voltage to the transistor at the base is V_{in} and the input current to the transistor at the base is I_b. Therefore if we divide V_{in} by I_b we obtain the input resistance r_{in} looking into the transistor at the base.

$$\boxed{r_{in} = (1 + \beta)r'_e} \tag{7-7}$$

The output voltage V_{out} is the IR drop across R_C.

$$V_{out} = I_c R_C$$

Replacing I_c by I_b and using the conversion from Table 4-2, we find that

$$I_c = \beta I_b$$

We have

$$V_{out} = \beta I_b R_C$$

The voltage gain A_v is the ratio V_{out}/V_{in}.

$$A_v = \frac{V_{out}}{V_{in}} = \frac{\beta I_b R_C}{(1 + \beta)I_b r'_e}$$

Dividing through by I_b, we find

$$A_v = \frac{\beta R_C}{(1 + \beta)r'_e}$$

If the value of β is as low as 24, the ratio of $\beta/(1+\beta)$ is 24/25 or 0.96. The assumption that this ratio is unity introduces an error of 0.04 or 4%. If the value of β is 49, the ratio of $\beta/(1+\beta)$ is 49/50 or 0.98. If we assume that this ratio is unity, we make an error of 0.02 or 2%. By accepting this small error, we can reduce the gain equation to a simple equation.

$$\boxed{A_v \approx \frac{R_C}{r'_e}} \tag{7-8a}$$

We can generalize the gain equation by replacing R_C in Eq. 7-8*a* with Z_L.

$$A_v \approx \frac{Z_L}{r'_e} \tag{7-8b}$$

The load on the amplifier Z_L can be any simple or complex arrangement of R and/or L and/or C.

The current gain of the circuit is β.

$$A_i = \frac{I_c}{I_b} = \beta \tag{7-8c}$$

Equation 7-3 relates A_v to A_e.

$$A_e = \frac{r'_{\text{in}}}{r'_{\text{in}} + R_s} A_v \tag{7-3}$$

If we compare Fig. 7-2 to the simplified model (Fig. 7-5*c*), it is evident that r'_{in} is the parallel combination of R_B and r'_{in}.

$$r'_{\text{in}} = \frac{R_B r_{\text{in}}}{R_B + r_{\text{in}}} \tag{7-9}$$

In most cases R_B is very much greater than r_{in} with the result that the parallel combination of R_B and r_{in} is effectively r_{in}.

$$\text{If} \quad R_B \gg r_{\text{in}}$$

$$r'_{\text{in}} \approx r_{\text{in}} \tag{7-10}$$

Example 7-4
The value of r'_e is 25 Ω and the value of β is 50 for a particular transistor. Determine the input resistance when it is used in a common-emitter amplifier circuit.

Solution
The input resistance to the common-emitter circuit is

$$r_{\text{in}} = (1+\beta)r'_e = 51 \times 25 = \mathbf{1275\ \Omega} \tag{7-7}$$

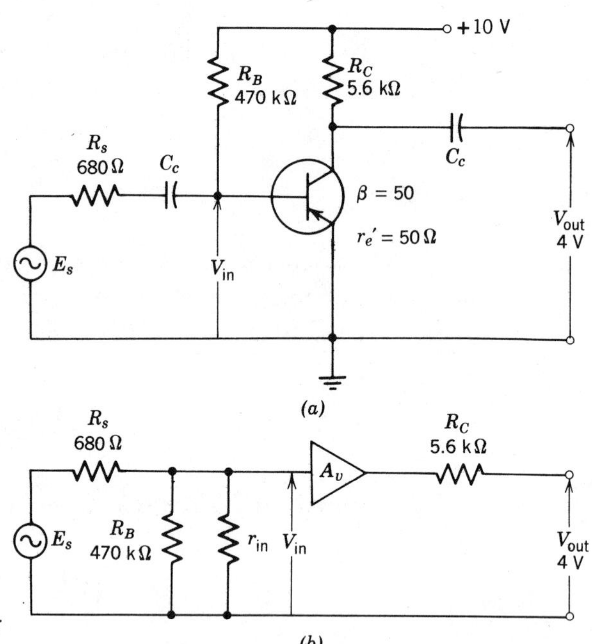

Circuit (*a*) and simplified model (*b*) for Example 7-5.

Example 7-5
Determine V_{in}, E_s, and A_e for the illustrated common-emitter amplifier.

Solution
The first step in determining the signal-level values is to draw the simplified model. Then numerical values are transferred from the circuit to the model.

R_s, R_B, R_C, and V_{out}

E_s, r_{in}, V_{in}, and A_v are placed on the model as unknown values at this point. Now we must determine these unknown values.

The input resistance r_{in} to the common-emitter amplifier is

$$r_{in} = (1+\beta)r'_e = 51 \times 50 = 2550\ \Omega \qquad (7\text{-}7)$$

The voltage gain A_v across the transistor is

$$A_v = \frac{R_C}{r'_e} = \frac{5600\ \Omega}{50\ \Omega} = 112 \qquad (7\text{-}8a)$$

Then V_{in} can be found from the definition of A_v.

$$A_v = \frac{V_{out}}{V_{in}} \qquad (7\text{-}1a)$$

$$112 = \frac{4\ \text{V}}{V_{in}}$$

$$V_{in} = 0.0357\ \text{V} = \mathbf{35.7\ mV}$$

It is obvious that R_B is so much larger than r_{in} that R_B can be neglected. Then

$$r'_{in} = r_{in} = 2550\ \Omega \qquad (7\text{-}10)$$

Using the action of the voltage divider in the input circuit of the model, we find that

$$V_{in} = \frac{r'_{in}}{r'_{in} + R_s} E_s \qquad (7\text{-}2)$$

$$35.7\ \text{mV} = \frac{2550\ \Omega}{2550\ \Omega + 680\ \Omega} E_s$$

$$E_s = \mathbf{45.2\ mV}$$

The overall amplification A_e of the circuit is

$$A_e = \frac{V_{out}}{E_s} = \frac{4\ \text{V}}{0.0452\ \text{V}} = \mathbf{88} \qquad (7\text{-}1b)$$

Problems All transistors are silicon, and r'_e is obtained from $25\ \text{mV}/I_E$ unless specified. The ac model is required for all problems.

7-3.1 R_s is 3600 Ω, R_B is 80 kΩ, and R_C is 3000 Ω. For the transistor,

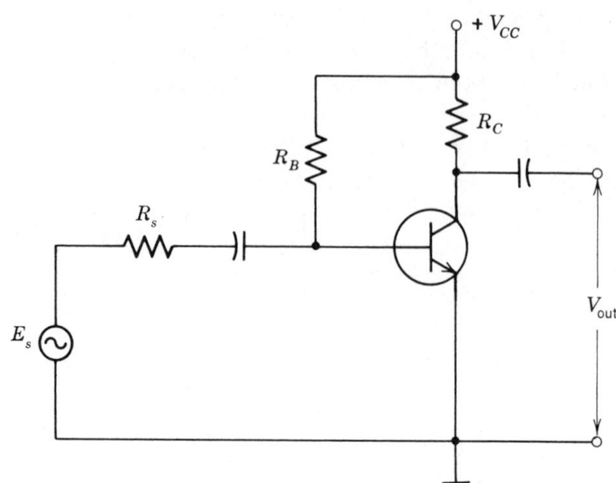

Circuit for Problems 7-3.1 through 7-3.3.

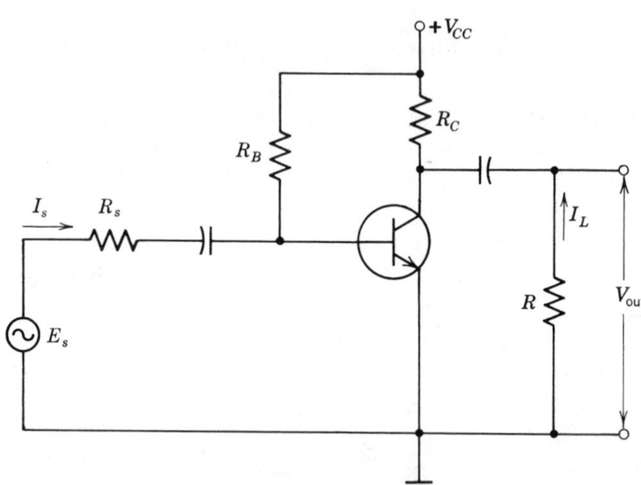

Circuit for Problem 7-3.4.

r_e' is 20 Ω and β is 100. Determine A_e, A_v, and the resistance that E_s sees.

7-3.2 R_s is 600 Ω, R_B is 75 kΩ, R_C is 2.0 kΩ, V_{CC} is +7.5 V, and β is 20. Determine the overall voltage gain of the circuit.

7-3.3 Repeat Problem 7-3.2 if β is 50.

7-3.4 R_s is 600 Ω, R_C is 4.7 kΩ, R is 4.3 kΩ, V_{CC} is +9.0 V, and β is 40. Determine the value of R_B for optimum bias that yields maximum peak-to-peak output voltage. Determine the maximum allowable value of E_s without clipping and determine the current gain I_L/I_s.

Section 7-4 The Common-Collector Amplifier Model

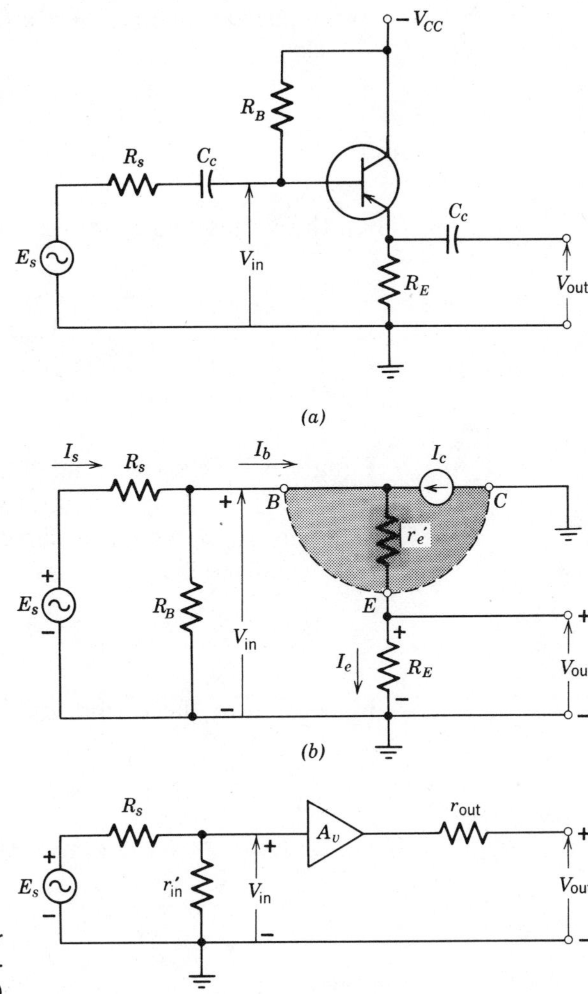

Figure 7-6 The common-collector or emitter-follower amplifier. (*a*) The complete circuit. (*b*) The formal model. (*c*) The simplified model.

The common-collector or emitter-follower amplifier circuit is given in Fig. 7-6*a*. The formal model (Fig. 7-6*b*) is readily drawn from the complete circuit. If the source E_s has the polarity as marked, the direction of I_s is toward the base. Then the base current is directed into the base. This direction of I_b determines the directions of I_c and I_e. The direction of I_e through R_E is downward yielding the positive polarity of V_{out}. Thus, the input voltage and the output voltage of the emitter follower are in phase.

By inspection, we can write the equations

$$V_{in} = I_e(r'_e + R_E)$$

and

$$V_{out} = I_e R_E$$

Then the voltage gain across the transistor A_v is

$$\boxed{A_v = \frac{V_{out}}{V_{in}} = \frac{R_E}{r'_e + R_E}} \tag{7-11}$$

Equation 7-11 shows that the voltage gain A_v must be less than 1.00. If R_E is much greater than r'_e, A_v is approximately 1.00. There is, however, a current gain through the transistor.

$$A_i = \frac{I_e}{I_b} = \frac{(1+\beta)I_b}{I_b} = 1 + \beta \tag{7-12}$$

Inspection of the formal model (Fig. 7-6*b*) shows that

$$V_{in} = I_e(r'_e + R_E)$$

If we replace I_e by the conversion from Table 4-2,

$$I_e = (1+\beta)I_b$$

we have

$$V_{in} = (1+\beta)(r'_e + R_E)I_b$$

Dividing both sides of this equation by I_b gives the ratio V_{in}/I_b, which is the input resistance r_{in} looking into the base of the transistor.

$$\boxed{r_{in} = (1+\beta)(r'_e + R_E)} \tag{7-13}$$

The simplified model (Fig. 7-6*c*) can now be formed. The

voltage gain A_v is given by Eq. 7-11. The resistive load r'_{in} on the source E_s and R_s is the parallel combination of r_{in} and R_B.

$$r'_{in} = \frac{r_{in}R_B}{r_{in} + R_B} \tag{7-14}$$

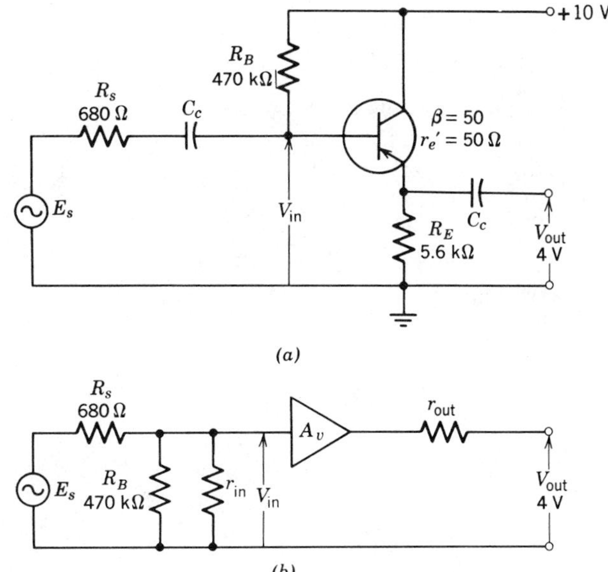

Circuit (*a*) and the simplified model (*b*) for Example 7-6.

Example 7-6
Determine V_{in}, E_s, and A_e for the common-collector amplifier.

Solution
The first step in determining the signal-level values is to draw the simplified model (*b*). Then numerical values are transferred from the circuit to the model.

$$R_s,\ R_B,\ \text{and}\ V_{out}$$

Then E_s, r_{in}, r_{out}, V_{in}, and A_v are placed on the model as unknowns at this point. Now we must determine these unknown values.

The input resistance r_{in} to the common-collector amplifier circuit is

$$r_{in} = (1 + \beta)(r'_e + R_E) = 51(50\ \Omega + 5600\ \Omega) = 288{,}000\ \Omega = 288\ \text{k}\Omega \tag{7-13}$$

The voltage gain A_v across the transistor is

$$A_v = \frac{R_E}{r'_e + R_E} = \frac{5600\ \Omega}{50\ \Omega + 5600\ \Omega} = 0.991 \approx 1 \qquad (7\text{-}11)$$

Since A_v is unity

$$V_{\text{in}} = \frac{V_{\text{out}}}{A_v} = \frac{4\text{ V}}{1} = \mathbf{4\ V} \qquad (7\text{-}1a)$$

The parallel combination of r_{in} and R_B is r'_{in}

$$r'_{\text{in}} = \frac{r_{\text{in}} R_B}{r_{\text{in}} + R_B} = \frac{288\text{ k}\Omega \times 470\text{ k}\Omega}{288\text{ k}\Omega + 470\text{ k}\Omega} = 177\text{ k}\Omega \qquad (7\text{-}14)$$

It is obvious that the small value of R_s (680 Ω) is negligible with respect to r'_{in} (177 kΩ). Therefore E_s and V_{in} are approximately equal.

$$E_s \approx V_{\text{in}} = \mathbf{4\ V}$$

The overall gain A_e of the amplifier is

$$A_e = \frac{V_{\text{out}}}{E_s} = \frac{4\text{ V}}{4\text{ V}} = \mathbf{1} \qquad (7\text{-}1b)$$

Now, let us replace E_s with a short circuit but keep the value of R_s in the circuit (Fig. 7-7*a*). Also, we will drive a signal E back into the output of the amplifier through a resistor R_1. The formal model for this is shown in Fig. 7-7*b*. The resulting current I_1 divides into two currents I_2 and I_e. This current division implies a parallel circuit. R_E is in parallel with a resistance r'_{out} that is obtained "looking back" into the emitter lead. The Kirchhoff voltage loop equation "looking back" into the emitter lead is

$$V_1 = I_e r'_e + I_b\left(\frac{R_s R_B}{R_s + R_B}\right)$$

We can replace I_b with I_e by using the conversion from Table 4-2.

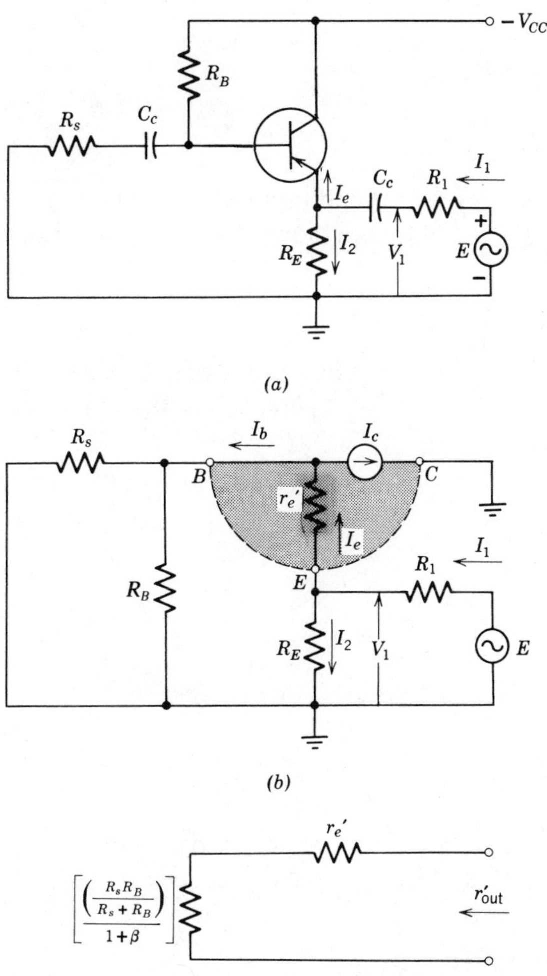

Figure 7-7 Determination of the output resistance of the emitter follower. (*a*) The complete test circuit. (*b*) The formal model. (*c*) The output resistance, r'_{out}.

$$I_b = \frac{I_e}{1+\beta}$$

Then

$$V_1 = I_e r'_e + \frac{I_e}{1+\beta}\left(\frac{R_s R_B}{R_s + R_B}\right)$$

When we divide this equation through by I_e, we have the equation for r'_{out}.

$$r'_{out} = r'_e + \frac{\left(\dfrac{R_s R_B}{R_s + R_B}\right)}{1+\beta} \tag{7-15}$$

If the input signal source E_s to the amplifier has zero internal impedance ($R_s = 0$), then Eq. 7-15 becomes simply

$$r'_{out} = r'_e \tag{7-16}$$

The output resistance of the whole circuit for Fig. 7-6*c* and for Example 7-6 is the parallel combination of R_E and r'_{out}.

$$r_{out} = \frac{R_E r'_{out}}{R_E + r'_{out}} \tag{7-17}$$

Example 7-7
Determine the value of r_{out} for the circuit of Example 7-6.

Solution
Using Eq. 7-15, we have

$$r'_{out} = r'_e + \frac{\left(\dfrac{R_s R_B}{R_s + R_B}\right)}{1+\beta} = 50\ \Omega + \frac{\dfrac{680\ \Omega \times 470{,}000\ \Omega}{680\ \Omega + 470{,}000\ \Omega}}{1+50}$$

$$\approx 50\ \Omega + \frac{680\ \Omega}{51} = 63.3\ \Omega$$

and using Eq. 7-17, we have

$$r_{out} = \frac{R_E r'_{out}}{R_E + r'_{out}} = \frac{63.3\ \Omega \times 5600\ \Omega}{63.3\ \Omega + 5600\ \Omega} = \mathbf{62.6\ \Omega}$$

Problems

7-4.1 R_B is selected to establish V_{CE} at 6 V. R is infinite and the value of β for the silicon transistor is 100. Find I_s and V_{out}. The value of r'_e is 50 mV/I_E.

7-4.2 Repeat Problem 7-4.1 if R is 3000 Ω.

7-4.3 Find I_s and V_{out}. The value of β is 50 and r'_e is 10 Ω.

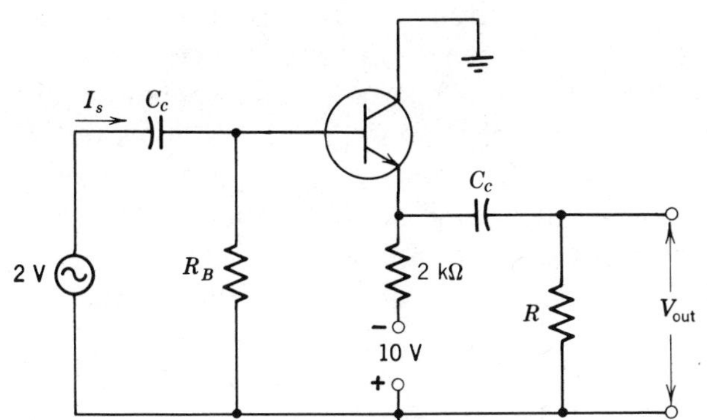

Circuit for Problems 7-4.1 and 7-4.2.

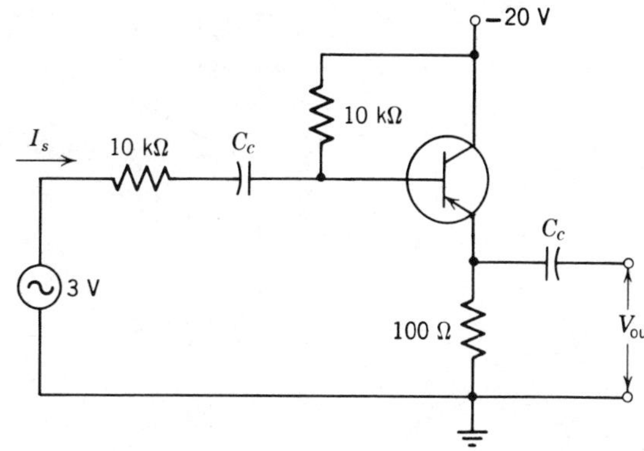

Circuit for Problem 7-4.3.

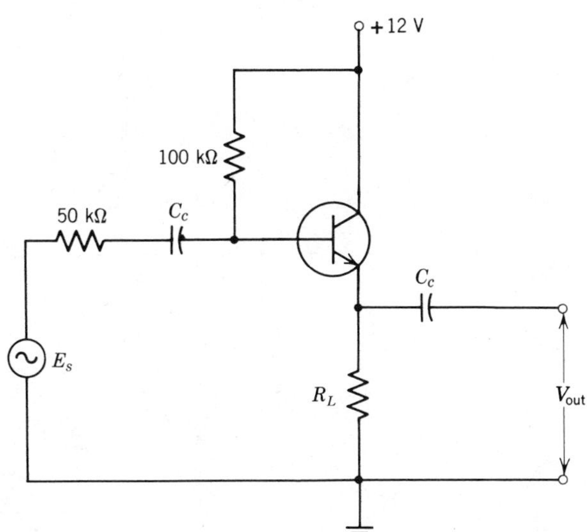

Circuit for Problem 7-4.4

7-4.4 The value of β for the silicon transistor is 100, and r_e' is 10 Ω. Then R_L is adjusted for the condition where optimum bias is obtained. Determine the value of R_L. What are the maximum peak-to-peak values of V_{out} and E_s without clipping?

7-4.5 Determine r'_{out} and r_{out} for the circuit of Problem 7-4.1.

7-4.6 Determine r'_{out} and r_{out} for the circuit of Problem 7-4.2.

7-4.7 Determine r'_{out} and r_{out} for the circuit of Problem 7-4.3.

7-4.8 Determine r'_{out} and r_{out} for the circuit of Problem 7-4.4.

Section 7-5 The Common-Base Amplifier Model

The common-base amplifier circuit is given in Fig. 7-8*a*. The formal model, Fig. 7-8*b*, is drawn from the complete circuit. If the source E_s has the polarity as marked, I_s is directed toward the emitter and I_e flows into the emitter. Then I_c flows out of the collector and into R_C. The polarity of V_{out} is the same as that of E_s. Thus, the input voltage and the output voltage of the common-base amplifier are in phase.

Inspection of the formal model shows that the output voltage is

$$V_{out} = I_c R_C$$

and the input voltage to the transistor is

$$V_{in} = I_e r'_e$$

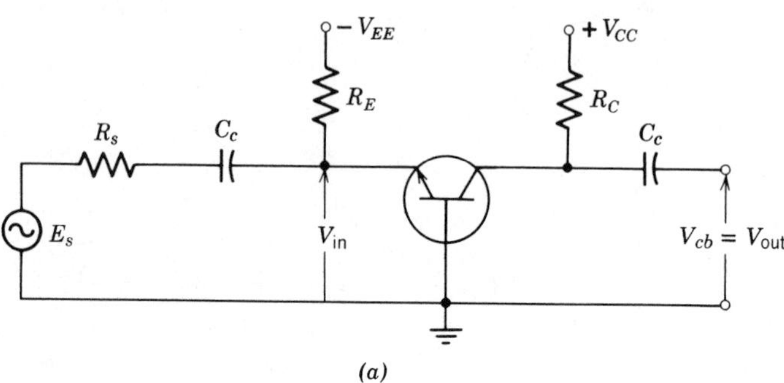

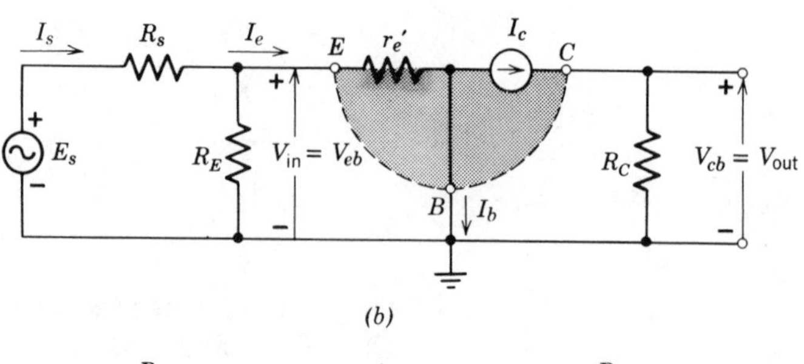

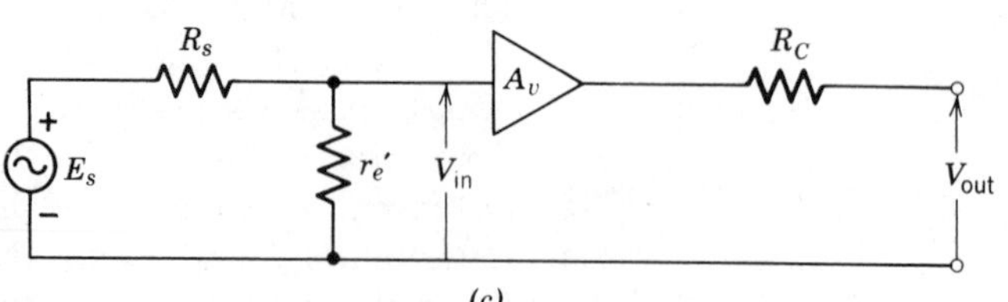

Figure 7-8 Models for the common-base amplifier circuit. (*a*) The complete circuit. (*b*) The formal model. (*c*) The simplified model.

If we substitute the conversions from Table 4-2, we have

$$V_{\text{out}} = \beta I_b R_C$$

and

$$V_{\text{in}} = (1+\beta) I_b r'_e$$

When we divide V_{out} by V_{in}, we have the voltage gain A_v.

$$A_v = \frac{V_{\text{out}}}{V_{\text{in}}} = \frac{\beta I_b R_C}{(1+\beta) I_b r'_e} \qquad (7\text{-}1a)$$

Dividing through by I_b and taking the ratio $\beta/(1+\beta)$ as unity

$$\boxed{A_v = \frac{R_C}{r'_e}} \qquad (7\text{-}18)$$

The current gain of the transistor is

$$A_i = \frac{I_c}{I_e} = \alpha \approx 1 \qquad (7\text{-}19)$$

Inspection of the formal model shows that the input resistance to the transistor is r'_e.

$$r_{\text{in}} = r'_e \qquad (7\text{-}20)$$

Since R_E is much greater than r'_e in a circuit, the parallel combination of R_E and r'_e is approximately r'_e.

$$\boxed{r'_{\text{in}} = r_{\text{in}} = r'_e} \qquad (7\text{-}21)$$

The fact that the input resistance to the common-base circuit is so low limits its application drastically. This circuit is only occasionally found in low-frequency applications; it is used primarily in radio-frequency applications.

The voltage gain equation for the common-emitter amplifier is

$$A_v = \frac{R_C}{r'_e} \qquad (7\text{-}8a)$$

The voltage gain equation for the emitter follower is

$$A_v = \frac{R_E}{r'_e + R_E} \tag{7-11}$$

The voltage gain equation for the common-base amplifier is

$$A_v = \frac{R_C}{r'_e} \tag{7-18}$$

These three gain equations are all the same equation if we define the voltage gain across a transistor as:

$$\boxed{A_v \equiv \frac{\text{the ac impedance of the load}}{\text{the ac impedance in the emitter}}} \tag{7-22}$$

The ac impedance in the emitter is r'_e plus any ac impedance between the emitter and ground.

We can apply Eq. 7-22 to the circuits in this chapter as well as to the amplifier circuits using field effect transistors (Chapter 8).

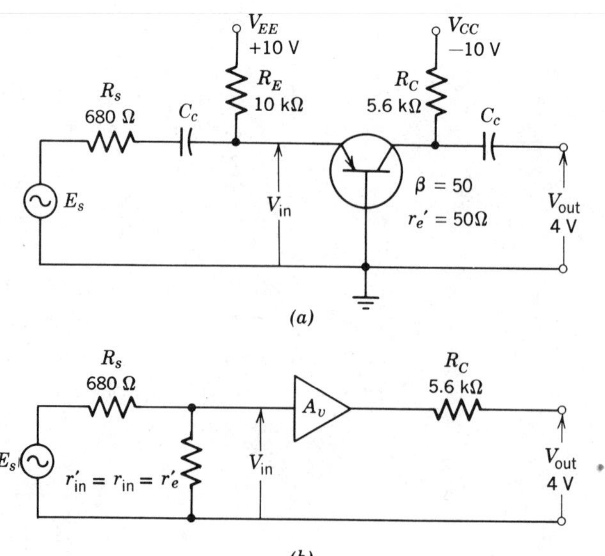

Circuit (*a*) and simplified model (*b*) for Example 7-8.

Example 7-8
Determine A_v, V_{in}, E_s, and A_e for the common-base amplifier.

Solution
The first step in determining the signal-level values is to draw the simplified model. Then numerical values are transferred from the circuit to the model.

$$R_s,\ R_E,\ R_C,\ \text{and}\ V_{out}$$

Then E_s, r'_{in}, V_{in}, and A_v are placed on the model as unknowns at this point. Now we must determine these unknown values.

The input resistance to the common-base amplifier circuit is

$$r'_{in} = r_{in} = r'_e = 50\ \Omega \tag{7-21}$$

Since R_E is 10 kΩ, the parallel combination of R_E and 50 Ω is 50 Ω with negligible error. The voltage gain A_v across the amplifier is

$$A_v = \frac{R_C}{r'_e} = \frac{5600\ \Omega}{50\ \Omega} = \mathbf{112} \tag{7-18}$$

V_{in} is found from the definition of A_v.

$$A_v = \frac{V_{out}}{V_{in}} \tag{7-1a}$$

$$112 = \frac{4\text{ V}}{V_{in}}$$

$$V_{in} = 0.0357\text{ V} = \mathbf{35.7\ mV}$$

The input resistance to the circuit is

$$r'_{in} = r_{in} = r'_e = 50\ \Omega \tag{7-21}$$

Using the concept of a voltage divider in the input circuit, we find

$$V_{in} = \frac{r'_{in}}{r'_{in} + R_s} E_s \tag{7-2}$$

$$37.5\text{ mV} = \frac{50\ \Omega}{50\ \Omega + 680\ \Omega} E_s$$

$$E_s = \mathbf{548\ mV}$$

The overall gain of the amplifier is

$$A_e = \frac{V_{out}}{E_s} = \frac{4\text{ V}}{0.548\text{ V}} = \mathbf{7.3} \tag{7-1b}$$

Problems

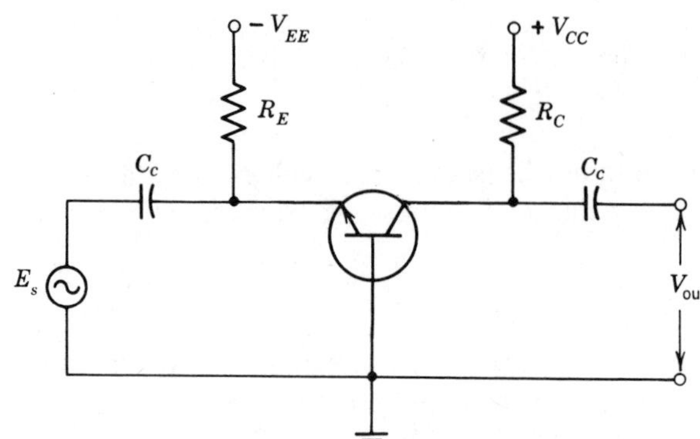

Circuit for Problem 7-5.1.

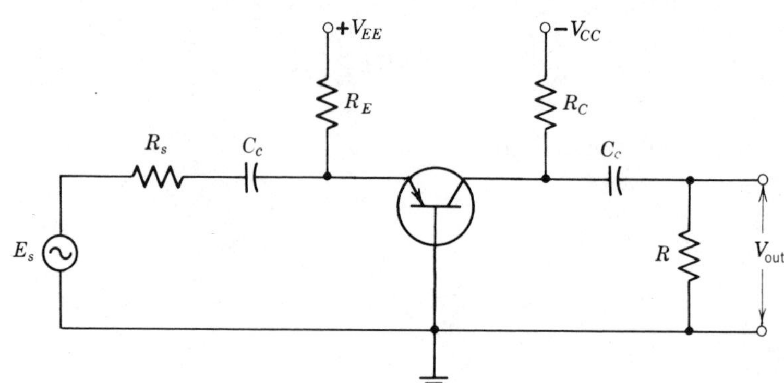

Circuit for Problems 7-5.2 through 7-5.5.

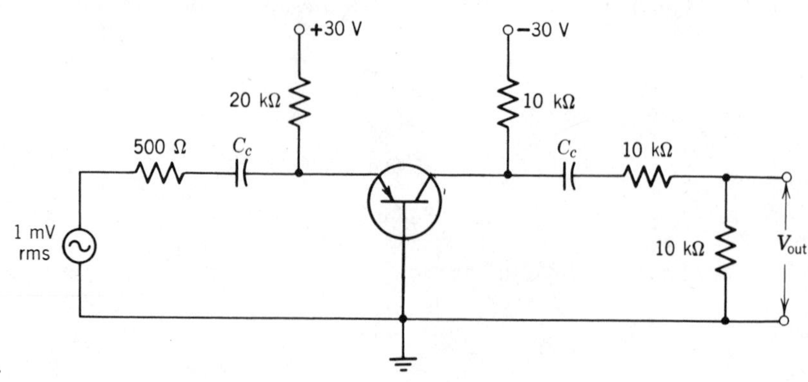

Circuit for Problem 7-5.6.

All transistors are silicon. Use $25\text{ mV}/I_E$ to find r'_e.

7-5.1 V_{EE} and V_{CC} are each 20 V with the proper polarity. R_E is 2 kΩ and R_C is 810 Ω. Determine the voltage gain of the circuit. What is the maximum allowable value of E_s without clipping?

7-5.2 V_{EE} is −20 V and V_{CC} is +150 V. R_s and R_E are each 2 kΩ and R_C is 7.5 kΩ. R is ∞Ω. Determine the operating point values and the voltage gain of the circuit. What is the maximum allowable value of E_s without clipping?

7-5.3 V_{EE} is +20 V and V_{CC} is −20 V. R_s is 20 Ω and R_E is 40 kΩ. R_C and R are each 20 kΩ. Determine the operating point and the voltage gain of the circuit.

7-5.4 V_{EE} is +20 V and V_{CC} is −20 V. R_E is 20 kΩ and R_C and R are each 10 kΩ. E_s is 5 mV rms and R_s varies from 0 to 1 kΩ. What are the minimum and maximum values of V_{out}?

7-5.5 E_s is 10 mV, V_{EE} is +4 V, R_E is 50 kΩ, V_{CC} is −4 V, R_C is 20 kΩ, and R is 30 kΩ. R_s is varied from 100 Ω to 1000 Ω. Determine the Q-point values of the circuit and determine the range of V_{out}.

7-5.6 Determine the value of V_{out}.

Section 7-6 The Common-Emitter Amplifier Model Using Emitter Feedback

The complete circuit for the common-emitter amplifier using emitter feedback is given in Fig. 7-9*a*. We can form the formal model, Fig. 7-9*b*, by following the same rules we used for the previous circuit. The only difference is that we now show the emitter resistance R_E positioned between the emitter terminal E and the common return. The input signal voltage to the transistor is V_{in}. We use Ohm's law to write

$$V_{\text{in}} = I_e(r'_e + R_E)$$

The output voltage V_{out} is

$$V_{\text{out}} = I_c R_C$$

The voltage gain across the transistor A_v is

$$A_v = \frac{V_{\text{out}}}{V_{\text{in}}} = \frac{I_c R_C}{I_e(r'_e + R_E)}$$

Substituting the conversions from Table 4-2

$$I_c = \beta I_b \quad \text{and} \quad I_e = (1+\beta)I_b$$

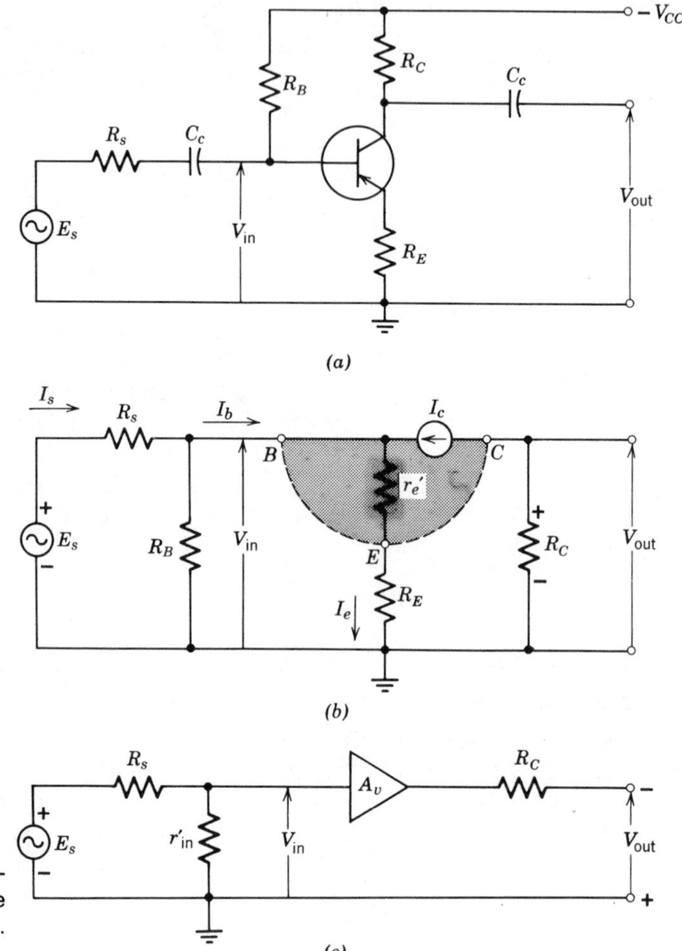

Figure 7-9 The common-emitter amplifier using emitter feedback. (*a*) The complete circuit. (*b*) The formal model. (*c*) The simplified model.

we have

$$A_v = \frac{\beta I_b R_C}{(1+\beta)I_b(r'_e + R_E)}$$

Simplifying by dividing through by I_b and by using unity for the ratio $\beta/(1+\beta)$, we have

$$\boxed{A_v = \frac{R_C}{r'_e + R_E}} \qquad (7\text{-}23)$$

The ac load impedance in the circuit is R_C and the ac impedance in the emitter circuit is $(r'_e + R_E)$. Thus this circuit satisfies the definition given by Eq. 7-19 in which A_v is the ratio of the ac impedance of the load to the ac impedance in the emitter.

Another approach we can use is to state that this circuit is an extension of the basic common-emitter amplifier but the value of r'_e is raised by the amount R_E to a new value $(r'_e + R_E)$.

The current gain of this circuit is the same as that in the basic common-emitter amplifier.

$$A_i = \frac{I_c}{I_b} = \beta \tag{7-24}$$

The input resistance r_{in} to the transistor can be determined from an inspection of the formal model.

$$r_{\text{in}} = \frac{V_{\text{in}}}{I_b} = \frac{I_e(r'_e + R_E)}{I_b}$$

Substituting $(1 + \beta)I_b$ for I_e and canceling I_b, we have

$$\boxed{r_{\text{in}} = (1 + \beta)(r'_e + R_E)} \tag{7-25}$$

The input resistance to the basic common-emitter amplifier, Section 7-3, is

$$r_{\text{in}} = (1 + \beta)r'_e \tag{7-7}$$

A comparison of Eq. 7-25 with Eq. 7-7 shows that the use of an unbypassed emitter resistor in the circuit materially raises the input resistance to the transistor. In a great many applications, the advantage of this increased input resistance materially outweighs the decrease in gain that results from an unbypassed R_E.

Since r_{in} is a large resistance value, we must form the parallel combination of R_B and r_{in} to obtain r'_{in} for the simplified model.

$$\boxed{r'_{\text{in}} = \frac{r_{\text{in}} R_B}{r_{\text{in}} + R_B}} \tag{7-26}$$

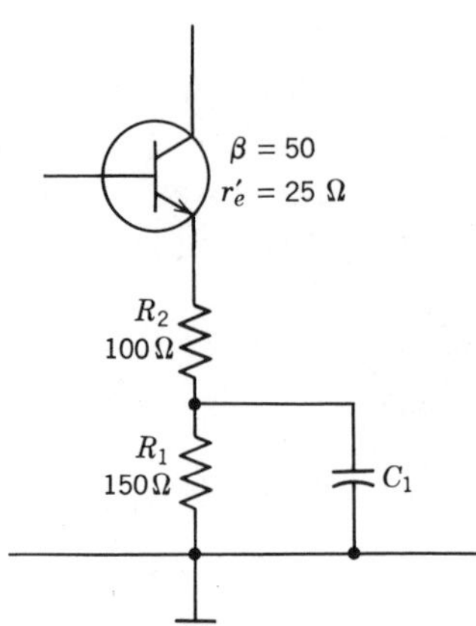

Figure 7-10 Emitter resistors with partial bypass.

In Fig. 7-10 a bypass capacitor C_1 is placed in parallel with R_1. If the bypass action is to be effective, the value of the reactance of C_1 must be numerically less than $0.1\,R_1$ at the lowest signal frequency in the circuit. In this example, the value of C_1 must be large enough to have a reactance no greater than $15\ \Omega$ at the lowest signal frequency.

The dc resistance from the emitter to ground is $R_2 + R_1$ or $250\ \Omega$. The ac resistance in the emitter circuit is $r'_e + R_2$ or $125\ \Omega$. The dc resistance value is used in Q-point calculations. The ac resistance value is used in ac gain calculations.

Example 7-9

Find the value of r_{in} if:

Case I The capacitor C_1 is placed across R_1 and R_2.
Case II The capacitor C_1 is placed across R_1.
Case III The capacitor is removed completely.

Solution

Case I. When C_1 is placed across both R_1 and R_2, we have the case of the amplifier acting as the basic common-emitter amplifier (Section 7-3), and the input resistance to the transistor involves r'_e only.

$$r_{\text{in}} = (1+\beta)r'_e = (1+50)\times 25\ \Omega = \mathbf{1275\ \Omega} \qquad (7\text{-}7)$$

Case II. C_1 is placed across R_1 as shown in Fig. 7-9. If R_1 is properly bypassed for the ac signal, we require that

$$X_{C_1} = \frac{1}{2\pi f C_1} \leq 0.1\,R_1$$

at the lowest signal frequency that will be processed by the amplifier.

R_1 and R_2 are both part of the dc circuit analysis to determine the dc operating point but only R_2 is considered in the ac circuit analysis. Therefore

$$r_{\text{in}} = (1+\beta)(r_e' + R_2) = (1+50)(25\ \Omega + 100\ \Omega) = \mathbf{6375\ \Omega} \quad (7\text{-}25)$$

Case III. When C_1 is removed from the circuit, R_1 and R_2 are considered both in the dc circuit analysis and in the ac circuit analysis. The input resistance to the transistor is

$$r_{\text{in}} = (1+\beta)(r_e' + R_1 + R_2) = (1+50)(25\ \Omega + 100\ \Omega + 150\ \Omega) = \mathbf{14\,025\ \Omega} \quad (7\text{-}25)$$

This example shows that the use of an unbypassed emitter resistance R_E materially increases the input resistance to the transistor.

Problems

7-6.1 Determine the input resistance to the circuit and the voltage gain of the circuit. The β for the transistor is 50. R is infinite.

7-6.2 Repeat Problem 7-6.1 if R is 10 kΩ.

7-6.3 Repeat Problem 7-6.1 if the source resistance R_s is 2000 Ω.

7-6.4 If V_{out} is 2 V, determine E_s and I_s. β is 100, and R is 5.6 kΩ.

7-6.5 R_s is 10 kΩ, R_C is 2000 Ω, R_E is 75 Ω, V_{CC} is +4 V, and β is 40. R_B is adjusted to set I_C at 1 mA. Determine the load resistance on the source and determine the voltage gain of the circuit.

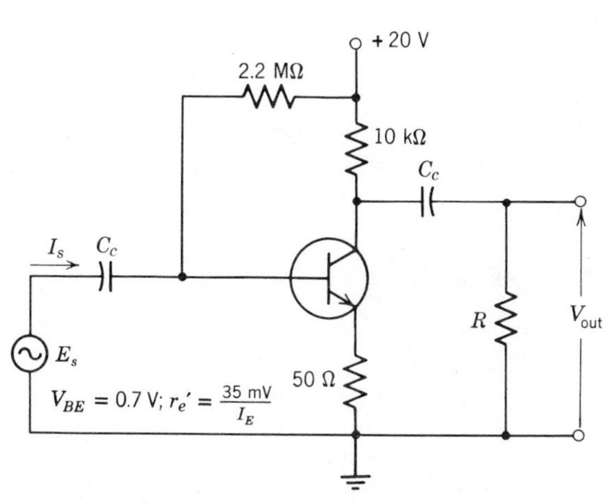

Circuit for Problems 7-6.1 to 7-6.4.

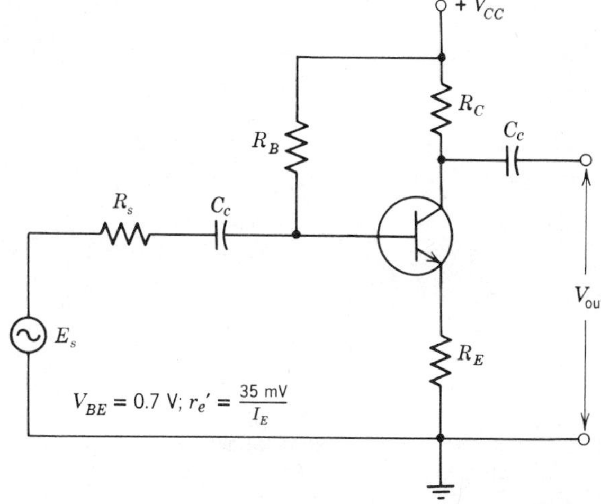

Circuit for Problems 7-6.5 to 7-6.7.

7-6.6 R_s is 10 kΩ, R_C is 12 kΩ, R_E is 3 kΩ, and V_{CC} is +8 V. Determine R_B to provide maximum peak-to-peak output voltage swing and determine the value of E_s that provides this swing. The value of β for the transistor is 60.

7-6.7 Repeat Problem 7-6.6 assuming that the emitter resistor R_E is adequately bypassed with a capacitor C_E at the signal frequency.

Section 7-7 The Common-Emitter Amplifier Model Using Collector-to-Base Feedback

The complete circuit for the common-emitter amplifier using collector-to-base feedback is given in Fig. 7-11*a*. The formal model (Fig. 7-11*b*) shows R_B now connected between the collector and the base.

The equation for the input voltage V_{in} looking into the base of the transistor is

$$V_{in} = I_e r'_e = (1+\beta) I_b r'_e$$

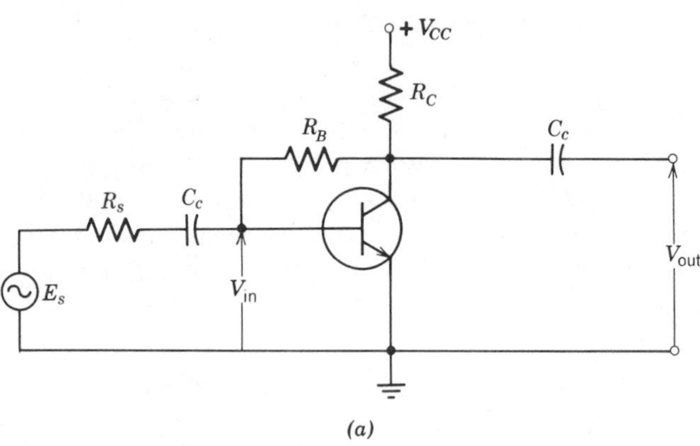

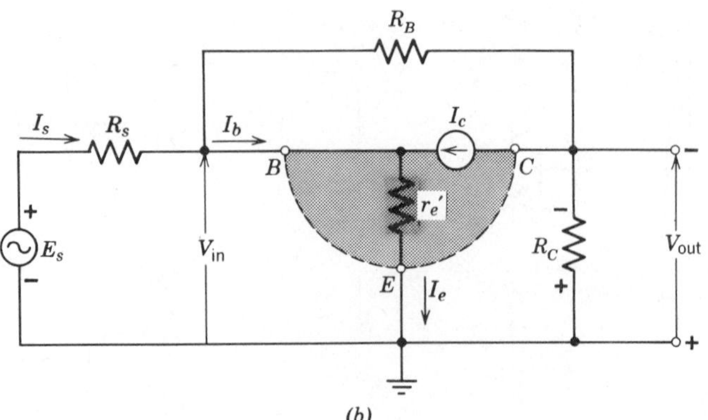

Figure 7-11 The common-emitter amplifier using collector-to-base feedback. (*a*) The complete circuit. (*b*) The formal model. (*c*) The modified formal model. (*d*) The simplified model.

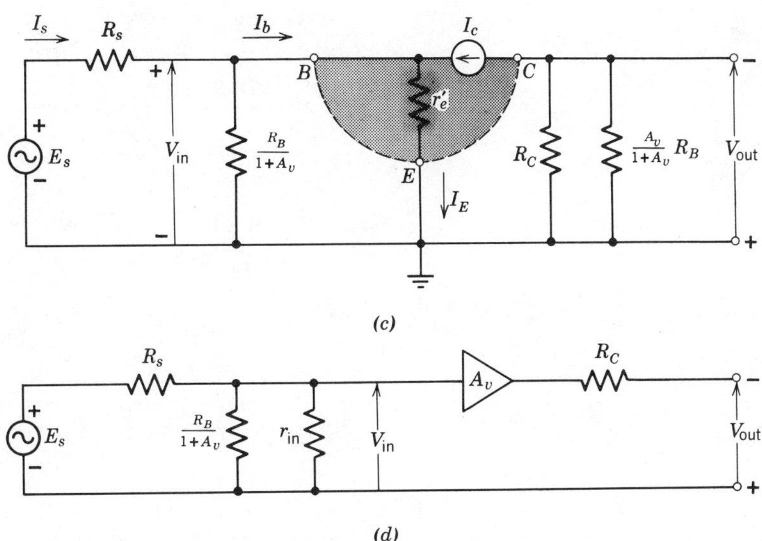

Figure 7-11 (Cont'd)

If we divide both sides of this equation by I_b, we have the resistance looking into the transistor r_{in}

$$\boxed{r_{in} = (1+\beta)r_e'} \tag{7-27}$$

Equation 7-27 is identical to Eq. 7-7, which was obtained for the basic common-emitter amplifier.

In order to determine how R_B is handled in forming the equations for gain and for input resistance, we must develop what is known as the *Miller theorem* or the *Miller effect*. In the formal model (Fig. 7-11*b*) V_{in} and V_{out} are 180° out of phase. We show this phase relation by placing polarity markings on V_{in} and on V_{out}. We redraw V_{in}, V_{out}, and R_B without the rest of the formal model in Fig. 7-12*a*.

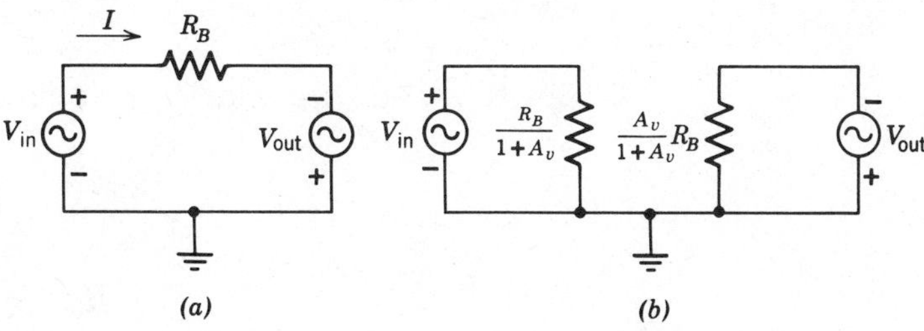

Figure 7-12 The Miller theorem. (*a*) Circuit. (*b*) Equivalent loading.

The voltage across R_B is V_{in} *plus* V_{out} since as far as R_B is concerned V_{in} and V_{out} are *series additive.* Thus, the total voltage across R_B is

$$(V_{in} + V_{out})$$

The current in R_B is shown as I in Fig. 7-12*a* in the direction indicated by the arrow. By Ohm's law, this current is

$$I = \frac{V_{in} + V_{out}}{R_B}$$

The resistance value R_{in} that V_{in} "sees" looking into R_B is

$$R_{in} = \frac{V_{in}}{I} = \frac{V_{in}}{\left(\dfrac{V_{in} + V_{out}}{R_B}\right)} = \frac{V_{in}}{V_{in} + V_{out}} R_B$$

Now, if we divide each term by V_{in} and replace V_{out}/V_{in} by A_v, we have

$$R_{in} = \frac{R_B}{1 + V_{out}/V_{in}} = \frac{R_B}{1 + A_v}$$

The resistance value R_{out} that V_{out} "sees" looking back into R_B is

$$R_{out} = \frac{V_{out}}{I} = \frac{V_{out}}{\left(\dfrac{V_{in} + V_{out}}{R_B}\right)} = \frac{V_{out}}{V_{in} + V_{out}} R_B$$

If we divide each term by V_{in} and replace V_{out}/V_{in} by A_v we have

$$R_{out} = \frac{A_v}{1 + A_v} R_B$$

If we add R_{in} and R_{out}, we have

$$R_{in} + R_{out} = \frac{R_B}{1 + A_v} + \frac{A_v}{1 + A_v} R_B = \frac{1 + A_v}{1 + A_v} R_B = R_B$$

$$R_{in} + R_{out} = R_B$$

Thus, the Miller theorem shows that a resistor (R_B) placed between the input terminal and the output terminal of an amplifier that has phase inversion can be broken into two parts:

The part placed across the input is

$$R_{\text{in}} = \frac{R_B}{1 + A_v} \tag{7-28}$$

The part placed across the output is

$$R_{\text{out}} = \frac{A_v}{1 + A_v} R_B \tag{7-29}$$

These results are shown in Fig. 7-12*b*. A modified formal model (Fig. 7-11*c*) shows how the two parts of R_B are included in the model.

The value of R_{out}, Eq. 7-29, is approximately equal to R_B. Since R_B is very much greater than R_C, we can consider that the load on the output of the transistor is R_C alone. Then we can write the voltage gain equation for the amplifier.

$$A_v = \frac{R_C}{r'_e} \tag{7-30}$$

The input resistance r_{in} given by Eq. 7-27 is placed in parallel with $R_B/(1 + A_v)$ given by Eq. 7-28 to form r'_{in} for the simplified model, Fig. 7-11*d*.

$$r'_{\text{in}} = \frac{r_{\text{in}}\left(\dfrac{R_B}{1 + A_v}\right)}{r_{\text{in}} + \dfrac{R_B}{1 + A_v}} \tag{7-31}$$

Example 7-10

A 1-MΩ resistor is connected from the output to the input of an amplifier as shown in the circuit (*a*). The amplifier has phase inversion. Determine the equivalent circuit (*b*) of the amplifier showing R_{in} and R_{out}.

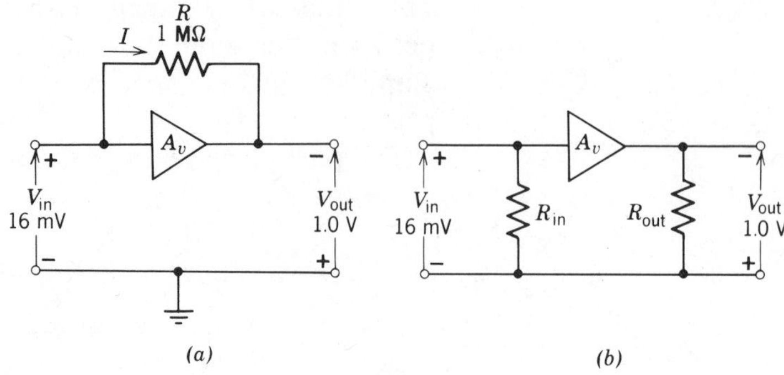

Circuit (*a*) and equivalent circuit (*b*) for Example 7-10.

Solution No. 1

The current I can be determined from Ohm's law as

$$I = \frac{V_{\text{in}} - (-V_{\text{out}})}{R} = \frac{0.016\text{ V} - (-1.00)\text{ V}}{1 \times 10^6\ \Omega}$$

$$= 1.016 \times 10^{-6}\text{ A} = 1.016\ \mu\text{A}$$

The input resistance determined by Ohm's law is

$$R_{\text{in}} = \frac{V_{\text{in}}}{I} = \frac{0.016\text{ V}}{1.016 \times 10^{-6}\ \Omega} = \mathbf{16\ k\Omega}$$

The output resistance determined by Ohm's law is

$$R_{\text{out}} = \frac{V_{\text{out}}}{I} = \frac{1.00\text{ V}}{1.016 \times 10^{-6}\ \Omega} = \mathbf{984\ k\Omega}$$

Solution No. 2

The magnitude of the voltage gain of the amplifier is

$$A_v = \frac{V_{\text{out}}}{V_{\text{in}}} = \frac{1.00\text{ V}}{0.016\text{ V}} = 62.5 \tag{7-1a}$$

The input resistance to the circuit is

$$R_{\text{in}} = \frac{R_B}{1 + A_v} = \frac{1 \times 10^6\ \Omega}{62.5} = \mathbf{16\ k\Omega} \tag{7-28}$$

and R_{out} can be found from

$$R_B = R_{\text{in}} + R_{\text{out}}$$

$$1\text{ M}\Omega = 16\text{ k}\Omega + R_{\text{out}}$$

$$R_{\text{out}} = \mathbf{984\ k\Omega}$$

or from

$$R_{out} = \frac{A_v}{1 + A_v} R_B = \frac{62.5}{1 + 62.5} 1000\ \text{k}\Omega = \mathbf{984\ k\Omega} \qquad (7\text{-}29)$$

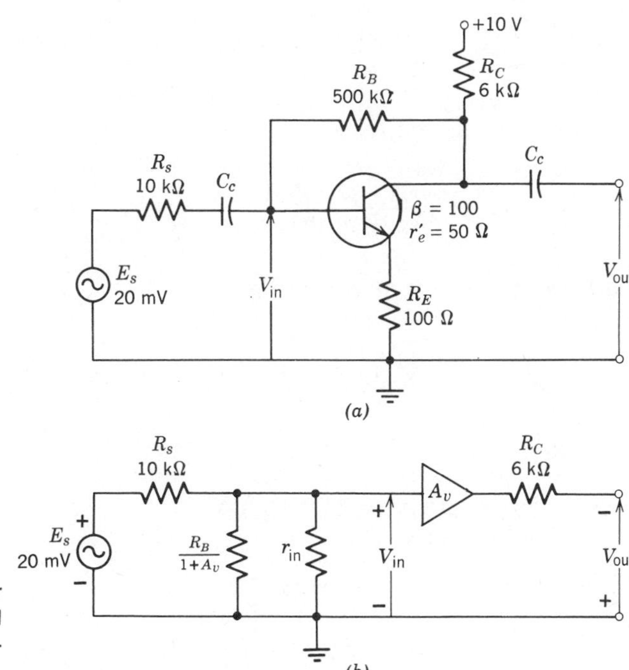

Figure 7-13 Common-emitter amplifier using collector-to-base feedback and emitter feedback. (*a*) The complete circuit. (*b*) The simplified model.

Example 7-11

Determine V_{in}, V_{out}, and A_e for the amplifier shown in Fig. 7-13*a*.

Solution

The first step in determining the signal-level values is to draw the simplified model, Fig. 7-13*b*. Then numerical values are transferred from the circuit to the model for R_s, R_C, and E_s. Then R_B is transferred to the model as $R_B/(1 + A_v)$. At this point $R_B/(1 + A_v)$, r_{in}, V_{in}, A_v, and V_{out} are placed on the model as unknowns. Now we must determine these unknown values.

The input resistance r_{in} to the common-emitter amplifier with emitter feedback is

$$r_{in} = (1 + \beta)(r'_e + R_E) = 101(50\ \Omega + 100\ \Omega) = 15{,}150\ \Omega \qquad (7\text{-}25)$$

The voltage gain A_v across the amplifier is obtained from

$$A_v = \frac{R_C}{r'_e + R_E} = \frac{6000\ \Omega}{50\ \Omega + 100\ \Omega} = 40 \tag{7-23}$$

The bias resistor R_B is transformed by the Miller theorem to

$$R_{\text{in}} = \frac{R_B}{1 + A_v} = \frac{500{,}000\ \Omega}{1 + 40} = 12{,}195\ \Omega \tag{7-28}$$

The parallel combination of r_{in} and R_{in} is r'_{in}.

$$r'_{\text{in}} = \frac{r_{\text{in}} R_{\text{in}}}{r_{\text{in}} + R_{\text{in}}} = \frac{15{,}150\ \Omega \times 12{,}195\ \Omega}{15{,}150\ \Omega + 12{,}195\ \Omega} = 6756\ \Omega \tag{7-31}$$

V_{in} can be found from the voltage divider in the input circuit.

$$V_{\text{in}} = \frac{r'_{\text{in}}}{r'_{\text{in}} + R_s} E_s = \frac{6756\ \Omega}{6756\ \Omega + 10{,}000\ \Omega} 20\text{ mV} = \mathbf{8\ mV} \tag{7-2}$$

and the output voltage V_{out} is

$$V_{\text{out}} = V_{\text{in}} A_v = 8\text{ mV} \times 40 = \mathbf{320\ mV} \tag{7-1a}$$

The overall gain is

$$A_e = \frac{V_{\text{out}}}{E_s} = \frac{320\text{ mV}}{20\text{ mV}} = \mathbf{16} \tag{7-1b}$$

Problems All transistors are silicon, and r'_e is determined from $50\text{ mV}/I_E$.

7-7.1 R_s is 1000 Ω, R_B is 1 MΩ, R_C is 10 kΩ, V_{CC} is 12 V, and R_E is 1000 Ω. The value of r'_e is 50 Ω. R is infinite and the value of β is 100. Determine the load on the source and find the voltage gain of the circuit.

7-7.2 E_s is 1 mV and R_s is 4.7 kΩ. R_B is 750 kΩ, and R_C and R are each 47 kΩ. Assume that r'_e is 100 Ω. β for the transistor varies between 40 and 100. What is the variation in V_{out}? R_E is adequately bypassed.

7-7.3 Determine R_B when R_B is adjusted to provide a maximum undistorted peak-to-peak output voltage. Determine the value of E_s required to drive the amplifier when it delivers the maximum undistorted peak-to-peak output voltage. What is the overall circuit gain?

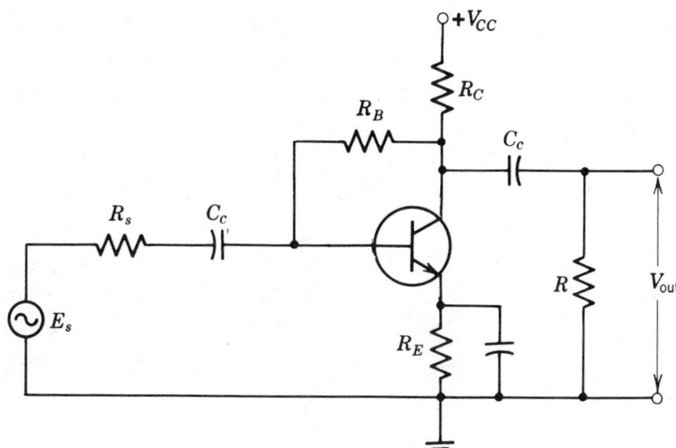

Circuit for Problems 7-7.1 and 7-7.2.

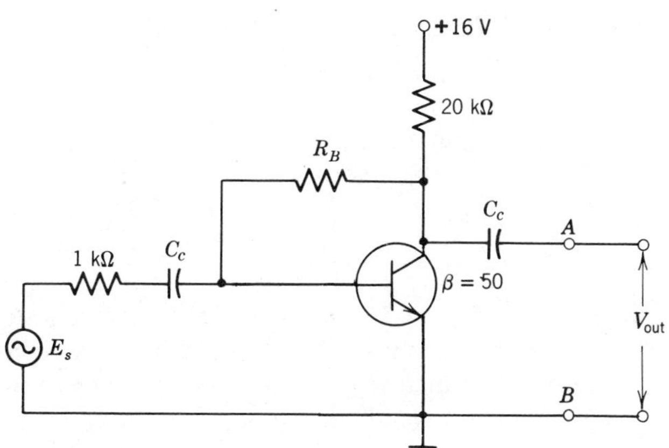

Circuit for Problems 7-7.3 and 7-7.4.

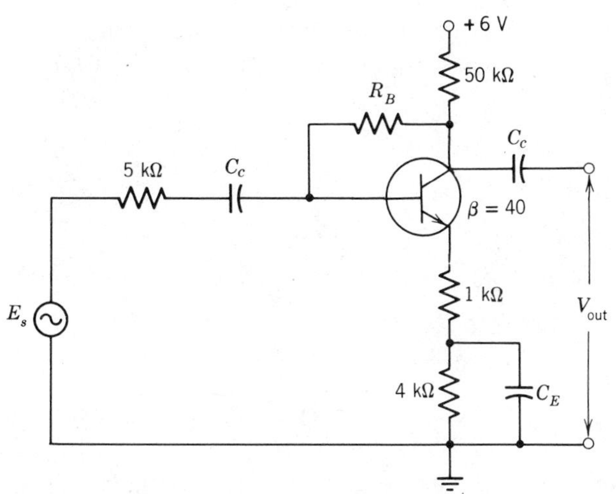

Circuit for Problems 7-7.5 and 7-7.6.

7-7.4 An additional load resistor of a value of 15 kΩ is placed between terminals *A* and *B*. What value must R_B have to establish an optimum *Q*-point for the circuit? Determine the value of E_s that develops a maximum peak-to-peak output voltage.

7-7.5 Determine R_B when R_B is adjusted to make V_{CE} 2 V. What is the maximum available undistorted peak-to-peak output voltage? Determine the overall gain of the circuit. What value of E_s delivers the maximum undistorted peak-to-peak output voltage?

7-7.6 The emitter bypass capacitor C_E is removed from the circuit. Recalculate Problem 7-7.5 for this new circuit.

Section 7-8 Cascaded Amplifiers

Example 7-12

A cascaded three-stage amplifier circuit with numerical values is shown in Fig. 7-14*a*. The signal levels are required at all points in the amplifier.

Solution

This complex circuit, Fig. 7-14*a*, can be simplified by "unsoldering" the coupling capacitors between amplifier stages. Now the circuit consists of three separate amplifier stages. Each of the three amplifier stages must be reduced to simplified models for each stage (Fig. 7-14*b*). For each stage, we must find r_{in}, A_v, and R_{out} (r'_{out}). When we determine these values, we can redraw Fig. 7-14*b* as shown in Fig. 7-14*c*. Now, when we close the gaps (*A* and *B*), we have the complete cascaded circuit that shows a succession of voltage dividers and amplifiers.

Stage 1, *Q*1

This stage is an emitter follower. The voltage gain A_{v_1} is

$$A_{v_1} = \frac{R_E}{r'_e + R_E} = \frac{4700\ \Omega}{15\ \Omega + 4700\ \Omega} \approx 1 \qquad (7\text{-}11)$$

The input resistance r_{in_1} to the transistor Q_1 is

$$r_{in_1} = (1+\beta)(r'_e + R_E) = (1+50)(15\ \Omega + 4700\ \Omega) = 240\ \text{k}\Omega$$

The input resistance $r_{in'_1}$ to the whole circuit is R_B in parallel with r_{in}

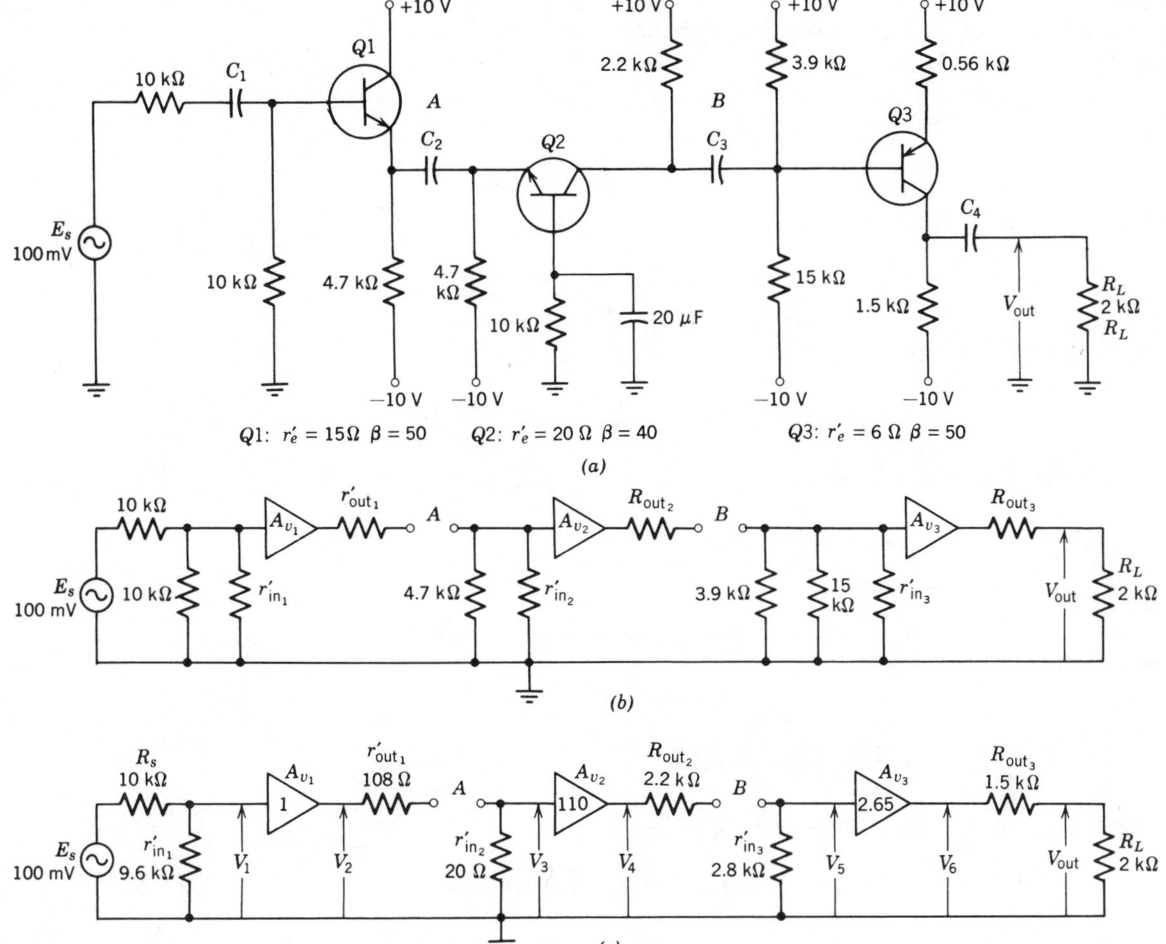

Figure 7-14 A cascaded amplifier. (*a*) Actual circuit. (*b*) Simplified circuit models for each stage. (*c*) Cascaded simplified models.

$$r_{in_1'} = \frac{R_B r_{in}}{R_B + r_{in}} = \frac{10\ \text{k}\Omega \times 240\ \text{k}\Omega}{10\ \text{k}\Omega + 240\ \text{k}\Omega} = 9.6\ \text{k}\Omega \qquad (7\text{-}14)$$

The resistance r_{out_1} seen by "looking back" into the emitter of $Q1$ is

$$r_{out_1} = r'_e + \frac{\left(\dfrac{R_s R_B}{R_s + R_B}\right)}{1+\beta} = 15\ \Omega + \frac{\left(\dfrac{10\ \text{k}\Omega \times 10\ \text{k}\Omega}{10\ \text{k}\Omega + 10\ \text{k}\Omega}\right)}{1+50} = 111\ \Omega \qquad (7\text{-}15)$$

This resistance r_{out_1} is in parallel with R_E.

$$r'_{out_1} = \frac{R_E r_{out_1}}{R_E + r_{out_1}} = \frac{4700\ \Omega \times 111\ \Omega}{4700\ \Omega + 111\ \Omega} = 108\ \Omega \qquad (7\text{-}17)$$

Stage 2, Q2

The second stage $Q2$ is a common-base amplifier. The voltage gain across the transistor A_{v_2} is

$$A_{v_2} = \frac{R_C}{r'_e} = \frac{2200\ \Omega}{20\ \Omega} = 110 \qquad (7\text{-}18)$$

The input resistance to the transistor is

$$r_{\text{in}} = r'_e = 20\ \Omega \qquad (7\text{-}20)$$

Since R_E is very much greater than r'_e,

$$r_{\text{in}'_2} = r_{\text{in}} = r'_e = 20\ \Omega \qquad (7\text{-}21)$$

The output resistance R_{out_2} is R_C

$$R_{\text{out}_2} = R_C = 2.2\ \text{k}\Omega$$

Stage 3, Q3

The third stage $Q3$ is a common-emitter amplifier using emitter feedback. The voltage gain, A_{v_3}, is

$$A_{v_3} = \frac{R_C}{r'_3 + R_E} = \frac{1500\ \Omega}{6\ \Omega + 560\ \Omega} = 2.65 \qquad (7\text{-}23)$$

The input resistance to the transistor r_{in} is

$$r_{\text{in}} = (1+\beta)(r'_e + R_E) = (1+50)(6\ \Omega + 560\ \Omega) = 28.9\ \text{k}\Omega \qquad (7\text{-}25)$$

The input resistance to the third stage $r_{\text{in}'_3}$ is determined from

$$\frac{1}{r_{\text{in}_3}} = \frac{1}{r_{\text{in}}} + \frac{1}{3.9\ \text{k}\Omega} + \frac{1}{15\ \text{k}\Omega} = \frac{1}{28.9\ \text{k}\Omega} + \frac{1}{3.9\ \text{k}\Omega} + \frac{1}{15\ \text{k}\Omega}$$

Solving for $r_{\text{in}'_3}$

$$r_{\text{in}'_3} = 2.8\ \text{k}\Omega$$

The output resistance R_{out_3} is R_C

$$R_{\text{out}_3} = 1.5\ \text{k}\Omega$$

The Cascaded Amplifier

The values calculated for each of the three stages are shown on Fig. 7-14*c*. The gaps, *A* and *B*, are closed and inspection of the circuit shows

$$V_1 = \frac{r'_{\text{in}_1}}{r'_{\text{in}_1} + R_s} E_s = \frac{9.6\text{ k}\Omega}{9.6\text{ k}\Omega + 10\text{ k}\Omega} 100\text{ mV} = \mathbf{49.0\text{ mV}} \qquad (7\text{-}2)$$

$$V_2 = V_1 A_{v_1} = 49.0\text{ mV} \times 1 = \mathbf{49.0\text{ mV}} \qquad (7\text{-}1a)$$

$$V_3 = \frac{r'_{\text{in}_2}}{r'_{\text{in}_2} + r'_{\text{out}_1}} V_2 = \frac{20\ \Omega}{20\ \Omega + 108\ \Omega} 49.0\text{ mV} = \mathbf{7.66\text{ mV}}$$

$$V_4 = V_3 A_{v_2} = 7.66\text{ mV} \times 110 = \mathbf{843\text{ mV}}$$

$$V_5 = \frac{r'_{\text{in}_3}}{r'_{\text{in}_3} + R_{\text{out}_2}} V_4 = \frac{2.8\text{ k}\Omega}{2.8\text{ k}\Omega + 2.2\text{ k}\Omega} 843\text{ mV} = \mathbf{472\text{ mV}}$$

$$V_6 = V_5 \times A_{v_3} = 472\text{ mV} \times 2.65 = \mathbf{1250\text{ mV}}$$

$$V_{\text{out}} = \frac{R_L}{R_L + R_{\text{out}_3}} V_6 = \frac{2\text{ k}\Omega}{2\text{ k}\Omega + 1.5\text{ k}\Omega} 1250\text{ mV} = \mathbf{714\text{ mV}}$$

The overall gain of the circuit can be written directly from an examination of the cascaded simplified models, Fig. 7-14*c*.

$$A_e = \left(\frac{9.6}{10 + 9.6}\right)(1)\left(\frac{20}{20 + 108}\right)(110)\left(\frac{2.8}{2.2 + 2.8}\right)(2.65)\left(\frac{2}{1.5 + 2}\right) = \mathbf{7.14}$$

Then

$$V_{\text{out}} = A_e E_s = 7.14 \times 100\text{ mV} = \mathbf{714\text{ mV}} \qquad (7\text{-}1b)$$

Problems

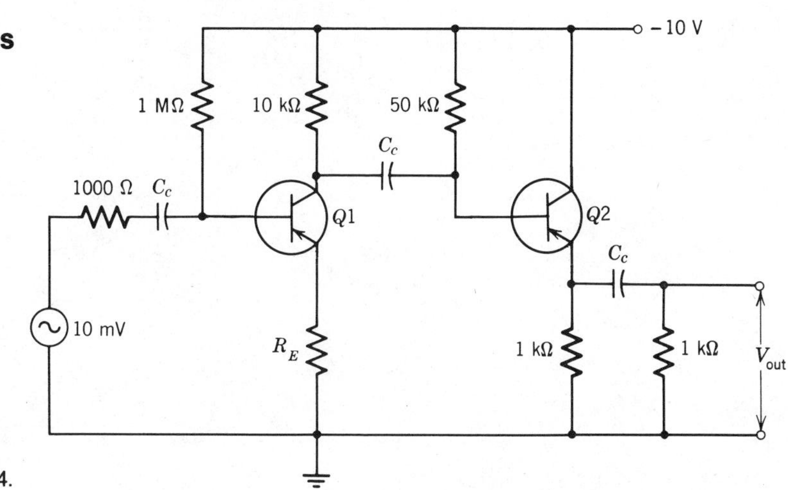

Circuit for Problems 7-8.1 through 7-8.4.

7-8.1 Determine V_{out} when R_E is zero. For each transistor β is 50 and r'_e is 30 Ω.

7-8.2 Repeat Problem 7-8.1 using a value of 500 Ω for R_E.

7-8.3 Repeat Problem 7-8.1 using a value of 1000 Ω for R_E.

7-8.4 Plot a curve showing V_{out} on the Y axis and R_E on the X axis. Use values of 0 Ω, 500 Ω, and 1000 Ω for R_E. Explain the shape of the curve.

Supplementary Problems

7-1 $E_s = 80$ mV; $R_s = 20$ kΩ; $R_B = 400$ kΩ; $R_C = 10$ kΩ; $r'_e = 100$ Ω; and $\beta = 70$. R is omitted from the circuit. Find V_{in}, V_{out}, A_v, and A_e.

7-2 $E_s = 20$ mV; $R_s = 10$ kΩ; $R_B = 100$ kΩ; $R_C = 10$ kΩ; $r'_e = 100$ Ω; and $\beta = 80$. R is omitted from the circuit. Find V_{in}, V_{out}, A_v, and A_e.

7-3 $E_s = 30$ mV; $R_s = 2.4$ kΩ; $R_B = 400$ kΩ; $R_C = 5.6$ kΩ; $r'_e = 75$ Ω; and $\beta = 65$. R is omitted from the circuit. Find V_{in}, V_{out}, A_v, and A_e.

7-4 $E_s = 50$ mV; $R_s = 1.5$ kΩ; $R_B = 300$ kΩ; $R_C = 8.2$ kΩ; $r'_e = 50$ Ω; and $\beta = 50$. R is omitted from the circuit. Find V_{in}, V_{out}, A_v, and A_e.

7-5 Use the data of Problem 7-1 but R is 24 kΩ. Find V_{in}, V_{out}, A_e, and A_v.

7-6 $E_s = 1$ V; $R_s = 50$ kΩ; $R_B = 600$ kΩ; $R_C = 6$ kΩ; $R = 8$ kΩ; $r'_e = 25$ Ω; and $\beta = 40$. Find V_{out} and A_v.

7-7 $E_s = 25$ mV; $R_s = 20$ kΩ; $R_B = 500$ kΩ; $R_C = 20$ kΩ; $R = 15$ kΩ; $r'_e = 100$ Ω; and $\beta = 60$. Find V_{in} and V_{out}.

7-8 $V_{out} = 2$ V; $R_s = 2.4$ kΩ; $R_B = 200$ kΩ; $R_C = 10$ kΩ; $R = 10$ kΩ; $r'_e = 120$ Ω; and $\beta = 65$. Find E_s and A_v.

7-9 $E_s = 20$ mV; $R_s = 10$ kΩ; $R_B = 485$ kΩ; $R_E = 100$ Ω; $R_C = 3.9$ kΩ; $r'_e = 15$ Ω; and $\beta = 65$. Find V_{in}, V_{out}, A_v, and A_e.

7-10 $E_s = 2$ mV; $R_s = 20$ kΩ; $R_B = 500$ kΩ; $R_E = 250$ Ω; $R_C = 5$ kΩ; $r'_e = 50$ Ω; and $\beta = 50$. Find V_{in}, V_{out}, and A_e.

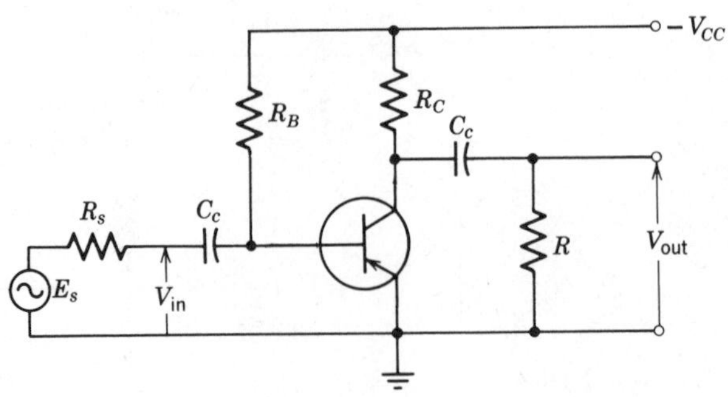

Circuit for Problems 7-1 through 7-8.

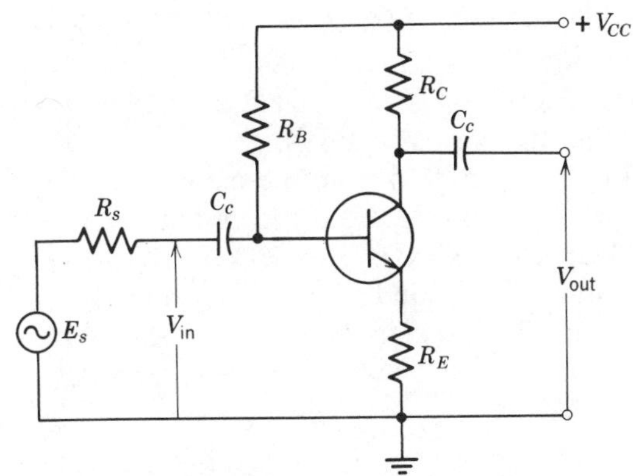

Circuit for Problems 7-9 through 7-12.

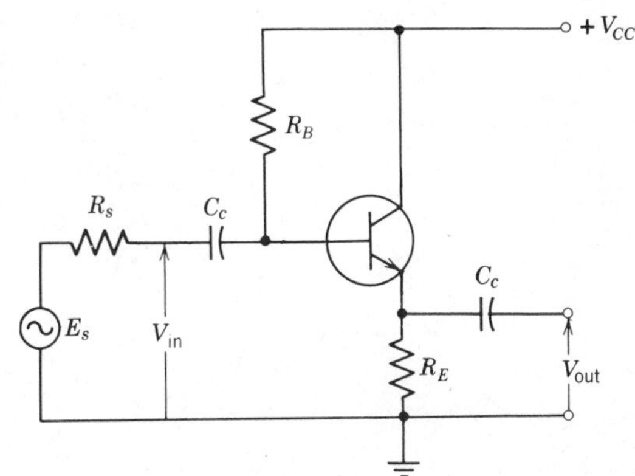

Circuit for Problems 7-13 through 7-16.

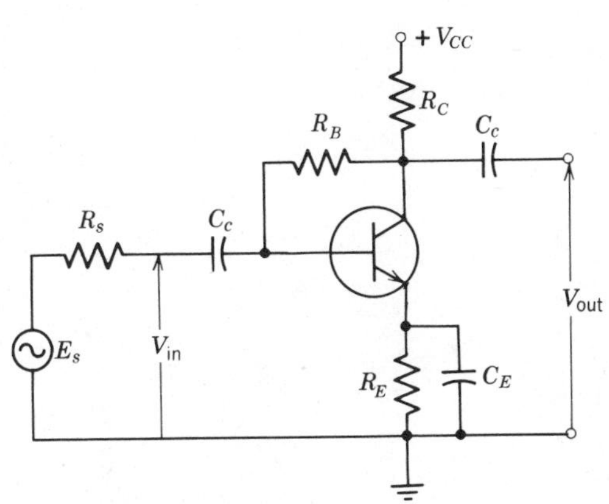

Circuit for Problems 7-17 through 7-20.

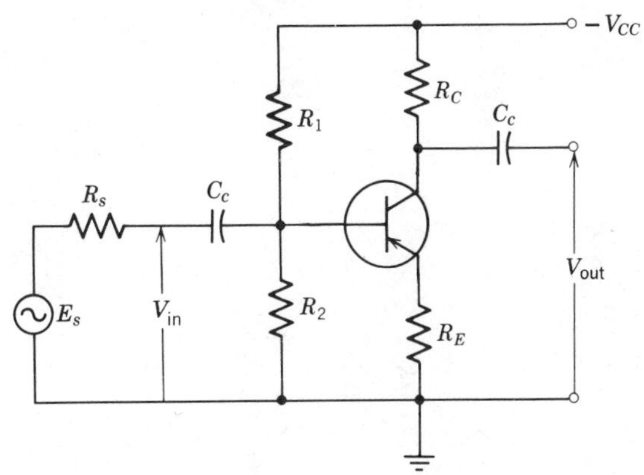

Circuit for Problems 7-21 and 7-22.

7-11 $E_s = 45$ mV; $R_s = 40$ kΩ; $R_B = 350$ kΩ; $R_E = 5$ kΩ; $R_C = 20$ kΩ; $r'_e = 50$ Ω; and $\beta = 45$. Find V_{in}, V_{out}, A_v, and A_e.

7-12 $E_s = 0.1$ V; $R_s = 100$ kΩ; $R_B = 300$ kΩ; $R_E = 1000$ Ω; $R_C = 12$ kΩ; $r'_e = 100$ Ω; and $\beta = 60$. Find V_{in}, V_{out}, A_v, and A_e.

7-13 $E_s = 5$ V; $R_s = 100$ Ω; $R_B = 3$ kΩ; $R_E = 10$ Ω; $r'_e = 1$ Ω; and $\beta = 50$. Find V_{in} and V_{out}.

7-14 $E_s = 6$ V; $R_s = 10$ kΩ; $R_B = 40$ kΩ; $R_E = 500$ Ω; $r'_e = 100$ Ω; and $\beta = 45$. Find V_{in} and V_{out}.

7-15 In the circuit for Problem 7-13, R_E is replaced by that value required for maximum power transfer. What is that value and what is V_{out} for the 5-V input?

7-16 In the circuit for Problem 7-14, R_E is replaced by that value required for maximum power transfer. What is that value and what is V_{out} for the 6-V signal?

7-17 $E_s = 100$ mV; $R_s = 20$ kΩ; $R_B = 200$ kΩ; $R_C = 4$ kΩ; $R_E = 200\ \Omega$; $r'_e = 50\ \Omega$; and $\beta = 35$. Find V_{in}, V_{out}, A_v, and A_e.

7-18 $E_s = 10$ mV; $R_s = 5$ kΩ; $R_B = 400$ kΩ; $R_C = 6$ kΩ; $r'_e = 50\ \Omega$; $\beta = 80$; and $R_E = 150\ \Omega$. Find V_{in}, V_{out}, A_v, and A_e.

7-19 Repeat Problem 7-17 if C_E is omitted from the circuit.

7-20 Repeat Problem 7-18 if C_E is omitted from the circuit.

7-21 $E_s = 60$ mV; $R_s = 20$ kΩ; $R_1 = 140$ kΩ; $R_2 = 20$ kΩ; $R_E = 1$ kΩ; $R_C = 8$ kΩ; $r'_e = 40\ \Omega$; and $\beta = 50$. Find V_{in}, V_{out}, A_v, and A_e.

7-22 $E_s = 300$ mV; $R_s = 6$ kΩ; $R_1 = 20$ kΩ; $R_2 = 20$ kΩ; $R_E = 2$ kΩ; $R_C = 8$ kΩ; $r'_e = 100\ \Omega$; and $\beta = 60$. Find V_{in}, V_{out}, A_v, and A_e.

Chapter 8 Field Effect Transistors

The operation of the junction field effect transistor is explained and the drain characteristic developed (Section 8-1). The transfer characteristic is derived from the drain characteristic (Section 8-2). The depletion-type MOSFET (Section 8-3) and the enhancement-type MOSFET (Section 8-4) are examined together with methods of determining the drain current and the transconductance.

Section 8-1 The Junction Field Effect Transistor

Conventional transistors, *PNP* or *NPN*, function both on hole current and on electron current. Consequently, they are referred to in the literature as *bipolar junction transistors* (*BJT*). The field effect transistor (FET) operates on *either* electron-current flow *or* on hole-current flow. In contrast to the BJT, the FET is a *unipolar transistor.*

The construction of an *N*-channel junction field effect transistor (JFET) is shown in Fig. 8-1*a*. Metallic contacts, the *source* (*S*) and the *drain* (*D*), are placed at opposite ends of the *channel.* The contact of the source to the channel and the contact of the drain to the channel are ohmic contacts and are not *P-N* junctions. A *gate* (*G*) is formed by placing a ring of *P* material around the center of the channel to form a *P-N* junction. Electrode voltages are placed on the JFET as shown in the circuit. The polarities of both V_{GG} and V_{DD} would be reversed if the JFET had a *P* channel. The dc source current is I_S, the dc drain current is I_D, and any gate current is I_G. The dc voltage between the gate and the source is V_{GS} and the dc voltage between the drain and the source is V_{DS}. Usually the source is the reference point in the JFET. Most of the literature refers to the JFET simply as an FET with the *J* understood. In this text, we will retain the *J* in order to distinguish the junction type from the other types of field effect transistors.

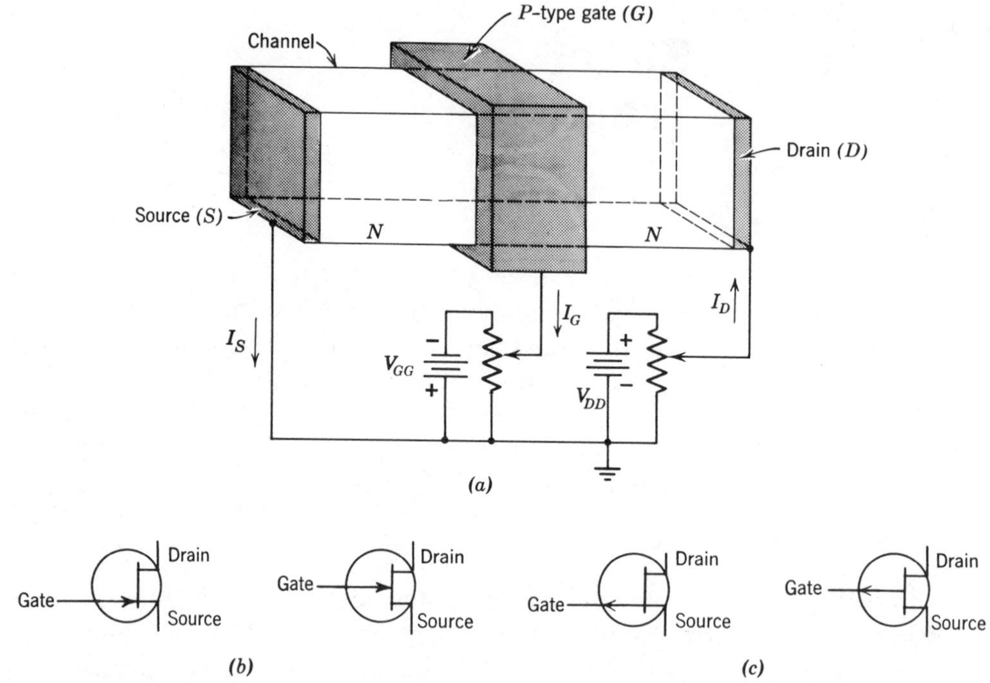

Figure 8-1 The junction field effect transistor (JFET). (*a*) Construction. (*b*) Symbol for an *N*-channel JFET. (*c*) Symbol for a *P*-channel JFET.

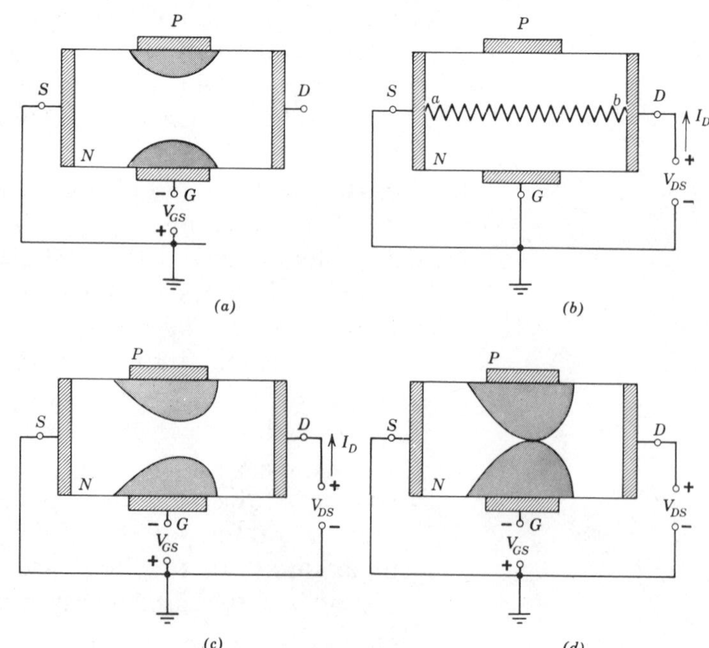

Figure 8-2 The electric field in a FET. (*a*) Depletion region caused by a reverse-biased gate. (*b*) Voltage drop in the channel. (*c*) Effect of drain voltage on depletion region with a negative gate. (*d*) Pinch-off.

The circuit symbols for an N-channel JFET are shown in Fig. 8-1*b*, and the circuit symbols for a P-channel JFET are shown in Fig. 8-1*c*.

The cross-sectional view of the JFET is shown in Fig. 8-2*a*. There is a P-N junction between the gate and the channel and, consequently, there is a depletion region in the channel surrounding the gate. Since the action of the JFET does not depend on the depletion region within the gate, the depletion region inside the gate is ignored. The gate is normally reverse biased and, as a result, I_G is zero.

In Fig. 8-2*b* the gate is connected to the source making V_{GS} zero. A positive voltage $+V_{DS}$ is placed between the drain and the source. Assume there is a uniform channel resistance between a and b in Fig. 8-2*b*. Then the drain current I_D produces a uniform voltage drop between a and b. The voltage at any point depends on the location of that point between a and b. The voltage at each point in the channel between a and b contributes to the reverse bias and to the depletion region between the channel and the gate. This condition could not occur if V_{DS} were negative. If we use a P-channel JFET, the supply voltage on the drain must be negative to obtain the necessary reverse bias between the channel and the gate.

When we have both V_{GS} and V_{DS} on the JFET, we have the depletion region condition shown in Fig. 8-2*c*. The depletion region acts like a choke or throttle to reduce the drain current. The deeper the penetration of the depletion region, the smaller the drain current. At some point, as the gate voltage is increased negatively, the depletion region extends completely across the channel (Fig. 8-2*d*). The drain current I_D is now zero. The particular gate-to-source voltage that produces *cutoff* for the drain current is called the ***pinch-off voltage*** V_P.

The drain current obtained when V_{GS} is zero is I_{DSS}. The *SS* in I_{DSS} indicates the gate is shorted to the source to insure that V_{GS} is zero.

We must be careful not to allow the gate to become forward biased with respect to the channel. In certain applications, we can allow the gate to become forward biased as long as the forward bias does not exceed the threshold voltage for the silicon P-N junction (about 0.6 V or 0.7 V at room temperature).

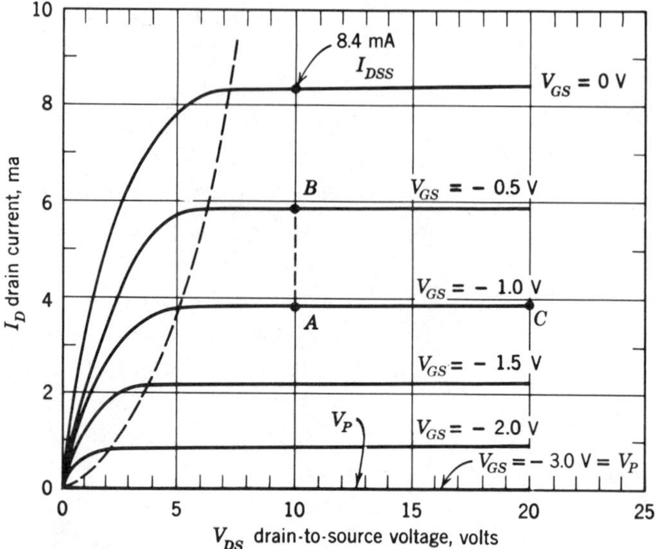

Figure 8-3 Drain characteristic for JFET.

The transistor (BJT) is a current-controlled device. For example, the base current in the common-emitter amplifier controls the collector current. This explanation of how the JFET works shows that the JFET is a voltage-controlled device. The gate-to-source voltage controls the drain current.

A typical set of curves called the *drain characteristic*, obtained by maintaining a fixed gate voltage and by varying V_{DS}, is given in Fig. 8-3. The value of V_P for this particular JFET is -3 V. The curves of the JFET are quite flat once V_{DS} exceeds the values shown by the dashed curve. The drain current is effectively independent of the drain voltage.

Methods of analysis using theoretical physics show that an equation can be developed for I_D in the flat portion of the drain characteristic as

$$I_D = I_{DSS}\left(1 - \frac{V_{GS}}{V_P}\right)^2 \tag{8-1}$$

If a slope is drawn to the drain characteristic at a point such as B on Fig. 8-3, the slope of the tangent defines the *ac drain resistance* r_d.

$$r_d \equiv \frac{\Delta V_{DS}}{\Delta I_{DS}} \tag{8-2}$$

Note that r_d is defined at a constant gate voltage.

Example 8-1
The value of I_{DSS} is 8.4 mA for the JFET used for Fig. 8-3. The change in current from point A to point C is 100 μA. Determine I_D and r_d at point A.

Solution
Inspection of Fig. 8-3 shows that V_P is -3.0 V and that V_{GS} is -1.0 V at point A. Then I_D at point A is

$$I_D = I_{DSS}\left(1 - \frac{V_{GS}}{V_P}\right)^2 = 8.4\ \text{mA}\left(1 - \frac{-1\ \text{V}}{-3\ \text{V}}\right)^2 = \mathbf{3.73\ mA} \qquad (8\text{-}1)$$

The drain resistance at point A is determined by using ΔI_D and ΔV_{DS} between point C and point A.

$$r_d = \frac{\Delta V_{DS}}{\Delta I_D} = \frac{20\ \text{V} - 10\ \text{V}}{100 \times 10^{-6}\ \text{A}} = \mathbf{100\ 000\ \Omega} \qquad (8\text{-}2)$$

The typical construction for the N-channel JFET on an integrated-circuit chip is shown in Fig. 8-4. The *linear gate* in the JFET shown in Fig. 8-4 has a rectangular form. The length of the gate is about 250 μm and its width is of the order of 20 to 30 μm. The depth of the channel under the gate is approximately 1 μm.

When the P substrate is connected to the source, the N channel is "floated" in a reverse bias depletion region to insulate the JFET electrically from other components in the chip. A disadvantage of the IC JFET is that I_{DSS} and V_p can vary as much as 5 to 1 from wafer to wafer. However, within the same wafer, the spread is much less. The circuit design can be adjusted to compensate for this variation.

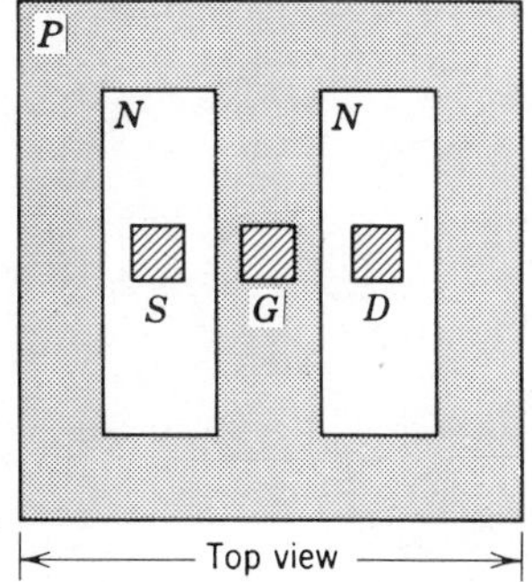

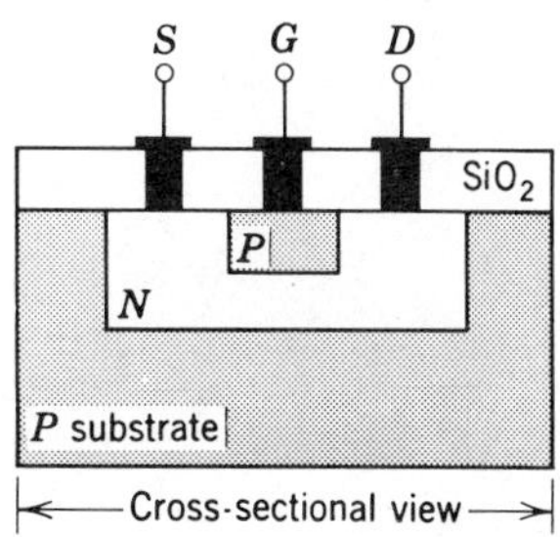

Figure 8-4 FET construction in integrated circuits.

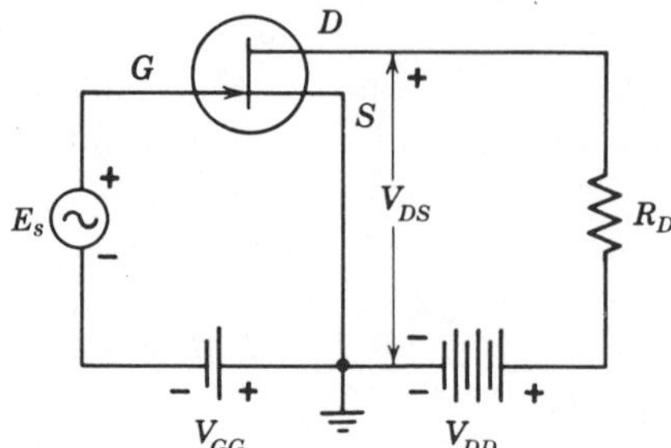

Figure 8-5 Simple JFET amplifier.

A simple amplifier circuit using an N-channel JFET is shown in Fig. 8-5. When E_s has the polarity shown, the sum of E_s and V_{GG} is less negative than V_{GG} alone. The depletion region decreases and I_D increases. The voltage drop $I_D R_D$ increases and the voltage from the drain to the source V_{DS}, which is positive, decreases. When the polarity of E_s reverses, the gate becomes more negative. The drain current decreases and the $I_D R_D$ voltage drop decreases. Consequently, V_{DS} increases in a positive direction. As a result of this action, there is a phase inversion between the input signal at the gate and the output signal at the drain in an FET amplifier. Usually the voltage gain in an FET amplifier is fairly low, of the order of 5 to 15.

The gate in the amplifier of Fig. 8-5 is negative at all times. Therefore, the gate current is zero at all times. Then, theoretically, the input resistance to a gate is infinite (an open circuit). The actual input resistance to the gate of a JFET is of the order of 10 MΩ. The basic transistor amplifiers have input resistance values very much less than 10 MΩ as we have seen. The wide use of the FET in applications results from this property of having a very high input resistance.

Problems **8-1.1** The equation for drain current in an N-channel JFET is

$$I_D = 8.4\left(1 - \frac{V_{GS}}{(-3.0)}\right)^2 \text{ mA} \qquad (V_{DS} = 10 \text{ V})$$

Find I_D at each of the following values of V_{GS}.

$$0, -0.5, -1.0, -1.5, -2.0, \text{ and } -3.0 \text{ V}$$

8-1.2 By using the results of Problem 8-1.1, plot the drain characteristic.

8-1.3 The following values are obtained for a P-channel JFET:

$$\text{when } V_{GS} = 2\text{ V},\ I_D = 7.2\text{ mA}$$

and $$\text{when } V_{GS} = 4\text{ V},\ I_D = 0.8\text{ mA}$$

Find V_P and I_{DSS}.

8-1.4 The following values are obtained for an N-channel JFET:

$$\text{when } V_{GS} = -1\text{ V},\ I_D = 6.75\text{ mA}$$

and $$\text{when } V_{GS} = -2\text{ V},\ I_D = 3.0\text{ mA}$$

Find V_P and I_{DSS}.

8.1.5 Using the data of Problem 8-1.3, calculate I_D when V_{GS} is -1 V.

8-1.6 Using the data of Problem 8-1.4, calculate I_D when V_{GS} is $+2$ V.

Section 8-2 Transfer Characteristics

If the points on the drain characteristic for a particular value of V_{DS} are plotted on a new set of axes (I_D, V_{GS}), the *transfer characteristic* is obtained. When this is done for the drain characteristic given in Fig. 8-3 for V_{DS} equal to 10 V, the transfer characteristic shown in Fig. 8-6 results. Points A and B are corresponding points on the two characteristics.

A tangent to the transfer characteristic at point B is drawn on Fig. 8-6, and its slope defines the *transconductance* g_m of

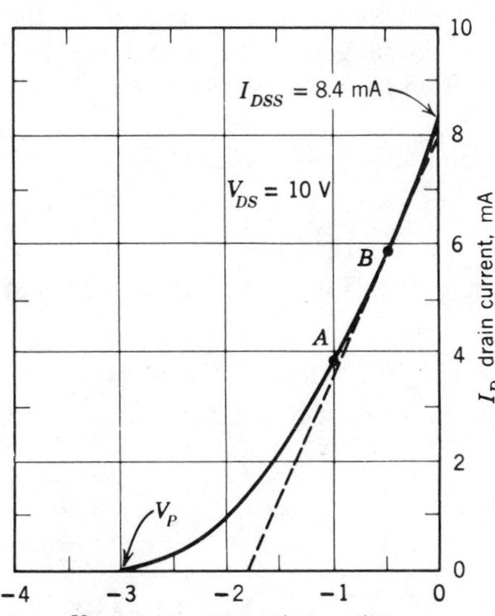

Figure 8-6 Transfer characteristic for JFET.

the JFET at point B as

$$g_m \equiv \frac{\Delta I_D}{\Delta V_{GS}} \text{ for a constant } V_{DS} \tag{8-3}$$

The transconductance measured at I_{DSS} is g_{mo}.

Transconductance is a term used in electronics. The *trans* signifies that the current is the current in the output circuit and that the voltage is the voltage in the input circuit when used in

$$G = \frac{I}{V}$$

The SI unit for conductance is the *siemens*. The unit symbol is G or g and the quantity symbol is S. In an FET g_m is expressed in *millisiemens* (*mS*) or in *microsiemens* (μS). The unit for conductance was formerly designated as the *mho* (℧).

Example 8-2

Determine the value for g_m at point B in Fig. 8-6.

Solution

The value of g_m is the value of the slope of the tangent to the transfer characteristic at point B. The tangent is the dashed straight line. The endpoints of the dashed line have the coordinates

$$V_{GS} = 0 \quad \text{at} \quad I_D = 8 \text{ mA and } V_{GS} = -1.8 \text{ V at } I_D = 0$$

Then

$$g_m = \frac{\Delta I_D}{\Delta V_{GS}} = \frac{8 \text{ mA} - 0}{0 - (-1.8 \text{ V})} = \mathbf{4.45\ mS = 4450\ \mu S} \tag{8-3}$$

If Eq. 8-1 is differentiated with respect to V_{GS}, a mathematical expression can be obtained for g_m.

$$I_D = I_{DSS}\left(1 - \frac{V_{GS}}{V_P}\right)^2$$

$$\frac{dI_D}{dV_{GS}} = 2I_{DSS}\left(1 - \frac{V_{GS}}{V_P}\right)\left(-\frac{1}{V_P}\right) = -\frac{2I_{DSS}}{V_P}\left(1 - \frac{V_{GS}}{V_P}\right)$$

$$g_m = -\frac{2I_{DSS}}{V_P}\left(1 - \frac{V_{GS}}{V_P}\right) \tag{8-4}$$

When V_{GS} is zero, g_m is g_{mo}

$$g_{mo} = -\frac{2I_{DSS}}{V_P} \tag{8-5}$$

And, substituting Eq. 8-5 into Eq. 8-4, we have

$$g_m = g_{mo}\left(1 - \frac{V_{GS}}{V_P}\right) \tag{8-6}$$

Example 8-3
Using values from Fig. 8-6, we have

$I_{DSS} = 8.4\text{ mA}$ $\quad V_p = -3.0\text{ V}$ and $V_{GS} = -0.5\text{ V}$ at point B

Determine g_{mo} and g_m at point B.

Solution No. 1
Substituting numerical values into Eq. 8-5, we have

$$g_{mo} = -\frac{2I_{DSS}}{V_P} = \frac{-2 \times 8.4\text{ mA}}{-3\text{ V}} = \mathbf{5.6\text{ mS} = 5600\mu S} \qquad (8\text{-}5)$$

and substituting numerical values into Eq. 8-6, we have

$$g_m = g_{mo}\left(1 - \frac{V_{GS}}{V_P}\right) = 5.6\left(1 - \frac{-0.5\text{ V}}{-3\text{ V}}\right) = \mathbf{4.67\text{ mS} = 4670\ \mu S} \qquad (8\text{-}6)$$

The value of g_m could have been obtained directly from Eq. 8-4.

Solution No. 2
Equation 8-6 plots as a straight line in the graph shown in Fig. 8-7. The endpoints of this straight line are g_{mo} and V_p. The intermediate values of g_m between zero and g_{mo} can be obtained by using proportions (ratios) derived from the triangle of Fig. 8-7. Using values of -3 V for V_P and 5600 μS for g_{mo}, we find that the value of

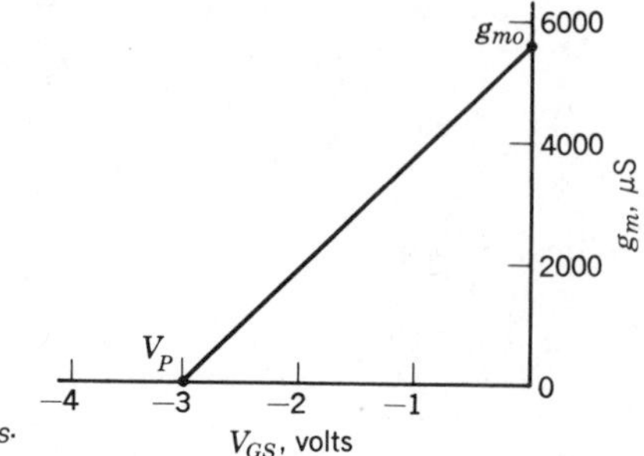

Figure 8-7 Variation of g_m with V_{GS}.

g_m at point B ($V_{GS} = -0.5$ V) is

$$\frac{g_m}{g_{mo}} = \frac{g_m}{5600\ \mu\text{S}} = \frac{V_P - V_{GS}}{V_P} = \frac{3\text{ V} - 0.5\text{ V}}{3\text{ V}} = \frac{2.5\text{ V}}{3\text{ V}}$$

Solving for g_m, we have

$$g_m = \frac{2.5\text{ V}}{3\text{ V}} 5600\ \mu\text{S} = \mathbf{4667\ \mu S}$$

Problems

8-2.1 The values for an N-channel JFET are −3.0 V for V_P and 8.4 mA for I_{DSS}. Determine g_{mo} and the equation for g_m. Plot g_m against V_{GS}.

8-2.2 The values for a P-channel JFET are +5 V for V_P and 20 mA for I_{DSS}. Determine g_{mo} and the equation for g_m. Plot g_m against V_{GS}.

8-2.3 The value of g_m is 500 μS when V_{GS} is −3 V for an N-channel JFET. When V_{GS} is −2 V, g_m is 1000 μS. Determine V_P and I_{DSS}.

8-2.4 For a P-channel JFET, g_m is 900 μS when V_{GS} is 3 V, and g_m is 1200 μS when V_{GS} is 2 V. Determine V_P and I_{DSS}.

Section 8-3 The Depletion-Type MOSFET

The construction of a different type of FET is shown in Fig. 8-8*a*. Here the gate is simply a metallic plate that has no P or N semiconductor property. The gate is insulated from the channel by a layer of silicon dioxide (SiO_2 glass). This device is called a *depletion-type metal oxide semiconductor field effect transistor* or *MOSFET*. An alternative terminology is *depletion-type insulated gate field effect transistor*. The circuit symbols are shown in Figs. 8-8*b*, 8-8*c*, and 8-8*d*. When the substrate connection is brought out externally, it is connected to the source externally.

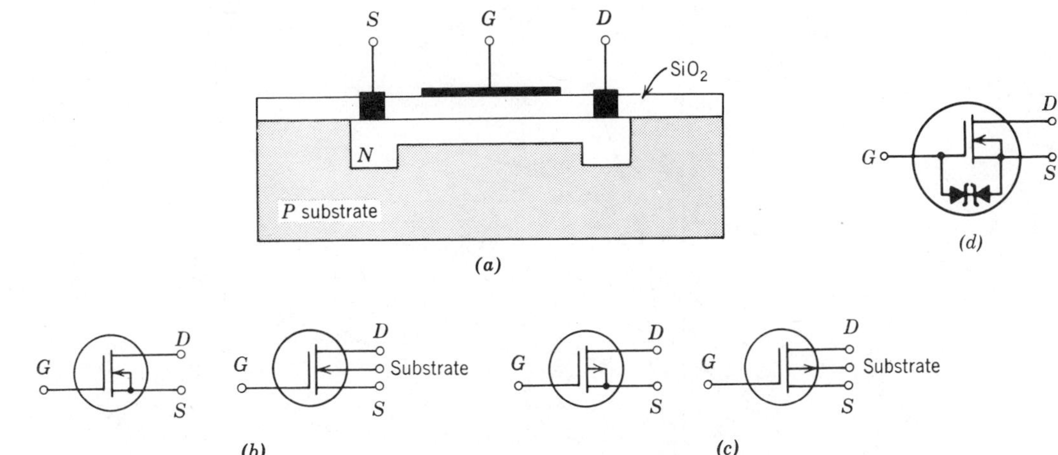

Figure 8-8 Depletion-type MOSFET. (*a*) Cross-sectional view. (*b*) Symbols for *N*-channel MOSFETS. (*c*) Symbols for *P*-channel MOSFETS. (*d*) MOSFET with internal Zener protection.

When the gate-to-source voltage V_{GS} is zero, the drain current is I_{DSS}. When a negative voltage is applied to the gate, the negative charge on the gate repels the electrons that are the current carriers in the *N* channel. The effect is to reduce or constrict the current flow to the drain. A sufficient negative voltage V_P on the gate produces a pinch-off in the channel, at which point I_D falls to zero. If a positive voltage is applied to the gate, more negative current carriers are drawn into the channel giving drain currents greater than I_{DSS}. When the gate is positive, the layer of insulation between the gate and the channel prevents any gate current flow. The MOSFET has the advantage of the insulated gate. The input resistance to the gate is of the order of 100 MΩ. The leakage current is of the order of 10^{-12} A (1 pA). The depletion-type MOSFET is usually operated at zero bias ($V_{GG} = 0$), since the gate voltage can normally swing either positive or negative.

It is imperative not to permit any stray or static voltage on the gate, or the SiO_2 layer between the gate and the channel will be destroyed. Even picking up the MOSFET can destroy it. Consequently, grounding rings must be used, and they are removed only after the MOSFET is securely wired into the circuit. Some MOSFET's have back-to-back Zener diodes internally formed in the monolithic structure to protect against these stray voltages (Fig. 8-8*d*).

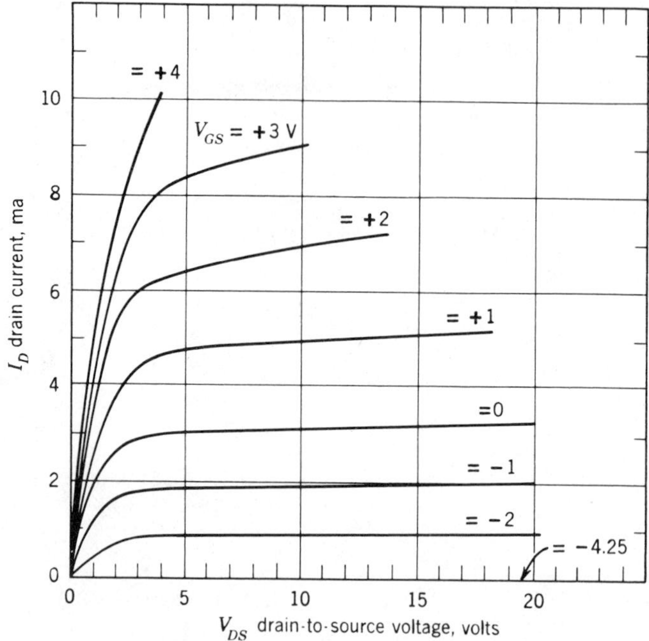

Figure 8-9 Drain characteristic for depletion-type MOSFET.

The drain characteristic for a typical N-channel depletion-type MOSFET is given in Fig. 8-9. The corresponding transfer characteristic is given in Fig. 8-10. The definitions for V_P, I_{DSS}, g_m, and r_d developed for the JFET apply to the depletion-type

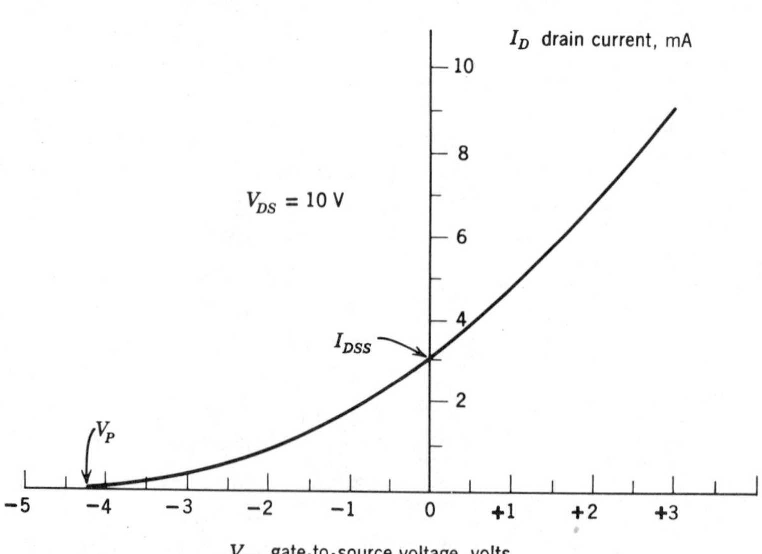

Figure 8-10 Transfer characteristic for depletion-type MOSFET.

MOSFET without restriction. Similarly, Eq. 8-1 to 8-6 are also valid. The equation for I_D for the MOSFET used for the curves of Figs. 8-9 and 8-10 is

$$I_D = I_{DSS}\left(1 - \frac{V_{GS}}{V_P}\right)^2 = 3.1\left(1 - \frac{V_{GS}}{-4.25}\right)^2 \text{ mA} \tag{8-1}$$

The MOSFET itself can be used in place of a resistor in an IC. The gate of the MOSFET is placed either to the supply (switch position 1) or to the ground return (switch position 2) in Fig. 8-11 as shown in the following example.

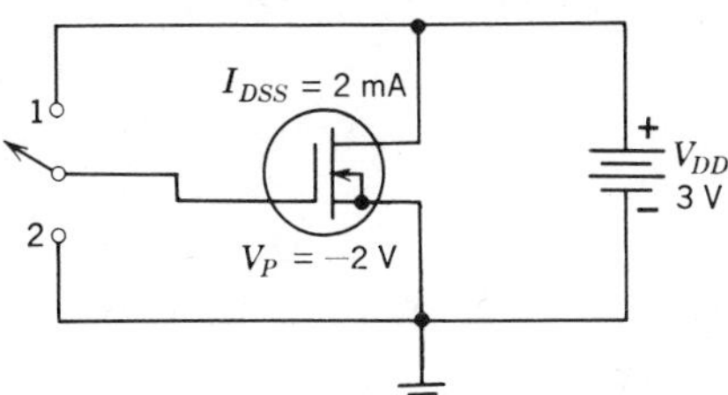

Figure 8-11 Depletion-type *N*-channel MOSFET used as a resistor.

Example 8-4

Determine the equivalent resistance of the circuit of Fig. 8-11 with the switch in position 1 and in position 2.

Solution

When the switch is in position 1, V_{GS} is +3 V. Then

$$I_D = I_{DSS}\left(1 - \frac{V_{GS}}{V_P}\right)^2 = 2\text{ mA}\left(1 - \frac{3\text{ V}}{-2\text{ V}}\right)^2 = 8.25\text{ mA} \tag{8-1}$$

and

$$R_1 = \frac{V_{DD}}{I_D} = \frac{3.0\text{ V}}{8.25\text{ mA}} = 0.364\text{ k}\Omega = \mathbf{364\ \Omega}$$

When the switch is in position 2, V_{GS} is zero and I_D is I_{DSS} or 2 mA. Then

$$R_2 = \frac{V_{DD}}{I_{DSS}} = \frac{3\text{ V}}{2\text{ mA}} = 1.5\text{ k}\Omega = \mathbf{1500\Omega}$$

Problems 8-3.1 A depletion-type *N*-channel MOSFET has a value of 4 mA for I_{DSS} and a value for V_P of −3 V. Determine I_D and g_m for the following values of V_{GS}.

−4 V, −2 V, 0 V, +2 V, and +4V

8-3.2 A depletion-type P-channel MOSFET has a value of 10 mA for I_{DSS} and a value for V_P of +4 V. Determine I_D and g_m for the following values of V_{GS}.

$$+6\text{ V}, +3\text{ V}, 0\text{ V}, -3\text{ V}, \text{and } -6\text{ V}$$

8-3.3 The MOSFET given in Problem 8-3.1 is used in the circuit of Fig. 8-11. V_{DD} is +5 V. Determine R_1 and R_2.

8-3.4 The MOSFET given in Problem 8-3.2 is used in the circuit of Fig. 8-11. V_{DD} is −4 V. Determine R_1 and R_2.

Section 8-4 The Enhancement-Type MOSFET

Another integrated-circuit structure is the *enhancement-type MOSFET.* This MOSFET is constructed without a channel. If the MOSFET shown in Fig. 8-12*a* had a channel, its channel would be N-type material. Therefore, this unit is called an *N-channel enhancement-type MOSFET.*

If a positive voltage is applied to the gate of the N-channel unit (Fig. 8-12*b*) at a sufficient positive voltage called the *threshold voltage* V_T, electrons are pulled into the substrate just below the layer of insulation beneath the gate. Now a *virtual N channel* exists between the source and the drain to permit a flow of drain current I_D. If the gate voltage is increased, the virtual channel deepens and I_D increases.

If the substrate is N-type material, the source P-type material, and the drain P-type material, the unit is called a

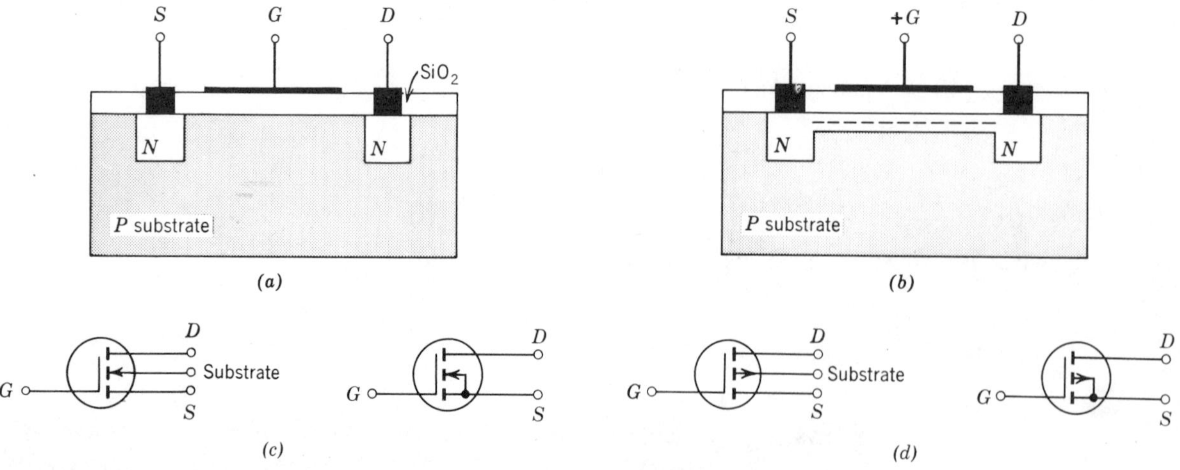

Figure 8-12 Enhancement-type MOSFET. (*a*) Cross-sectional view. (*b*) Formation of virtual channel. (*c*) Symbols for N-channel enhancement-type MOSFET. (*d*) Symbols for P-channel enhancement-type MOSFET.

P-channel enhancement-type MOSFET. A negative voltage on the gate repels electrons from the substrate just under the gate and a *virtual P channel* is formed. The threshold voltage V_T at which drain current just starts to flow is now a negative voltage.

The circuit symbols for the enhancement-type MOSFETs are shown in Fig. 8-12*c* and Fig. 8-12*d*. A set of typical drain characteristics is shown in Fig. 8-13. The corresponding transfer characteristic is given in Fig. 8-14.

Naturally, I_{DSS} and V_P have no meaning for the enhancement-type MOSFET. A new equation must be used for I_D.

$$I_D = K(V_{GS} - V_T)^2 \tag{8-7}$$

If Eq. 8-7 is differentiated with respect to V_{GS}, we obtain an equation for g_m.

$$I_D = K(V_{GS} - V_T)^2$$

$$\frac{dI_D}{dV_{GS}} = 2K(V_{GS} - V_T)$$

$$g_m = 2K(V_{GS} - V_T) \tag{8-8}$$

Example 8-5

The value of K for the MOSFET shown in Fig. 8-13 and Fig. 8-14 is $0.445\ \text{mA/V}^2$. Determine the value of I_D and g_m at point A.

Solution No. 1

The value of V_{GS} at point A is +3.0 V and the value of V_T is +0.2 V. Then

$$I_D = K(V_{GS} - V_T)^2 = 0.445\ (3.0 - 0.2)^2 = \mathbf{3.49\ mA} \tag{8-7}$$

and the transconductance is

$$g_m = 2K(V_{GS} - V_T) = 2 \times 0.445(3.0 - 0.2)$$

$$= \mathbf{2.5\ mS = 2500\ \mu S} \tag{8-8}$$

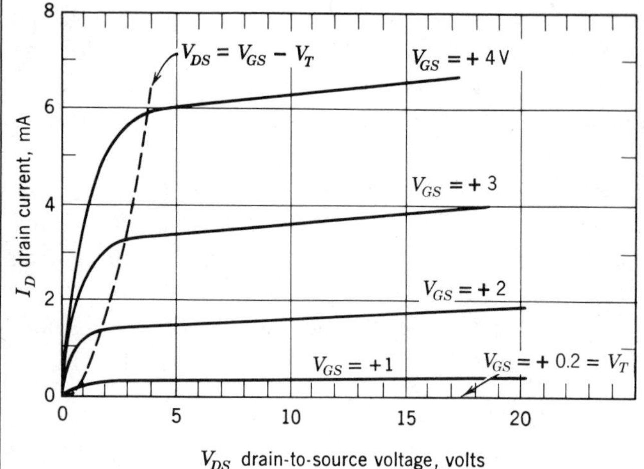

Figure 8-13 Drain characteristic for enhancement-type MOSFET.

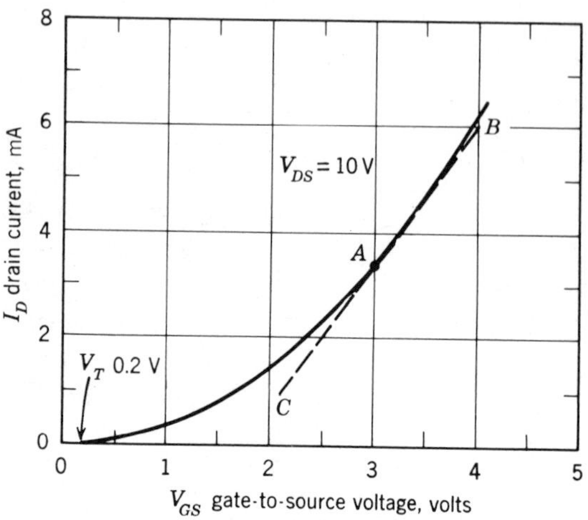

Figure 8-14 Transfer curve for enhancement-type MOSFET.

Solution No. 2

The value of g_m can be obtained from a tangent drawn to the transfer curve at point A. The coordinates of points B and C are

$$I_D = 6 \text{ mA} \quad \text{at} \quad V_{GS} = 4.0 \text{ V}$$

and

$$I_D = 1 \text{ mA} \quad \text{at} \quad V_{GS} = 2 \text{ V}$$

Then by the definition of g_m, Eq. 8-3,

$$g_m = \frac{\Delta I_D}{\Delta V_{GS}} = \frac{6-1}{4-2} = \frac{5 \text{ mA}}{2 \text{ V}} = \mathbf{2.5 \text{ mS} = 2500 \; \mu S} \tag{8-3}$$

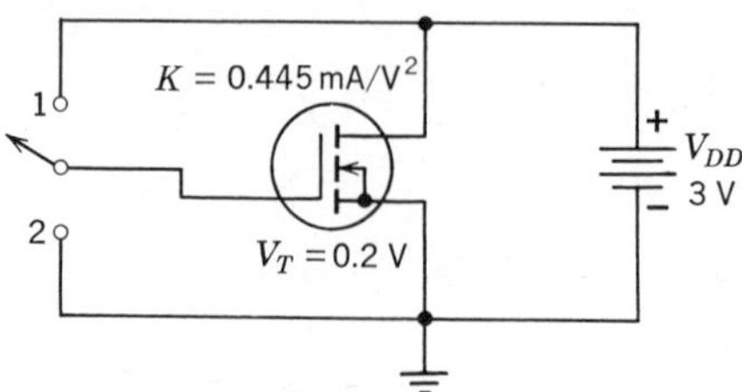

Figure 8-15 Enhancement-type N-channel MOSFET used as a resistor.

The enhancement-type MOSFET is widely used in digital circuits as a resistor or as an ON–OFF switch (Fig. 8-15)

Example 8-6

Determine the equivalent resistance of the circuit of Fig. 8-15 with the switch in position 1 and in position 2.

Solution

When the switch is in position 1, V_{GS} is +3 V. The value of I_D is

$$I_D = K(V_{GS} - V_T)^2 = 0.445(3 - 0.2)^2 = 3.49 \text{ mA} \qquad (8\text{-}7)$$

and

$$R_1 = \frac{V_{DD}}{I_D} = \frac{3 \text{ V}}{3.49 \text{ mA}} = 0.86 \text{ k}\Omega = \mathbf{860\ \Omega}$$

When the switch is in position 2, V_{GS} is zero. Unless V_{GS} is at least the value of V_T (+0.2 V), the current I_D is zero. Therefore the MOSFET acts as an open circuit.

$$R_2 = \frac{V_{DD}}{I_D} = \frac{3 \text{ V}}{0 \text{ mA}} = \mathbf{\infty\ \Omega}$$

Problems

8-4.1 The equation for an enhancement-type N-channel MOSFET is

$$I_D = 0.6(V_{GS} - 0.6)^2 \text{ mA} \qquad (V_{DS} = 10\ V)$$

Find I_D at each of the following values of V_{GS}.

1.0 V, 2.0 V, 3.0 V, and 4.0 V

Sketch the drain characteristic. Sketch the transfer characteristic.

8-4.2 The equation for an enhancement-type P-channel MOSFET is

$$I_D = 0.4(V_{GS} + 0.8)^2 \text{ mA} \qquad (V_{DS} = -10 \text{ V})$$

Find I_D at each of the following values of V_{GS}.

−1.0 V, −2.0 V, −3.0 V, and −4.0 V

Sketch the drain and transfer characteristics.

8-4.3 Determine g_m for each value of V_{GS} in Problem 8-4.1. Plot a curve of g_m against V_{GS}.

8-4.4 Determine g_m for each value of V_{GS} in Problem 8-4.2. Plot a curve of g_m against V_{GS}.

8-4.5 A circuit using the MOSFET of Problem 8-4.1 uses a V_{DD} of +5 V. The gate is switched from V_{DD} to the ground return. What is the circuit resistance in each case?

8-4.6 A circuit using the MOSFET of Problem 8-4.2 uses a V_{DD} of −4 V. The gate is switched from V_{DD} to the ground return. What is the circuit resistance in each case?

Supplementary Problems

8-1 The value for I_{DSS} is 20 mA and the value for V_P is -4 V for an N-channel JFET. Determine I_D and g_m when V_{GS} is -1 V.

8-2 The value of I_{DSS} is 15 mA and the value for V_P is $+3$ V for a P-channel JFET. Determine I_D for $+1.4$ V for V_{GS} and also for $+1.5$ V for V_{GS}. Find g_m by using Eq. 8-3 and also by using Eq. 8-4. What is the difference in the results in percent?

8-3 An N-channel depletion-type MOSFET has a value for I_{DSS} of 15 mA and a value for V_P of -3 V. If the maximum allowable current in the MOSFET is 40 mA, what is the maximum permissible voltage on the gate?

8-4 A P-channel depletion-type MOSFET has

$$I_{DSS} = 6 \text{ mA} \quad \text{and} \quad V_P = 3 \text{ V}$$

Determine V_{GS} and g_m when I_D is 4 mA.

8-5 Determine the value of R required to adjust V_{DS} to $+4$ V.

8-6 The JFET in Problem 8-5 is replaced with an N-channel depletion-type MOSFET having the same values of I_{DSS} and V_P. Determine the value of R required to adjust V_{DS} to $+4$ V.

8-7 The JFET in Problem 8-5 is replaced with an N-channel enhancement-type MOSFET that has a value for V_T of $+0.6$ V and a value for K of 1.2 mA/V^2. Determine the value of R required to adjust V_{DS} to $+4$ V.

8-8 The MOSFET of Problem 8-7 is to serve as the equivalent of a 3000-Ω resistor. What is the value of R?

8-9 Find the resistance of the circuit when the switch is in position A. Repeat for position B.

8-10 Repeat Problem 8-9 if the JFET is replaced with a depletion-type MOSFET.

8-11 Repeat Problem 8-9 is the JFET is replaced with an N-channel enhancement MOSFET that has a value of $+0.5$ V for V_T and a value for K of 2 mA/V^2.

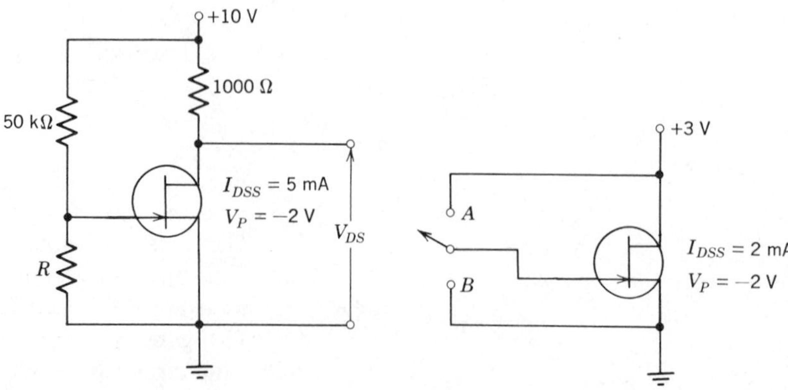

Circuit for Problems 8-5 through 8-8. Circuit for Problems 8-9 through 8-11.

Chapter 9 FET Bias, Load Lines, and Amplifiers

A dc bias establishes the quiescent operating point on the drain characteristic of the FET. A graphical approach is taken to determine the operating point with a self-bias arrangement. Often the bias is obtained by a combination of self-bias in the source and a voltage divider placed across the gate (Section 9-1). The load line drawn on the drain characteristic is used to obtain a dynamic transfer characteristic for an FET amplifier (Section 9-2). Since the dynamic transfer characteristic is nonlinear, the FET is used as a small-signal amplifier. The formal ac model can be reduced to the same form that was used for the transistor amplifier (Section 9-3). We also show that the reciprocal of g_m yields a resistance value that is equivalent to the r'_e that is used in transistor amplifier calculations. The source follower (Section 9-4) is equivalent to the emitter follower.

Section 9-1 FET Bias Methods

The drain characteristics for two FETs are shown in Fig. 9-1. The FET of Fig. 9-1*a* is an *N*-channel JFET. The characteristics of the JFET are shown for a range of V_{GS} from 0 V to −4 V, the pinch-off voltage. The characteristics of a comparable *N*-channel depletion-type MOSFET are given in Fig. 9-1*b*. Here the range of V_{GS} is from +2 V to −4 V, the pinch-off voltage.

The basic circuit used for the JFET is given in Fig. 9-2*a*. The gate for the *N*-channel JFET must be kept negative at all times. Therefore a bias source V_{GG} must be used to provide

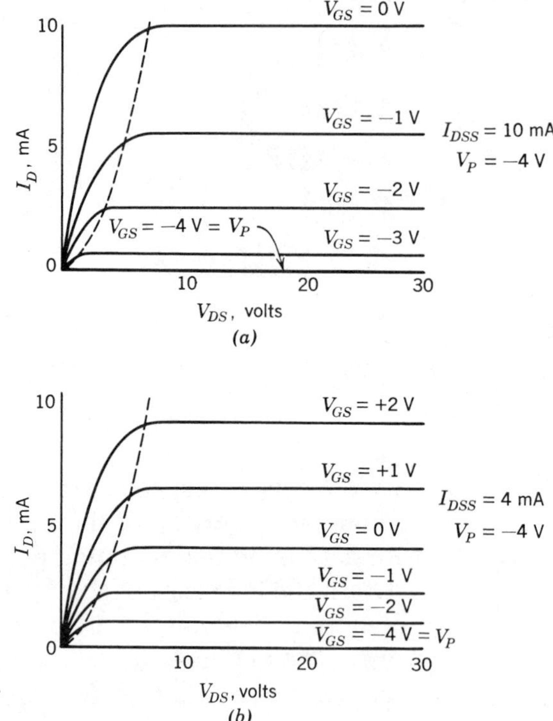

Figure 9-1 Typical drain characteristics. (*a*) *N*-channel junction FET. (*b*) *N*-channel depletion-type MOSFET.

the required negative bias voltage. The value of V_{GG} determines the value of I_D. The equation for I_D to the right of the dashed line is given by

$$I_D = I_{DSS}\left(1 - \frac{V_{GS}}{V_P}\right)^2 \tag{8-1}$$

When the value of V_{GG} is used for V_{GS}, Eq. 8-1 yields the resulting value of I_D directly.

It should be noted that this simple bias method has the disadvantage of requiring two different supply voltages, one positive and the other negative.

The biasing method for the MOSFET (Fig. 9-2*b*) is simple if the circuit is operated at zero voltage on the gate ($V_{GG} = 0$). The drain current then is I_{DSS}. The resistor R_G is required to provide a dc path to the common return (ground) in order to prevent a buildup of a charge on the gate. If R_G is omitted or open, the gate is *floating*. An accumulation of a charge on the

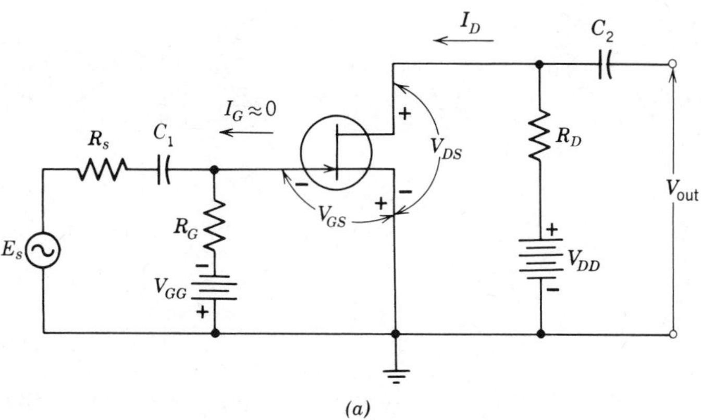

(a)

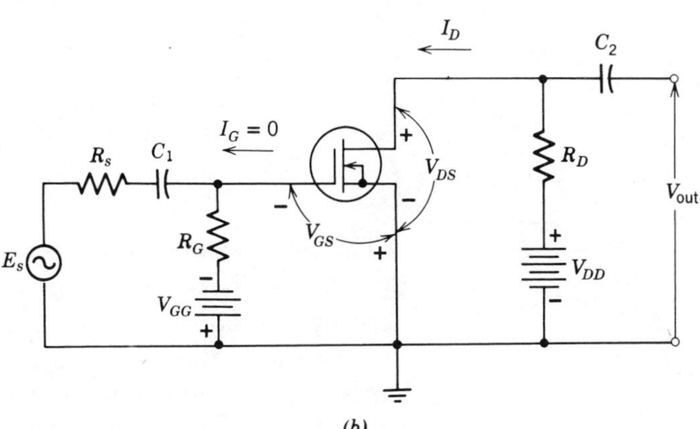

(b)

Figure 9-2 Basic FET bias circuits. (*a*) JFET circuit. (*b*) MOSFET circuit.

gate could produce a negative voltage equal to or greater than V_P. Then the MOSFET would be *cut off*. If a bias supply voltage V_{GG} is used for the MOSFET, we use Eq. 8-1 to obtain I_D. In Fig. 9-2*b*, if there is a dc path through the source, C_1 and R_G may be omitted from the circuit when V_{GG} is zero.

The Kirchhoff's voltage loop equation through the drain circuit is

$$V_{DD} = R_D I_D + V_{DS} \tag{9-1}$$

When the value of I_D is obtained either from the drain characteristic or from Eq. 8-1, the value of I_D can be substituted into Eq. 9-1 to yield a numerical value for V_{DS} or for R_D.

Example 9-1
In the circuit of Fig. 9-2*a*, R_G is 1 MΩ, V_{GG} is −2 V, and V_{DD} is 12 V. Then I_{DSS} is 9 mA and V_P is −3 V for the JFET. Find the value of R_D that sets V_{DS} to 7 V.

Solution
We find I_D from

$$I_D = I_{DSS}\left(1 - \frac{V_{GS}}{V_P}\right)^2 = 9\text{ mA}\left(1 - \frac{-2\text{ V}}{-3\text{ V}}\right)^2 = 1\text{ mA} \qquad (8\text{-}1)$$

Using this value of I_D in the Kirchhoff's voltage loop equation through the drain, we have

$$V_{DD} = R_D I_D + V_{DS} \qquad (9\text{-}1)$$

Substituting values, we have

$$12\text{ V} = R_D \times 1\text{ mA} + 7\text{ V}$$

$$R_D = \mathbf{5\ k\Omega}$$

The use of a second supply voltage source can be avoided by using a *self-bias* circuit arrangement (Fig. 9-3). The dc current through the JFET, I_D, also flows through the resistor in series with the source R_S. The polarity of the voltage drop across R_S is shown on the circuit diagram. Since the gate current is zero, the voltage drop across R_G is zero. Thus the gate is effectively connected to the negative side of the voltage drop across R_S. Thus, *the gate is negative with respect to the source* by the amount of the dc voltage drop across R_S.

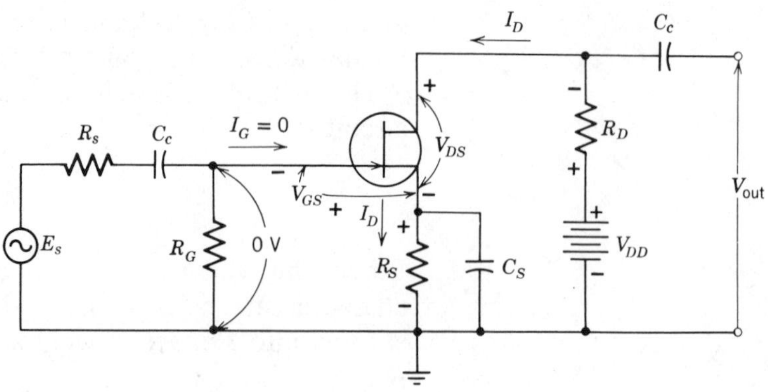

Figure 9-3 Amplifier circuit using self-bias.

$$V_{GS} = -I_D R_S$$

The transfer curve for the JFET used in the self-bias circuit (Fig. 9-3) is shown in Fig. 9-4. The data for this transfer curve is obtained from the drain characteristic given in Fig. 9-1*a*. Bias lines are drawn from the origin to point A', to point B', and to point C'. The resistance value for the bias line to point A' is obtained by dividing the voltage value at point A' by the current value at point A'.

The Kirchhoff's voltage loop equation through the drain circuit is

$$V_{DD} = R_D I_D + V_{DS} + R_S I_D \tag{9-2}$$

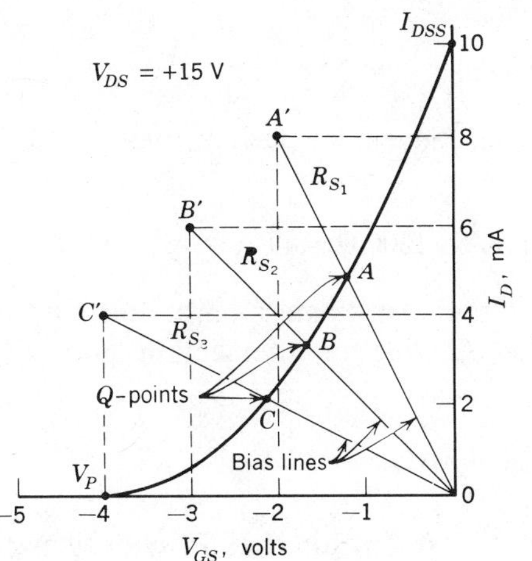

Figure 9-4 Transfer curve for the JFET used in the circuit of Fig. 9-3.

Example 9-2

Construct bias lines on the transfer characteristic given in Fig. 9-4 for the following values of R_S.

Case I. $R_{S_1} = 250\ \Omega$
Case II. $R_{S_2} = 500\ \Omega$
Case III. $R_{S_3} = 1000\ \Omega$

Find the Q-point values of I_D and V_{GS} for each case.

Solution

Case I. If we assume a convenient value of current as 8 mA, we find by Ohm's law

$$V = IR_{S_1} = 0.008\ \text{mA} \times 250\ \Omega = 2.0\ \text{V}$$

Point A' is located by the coordinates 8 mA and −2 V. A straight line, the bias line, is drawn from point A' to the origin. The intersection, point A, of the bias line with the transfer characteristic gives the operating or Q point.

$$I_D = \mathbf{4.8\ mA} \quad \text{and} \quad V_{GS} = \mathbf{-1.2\ V}$$

Case II. If we assume 6 mA as a convenient current, we have by Ohm's law

$$V = IR_{S_2} = 0.006\ \text{mA} \times 500\ \Omega = 3.0\ \text{V}$$

Point B' is located at the coordinates 6 mA and −3 V. Drawing the bias line as before, we find that the intersection, point B, the Q-point, is

$$I_D = \mathbf{3.3\ mA} \quad \text{and} \quad V_{GS} = \mathbf{-1.65\ V}$$

Case III. If we assume 4 mA as a convenient current, we have by Ohm's law

$$V = IR_{S_3} = 0.004\ \text{mA} \times 1000\ \Omega = 4.0\ \text{V}$$

Point C' is located at the coordinates 4 mA and −4 V. Drawing the bias line and finding the intersection, point C, we have the Q-point

$$I_D = \mathbf{2.2\ mA} \quad \text{and} \quad V_{GS} = \mathbf{-2.2\ V}$$

A circuit that is widely used to bias the FET is shown in Fig. 9-5. There is a voltage from the gate to the common

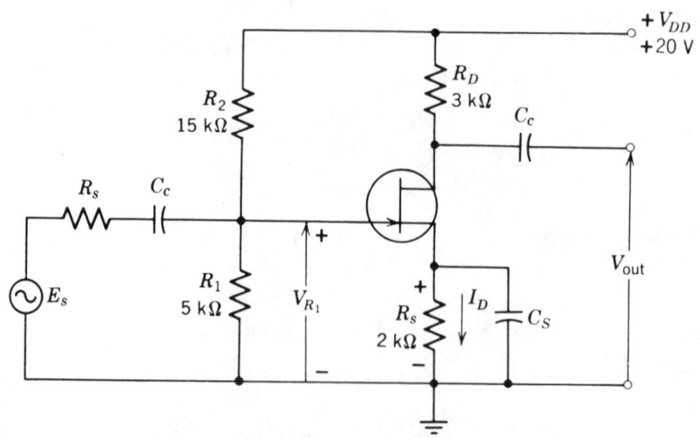

Figure 9-5 Circuit deriving bias both from self-bias and from a voltage divider.

return (ground) produced by the action of the voltage divider formed by R_1 and R_2. There is a voltage from the source to the common return (ground) produced by the voltage drop $I_D R_S$ across R_S. The polarities of these voltages are shown on the circuit diagram. The bias on the FET, V_{GS}, is the difference between these two voltages.

The voltage across the voltage divider measured from the gate to ground is

$$V_{R_1} = \frac{R_1}{R_1 + R_2} V_{DD} \tag{9-3}$$

This voltage (V_{R_1}) is located on the transfer characteristic (Fig. 9-6) at point A. The voltage produced by the voltage divider acts as an *offset voltage* for the bias line. Now we must draw the bias line from point A whereas in the previous circuit arrangement we started the bias line at the origin.

A convenient value of current I is assumed and Ohm's law is used to determine

$$V = IR_S$$

Now voltage V is the total distance we go to the left from point A to point B on Fig. 9-6. At point B, we go vertically the amount of the assumed current I to locate point C. The bias line is drawn from point C to point A. The intersection of the

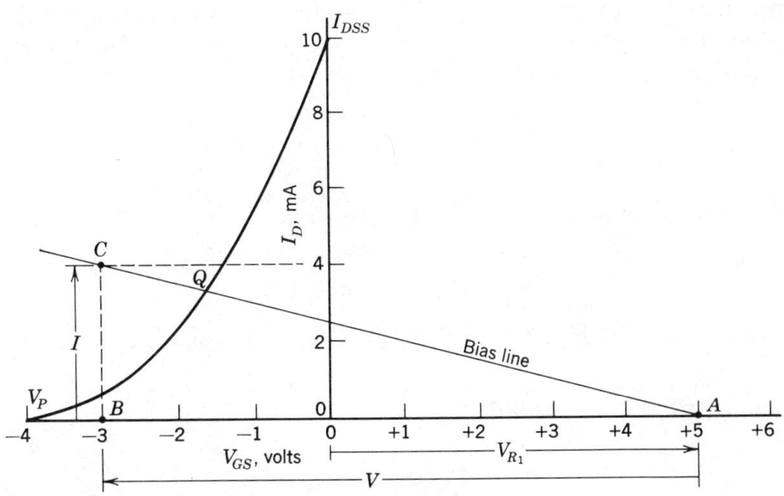

Figure 9-6 Bias line for the circuit given in Fig. 8-15.

bias line with the transfer characteristic gives the operating point.

The Kirchhoff's voltage loop equation through the drain circuit is

$$V_{DD} = R_D I_D + V_{DS} + R_S I_D \tag{9-2}$$

Example 9-3

Find I_D, V_{GS}, and V_{DS} for the circuit shown in Fig. 9-5. The transfer characteristic for the FET is given in Fig. 9-6.

Solution

The bias offset voltage V_{R_1} is found from the voltage divider rule

$$V_{R_1} = \frac{R_1}{R_1 + R_2} V_{DD} = \frac{5000\ \Omega}{5000\ \Omega + 15{,}000\ \Omega} 20\ \text{V} = 5.0\ \text{V} \tag{9-3}$$

V_{R_1} is the offset value that locates point A on Fig. 9-6.
Let us assume 4 mA as a convenient current value for I and using Ohm's law

$$V = IR_S = 0.004\ \text{mA} \times 2000\ \Omega = 8.0\ \text{V}$$

We move to the left from point A (+5 V) a total distance of 8 V to locate point B. Point B is located at −3.0 V. Now from point B we move up 4 mA to locate point C. The bias line is drawn from point C to point A. The intersection of the bias line with the transfer characteristic is the Q-point (Q).

$$I_D = \mathbf{3.32\ mA} \quad \text{and} \quad V_{GS} = \mathbf{-1.64\ V}$$

Using these values in Eq. 9-3, we have

$$V_{DD} = R_D I_D + V_{DS} + R_S I_D$$
$$20\ \text{V} = 3000\ \Omega \times 0.00332\ \text{A} + V_{DS} + 2000\ \Omega \times 0.00332\ \text{A}$$

Solving for V_{DS}, we find that

$$V_{DS} = \mathbf{3.40\ V}$$

Problems **9-1.1** The JFET has the values

$$I_{DSS} = 8\text{ mA} \quad \text{and} \quad V_P = -3\text{ V}$$

If the circuit is biased at -1 V (V_{GG}), determine the operating point and show this operating point on the drain characteristic.

9-1.2 Repeat Problem 9-1.1 for

$$I_{DSS} = 8\text{ mA} \qquad V_P = -5\text{ V} \quad \text{and} \quad V_{GG} = -2\text{ V}$$

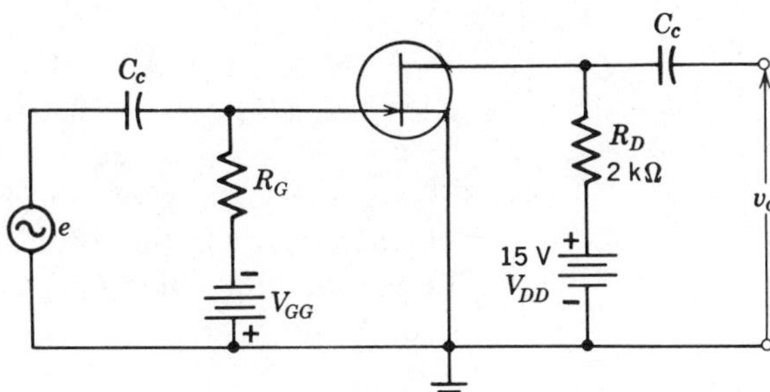

Circuit for Problems 9-9.1 and 9-1.2.

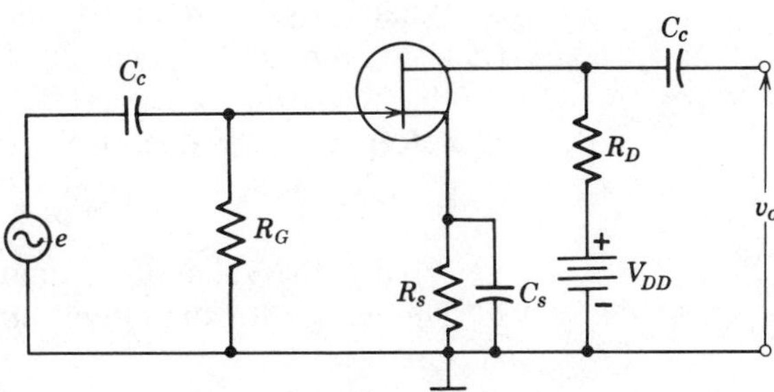

Circuit for Problems 9-1.3 through 9-1.6.

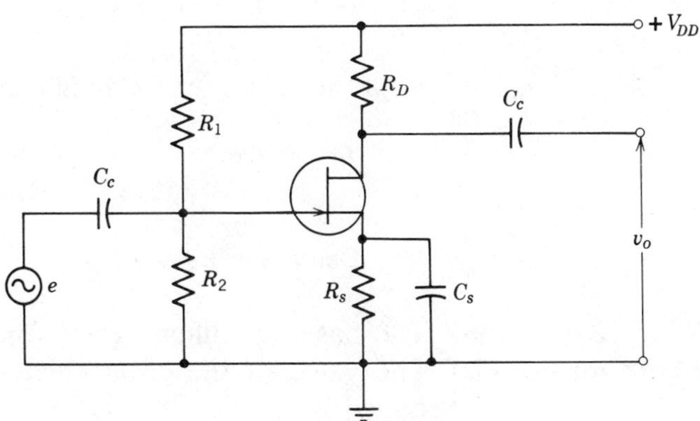

Circuit for Problems 9-1.7 and 9-1.8.

9-1.3 The values for the circuit are

$$I_{DSS} = 8\text{ mA} \qquad V_P = -3\text{ V} \qquad V_{DD} = 30\text{ V} \quad \text{and} \quad (R_D + R_S) = 4000\ \Omega$$

Determine R_S to obtain the operating point.

$$I_{DQ} = 2\text{ mA}$$

Show this operating point on the load line.

9-1.4 Repeat Problem 9-1.3 for

$$I_{DSS} = 10\text{ mA} \qquad V_P = -4\text{ V} \qquad V_{DD} = 20\text{ V}$$
$$(R_D + R_S) = 2000\ \Omega \quad \text{and} \quad I_{DQ} = 5\text{ mA}$$

9-1.5 Using the data given in Problem 9-1.3, draw the transfer characteristic of the FET. By using the graphical approach, determine the value of R_S to establish the operating point at

$$V_{GS} = -2.5\text{ V}$$

What is the value of I_{DQ} and what is the value of V_{DS}?

9-1.6 Using the data given in Problem 9-1.4, draw the transfer characteristic of the FET. By using the graphical approach, determine the value of R_S to establish the operating point at

$$V_{GS} = -1.5\text{ V}$$

What is the value of I_{DQ} and what is the value of V_{DS}?

9-1.7 The values for the circuit are

$$I_{DS} = 10\text{ mA} \qquad V_P = -4\text{ V} \qquad V_{DD} = +33\text{ V} \qquad R_1 = 470\ \Omega$$
$$R_2 = 47\text{ k}\Omega \qquad R_S = 2500\ \Omega \qquad V_{DS} = 10\text{ V}$$

Determine I_{DQ} and R_D.

9-1.8 The values for the circuit are

$$I_{DSS} = 20\text{ mA} \qquad V_P = -4\text{ V} \qquad V_{DD} = +25\text{ V} \qquad R_1 = 300\text{ k}\Omega$$
$$R_2 = 56\text{ k}\Omega \qquad R_D = 1500\ \Omega \qquad I_{DQ} = 4\text{ mA}$$

Determine R_S and V_{DS}.

Section 9-2 The Load Line for the FET

The basic amplifier circuit for the FET is shown in Fig. 9-7. The value of the drain current at the Q-point can be determined from

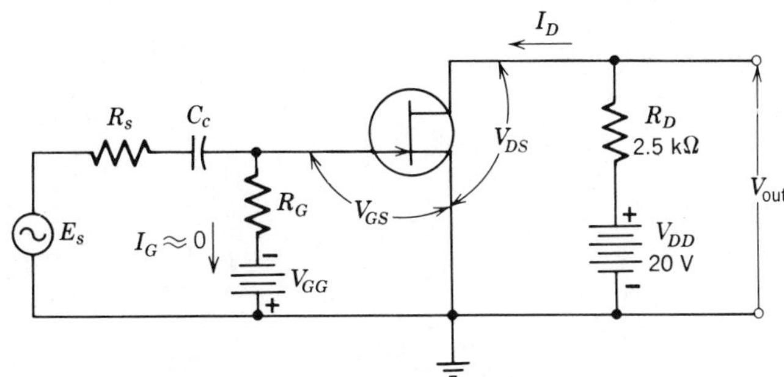

Figure 9-7 Amplifier circuit using an *N*-channel JFET.

$$I_D = I_{DSS}\left(1 - \frac{V_{GS}}{V_P}\right)^2 \tag{8-1}$$

or from

$$I_D = K(V_{GS} - V_T)^2 \tag{8-7}$$

for the enhancement-type MOSFET. The Kirchhoff's voltage loop equation through the drain circuit is

$$V_{DD} = R_D I_D + V_{DS} \tag{9-1}$$

Solving this equation for I_D, we have

$$I_D = -\frac{1}{R_D}V_{DS} + \frac{V_{DD}}{R_D} \tag{9-4}$$

Equation 9-4 has the form

$$y = mx + b$$

that we used previously to show that the load line for a transistor was a straight line on the collector characteristic. Thus, Eq. 9-4 shows that a load line can be drawn as a straight line for the FET on a drain characteristic (Fig. 9-8). The endpoints of the load line are V_{DD}/R_D and V_{DD}.

The intersection points of the load line with the V_{GS} curves are numbered *points 1 through 6*. These are possible *Q*-points for circuit operation. The particular *Q*-point is determined by the value of V_{GS} used for the circuit.

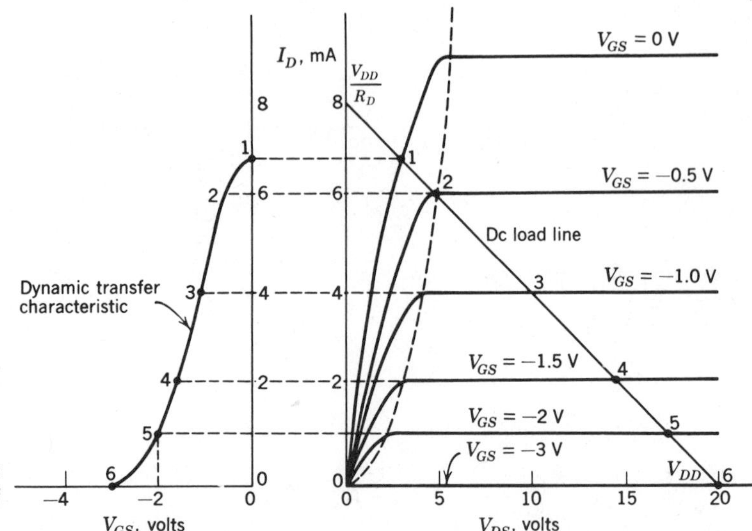

Figure 9-8 Load line and dynamic transfer characteristic for the circuit of Fig. 9-7.

Let us place a new set of axes to the left of the drain characteristic. The vertical axis is I_D to the same scale as the drain characteristic. We show values of V_{GS} increasing in magnitude to the left of the origin. We extend this axis to the left both for an N-channel FET and for a P-channel FET. Now we locate points 1 through 6 on this new set of axes. These points are now connected with a smooth curve. This new curve represents *all* the points on the load line on a new set of axes. Obviously, we do not have a straight line now although we did prove that the load line must be a straight line on the drain characteristic. We call this new curve the *dynamic transfer characteristic* (Fig. 9-8).

We use the word *dynamic* to show that the points come from a specific load line as contrasted with the ordinary transfer curves we used for the FET in Section 8-2. This dynamic transfer characteristic shows the input-voltage-to-output-current relationship for an FET amplifier circuit having specific values for R_D and for V_{DD}. If we change either R_D or V_{DD}, a new dynamic transfer characteristic results.

Example 9-4

Develop the dynamic transfer characteristic for the JFET circuit shown in Fig. 9-7 using the drain characteristic given in Fig. 9-8. Draw the load line for a supply voltage of 20 V for V_{DD} and a load resistance of 2.5 kΩ for R_D. Plot the dynamic transfer characteristic.

Solution

The endpoints of the load line are V_{DD} (20 V) and V_{DD}/R_D (20/2.5 = 8 mA). The load line is drawn on the drain characteristic. The intersections of the load line with the V_{GS} curves are labeled *points 1, 2, 3, 4, 5,* and *6.* Each point corresponds to a different value of V_{GS}.

Point 5 is projected over to the left through the duplicated I_D vertical axis used for the transfer curve to the value of V_{GS}, in this case −2 V. This new point corresponds to point 5 on the load line and is also marked *5.* Each of the other points is projected over. Then, all the points are connected by a smooth curve. This resultant curve is the desired dynamic transfer characteristic.

It is obvious that this dynamic transfer characteristic is not a straight line. If we use only a very small segment, we can say that this small segment is a straight line. Therefore, when the input signal to the FET amplifier is very small, the amplifier is a *linear amplifier.* In a linear amplifier, the distortion produced in the output is negligible. If we use a large segment of the dynamic transfer characteristic, we have a noticeable curvature. Now the output current is not proportional to the input signal. Consequently, a large amount of signal distortion results.

Problems **9-2.1** The supply voltage is 20 V and the load resistance R_D is 5000 Ω. Determine the *Q*-point and the voltage gain when the circuit is biased at −2.0 V and the input signal has a peak value of 0.5 V.

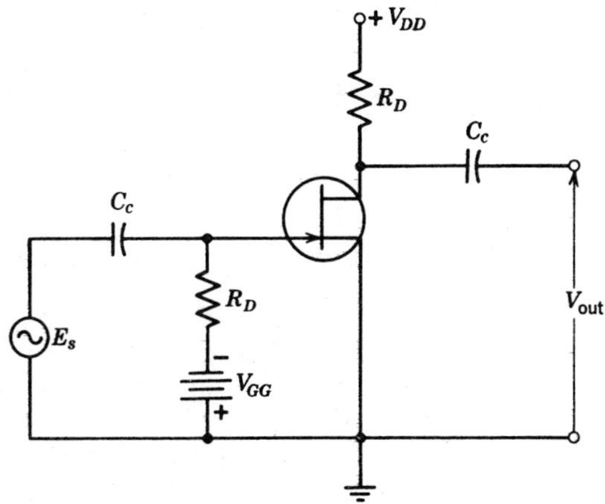

Circuit and characteristic for Problems 9-2.1 to 9-2.5.

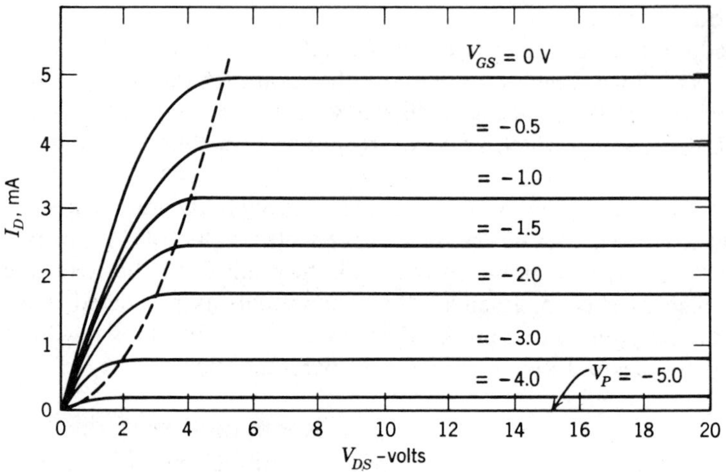

9-2.2 Repeat Problem 9-2.1 for a supply voltage of 15 V and a load resistance of 3000 Ω.

9-2.3 Repeat Problem 9-2.1 for a supply voltage of 20 V and a load resistance of 20 kΩ.

9-2.4 Construct the dynamic transfer characteristic for Problem 9-2.1.

9-2.5 Construct the dynamic transfer characteristic for Problem 9-2.2.

Section 9-3 The FET Amplifier Model

The equation obtained in Chapter 8 for the drain current in the FET of the amplifier shown in Fig. 9-9 is

$$I_D = I_{DSS}\left(1 - \frac{V_{GS}}{V_P}\right)^2 \tag{8-1}$$

If a small signal is applied to the gate, the gate voltage is

$$v_{GS} = V_{GS} + v_{gs}$$

and the resulting drain current is

$$i_D = I_D + i_d$$

The symbols used in these two equations are:

V_{GS} The dc voltage between gate and source—in this case, V_{GG}.

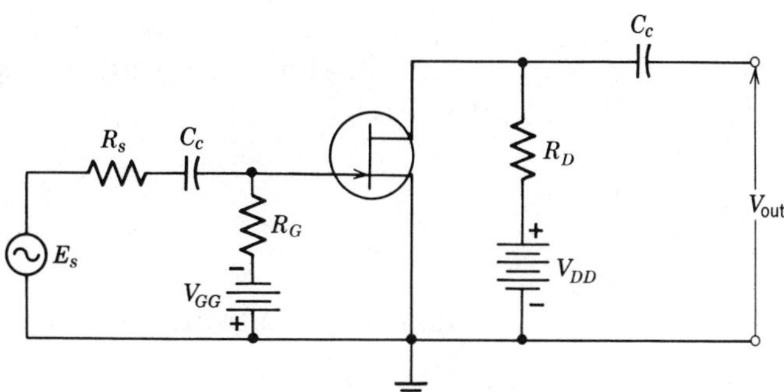

Figure 9-9 The FET amplifier.

v_{gs} The instantaneous value of the ac signal voltage between gate and source.

v_{GS} The instantaneous total voltage between gate and source, which is the sum of V_{GS} and v_{gs}.

I_D The dc current in the drain.

i_d The instantaneous value of the ac signal current in the drain.

i_D The instantaneous total current in the drain, which is the sum of I_D and i_d.

Substituting v_{GS} for V_{GS} and i_D for I_D in Eq. 8-1, we have

$$I_D + i_d = i_D = I_{DSS}\left(1 - \frac{V_{GS} + v_{gs}}{V_P}\right)^2$$

Rearranging, we have

$$I_D + i_d = i_D = I_{DSS}\left[\left(1 - \frac{V_{GS}}{V_P}\right) - \frac{v_{gs}}{V_P}\right]^2$$

Expanding, we have

$$I_D + i_d = I_{DSS}\left(1 - \frac{V_{GS}}{V_P}\right)^2 - 2I_{DSS}\left(1 - \frac{V_{GS}}{V_P}\right)\frac{v_{gs}}{V_P} + I_{DSS}\left(\frac{v_{gs}}{V_P}\right)^2$$

Subtracting Eq. 8-1 from this result removes the dc component (I_D) from the output:

$$i_d = -\frac{2I_{DSS}}{V_P}\left(1 - \frac{V_{GS}}{V_P}\right)v_{gs} + I_{DSS}\left(\frac{v_{gs}}{V_P}\right)^2$$

This expression contains signal components *only*.

In Chapter 8, we showed that the transconductance g_m for the FET is

$$g_m = -\frac{2I_{DSS}}{V_P}\left(1 - \frac{V_{GS}}{V_P}\right) \tag{8-4}$$

Making this substitution, we have

$$i_d = g_m v_{gs} + I_{DSS}\left(\frac{v_{gs}}{V_P}\right)^2 \tag{9-5}$$

The second term of Eq. 9-5 represents the distortion produced in the output of the FET caused by the curvature of the dynamic transfer characteristic. If v_{gs} is small as compared with V_P, this term can be neglected. Equation 9-5, can, then, be reduced to

$$i_d = g_m v_{gs} \quad \text{or} \quad I_d = g_m V_{gs} \tag{9-6}$$

From Fig. 9-9, the instantaneous total voltage at the drain is

$$\begin{aligned} v_{DS} = V_{DD} - R_D i_D &= V_{DD} - R_D(I_D + i_d) \\ &= V_{DD} - R_D I_D - R_D i_d \end{aligned}$$

If we remove the dc component $(V_{DD} - R_D I_D)$, we have v_{ds}, which is the output voltage v_{out}.

$$v_{ds} = v_{\text{out}} = -R_D i_d \tag{9-7a}$$

and substituting Eq. 9-6 into this last equation, we have

$$v_{\text{out}} = -g_m R_D v_{gs} \tag{9-7b}$$

or

$$V_{\text{out}} = -g_m R_D V_{gs} \tag{9-7c}$$

The negative sign in Eq. 9-7*c* *proves* that phase inversion occurs in this FET amplifier circuit just as we have phase inversion in the common-emitter transistor amplifier circuit.

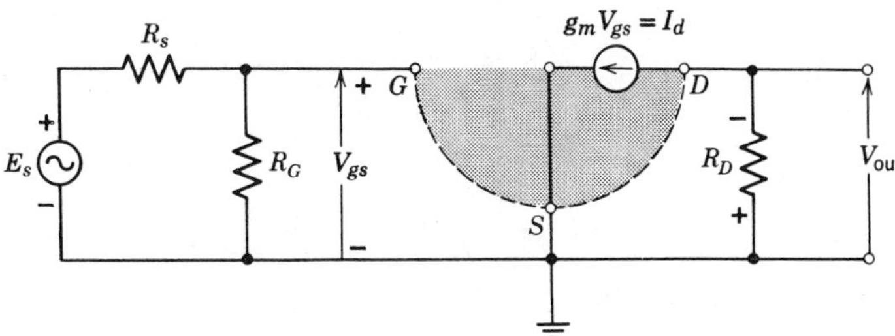

Figure 9-10 The formal model for the FET amplifier.

We use this fact when we form the formal model for the FET amplifier.

The formal model for the FET signal amplifier (Fig. 9-10) shows how Eq. 9-6, Eq. 9-7*a*, and Eq. 9-7*c* are related in a circuit diagram.

The current generator in the formal model for the transistor (BJT) amplifier has the label βI_b. The current generator in the formal model for the FET amplifier has the label $g_m V_{gs}$. Thus, the transistor (BJT) is a *current-controlled* device. The collector then is a *current-controlled current source.* The FET is a *voltage-controlled* device. The drain is a *voltage-controlled current source.*

The input resistance to the FET is sufficiently high that the gate is shown as a terminal on the model without a connection to the FET internally. Ordinarily R_s is very much smaller than R_G. As a result, for most FET amplifier circuits

$$V_{gs} = E_s \tag{9-8}$$

and

$$A_v = A_e \tag{9-9}$$

We recall that the negative sign merely indicates phase inversion. Dividing both sides of Eq. 9-7*c* by V_{gs}, we have the equation for voltage gain.

$$\boxed{A_v = A_e = g_m R_D} \tag{9-10}$$

The simplified model is shown in Fig. 9-11. This arrangement is the same as the one we used for the transistor amplifier circuits. Since this model has the same form as the simplified model for transistors, we find it convenient to

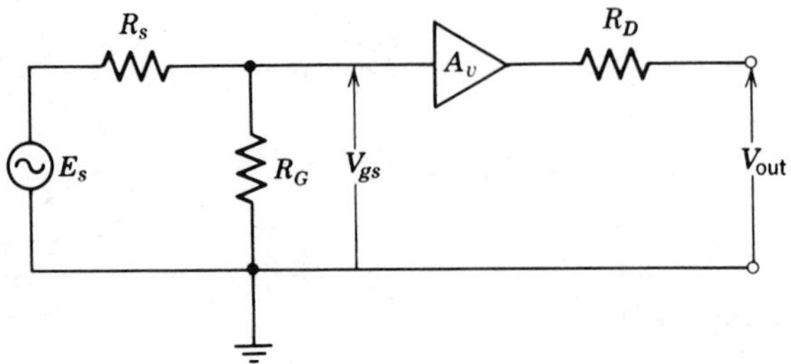

Figure 9-11 The simplified model for the FET amplifier.

convert Eq. 9-10 to the same form that we used for the common-emitter transistor amplifier circuits:

$$A_v = \frac{R_C}{r'_e} \tag{7-8a}$$

and

$$A_V = \frac{R_C}{r'_e + R_E} \tag{7-23}$$

in which R_E is the unbypassed emitter-feedback resistor.

If we compare Eq. 9-10 with Eq. 7-8a, we can define a new term r'_s.

$$r'_s \equiv \frac{1}{g_m}\,\Omega \quad \text{or} \quad g_m = \frac{1}{r'_s}\,S \tag{9-11}$$

Corresponding equations for the FET amplifier can be written

$$A_e = A_v = g_m R_D = \frac{R_D}{r'_s} \tag{9-12}$$

and

$$A_e = A_v = g'_m R_D = \frac{R_D}{r'_s + R_S} \tag{9-13}$$

in which

$$g'_m = \frac{1}{r'_s + R_S} = \frac{1}{\dfrac{1}{g_m} + R_S} \tag{9-14}$$

Example 9-5
Determine the gain and the output voltage for the JFET amplifier shown in Fig. 9-12.

Case I. With the switch closed.
Case II. With the switch opened.

Solution
Case I. The operating point for the circuit must be determined graphically. We must proceed using the procedures of Section 8-2. First, we require a transfer characteristic ($I_D - V_{GS}$). This data can be obtained from Eq. 8-1.

$$I_D = I_{DSS}\left(1 - \frac{V_{GS}}{V_P}\right)^2 = 10\left(1 - \frac{V_{GS}}{-4}\right)^2 \text{ mA} \qquad (8\text{-}1)$$

Using this equation, we obtain the values for I_D shown in Table 9-1.

Table 9-1
Values for Transfer Characteristic

V_{GS}, V	0	−0.5	−1.0	−1.5	−2.0	−3.0	−4.0
I_D, mA	10	7.7	5.6	3.9	2.5	0.6	0

These data points are plotted on the curve of Fig. 9-13. A bias line for 2000 Ω must be drawn at an offset value of +3 V (V_{GG}) at point *A*. A convenient current value *I* is assumed. If we take 3 mA for *I*, the voltage *V* across the bias resistor R_s is

$$V = IR_s = 0.003 \times 2000 = 6.0 \text{ V}$$

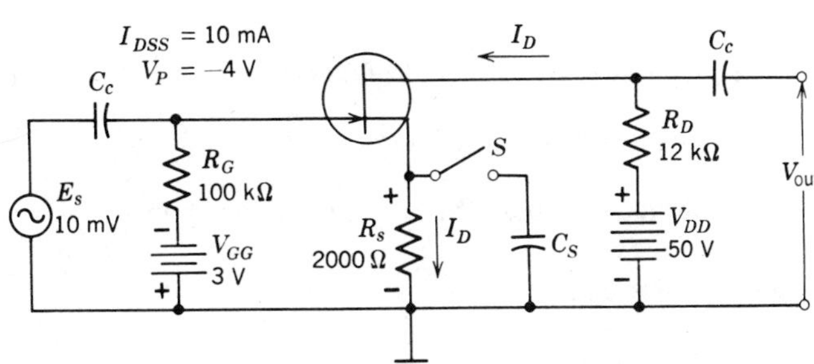

Figure 9-12 FET amplifier circuit with source feedback.

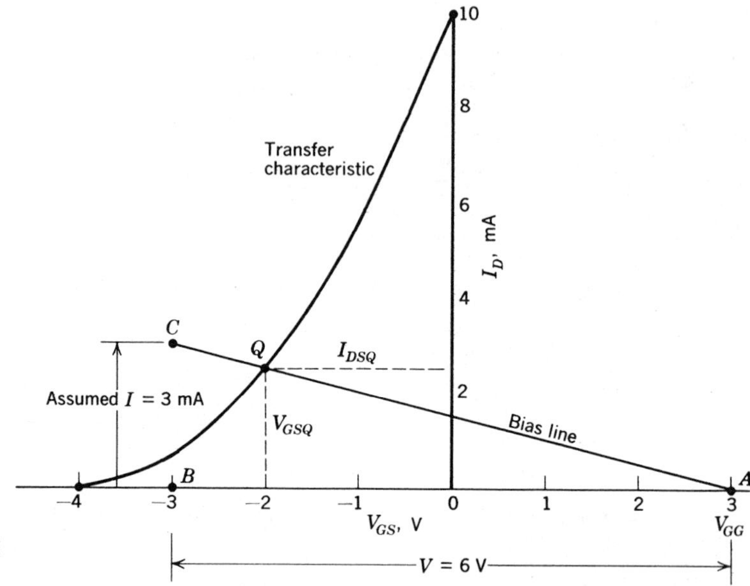

Figure 9-13 Transfer characteristic and bias line.

We move to the left 6 V (V) from point A to point B. From point B, we move up 3 mA (I) to point C. The bias line is drawn from point C to point A.

The operating point Q is the intersection of the bias line with the transfer characteristic.

$$V_{GSQ} = -2.0 \text{ V} \quad \text{and} \quad I_{DQ} = 2.5 \text{ mA}$$

The value of g_m at the operating point is obtained from Eq. 8-4.

$$g_m = -\frac{2I_{DSS}}{V_P}\left(1 - \frac{V_{GSQ}}{V_P}\right) = -\frac{2 \times 10 \text{ mA}}{-4 \text{ V}}\left(1 - \frac{-2 \text{ V}}{-4 \text{ V}}\right) = 2.5 \text{ mS} \tag{8.4}$$

The output voltage of the amplifier is

$$V_{\text{out}} = -g_m R_D V_{gs} = -(2.5 \times 10^{-3} \text{ S})(12{,}000 \ \Omega)10 \text{ V}$$
$$= -0.3 \text{ V} = \mathbf{-300 \ mV} \tag{9-7c}$$

The negative sign indicates that the circuit gives phase inversion.

The voltage gain of the circuit is

$$A_e = A_v = g_m R_D = (2.5 \times 10^{-3} \text{ S})(12{,}000 \ \Omega) = \mathbf{30} \tag{9-10}$$

Solution

Case II. We can use Eq. 9-11 to obtain r'_s, the equivalent resistance of g_m.

$$r'_s = \frac{1}{g_m} = \frac{1}{2.5 \times 10^{-3}\ \text{S}} = 400\ \Omega \qquad (9\text{-}11)$$

This 400-Ω resistance is in series with R_S when the switch is open. The equivalent transconductance produced by this series combination is

$$g'_m = \frac{1}{r'_s + R_S} = \frac{1}{400\ \Omega + 2000\ \Omega} = 4.17 \times 10^{-4}\ \text{S}$$

$$= 0.417\ \text{mS} \qquad (9\text{-}14)$$

Now the voltage gain of the circuit is

$$A_e = A_v = g'_m R_D = (4.17 \times 10^{-4}\ \text{S})(12{,}000\ \Omega) = \mathbf{5} \qquad (9\text{-}13)$$

and the output voltage is

$$V_{\text{out}} = A_e V_{\text{in}} = 5 \times 10\ \text{mV} = \mathbf{50\ mV} \qquad (7\text{-}1a)$$

There is one additional topic that must be considered for the FET amplifier. The drain characteristics show a significant rise as V_{DS} is increased (Fig. 9-14). The value of r_d is often not negligible. The drain resistance r_d is defined at point Q as

$$r_d \equiv \frac{\Delta V_{DS}}{\Delta I_D} \qquad (9\text{-}15)$$

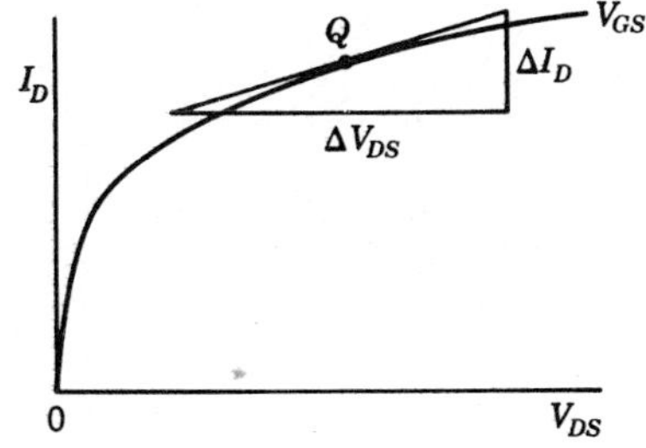

Figure 9-14 Determination of r_d, drain resistance.

The value of r_d is given in the specification sheet issued by the manufacturer. Some specifications give values for y_{os}. The reciprocal of the real part of y_{os} is r_d.

The value of r_d appears in the formal model as a resistor connected between D and S in Fig. 9-10. Now r_d is in parallel with R_D. Therefore, if we form the parallel combination R_D' of r_d and R_D

$$R_D' = \frac{r_d R_D}{r_d + R_D} \tag{9-16}$$

Equations 9-10, 9-12, and 9-13 are used with this new value of R_D'.

Problems The model is required for all problems.

9-3.1 The values for the circuit are

$$V_{GG} = -2\text{ V} \qquad V_{DD} = +20\text{ V} \qquad I_{DSS} = 6\text{ mA}$$
$$V_P = -4\text{ V} \qquad V_{DS} = 10\text{ V}$$

Determine R_D and V_{out}.

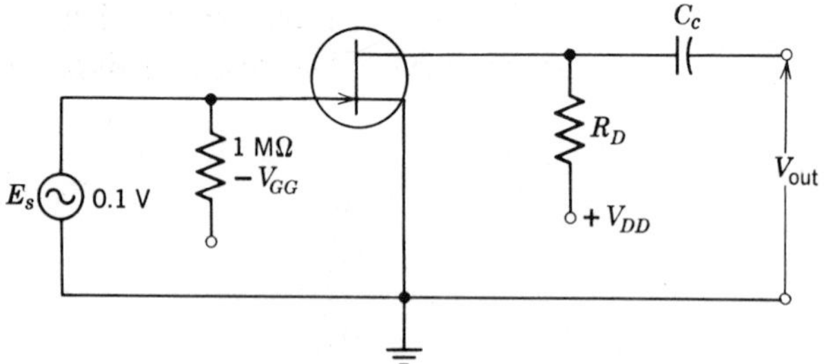

Circuit for Problems 9-3.1 and 9-3.2.

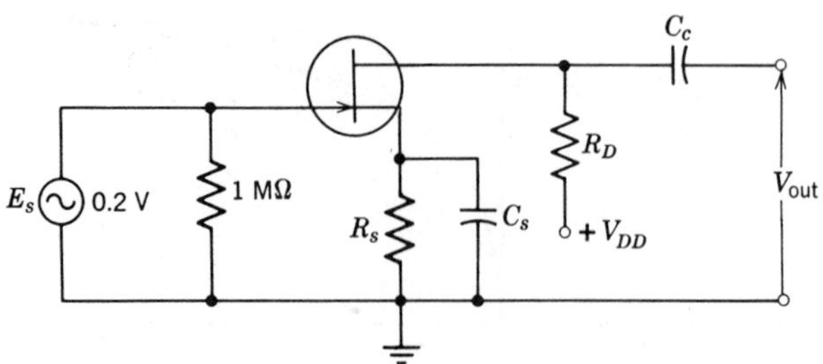

Circuit for Problems 9-3.3 and 9-3.4.

9-3.2 The values for the circuit are

$$V_{GG} = -3\text{ V} \qquad V_{DD} = +10\text{ V} \qquad I_{DSS} = 10\text{ mA}$$
$$V_P = -4\text{ V} \qquad V_{DS} = 6\text{ V}$$

Determine R_D and V_{out}.

9-3.3 The values for the JFET are

$$V_P = -5\text{ V} \quad \text{and} \quad I_{DSS} = 20\text{ mA}$$

Determine the value of R_S to establish I_{DQ} at 10 mA. The value of V_{DD} is +20 V and R_D is 1000 Ω. What is V_{out}? What is V_{out} if C_S is removed from the circuit?

9-3.4 In Problem 9-3.3, the JFET fails and is replaced by a JFET having a V_P of −4 V and an I_{DSS} of 16 mA. The other component values remain unchanged. Now, what are the values of V_{out} with and without C_S?

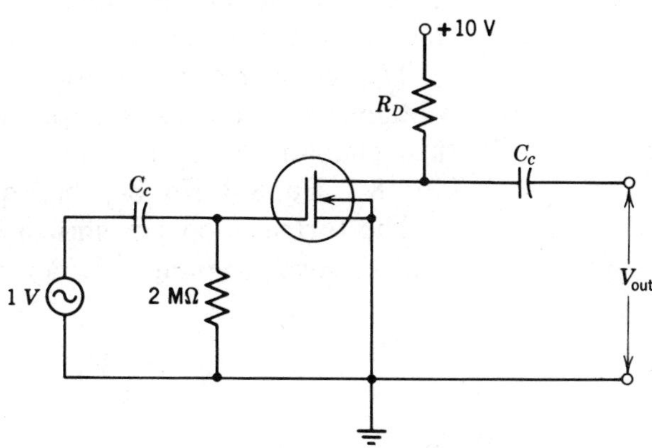

Circuit for Problems 9-3.5 and 9-3.6.

9-3.5 The equation for the current in the N-channel depletion-type MOSFET is

$$I_D = 2.0\left(1 - \frac{V_{GS}}{V_P}\right)^2 \text{ mA}$$

R_d is 2.5 kΩ, V_P is −4 V, and r_d is 10 kΩ. Find V_{DS}, g_m, and V_{out}.

9-3.6 The MOSFET of Problem 9-3.5 is operated in a circuit with a self-bias arrangement of R_S suitably bypassed with a capacitor C_S to operate the circuit at V_{GS} equal to −2.0 V. What is the value of R_S? What is the value of V_{DS}? Determine the circuit gain. Draw the circuit. R_D and r_d are each 10 kΩ.

9-3.7 An N-channel enhancement-type MOSFET is used in an amplifier with a 10 kΩ load. The supply voltage is +20 V and the signal input is 200 mV. The equation for the drain current for the MOSFET is

$$I_D = 1.2(V_{GS} - V_T)^2 \text{ mA}$$

where V_T is 1.0 V.
The amplifier is operated at V_{GS} equal to +2.0 V.
Determine V_{DS}, g_m, and the gain of the circuit. Draw the circuit.

9-3.8 Repeat Problem 9-3.7 if r_d is 15 kΩ.

9-3.9 Repeat Problem 9-3.7 if the MOSFET is operated at a quiescent point of I_D equal to 0.8 mA.

Section 9-4 The Source Follower

The circuit for the *source follower* is shown in Fig. 9-15*a*. This circuit is also referred to as the *common-drain amplifier*. The configuration is equivalent to the transistor version, the emitter follower (the common-collector amplifier).

The operating point for the circuit (Fig. 9-15*a*) can be obtained by the method explained in developing Fig. 8-4 in the last chapter. Then the value for g_m can be determined from Eq. 8-3, Eq. 8-4, Eq. 8-6, or Eq. 8-8.

The formal model is shown in Fig. 9-15*b*. The resistors R_G and R_s form a voltage divider across E_s. Then

$$V_{\text{in}} = \frac{R_G}{R_s + R_G} E_s$$

The output voltage is the signal voltage drop across R_S.

$$V_{\text{out}} = I_d R_S = g_m R_S V_{gs}$$

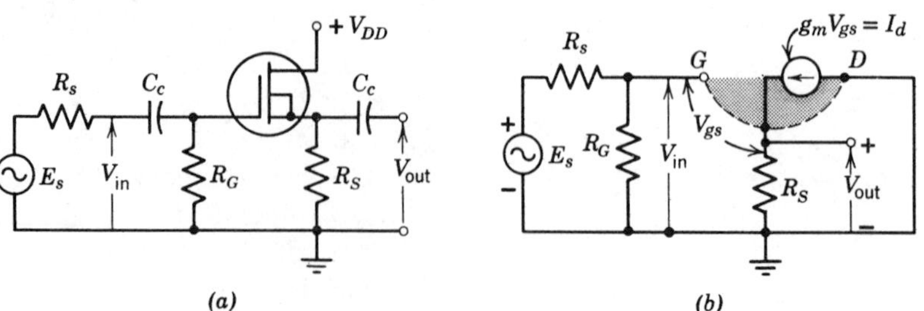

Figure 9-15 The source follower. (a) Actual circuit. (b) The formal model.

From inspection of the formal model, we see that

$$V_{\text{in}} = V_{gs} + V_{\text{out}}$$

Substituting, we find

$$V_{\text{in}} = V_{gs} + g_m R_S V_{gs} = (1 + g_m R_S) V_{gs}$$

The in-phase voltage gain across the FET is

$$A_v = \frac{V_{\text{out}}}{V_{\text{in}}} = \frac{g_m R_S V_{gs}}{(1 + g_m R_S) V_{gs}}$$

$$\boxed{A_v = \frac{g_m R_S}{1 + g_m R_S} < 1} \tag{9-17}$$

and

$$A_e = \frac{R_G}{R_s + R_G} A_v = \frac{R_G}{R_s + R_G} \times \frac{g_m R_S}{1 + g_m R_S} \tag{9-18}$$

The characteristics of this circuit, its advantages and its disadvantages, are very similar to the properties of the emitter follower.

Problems **9-4.1** R_S is 500 Ω. Determine I_D and V_{out}.
9-4.2 R_S is 1000 Ω. Determine I_D and V_{out}.
9-4.3 R_S is 500 Ω. Determine I_D and V_{out}.
9-4.4 R_S is 1000 Ω. Determine I_D and V_{out}.

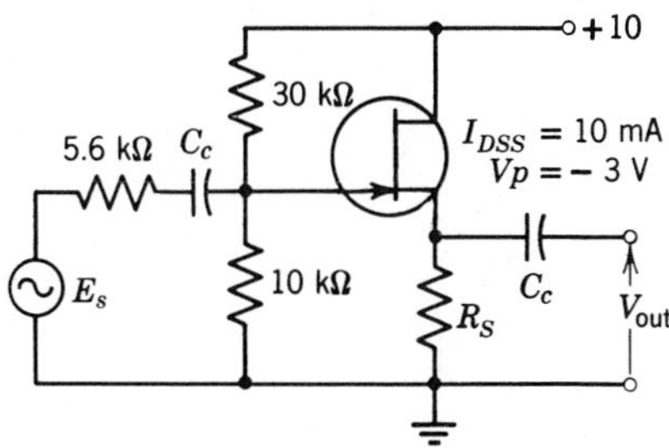

Circuit for Problems 9-4.1 and 9-4.2.

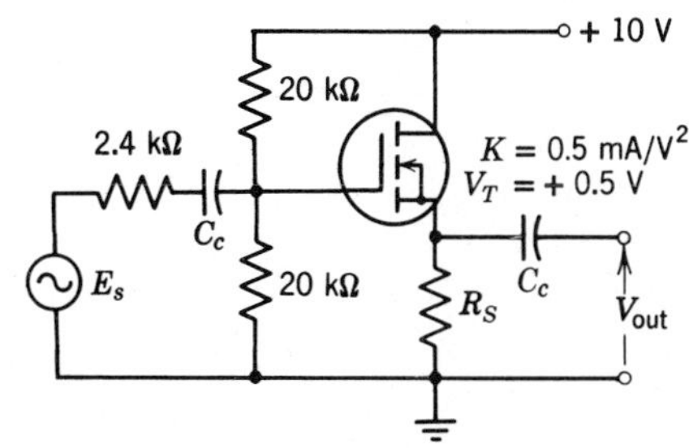

Circuit for Problems 9-4.3 and 9-4.4.

Supplementary Problems

9-1 If V_{GG} is 2 V and R_S is zero, what are the values for I_D and V_{DS}?

9-2 If R_S is zero and V_{DS} is 12 V, what are the values for I_D and V_{GG}?

9-3 If V_{GG} is zero and R_S is 1000 Ω, what are the values for I_D and V_{DS}?

9-4 If V_{GG} is zero and R_S is 2000 Ω, what are the values for I_D and V_{DS}?

9-5 R_S is 500 Ω. What are the values for I_D and V_{DS}?

9-6 If R_S is 1000 Ω, what are the values for I_D and V_{DS}?

9-7 What is A_e for the circuit of Problem 9-1?

9-8 What is A_e for the circuit of Problem 9-3?

9-9 What is A_e for the circuit of Problem 9-3 if C_S is removed?

9-10 What is A_e for the circuit of Problem 9-4 if C_S is removed?

9-11 What is A_e for the circuit of Problem 9-6?

9-12 What is A_e for the circuit of Problem 9-6 if C_S is removed?

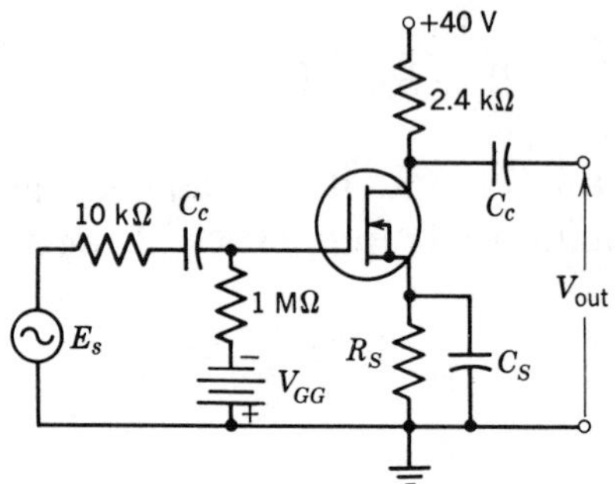

I_{DSS} = 20 mA; V_P = − 6 V; r_d = 10 kΩ

Circuit for Problems 9-1 through 9-4 and 9-7 through 9-10.

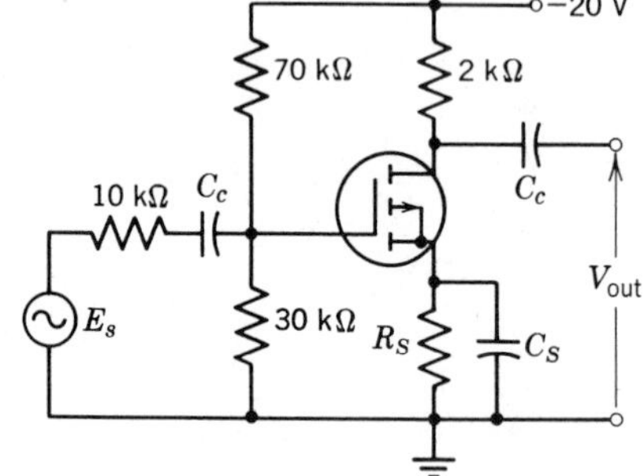

V_T = − 0.8 V; K = 2 mA/V²; r_d = 10 kΩ

Circuit for Problems 9-5, 9-6, 9-11, and 9-12.

Chapter 10 Stability and Compensation

The reasons and needs for considering the effect of a variation between the characteristics of different transistors that have the same type number are explained. A definition is formed for beta stability, K (Section 10-1). A method for determining K is developed for the transistor circuits considered in Chapter 5 and in Chapter 7 (Section 10-2). Leakage current in a transistor is materially increased with an increase in temperature (Section 10-3). The temperature sensitivity S is defined to show the effect of an increase in leakage current on the collector current. A method of determining S is developed for the transistor circuits used in Chapter 5 and in Chapter 7 (Section 10-4). The FET amplifier can be made less dependent on variations in I_{DSS} and in V_P (Section 10-5). Many circuits, especially integrated circuits and operational amplifiers, use diodes to bias a circuit and, at the same time, to compensate for temperature effects (Section 10-6).

Section 10-1 General Concepts of Beta Stability

The mass production methods of the manufacture of electronics equipment for the home consumer market, in particular, can easily result in the construction of many thousands of copies of the same circuit. An electronic item used in the automotive industry could approach a production run of a million units. Let us assume that two transistor types are available for the same circuit application. One transistor has its variation of β held to $100 \pm 10\%$ and costs 30¢ each in lots of 10,000. The other has a spread in β from 50 to 150 and costs 6¢ each in lots of 10,000. A method that can utilize the cheaper transistor results in the savings of \$24,000 in the cost of one transistor in a production run of 100,000 units. If the circuit has many transistors, a considerable sum of money is involved.

This same concept extends to servicing. If a transistor is replaced, the overall operation of the equipment should not change radically when the replacement transistor has a different value of β.

When the application becomes very critical—for instance, in space or military equipment—the design often requires semiconductors that are controlled within very close tolerances. Also, the production runs are usually not very large. Consequently, the cost factor takes a place that is secondary to performance and reliability.

Consider the simple amplifier that is shown in Fig. 10-1*a*. The base current is found from

$$V_{BB} = R_B I_B + V_{BE}$$

(*a*)

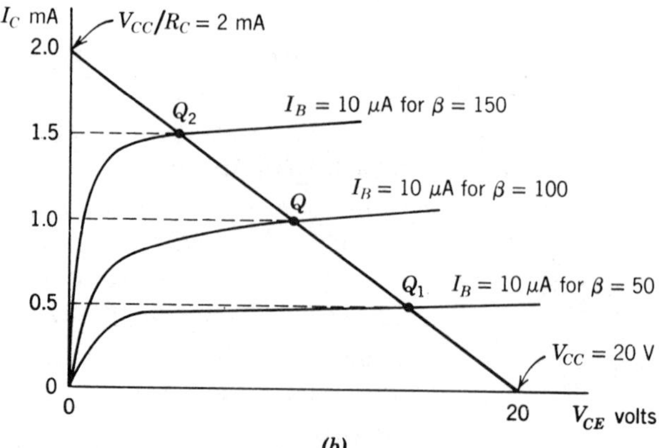

(*b*)

Figure 10-1 Effect of change in β. (*a*) Circuit. (*b*) Load line.

If we neglect V_{BE}

$$V_{BB} = R_B I_B$$

and the base current is given by Ohm's law.

$$I_B = \frac{V_{BB}}{R_B} = \frac{20\text{ V}}{10\text{ M}\Omega} = 10\ \mu\text{A}$$

If V_{BE} is not neglected and is 0.7 V, the base current would be

$$I_B = \frac{V_{BB} - V_{BE}}{R_B} = \frac{20\text{ V} - 0.7\text{ V}}{2\text{ M}\Omega} = 9.65\ \mu\text{A}$$

This calculation shows that the maximum conceivable change of I_B in this circuit caused by a variation of V_{BE} is 0.35 μA or about ±2% from a center value. We are concerned with a transistor that has a variation in β from 50 to 150. We are not concerned with the contribution of any small variation in V_{BE} that may occur from transistor to transistor.

Now let us assume that the nominal value of β for the transistor is 100. The nominal collector current is

$$I_C = \beta I_B = 100 \times 10 = 1000\ \mu\text{A} = 1\text{ mA}$$

This value of I_C locates the operating point of the circuit at Q (Fig. 10-1*b*). The least value for β expected in a sample of this particular transistor type is 50. The resulting collector current is

$$I_C = \beta I_B = 50 \times 10 = 500\ \mu\text{A} = 0.5\text{ mA}$$

which is Q_1 on the load line.

The maximum expected value of β is 150, and the resulting collector current is

$$I_C = \beta I_B = 150 \times 10 = 1500\ \mu\text{A} = 1.5\text{ mA}$$

which is Q_2 on the load line.

When the β of a transistor decreases, the family of curves *shrinks* toward the horizontal axis. When β is high, the family of curves *spreads* upward. The operating point of the circuit shifts materially along the load line. What is an acceptable signal swing at the mid-value Q, obviously, could produce cutoff at Q_1 and saturation at Q_2.

The objective of this analysis is to investigate this shift of operating point. From our numerical values, we take β as the nominal mid-value. The $+\Delta\beta$ is the upward increase in β that produces an upward increase $+\Delta I_C$ in collector current. Then $-\Delta\beta$ is the downward change in β that results in a decrease $-\Delta I_C$ in collector current.

The definition of *beta stability* K is

$$\boxed{K \equiv \left(\frac{\Delta I_C}{I_C}\right) \Big/ \left(\frac{\Delta\beta}{\beta}\right)} \tag{10-1}$$

where

$$0 \le K \le 1$$

Equation 10-1 can be rearranged to the form

$$\boxed{\left(\frac{\Delta I_C}{I_C}\right) = K\left(\frac{\Delta\beta}{\beta}\right)} \tag{10-2a}$$

Equation 10-2*a* can be expressed in words:

The percentage change in I_C is K times the percentage change in β (10-2*b*)

If K is zero, a change in β produces no change in I_C. This is ideal. The worst case is a value for K of unity. Then a particular percentage change in β produces the same percentage change in I_C.

In the example we used for Fig. 10-1, we have

$$\beta = 100 \qquad I_C = 1.0\text{ mA}$$

and

$$\Delta\beta = \pm 50 \qquad \Delta I_C = \pm 0.5\text{ mA}$$

Substituting into Eq. 10-2*a*, we have

$$\frac{\pm 0.5}{1.0} = K\frac{\pm 50}{100}$$

or

$$K = 1$$

The circuit shown in Fig. 10-1*a* is the "worst-case" condition for beta stability. We will next proceed to examine other amplifier circuits to show how we can obtain values of K that are less than unity.

Problems

10-1.1 In Fig. 10-1, R_B is 300 kΩ, R_C is 2 kΩ, and β is 50. The transistor is silicon and the supply voltage is −20 V. What is the Q-point and what is the new Q-point if β is doubled? Sketch the shift on the load line. Show that K is unity.

10-1.2 In Fig. 10-1, R_B is 10 kΩ, R_C is 75 Ω, and β is nominally 100. The transistor is an *NPN* unit and the supply voltage is +3 V. If β varies from 50 to 150, what is the shift of the Q-point from the nominal value? Show this shift on a sketch on the load line for the silicon transistor. Show that K is unity.

Section 10-2 Beta Stability Circuit Analysis

Let us investigate the beta stability of the circuit shown in Fig. 10-2. *We are concerned with the stability of the operating point.* The operating point of the circuit is determined by the

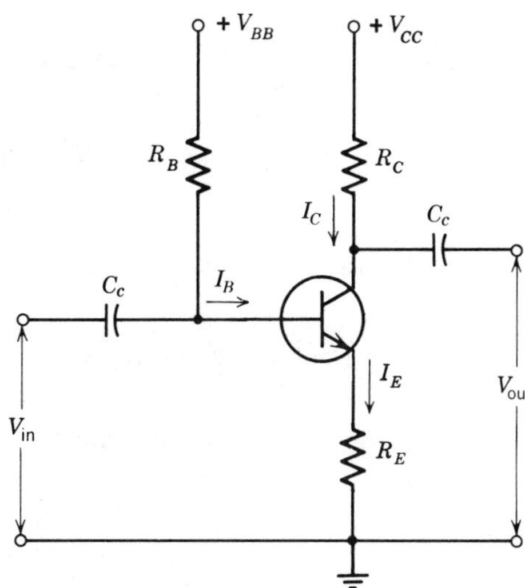

Figure 10-2 Circuit using emitter resistor.

dc circuit analysis. A bypass capacitor placed in parallel with R_E will not change the operating point. The use of a bypass capacitor changes the input impedance and the gain of the amplifier, which are factors in the ac circuit analysis. Thus, if a circuit has a bypass capacitor, it is ignored in a calculation for beta stability.

The Kirchhoff's voltage loop equation through the base resistor R_B is

$$V_{BB} = I_B R_B + V_{BE} + I_E R_E$$

Rearranging and substituting $(I_B + I_C)$ for I_E, we have

$$V_{BB} - V_{BE} = I_B R_B + (I_B + I_C)R_E = (R_B + R_E)I_B + R_E I_C$$

Substituting I_C/β for I_B, we find that

$$V_{BB} - V_{BE} = (R_B + R_E)\frac{I_C}{\beta} + R_E I_C \qquad (10\text{-}3a)$$

If the transistor is replaced with a unit that has a new value of beta $(\beta + \Delta\beta)$ the collector current becomes $(I_C + \Delta I_C)$. Substituting these new values into Eq. 10-3*a*, we have

$$V_{BB} - V_{BE} = (R_B + R_E)\frac{I_C + \Delta I_C}{\beta + \Delta\beta} + R_E(I_C + \Delta I_C) \quad (10\text{-}3b)$$

Subtract Eq. 10-3*a* from Eq. 10-3*b*.

$$(R_B + R_E)\frac{I_C + \Delta I_C}{\beta + \Delta\beta} - (R_B + R_E)\frac{I_C}{\beta} + (R_E \Delta I_C) = 0$$

Clearing fractions and collecting terms yields

$$[\beta(R_B + R_E) + \beta R_E(\beta + \Delta\beta)]\Delta I_C = (R_B + R_E)I_C \Delta\beta$$

Solving for $\Delta I_C/I_C$, we have

$$\frac{\Delta I_C}{I_C} = \frac{R_B + R_E}{R_B + R_E + \beta R_E + \Delta\beta R_E}\left(\frac{\Delta\beta}{\beta}\right)$$

Dividing through by $(R_B + R_E)$, we have

$$\frac{\Delta I_C}{I_C} = \frac{1}{1 + (\beta + \Delta\beta)\dfrac{R_E}{R_E + R_B}} \left(\frac{\Delta\beta}{\beta}\right) \tag{10-4}$$

If we compare Eq. 10-4 with Eq. 10-2a, we see that

$$\boxed{K = \frac{1}{1 + (\beta + \Delta\beta)\dfrac{R_E}{R_E + R_B}}} \tag{10-5}$$

If we were to make formal algebraic derivations for K for all the transistor circuits we use, we would devote quite a few pages for the purpose. Instead, we will summarize the results of the lengthy derivations.

1. In the circuit shown in Fig. 10-1, there is no external emitter resistance in the circuit. Consequently R_E is zero and the substitution of zero for R_E into Eq. 10-5 yields a value of 1 for K.
2. In an emitter-follower amplifier circuit, we use the values of R_B and R_E directly into Eq. 10-5.
3. In a circuit that uses a voltage divider to obtain the bias (Fig. 10-3a) we use Thévenin's theorem to obtain a value for R_B.

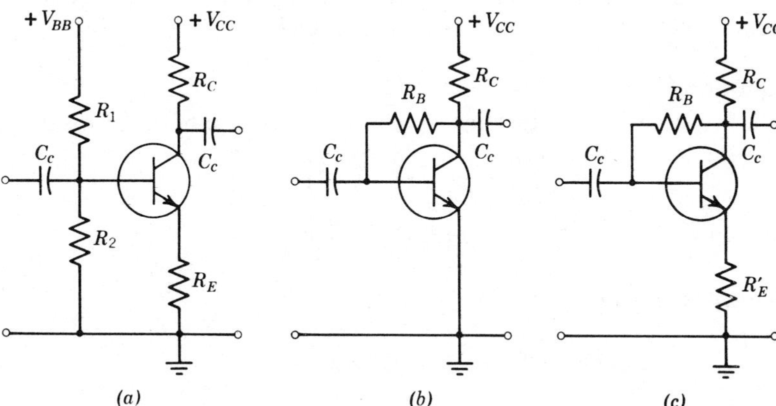

Figure 10-3 Transistor circuit arrangements. (a) Bias obtained from voltage divider. (b) Collector-to-base feedback. (c) Collector-to-base feedback with emitter feedback.

$$R_B = \frac{R_1 R_2}{R_1 + R_2} \qquad (10\text{-}6a)$$

4. In a circuit that uses collector-to-base feedback (Fig. 10-3*b*) use

$$R_E = R_C \qquad (10\text{-}6b)$$

5. In a circuit that uses collector-to-base feedback in addition to emitter feedback (Fig. 10-3*c*) use

$$R_E = R_C + R_E' \qquad (10\text{-}6c)$$

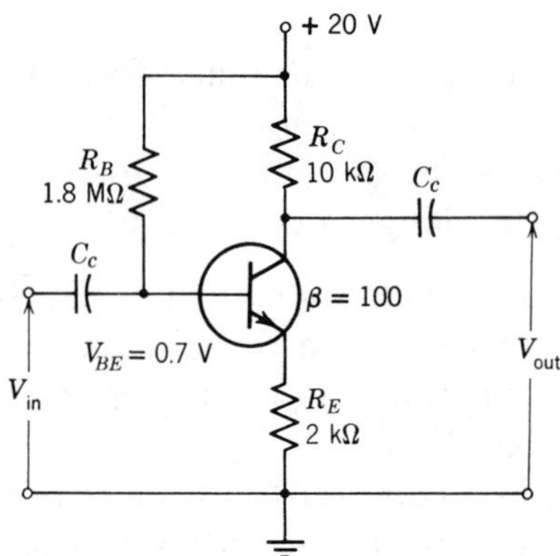

Circuit for Example 10-1.

Example 10-1

Determine the operating point for the circuit. A transistor with a beta of 150 is substituted. Determine the K and the new operating point. Determine the percentage change in I_C.

Solution

The Kirchhoff's voltage loop equation through the base circuit is

$$V_{BB} = R_B I_B + V_{BE} + R_E I_E$$

or

$$V_{BB} = R_B I_B + V_{BE} + R_E(1 + \beta) I_B$$

Substituting values, we have

$$20\text{ V} = 1800\text{ k}\Omega \times I_B + 0.7\text{ V} + 2\text{ k}\Omega \times (101) \times I_B$$

Solving, we find

$$I_B = 9.64\ \mu\text{A}$$

then

$$I_C = \beta I_B = 0.964\text{ mA}$$

and

$$I_E = (1+\beta)I_B = 0.974\text{ mA}$$

The Kirchhoff's voltage loop equation through the collector circuit is

$$V_{CC} = R_C I_C + V_{CE} + R_E I_E$$

Substituting values, we have

$$20\text{ V} = 10\text{ k}\Omega \times 0.964\text{ mA} + V_{CE} + 2\text{ k}\Omega \times 0.974\text{ mA}$$

$$V_{CE} = 8.4\text{ V}$$

Therefore the operating point using the transistor with a beta of 100 is

$$\mathbf{I_{CQ} = 0.964\text{ mA}} \quad \text{and} \quad \mathbf{V_{CEQ} = 8.4\text{ V}}$$

The new beta of 150 compared to the original beta of 100 gives a value of 50 for $\Delta\beta$. Referring to Fig. 10-3*b* and using Eq. 10-5, we have

$$K = \frac{1}{1+(\beta+\Delta\beta)\left(\dfrac{R_E}{R_E+R_B}\right)} = \frac{1}{1+150\left(\dfrac{2\text{ k}\Omega}{2\text{ k}\Omega+1800\text{ k}\Omega}\right)} = \mathbf{0.857} \tag{10-5}$$

By the definition of K, Eq. 10-1

$$\frac{\Delta I_C}{I_C} = K\frac{\Delta\beta}{\beta} \tag{10-1}$$

Substituting values, we find

$$\frac{\Delta I_C}{0.964} = 0.857\frac{50}{100}$$

Solving, we have

$$\Delta I_C = 0.413\text{ mA}$$

Then the new operating point, I'_{CQ} is

$$I'_{CQ} = I_C + \Delta I_C = 0.964\text{ mA} + 0.413\text{ mA} = 1.377\text{ mA}$$

The new value of V'_{CE} can be obtained from

$$V_{CC} = R_C I_C + V'_{CE} + R_E I_E$$

or

$$V_{CC} = R_C I_C + V'_{CE} + R_E \frac{1+\beta}{\beta} I_C$$

Substituting numerical values, we find

$$20 \text{ V} = 10 \text{ k}\Omega \times 1.377 \text{ mA} + V'_{CE} + 2 \text{ k}\Omega \times \left(\frac{151}{150}\right) 1.377 \text{ mA}$$

$$V'_{CE} = 3.46 \text{ V}$$

Then the new operating point using a transistor with a beta of 150 is

$$\mathbf{I'_{CQ} = 1.377 \text{ mA}} \quad \text{and} \quad \mathbf{V'_{CEQ} = 3.46 \text{ V}}$$

The percentage change in I_C is

$$\frac{I'_{CQ} - I_{CQ}}{I_{CQ}} 100 = \frac{1.377 \text{ mA} - 0.964 \text{ mA}}{0.964 \text{ mA}} 100 = \mathbf{42.8\%}$$

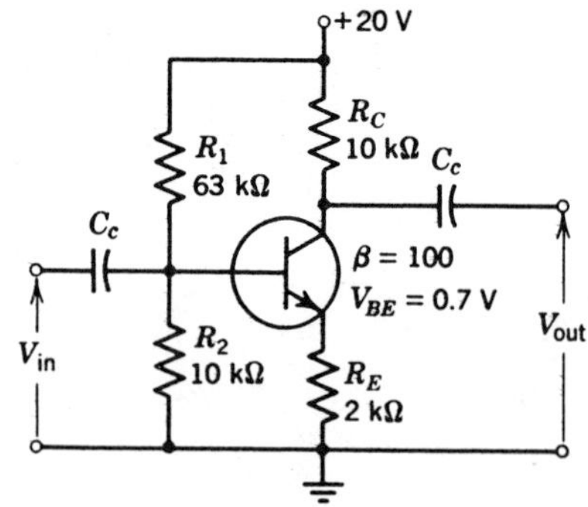

Circuit for Example 10-2.

Example 10-2

A transistor with a beta of 150 is substituted into the circuit. Determine the value of K and the new operating point. Determine the percentage change in I_C.

Solution

The values for R_1 and for R_2 have been selected to give the same operating point for a transistor with a beta of 100 that we determined in Example 10-1.

$$I_{CQ} = 0.964\text{ mA} \quad \text{and} \quad V_{CEQ} = 8.4\text{ V}$$

The equivalent value of R_B using Thévenin's theorem for the voltage divider is

$$R_B = \frac{R_1 R_2}{R_1 + R_2} = \frac{63\text{ k}\Omega \times 10\text{ k}\Omega}{63\text{ k}\Omega + 10\text{ k}\Omega} = 8.63\text{ k}\Omega \qquad (10\text{-}6a)$$

Using this value for R_B and a value of 50 for $\Delta\beta$, we have

$$K = \frac{1}{1 + (\beta + \Delta\beta)\dfrac{R_E}{R_E + R_B}} = \frac{1}{1 + 150\dfrac{2\text{ k}\Omega}{2\text{ k}\Omega + 8.63\text{ k}\Omega}} = \mathbf{0.034} \qquad (10\text{-}5)$$

By the definition of K, Eq. 10-1, we have

$$\frac{\Delta I_C}{I_C} = K\frac{\Delta\beta}{\beta} \qquad (10\text{-}1)$$

Substituting values, we find

$$\frac{\Delta I_C}{0.964\text{ mA}} = 0.034\frac{50}{100}$$

Solving, we have

$$\Delta I_C = 0.016\text{ mA}$$

Then the new operating point I'_{CQ} is

$$I'_{CQ} = I_C + \Delta I_C = 0.964 + 0.016 = 0.980\text{ mA}$$

The new value of V'_{CE} can be obtained from

$$V_{CC} = R_C I_C + V'_{CE} + R_E \frac{1+\beta}{\beta} I_C$$

$$20\text{ V} = 10\text{ k}\Omega \times 0.980\text{ mA} + V'_{CE} + 2\text{ k}\Omega\left(\frac{151}{150}\right)0.980\text{ mA}$$

$$V'_{CE} = 8.23\text{ V}$$

The new operating point using a transistor with a beta of 150 is

$$\mathbf{I'_{CQ} = 0.980\text{ mA}} \quad \text{and} \quad \mathbf{V'_{CEQ} = 8.23\text{ V}}$$

The original operating point for a beta of 100 was

$$I_{CQ} = 0.964 \text{ mA} \quad \text{and} \quad V_{CEQ} = 8.4 \text{ V}$$

The percentage change in I_C is

$$\frac{I'_{CQ} - I_{CQ}}{I_{CQ}} 100 = \frac{0.980 - 0.964}{0.964} 100 = \mathbf{1.7\%}$$

When we used the simple amplifier circuit of Fig. 10-1, a change of 50% in β caused the operating point I_{CQ} to shift 50% since K is unity. When we added the emitter-feedback resistor used in Example 10-1, a 50% change in β reduced the change in the operating point I_{CQ} to 42.8%. When we used a voltage divider arrangement in the base circuit, Example 10-2, a 50% change in β results in the very low change of 1.7% in the operating point I_{CQ}.

A comparison of these three sets of operating-point data clearly shows the importance of considering beta stability K in a circuit that is to be used for a production run. It also shows that, when we change a transistor in an amplifier, we may find that the characteristics of the amplifier have changed considerably.

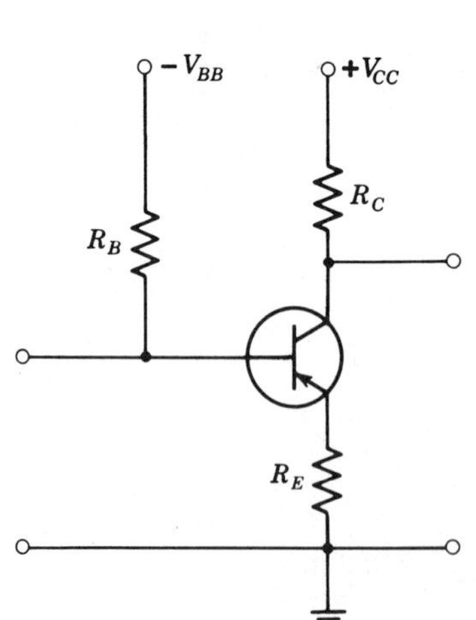

Circuit for Problems 10-2.3 and 10-2.4.

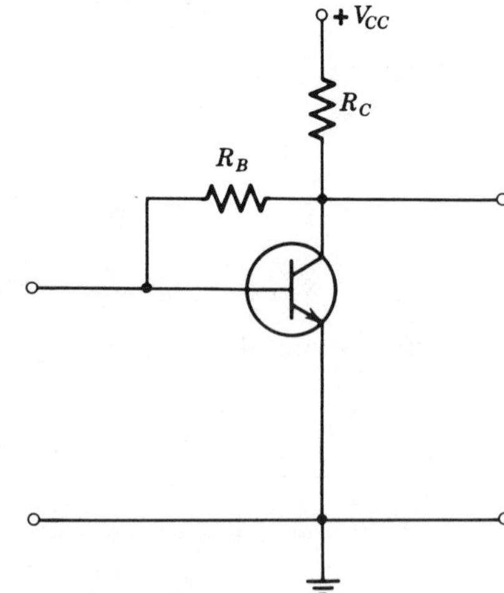

Circuit for Problems 10-2.6 and 10-2.7.

Problems **10-2.1** Repeat Problem 10-1.1 making use of Eq. 10-1.

10-2.2 Repeat Problem 10-1.2 making use of Eq. 10-1.

10-2.3 R_B is 300 kΩ, R_C is 2000 Ω, and R_E is 1000 Ω. The silicon

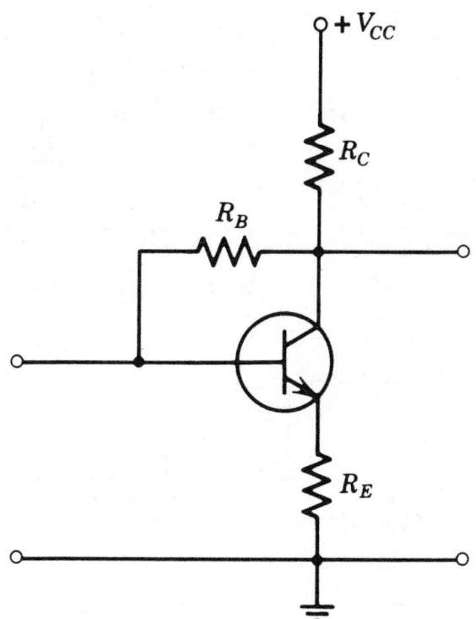

Circuit for Problems 10-2.8 and 10-2.9.

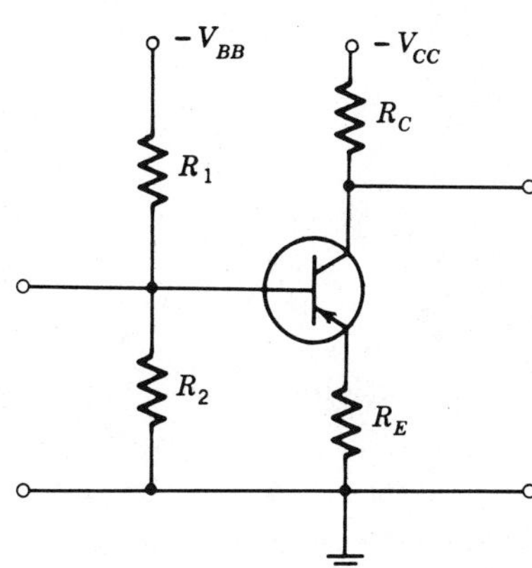

Circuit for Problems 10-2.10 and 10-2.11.

transistor has a β of 50, and the supply voltage is 20 V. Determine the operating point and K. If the range of variation in β is from 30 to 75, what is the change in I_C and what is the change in V_{CE}?

10-2.4 R_B is 750 kΩ, R_C is 3.6 kΩ, and R_E is 2000 Ω. The silicon transistor has a β of 100 and the supply voltage is 10 V. Calculate K and the change in I_C for a variation in β of ±20%.

10-2.5 Use the data given in Problem 10-2.4. If the maximum allowable shift in operating current for a particular application is ±20%, what range of β is acceptable for the transistor?

10-2.6 V_{CC} is 20 V, R_C is 3.9 kΩ, R_B is 390 kΩ, and β is 100 for the silicon transistor. Determine the Q-point and K. If a transistor with a β of 150 is substituted, what is the new value of I_C?

10-2.7 If a transistor with a β of 50 is used in the circuit of Problem 10-2.6, what is the new value of I_C?

10-2.8 V_{CC} is 10 V, V_{CE} is 4 V, R_E is 1500 Ω, and I_C is 1 mA. V_{BE} is 0.7 V, and β is 100. Find R_C and R_B. What is K for the circuit? If a transistor having a β of 80 is substituted, what is the new value of I_C and what is the new value of V_{CE}?

10-2.9 V_{CC} is 10 V, R_C is 4 kΩ, R_B is 750 kΩ, R_E is 2000 Ω, β is 100, and V_{BE} is 0.7 V. Find I_C and K. If the β of the transistor varies from 50 to 150, what is the range of I_C?

10-2.10 R_2 is 200 kΩ, R_C is 3000 Ω, R_E is 2000 Ω, and β is 100 for the silicon transistor. The supply voltage is 10 V. The R_1 is adjusted to set V_{CE} to 5 V. Determine R_1 and I_C. What is the variation of I_C if β varies from 60 to 140? What is K?

10-2.11 R_1 is 75 kΩ, R_2 is 33 kΩ, R_C is 4.7 kΩ, and R_E is 1800 Ω. The supply voltage is 20 V, and the β of the silicon transistor is 30. Find I_C. What is I_C when β is 20? What is K?

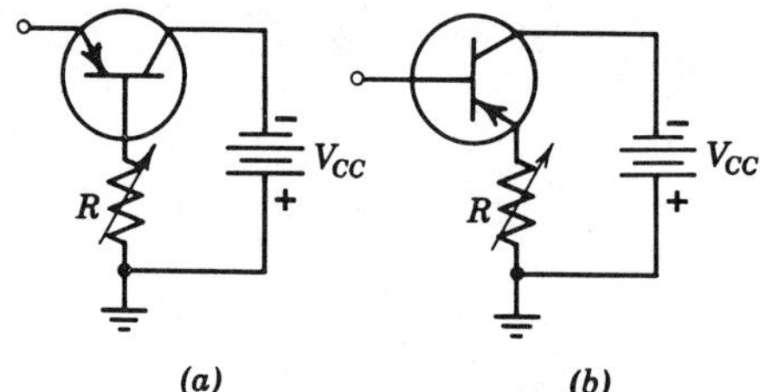

Figure 10-4 Leakage current. (*a*) I_{CBO}. (*b*) I_{CEO}.

Section 10-3 Leakage Currents

Let us perform an experiment on a *PNP* transistor (Fig. 10-4*a*). The emitter is left "floating"; that is, no connection is made to the emitter lead. A normal reverse bias voltage V_{CC} is applied to the collector. The value of R is raised from zero until we can read a dc voltage drop of the order of millivolts across R. When we have a suitable reading, we use Ohm's law to determine the current in the resistor.

The current we measure is I_{CBO}, the *leakage current*. The subscripts *CBO* are interpreted as "the current from collector to base with the emitter open."

The collector-to-base junction is reverse biased. Theoretically, the current should be zero but we actually find a small current I_{CBO}. In order to have a current flow across this junction, the current flow must come about as a result of a forward bias. A forward bias current requires that holes are present in the *N* material of the base and that electrons are present in the *P* material of the collector. This situation arises because of broken covalent bonds both in the base and in the collector. Thus I_{CBO} is a measure of the number of broken covalent bonds in the *N* material and in the *P* material. Whenever the temperature of the *PN* junction is greater than absolute zero, we have these *minority current carriers* present in semiconductors.

If we make the measurement we described for Fig. 10-4*a* and then hold the transistor with our fingers, we find that I_{CBO}

increases. The increase in temperature produced by our fingers breaks more covalent bonds and produces more leakage current. If the temperature of a transistor becomes high enough, the leakage currents can swamp out the normal operating current I_C.

Now let us repeat the experiment using the circuit shown in Fig. 10-4*b*. The base is "floating" in this circuit and the current we measure is I_{CEO}. We find that I_{CEO} is much higher than I_{CBO}. In the circuit of Fig. 10-4*b* we still have the leakage current from the collector to base I_{CBO}. However transistor action in this common-emitter circuit now multiplies I_{CBO} by β. The total current in the collector and in the emitter is the original leakage current I_{CBO} plus this leakage current multiplied by β.

$$I_{CEO} = I_{CBO} + \beta I_{CBO}$$

$$\boxed{I_{CEO} = (1 + \beta)I_{CBO}} \qquad (10\text{-}7)$$

Although the common-emitter amplifier circuit is used almost to the exclusion of the common-base amplifier circuit in practice, the manufacturers list I_{CBO} measured usually at 25°C in their specification sheets. Occasionally I_{CBO} is abbreviated to I_{CO}. Some typical values are listed in Table 10-1.

Minority current carriers must be present in any transistor. The more highly refined types, which consequently are the most expensive, have much lower values of the I_{CBO} than the

Table 10-1
Typical Leakage Current Values

I_{CBO}	$I_{C,max}$	**Transistor**
10 μA	50 mA	Germanium *PNP* for audio service
3 mA	3 A	Germanium *PNP* audio power amplifier
12 μA	10 mA	Germanium *PNP* for broadcast receivers
at 25°C 0.01 μA; at 150°C 1 μA	1.5 A	Silicon *NPN* power amplifier to 150 MHz
20 nA	8 mA	Silicon *NPN* small-signal amplifiers to 500 MHz
50 nA	200 mA	Silicon *NPN* for critical industrial amplifiers to 10 MHz

cheaper units. Also, the value of I_{CBO} in a silicon transistor is much less than the value of I_{CBO} in a germanium transistor.

Methods of modern physics show that the following "rules of thumb" are valid.

> 1. *I_{CBO} doubles for each 10°C rise in germanium transistors.*
> 2. *I_{CBO} doubles for each 6°C rise in silicon transistors.*

These two rules should be memorized.

If the temperature rise is ΔT in °C, the number of times I_{CBO} doubles is N.

$$N = \frac{\Delta T}{10} \quad \text{for germanium} \tag{10-8a}$$

and

$$N = \frac{\Delta T}{6} \quad \text{for silicon} \tag{10-8b}$$

and the leakage current at the higher temperature is

$$\boxed{I'_{CBO} = 2^N I_{CBO}} \tag{10-8c}$$

Equation 10-7 can be substituted into Eq. 10-8c to give the leakage current I'_{CEO} at an elevated temperature.

$$\boxed{I'_{CEO} = (1+\beta)2^N I_{CBO}} \tag{10-9}$$

Example 10-3

The leakage current I_{CBO} in a transistor is 2 μA. If the ambient temperature rise is 90°C, what is the leakage current for the transistor if it is germanium? If it is silicon?

Solution

For the germanium transistor, the leakage current doubles for each 10°C rise in ambient temperature. Therefore

$$N = \frac{\Delta T}{10} = \frac{90}{10} = 9 \tag{10-8a}$$

and the leakage current at high temperature is

$$I'_{CBO} = 2^N I_{CBO} = 2^9 \times 2\ \mu\text{A} = 1024\ \mu\text{A} = \mathbf{1.0\ mA} \qquad (10\text{-}8c)$$

For the silicon transistor, the leakage current doubles for each 6°C rise in ambient temperature. Therefore

$$N = \frac{\Delta T}{6} = \frac{90}{6} = 15 \qquad (10\text{-}8b)$$

and the leakage current at high temperature is

$$I'_{CBO} = 2^N I_{CBO} = 2^{15} \times 2\ \mu\text{A} = 65{,}536\ \mu\text{A} = \mathbf{65.5\ mA} \qquad (10\text{-}8c)$$

Example 10-4
The leakage current I_{CBO} in a silicon transistor is 25 nA. The value of β is 70. If the ambient temperature rise is 80°C, determine I'_{CEO}.

Solution
The leakage current doubles for each 6°C rise in ambient temperature. Therefore

$$N = \frac{\Delta T}{6} = \frac{80}{6} = 13.33 \qquad (10\text{-}8b)$$

Then

$$I'_{CEO} = (1+\beta)2^N I_{CBO} = 71 \times 2^{13.33} \times 25\ \text{nA}$$
$$= 1.83 \times 10^7\ \text{nA} = \mathbf{18.3\ mA} \qquad (10\text{-}9)$$

Table 10-1 shows that the values of I_{CBO} are much smaller for silicon units than for germanium units. Although I_{CBO} doubles for each 6°C rise for silicon transistors against a doubling for each 10°C rise for germanium transistors, the initial smaller value of I_{CBO} of silicon is the critical factor. As a result, high-temperature applications are limited to silicon transistors. Similarly, certain semiconductor devices, such as the SCR and the triac, must be made from silicon because of the requirement of a very low initial value of leakage current.

Problems **10-3.1 through 10-3.6** For each transistor listed in Table 10-1, assume that the leakage current is established at 25°C. For each listed transistor determine the temperature at which I_{CEO} equals the listed value of I_C. Assume β is 49 for each unit.

10-3.7 The maximum value of I_{CBO} for a particular transistor at room temperature is 15 nA. If β can vary from 150 to 240 for this transistor, what is the corresponding range for I_{CEO}?

10-3.8 The leakage current I_{CEO} is 75 μA and β is 135. What is I_{CBO}?

Section 10-4 Temperature Sensitivity

In the previous section we showed that a leakage current that is negligibly small at room temperature can become a large value at a high ambient temperature. The leakage current adds to the collector current and causes a shift ΔI_C in I_C. If the collector current without leakage is I_C, the collector current with leakage is $(I_C + \Delta I_C)$.

In the common-base amplifier a change in the leakage current (ΔI_{CBO}) reflects directly into the collector current as a change in collector current.

$$\Delta I_C = \Delta I_{CBO}$$

Therefore the ratio of $\Delta I_C/\Delta I_{CBO}$ is unity.

In the basic common-emitter amplifier, as used to develop Eq. 10-8 in the previous section, the change in the collector current is

$$\Delta I_C = \Delta I_{CEO} = (1+\beta)\Delta I_{CBO} \qquad (10\text{-}10)$$

Now ΔI_C is $(1+\beta)$ times ΔI_{CBO}. Then the ratio of $\Delta I_C/\Delta I_{CBO}$ is $(1+\beta)$.

These two cases give the limiting values 1.0 and $(1+\beta)$, for the *temperature sensitivity*, S

$$\boxed{S \equiv \frac{\Delta I_C}{\Delta I_{CBO}}} \qquad (10\text{-}11a)$$

where

$$1 \leq S \leq (1+\beta)$$

Equation 10-11a can be rearranged to

$$\boxed{\Delta I_C = S \times \Delta I_{CBO}} \qquad (10\text{-}11b)$$

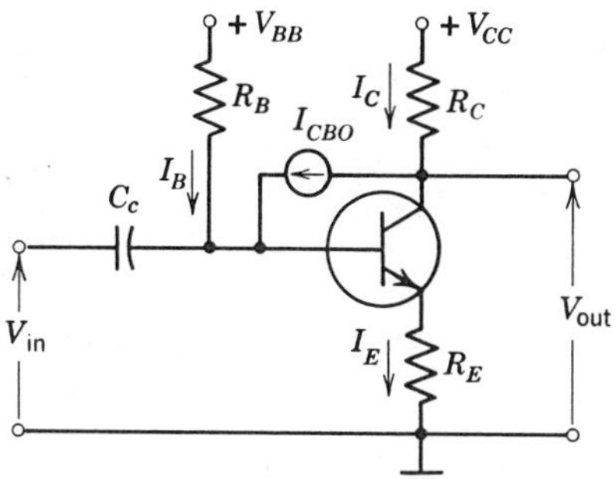

Figure 10-5 Common-emitter amplifier with emitter resistor.

Equation 10-11*b* shows that a rise in leakage current is multiplied by S to give the corresponding rise in collector current. For the ideal case, S is 1. For the "worst case" S is $(1+\beta)$. From the viewpoint of a circuit, S should be made as small as possible without sacrificing the other features of the circuit. The compromise result is termed a *trade-off*.

We will analyze the same circuit used in Section 10-2 for beta sensitivity K. This circuit is redrawn in Fig. 10-5 with one modification. The I_{CBO} is shown as a current generator externally connected from the collector to the base.

The Kirchhoff's voltage loop equation through the base is

$$V_{BB} = R_B I_B + V_{BE} + R_E I_E$$

The emitter current has two terms.

$$I_E = (1+\beta)I_B + (1+\beta)I_{CBO}$$

Substituting this value of I_E into the Kirchhoff's voltage loop equation, we have

$$V_{BB} = R_B I_B + V_{BE} + (1+\beta)R_E I_B + (1+\beta)R_E I_{CBO}$$

Solving for I_B, we have

$$I_B = \frac{V_{BB} - V_{BE}}{R_B + (1+\beta)R_E} - \frac{(1+\beta)R_E}{R_B + (1+\beta)R_E} I_{CBO}$$

The collector current is

$$I_C = \beta I_B + (1+\beta)I_{CBO}$$

Substituting the value for I_B into the equation for I_C, we have

$$I_C = \beta \frac{V_{BB} - V_{BE}}{R_B + (1+\beta)R_E} - \frac{\beta(1+\beta)R_E}{R_B + (1+\beta)R_E} I_{CBO} + (1+\beta)I_{CBO}$$

Now let I_{CBO} increase to $I_{CBO} + \Delta I_{CBO}$, causing I_C to rise to $I_C + \Delta I_C$.

$$I_C + \Delta I_C = \beta \frac{V_{BB} - V_{BE}}{R_B + (1+\beta)R_E} - \frac{\beta(1+\beta)R_E}{R_B + (1+\beta)R_E}(I_{CBO} + \Delta I_{CBO})$$
$$+ (1+\beta)(I_{CBO} + \Delta I_{CBO})$$

From this subtract the expression for I_C to obtain ΔI_C

$$\Delta I_C = \left[(1+\beta) - \frac{\beta(1+\beta)R_E}{R_B + (1+\beta)R_E}\right]\Delta I_{CBO}$$
$$= \left[\frac{(1+\beta)R_B + (1+\beta)R_E}{R_B + (1+\beta)R_E}\right]\Delta I_{CBO}$$

dividing numerator and denominator by $(1+\beta)$

$$\Delta I_C = \frac{R_E + R_B}{R_E + R_B/(1+\beta)}\Delta I_{CBO}$$

Comparison of this result with Eq. 10-11*b* shows that the temperature sensitivity S is

$$\boxed{S = \frac{R_E + R_B}{R_E + \dfrac{R_B}{1+\beta}}} \qquad (10\text{-}12)$$

In the discussion for beta stability K we did not perform derivation for other amplifier circuits but presented the final results. If these derivations were made, we would find that the rules used for K also apply to S.

1. In the circuit shown in Fig. 10-6*a*, there is no external emitter resistance in the circuit. Then, R_E is zero and Eq. 10-12 reduces to $(1+\beta)$, the "worst case."

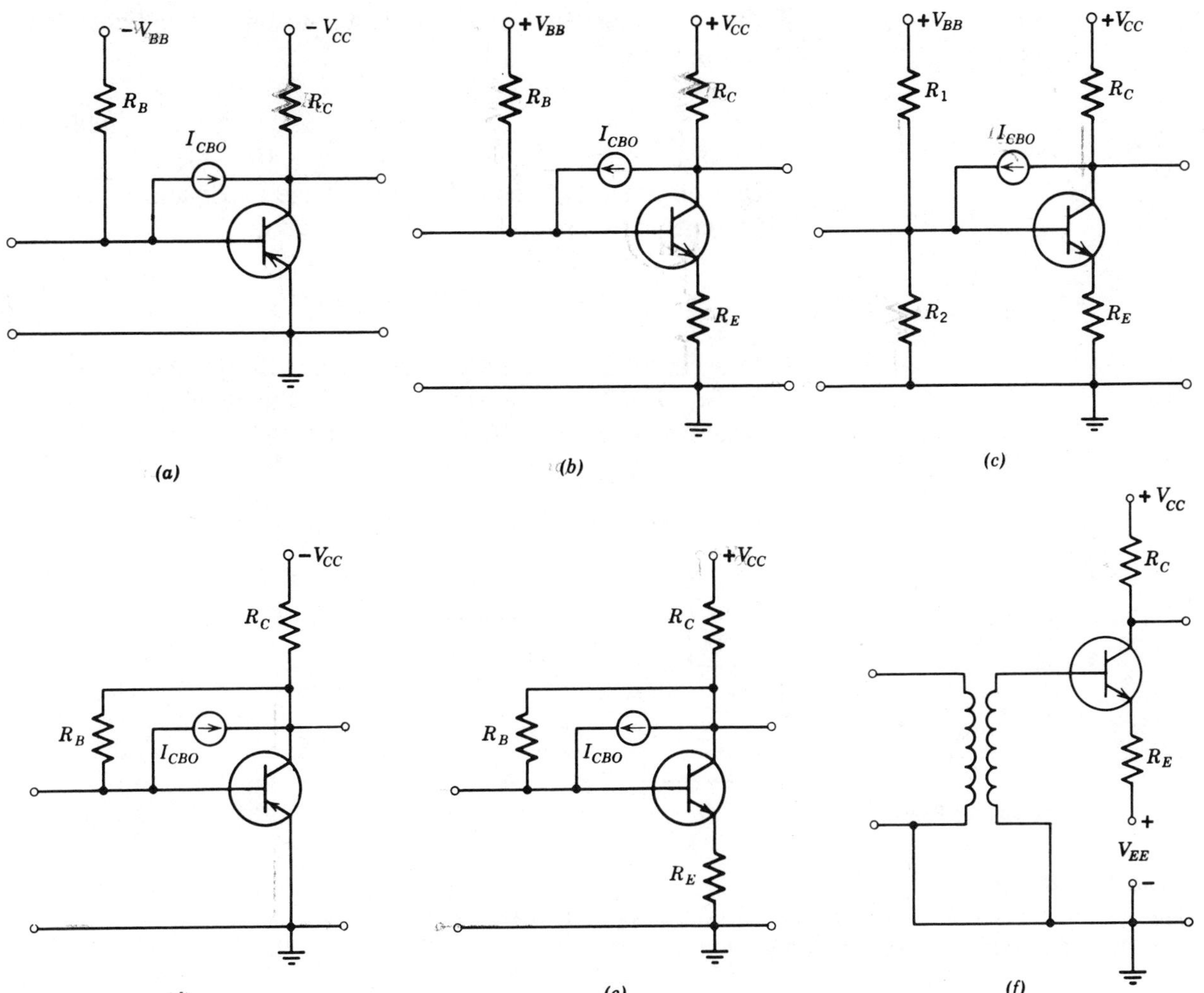

Figure 10-6 Circuits showing the method of determining values for *S*.

2. In an emitter follower amplifier circuit (Fig. 10-6*b*), we use the values of R_B and R_E directly into Eq. 10-12.
3. In a circuit that uses a voltage divider to obtain the bias (Fig. 10-6*c*), we use Thévenin's theorem to obtain a value for R_B.

$$R_B = \frac{R_1 R_2}{R_1 + R_2} \tag{10-13a}$$

4. In a circuit that uses collector-to-base feedback (Fig. 10-6*d*), use

$$R_E = R_C \tag{10-13b}$$

5. In a circuit that uses collector-to-base feedback in addition to emitter feedback (Fig. 10-6*e*), use

$$R_E = R_C + R_E' \tag{10-13c}$$

6. In a circuit in which R_B is zero (Fig. 10-6*f*), the value of S in Eq. 10-12 reduces to 1. This is the ideal condition, in which I_C rises only by the amount that I_{CBO} rises.

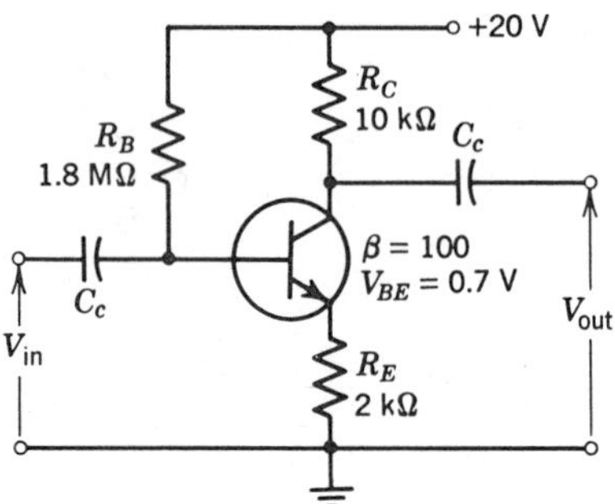

Circuit for Example 10-5.

Example 10-5
The silicon transistor has a leakage current I_{CBO} of 10 nA at 20°C. Find the operating point of the circuit at 75°C.

Solution
This is the circuit that was used to illustrate beta sensitivity K in Example 10-1. In Example 10-1, we found the operating point values:

$$I_{CQ} = 0.964 \text{ mA} \quad \text{and} \quad V_{CEQ} = 8.4 \text{ V}$$

To determine the value of I_{CBO} at 75°C, the number of times that I_{CBO} doubles is

$$N = \frac{\Delta T}{6} = \frac{75 - 20}{6} = 9.167 \tag{10-8b}$$

Then

$$I'_{CBO} = 2^N I_{CBO} = 2^{9.167} \times 10 = 5750 \text{ nA} = 5.75\ \mu\text{A} \tag{10-8c}$$

We use I'_{CBO} as ΔI_{CBO} in the definition of temperature sensitivity S in Eq. 10-11*a*.
S is obtained from Eq. 10-12.

$$S = \frac{R_E + R_B}{R_E + \dfrac{R_B}{1+\beta}} = \frac{2\ \text{k}\Omega + 1800\ \text{k}\Omega}{2\ \text{k}\Omega + \dfrac{1800\ \text{k}\Omega}{1+100}} = 90.9 \qquad (10\text{-}12)$$

The definition of temperature sensitivity S is

$$S = \frac{\Delta I_C}{\Delta I_{CBO}} \qquad (10\text{-}11a)$$

Substituting numerical values, we have

$$90.9 = \frac{\Delta I_C}{5.75\ \mu\text{A}}$$

$$\Delta I_C = 522\ \mu\text{A} = 0.522\ \text{mA}$$

The new value of I_C is

$$I_C + \Delta I_C = 0.964 + 0.522 = 1.486\ \text{mA}$$

Substituting this new value of I_C into the Kirchhoff's voltage loop equation through the collector circuit, we have

$$V_{CC} = R_C I_C + V_{CE} + R_E I_E$$

$$V_{CC} = R_C I_C + V_{CE} + R_E \frac{1+\beta}{\beta} I_C$$

or

$$20\ \text{V} = 10\ \text{k}\Omega \times 1.486\ \text{mA} + V_{CE} + 2\ \text{k}\Omega \times \frac{101}{100} \times 1.486\ \text{mA}$$

$$V_{CE} = 2.14\ \text{V}$$

The operating point at 75°C is

$$\mathbf{I'_{CQ} = 1.49\ mA} \quad \text{and} \quad \mathbf{V'_{CEQ} = 2.14\ V}$$

Example 10-6
The silicon transistor has a leakage current I_{CBO} of 10 nA at 20°C. Find the operating point of the circuit at 75°C

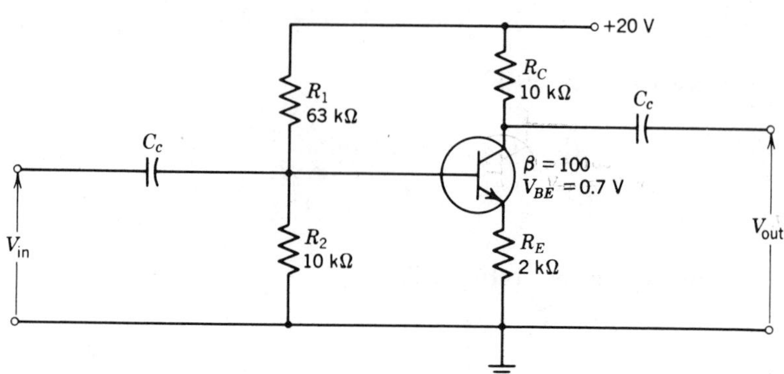

Circuit for Example 10-6.

Solution

This is the circuit that was used to illustrate beta stability K in Example 10-2. Also the operating point at room temperature is the same as the operating point that we used in Example 10-5.

$$I_{CQ} = 0.964 \text{ mA} \quad \text{and} \quad V_{CEQ} = 8.4 \text{ V}$$

Also, the value of ΔI_{CBO} is the same value obtained in Example 10-5.

$$\Delta I_{CBO} = I'_{CBO} = 5.75 \ \mu\text{A}$$

The equivalent resistance of the base-bias voltage divider given by Thévenin's theorem is

$$R_B = \frac{R_1 R_2}{R_1 + R_2} = \frac{63 \text{ k}\Omega \times 10 \text{ k}\Omega}{63 \text{ k}\Omega + 10 \text{ k}\Omega} = 8.63 \text{ k}\Omega \qquad (10\text{-}6a)$$

Then, by Eq. 10-12

$$S = \frac{R_E + R_B}{R_E + \dfrac{R_B}{1+\beta}} = \frac{2 \text{ k}\Omega + 8.63 \text{ k}\Omega}{2 \text{ k}\Omega + \dfrac{8.63 \text{ k}\Omega}{1 + 100}} = 5.10 \qquad (10\text{-}12)$$

The definition of temperature sensitivity S is

$$S = \frac{\Delta I_C}{\Delta I_{CBO}} \qquad (10\text{-}11a)$$

Substituting numerical values, we find

$$5.10 = \frac{\Delta I_C}{5.75\ \mu\text{A}}$$

$$\Delta I_C = 29\ \mu\text{A} = 0.029\ \text{mA}$$

The new value of I_C is

$$I_C + \Delta I_C = 0.964 + 0.029 = 0.993\ \text{mA}$$

Substituting into the Kirchhoff's voltage loop equation through the collector circuit, we have

$$V_{CC} + R_C I_C + V_{CE} + R_E\left(\frac{1+\beta}{\beta}\right) I_C$$

$$20\ \text{V} = 10\ \text{k}\Omega \times 0.993\ \text{mA} + V_{CE} + 2\ \text{k}\Omega\left(\frac{101}{100}\right) 0.993\ \text{mA}$$

$$V_{CE} = 8.06\ \text{V}$$

The operating point at 75°C is

$$\mathbf{I'_{CQ} = 0.993\ mA} \quad \text{and} \quad \mathbf{V'_{CEQ} = 8.06\ V}$$

Recalling that the operating point at room temperature (20°C) was

$$I_{CQ} = 0.964\ \text{mA} \quad \text{and} \quad V_{CEQ} = 8.4\ \text{V}$$

we see that the leakage current causes the collector current to rise from 0.964 mA to 0.993 mA. This is an increase in collector current of 3%.

In Example 10-5 the same leakage current rise caused the collector current to rise from 0.964 mA to 1.486 mA. This is a change of 54%.

A comparison of these two results clearly shows the importance of considering the effect of a rise in ambient temperature on the circuit.

In a general sense, it is obvious that a simultaneous improvement can be made in beta stability and in temperature sensitivity. An improvement in one also results in an improvement in the other.

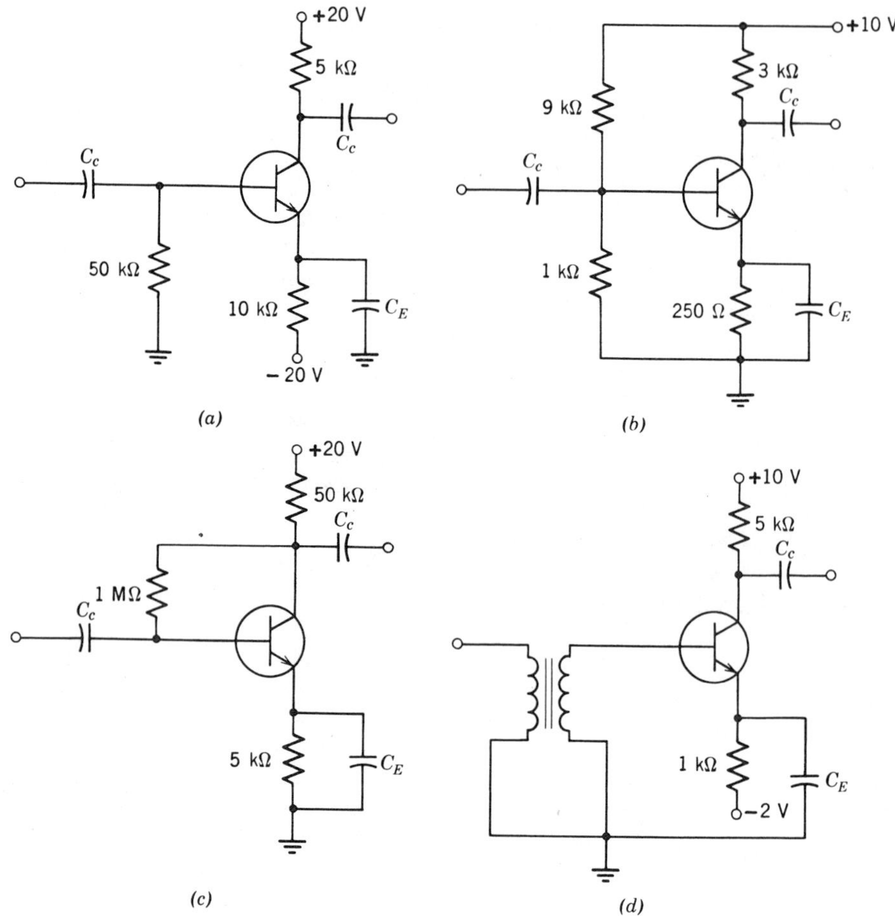

Circuits for Problem 10-4.10. (*a*) Circuit A. (*b*) Circuit B. (*c*) Circuit C. (*d*) Circuit D.

Problems

10-4.1 In Fig. 10-6*a*, R_B is 200 kΩ, R_C is 2 kΩ, and β is 50. The *PNP* transistor is germanium and the supply voltage is −20 V. At room temperature I_{CBO} is 0.1 μA, and its effect on I_C is negligible. At what elevated temperature will it cause I_C to increase by 50%?

10-4.2 In Fig. 10-6*a*, R_B is 10 kΩ, R_C is 75 Ω, and β is 60. The *NPN* transistor is silicon, and the supply voltage is +3 V. At room temperature I_{CBO} is 50 nA, and its effect on I_C is negligible. At what elevated temperature will it cause I_C to rise by 40%?

10-4.3 In Fig. 10-6*b*, R_B is 300 kΩ, R_C is 2000 Ω, R_E is 1000 Ω, and β is 75. The transistor is silicon and the supply voltage is 20 V. At room temperature I_{CBO} is 20 nA, and its effect on I_C is negligible. At what elevated temperature will I_C increase by 50%?

10-4.4 In Fig. 10-6*b*, R_B is 750 kΩ, R_C is 3.9 kΩ, R_E is 2000 Ω, and β is 100. The transistor is germanium and the supply voltage is 12 V. At room temperature I_{CBO} is 0.1 μA, and its effect on I_C at room temperature is negligible. At what elevated temperature will it cause I_C to increase 30%?

10-4.5 In Fig. 10-6*c*, R_2 is 100 kΩ, R_C is 3000 Ω, R_E is 1000 Ω, and β is 150 for the silicon transistor. R_1 is adjusted to set V_{CE} to 4 V with a 10-V supply. At room temperature I_{CBO} is 5 nA, and its effect on I_C at room temperature is negligible. At what elevated temperature will it cause I_C to rise 20%?

10-4.6 In Fig. 10-6*d*, R_C is 5 kΩ, and β is 60 for the germanium transistor. R_B is adjusted to set V_{CE} to 2 V with a −4-V supply. At room temperature I_{CBO} is 0.1 μA, and its effect is negligible on I_C. At what elevated temperature will I_C be increased by 15%?

10-4.7 In Fig. 10-6*d*, R_C is 39 kΩ and β is 200 for the silicon transistor. R_B is selected to set V_{CE} to 10 V with a 20-V supply. At room temperature I_{CBO} is 30 nA, and its effect on I_C is negligible. At what elevated temperature will I_C be increased 20%?

10-4.8 Use a value of 300 Ω for R_E in Fig. 10-6*e* and use the other data given in Problem 10-4.6. What is I_C at 80°C?

10-4.9 Use a value of 300 Ω for R_E in Fig. 10-6*e* and use the other data given in Problem 10-4.7. What is I_C at 60°C?

10-4.10 The transistor is silicon and has a value of 0.7 V for V_{BE} and a value of 100 for β. At 25°C, I_{CBO} is 1 μA, and its effect on I_C is negligible. The maximum allowable ambient temperature for the transistor is 67°C. The maximum permissible shift in I_C is 50%. What is the maximum allowable operating ambient temperature for each of the four circuits? Use 25°C for room temperature.

Section 10-5 FET Stabilization

The collector current in a transistor (BJT) is a current-controlled current source that depends both on the current amplification factor (beta) and on the leakage current. The drain current in a FET is a voltage-controlled current source with the feature that the gate current is zero. Thus, we do not become involved either in a concept for *K* or in a concept for *S* for the FET.

However, we do find that there is a wide range of variations

Table 10-2
Range of FET Parameters

	Maximum Value	Nominal Value	Minimum Value
I_{DSS}	13 mA	9 mA	4.5 mA
V_P	−5.6 V	−4.0 V	−3.1 V

in the specification between the maximum and minimum values of I_{DSS} that can be expected in a large lot. Table 10-2 lists the expected range of variation for one particular FET type.

Three circuit variations for an FET amplifier are shown in Fig. 10-7. The corresponding load lines and transfer characteristics are given in Fig. 10-8. Each load line for the drain circuit is drawn for a supply voltage V_{DD} of 13 V and for a total dc resistance of 1625 Ω. The intercept of the load line on the I_D axis is 13 V/1625 Ω or 8 mA. Also the circuits are designed so that the operating point for the nominal value is the same for all three circuits.

In Circuit No. 1, the bias on the gate is derived from a bias source V_{GG} set at −1.5 V. The operating points are found by drawing a vertical line at $V_{GS} = V_{GG} = -1.5$ V. The intersections with the transfer curves are points C, A, and B. These three points are projected over to the load line to give the operating points Q_2, Q, and Q_1. These operating-point currents, I_{DQ2} (the minimum value), I_{DQ} (the nominal value), and I_{DQ1} (the maximum value) are listed in Table 10-3. The operating-point values for V_{DS} are calculated from the Kirchhoff's voltage loop equation through the drain circuit.

$$V_{DD} = R_D I_{DQ} + V_{DSQ} + R_S I_{DQ} \tag{10-14}$$

Table 10-3
FET Amplifier Operating Points

Circuit	R_D (ohms)	R_S (ohms)	I_{DQ1} (mA)	I_{DQ} (mA)	I_{DQ2} (mA)	V_{DSQ1} (V)	V_{DSQ} (V)	V_{DSQ2} (V)
No. 1	1625	0	7.0	3.5	1.2	1.6	7.3	11.1
No. 2	1196	429	5.1	3.5	2.1	4.7	7.3	9.6
No. 3	625	1000	4.3	3.5	2.8	6.0	7.3	8.5

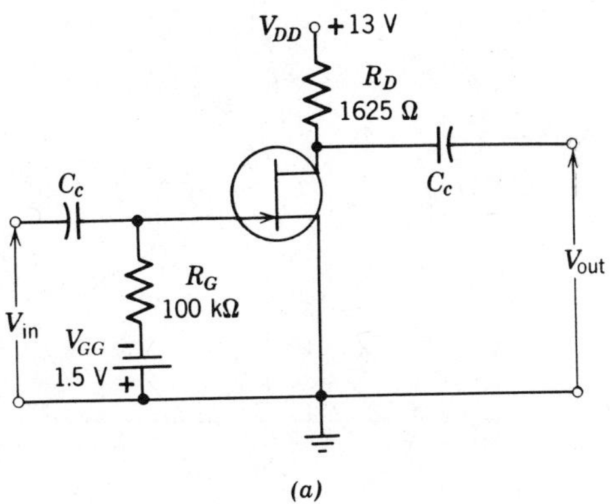

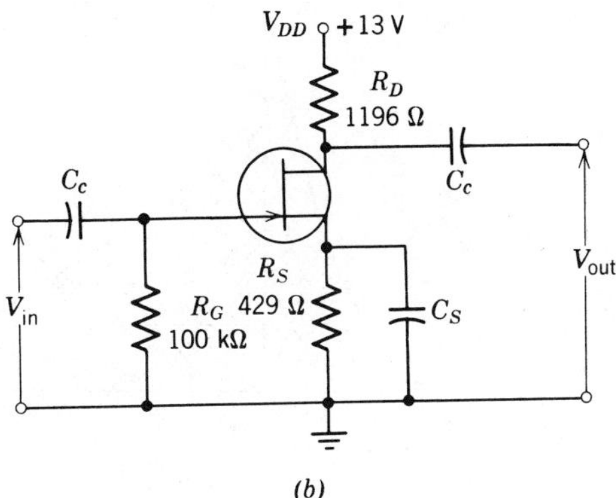

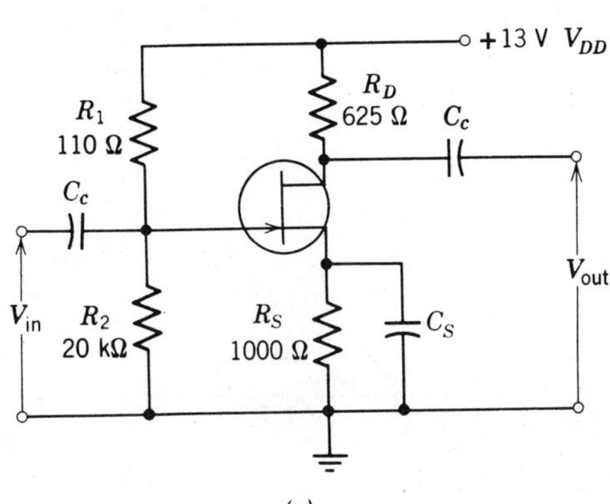

Figure 10-7 FET amplifier circuits with different bias arrangements. (*a*) Circuit No. 1. (*b*) Circuit No. 2. (*c*) Circuit No. 3.

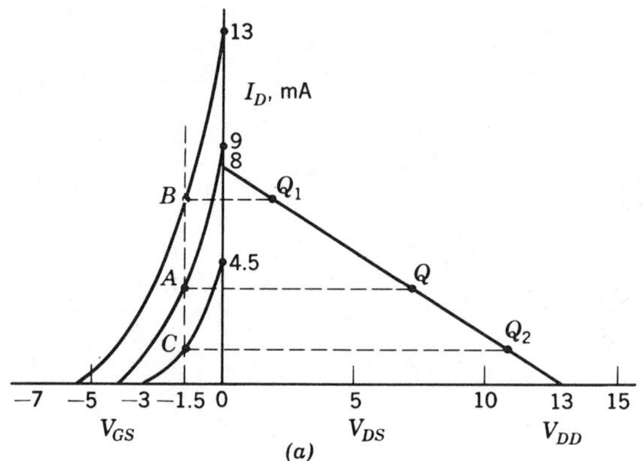

(a)

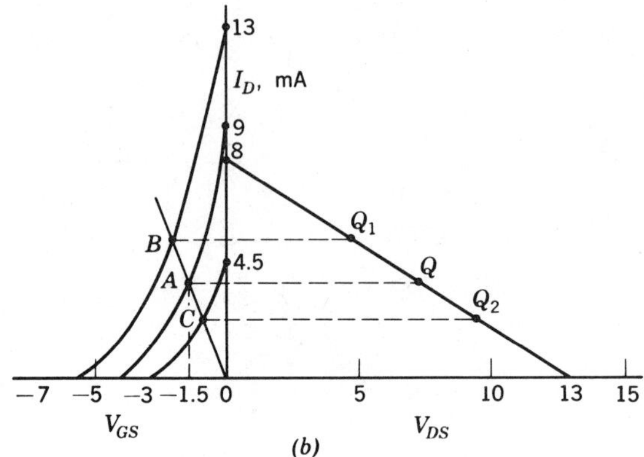

(b)

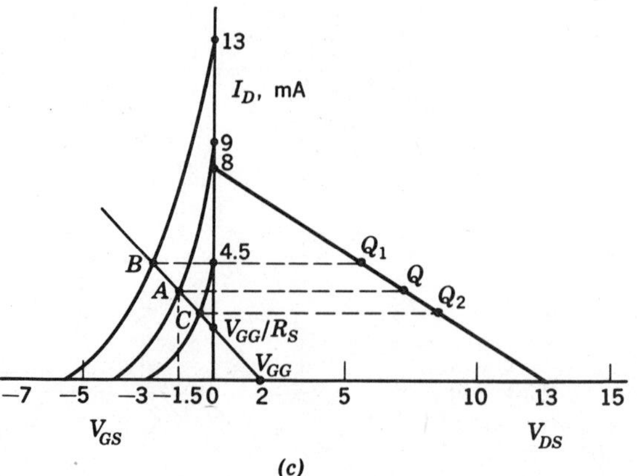

(c)

Figure 10-8 Load lines for different bias circuits. (*a*) Load line for Circuit No. 1. (*b*) Load line for Circuit No. 2. (*c*) Load line for Circuit No. 3.

The values for V_{DSQ2}, V_{DSQ}, and V_{DSQ1} are also listed in Table 10-3.

In Circuit No. 2, the bias is derived from the dc voltage drop across the source resistor R_S.

$$V_{GG} = V_{GS} = R_S I_{DS} \tag{10-15}$$

The operating points are found by constructing a *bias line.* The bias line is drawn on the I_{DS}–V_{GS} axis as

$$V_{GS} = -R_S I_{DS} \tag{10-16}$$

In this case, the bias line that is required must pass through the operating point for the nominal values ($I_{DQ} = 3.5$ mA; $V_{GS} = -1.5$ V). Then

$$R_S = \frac{V_{GS}}{I_{DQ}} = \frac{1.5\text{ V}}{0.0035\text{ A}} = 429\ \Omega$$

Since

$$R_S + R_D = 1625\ \Omega$$

$$R_D = 1625\ \Omega - 429\ \Omega = 1196\ \Omega$$

The operating points on the bias line are *C*, *A*, and *B*. By projecting these points over to the load line, we have the operating points I_{DQ2}, I_{DQ}, and I_{DQ1}. Corresponding values of V_{DSQ2}, V_{DSQ}, and V_{DSQ1} are calculated from Eq. 10-14 and the results entered into Table 10-3.

In Circuit No. 3, there is a fixed dc bias obtained from the divider network of R_1 and R_2. This bias is V_{GG} and is found from

$$V_{GG} = \frac{R_2}{R_1 + R_2} V_{DD} = \frac{20\text{ k}\Omega}{110\text{ k}\Omega + 20\text{ k}\Omega} 13\text{ V} = 2\text{ V} \tag{10-17}$$

This bias voltage V_{GG} is positive whereas the net bias for this FET must be negative. The value of V_{GG} is an *offset* for the bias line as shown on the load line for Circuit No. 3 on Fig. 10-8. Now the equation for the bias line for R_S is

$$V_{GG} - (-V_{GS}) = R_S I_D \tag{10-18}$$

At the point where V_{GS} is zero, we have the current value V_{GG}/R_S. This is the current value where the bias line crosses the I_D axis.

For our circuit, the bias line must pass through the operating point for the nominal-value FET, which is 3.5 mA at −1.5 V. Then

$$2\text{ V} - (-1.5\text{ V}) = 0.0035\text{ A} \times R_S$$

or

$$R_S = \frac{3.5\text{ V}}{0.0035\text{ A}} = 1000\ \Omega$$

Since we require that

$$R_S + R_D = 1625\ \Omega$$

we have

$$1000\ \Omega + R_D = 1625\ \Omega$$

$$R_D = 625\ \Omega$$

The operating points on the bias line are C, A, and B. These points are projected over to the load line to Q_2 (I_{DQ2}), to Q (I_{DQ}), and to Q_1 (I_{DQ1}). The corresponding values for V_{DSQ2}, V_{DSQ}, and V_{DSQ1} are calculated from Eq. 10-14 and the results entered into Table 10-3.

Inspection of the operating points from Circuit No. 1 shows that Q_1 is not far from saturation and that Q_2 is not far from cutoff. The use of the self-bias resistor R_S moves both Q_1 and Q_2 closer to Q. When both a voltage divider and R_S are used, Q_1 and Q_2 move still closer to Q.

It should be recognized that, as we proceed from Circuit No. 1 to Circuit No. 2 to Circuit No. 3, the values of R_D decrease. Since the value of g_m at Q is fixed for all three circuits, the voltage gain (g_mR_D) of the stage is reduced as a *trade-off* for obtaining less variation in the operating point.

Since the FET is a nonlinear device, a mathematical approach to this problem is very complicated so a graphical approach is used. Calculations of percentage changes and gain variations are left to the problem set.

Problems For problems 10-5.1 through 10-5.9 use the data given in Table 10-2, Table 10-3, circuits of Fig. 10-7, and the load lines of Fig. 10-8.

10-5.1 What are the percentage variations in I_{DQ} and in V_{DSQ} for Circuit No. 1? Use the nominal values as the reference values.

10-5.2 Repeat Problem 10-5.1 for Circuit No. 2.

10-5.3 Repeat Problem 10-5.1 for Circuit No. 3.

10-5.4 Determine the equation for g_m for the FET having the maximum value specification.

10-5.5 Determine the equation for g_m for the FET having the nominal value specification.

10-5.6 Determine the equation for g_m for the FET having the minimum value specification.

10-5.7 Calculate the circuit gains for Circuit No. 1 at Q_2, Q, and Q_1. What is the percentage variation in gain using the nominal value as reference?

10-5.8 Repeat Problem 10-5.7 for Circuit No. 2.

10-5.9 Repeat Problem 10-5.7 for Circuit No. 3.

10-5.10 A FET has a nominal value for I_{DSS} of 10 mA with V_P rated at −4 V. The minimum expected value for I_{DSS} is 5 mA with a V_P of −2 V. The maximum expected value for I_{DSS} is 15 mA with a V_P of −6 V. The FET is used in Circuit No. 2 of Fig. 10-7 that has the values $V_{DD} = 24$ V, $R_G = 100$ kΩ, $R_S = 500$ Ω, and $R_D = 1500$ Ω. Construct a graph similar to the load line for Circuit No. 2, Fig. 10-7, giving the location of the operating points Q_1, Q, and Q_2 that are the limits for this FET. What are the voltage gains of the circuit at the operating points?

10-5.11 The FET used in Problem 10-5.10 is now used in Circuit No. 3, Fig. 10-7. Now, $R_1 = 100$ kΩ, $R_2 = 20$ kΩ, $R_S = 1000$ Ω, and $R_D = 2400$ Ω for a supply voltage of 24 V. Draw the curves similar to the load line for Circuit No. 3, Fig. 10-7, and determine the operating points Q_1, Q, and Q_2. What is the voltage gain of the circuit at each operating point?

Section 10-6 Diode Biasing and Compensation

The value of V_{BE} is taken as 0.3 V for a germanium junction and 0.7 V for a silicon junction. These values are valid for room temperature but must be corrected for other temperatures. In germanium units V_{BE} decreases 1.6 mV/°C and in silicon units V_{BE} decreases 2.0 mV/°C. Corrections need only be made when V_{BB} or V_{EE} are very low values.

The amplifier shown in Fig. 10-9*a* has a diode *D*1 placed in the emitter circuit that compensates against changes in V_{BE} that occur as the ambient temperature changes. In order to

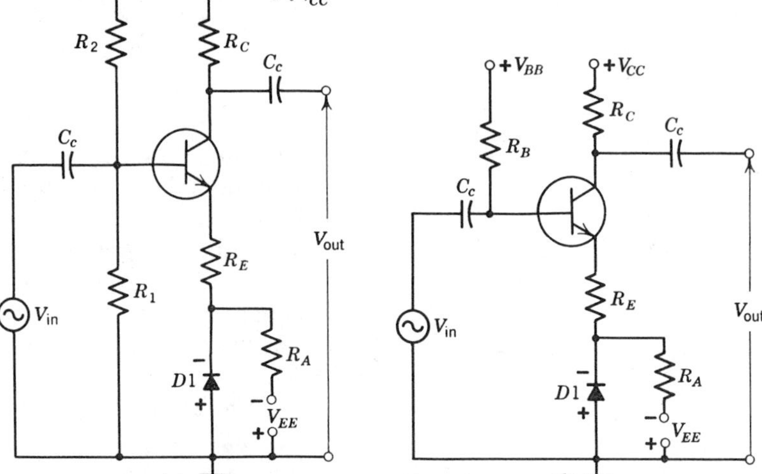

Figure 10-9 Diode compensation for variations in V_{BE}. (*a*) Actual circuit. (*b*) Modified circuit for analysis.

facilitate analysis, the bias circuit comprising V_{CC}, R_2, and R_1 is converted by Thévenin's theorem to an equivalent V_{BB} and R_B as shown in Fig. 10-9*b*. The Kirchhoff's voltage loop equation through the base is

$$V_{BB} = R_B I_B + V_{BE} + R_E I_E - V_{D1}$$

where V_{D1} is the voltage drop across the diode $D1$, developed by the value of V_{EE} and R_A. Then

$$V_{BB} = R_B I_B + R_E I_E + (V_{BE} - V_{D1})$$

if

$$V_{BE} = V_{D1}$$

then

$$V_{BB} = R_B I_B + R_E I_E$$

Thus, when a diode is selected that will have the identical characteristics of variation of V_{D1} that V_{BE} has, the equation is independent of V_{BE} and perfect compensation is achieved. The practical difficulty that arises is in the selection of a diode that has the exact variation required for compensation. Therefore, a compromise is made to get the compensation as close as possible. It is possible to use two diodes in series and to obtain a degree of compensation for changes in β also.

The circuit shown in Fig. 10-10 uses a diode to compensate for

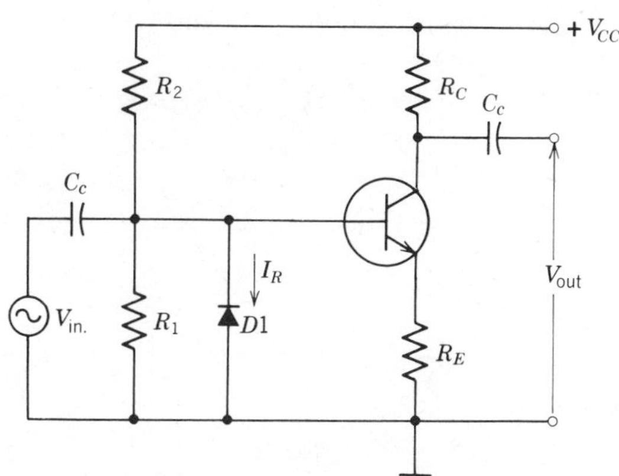

Figure 10-10 Diode compensation for I_{CBO}.

a shift in the operating point caused by a change in I_{CBO}. The compensation is based on the fact that there is a reverse current I_R in the diode.

A lengthy algebraic process yields a conclusion that is quite simple: the compensation for I_{CBO} is achieved when a diode $D1$ is selected which has a reverse current that is identical to the I_{CBO} of the transistor over the expected range of the ambient temperature variation. Again, in a practical situation, a compromise must be accepted.

A circuit that shows both methods of diode compensation is given in Fig. 10-11. An adjustable resistor (a potentiometer)

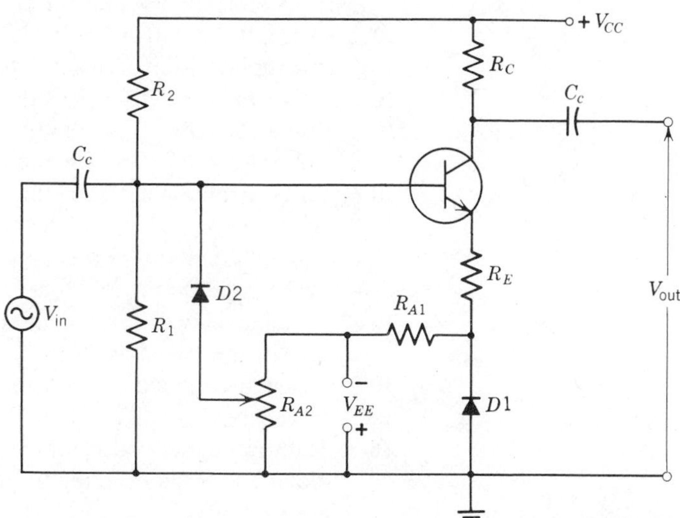

Figure 10-11 Circuit using diodes to compensate for I_{CBO} and for variations in V_{BE}.

used for R_{A2} permits individual circuit compensations to be made.

Very often, *thermistors* are used in place of diodes. A *thermistor* is a resistor that has a negative temperature coefficient. Thermistors are available over a wide range of resistance and current ranges.

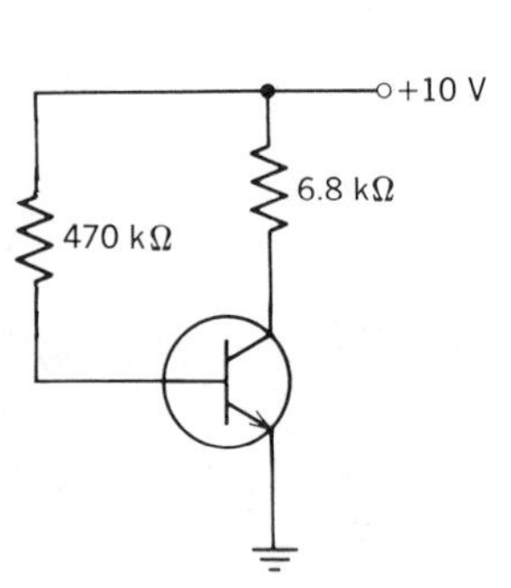

Circuit for Problems 10-1, 10-2, 10-7, 10-8, and 10-9.

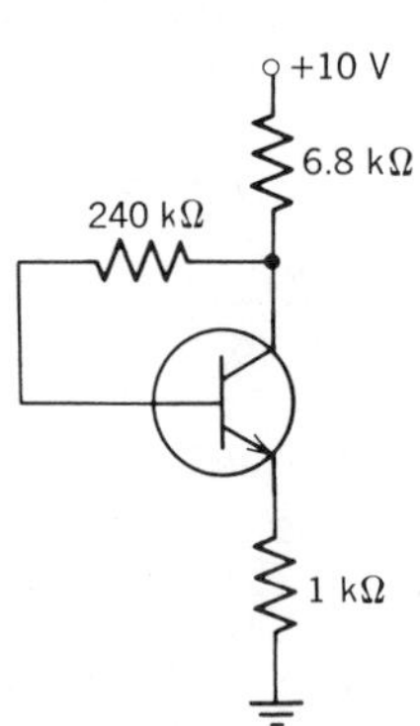

Circuit for Problems 10-3, 10-4, 10-10, 10-11, and 10-12.

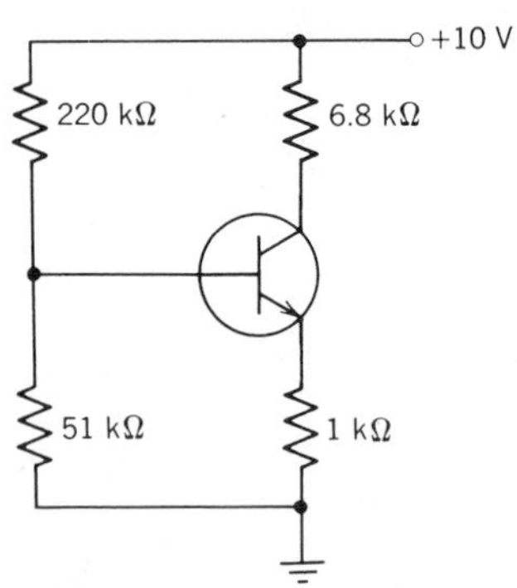

Circuit for Problems 10-5, 10-6, 10-13, and 10-14.

Supplementary Problems

For all circuits, $V_{BE} = 0.7$ V (silicon); $\beta = 50$, and $I_{CBO} = 20$ nA at 20°C.

10-1 Find I_{CQ} and V_{CEQ}. Determine K. Using this value of K, determine what I_{CQ} is and what V_{CEQ} is if a replacement transistor has a value for β of 75.

10-2 Using the data for Problem 10-1, what is I_{CQ} and what is V_{CEQ} if a replacement transistor has a value for β of 30?

10-3 Solve Problem 10-1 for the new circuit.

10-4 Solve Problem 10-2 for the new circuit.

10-5 Solve Problem 10-1 for the new circuit.

10-6 Solve Problem 10-2 for the new circuit.

10-7 In Problem 10-1 we neglected the effect of I_{CBO}. If we consider I_{CBO}, what is the percentage error we make if we neglect I_{CBO} at room temperature?

10-8 At what elevated ambient temperature will I_{CBO} cause saturation in the transistor?

10-9 At what elevated ambient temperature will I_{CBO} cause I_{CQ} to be increased by 20%?

10-10 Determine the value of S for the circuit. Using S, at what elevated ambient temperature will I_{CBO} cause I_{CQ} to be increased 10%?

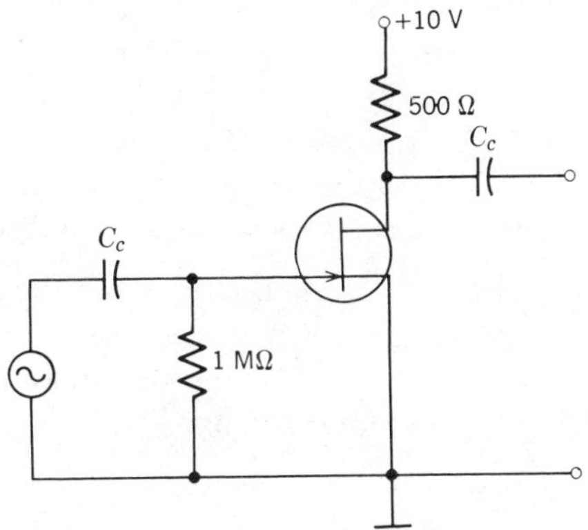

Circuit for Problem 10-15.

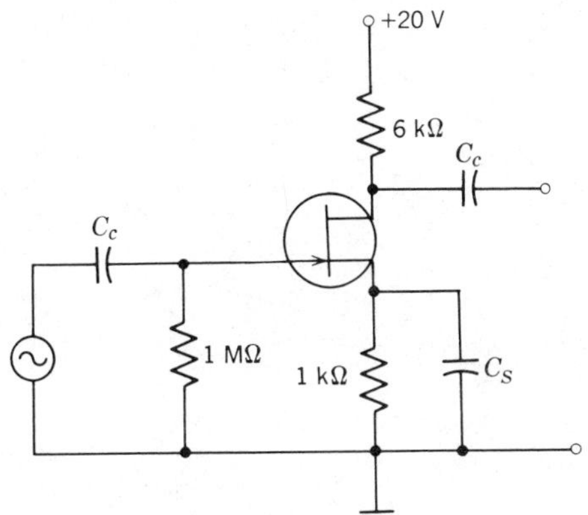

Circuit for Problem 10-16.

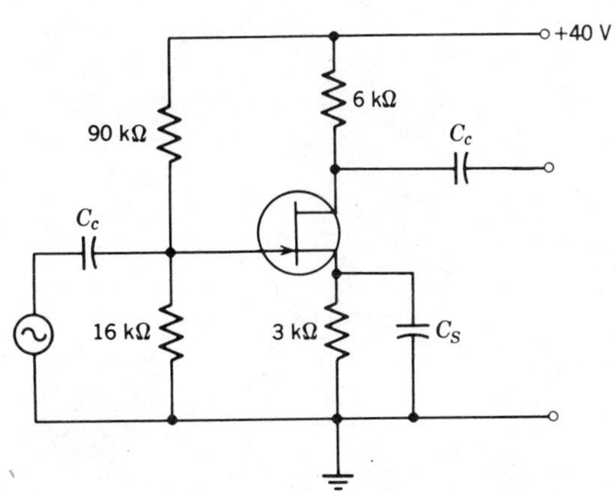

Circuit for Problem 10-17.

10-11 Determine the value of S for the circuit. Using S, at an ambient temperature of 100°C, determine I_{CQ}.
10-12 Solve Problem 10-11 for an ambient temperature of 60°C.
10-13 Determine the value of S for the circuit. Using S, at an ambient temperature of 60°C, determine I_{CQ}.
10-14 Solve Problem 10-13 for an ambient temperature of 100°C

The typical values ($Q1$) for the FET are 7.2 mA for I_{DSS} and -4 V for V_P. The maximum values ($Q2$) for the FET are 10.8 mA for I_{DSS} and -6 V for V_P. The minimum values ($Q3$) for the FET are 3.6 mA for I_{DSS} and -2 V for V_P.

10-15 What are the values for I_D and V_{DS} if each of $Q1$, $Q2$, and $Q3$ is used in the circuit?
10-16 What are the values for I_D and V_{DS} if each of $Q1$, $Q2$, and $Q3$ is used in the circuit?
10-17 What are the values for I_D and V_{DS} if each of $Q1$, $Q2$, and $Q3$ is used in the circuit?

Chapter 11 Decibels

Human vision and hearing require a nonlinear system of measurement (Section 11-1). The decibel (Section 11-2) is devised to meet this measurement need but, at the same time, retains the decimal system in the definition. A short review of logarithms is given (Section 11-3). In practice, decibels are usually determined from measured values of voltage and resistance (Section 11-4).

Section 11-1 The Need for a Nonlinear System of Measurement

Human sensory response is nonlinear. As an example showing this nonlinearity, a single match, when suddenly ignited in a dark room, produces a lasting glare. In bright sunlight, the same-size match, when struck, does not give off noticeable light. As another example, the noise of an insect can disrupt the calm of a still summer's night. On the other hand, it would take millions of these insects to be heard over the roar of a passing railroad train. In a dark room, two lighted matches give twice the effect of one match on the response of the human eye. In broad daylight, it would take two suns to give twice the effect of one on human vision. These facts would indicate that a true response would be of the order:

Steps of equal response	1	2	3	4	5	6	7	8	9
Quantity of cause	$\frac{1}{16}$	$\frac{1}{8}$	$\frac{1}{4}$	$\frac{1}{2}$	1	2	4	8	16

Each successive step in cause doubles the previous quantity, but the change in response is linear.

A further indication of the usefulness of such a scheme is given by the system used in music. In music, an increase in one *octave* doubles the pitch or frequency. The reference frequency used is A above middle C at 440 Hz. If the relative pitch is plotted on a linear axis corresponding to the keys on a piano, as in Fig. 11-1, we see that the frequency scale is nonlinear.

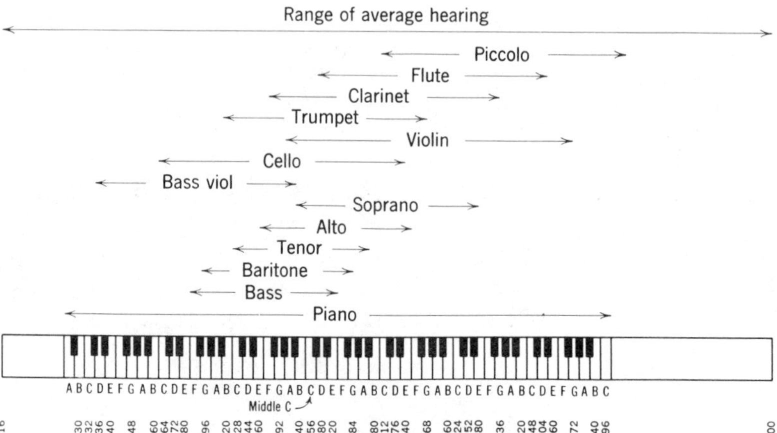

Figure 11-1 Range of frequencies in music.

In mathematics the process of taking logarithms of numbers converts a nonlinear scale, such as the musical scale, into a linear scale. Since each octave in the musical scale is a multiplication of the frequency of the preceding octave by two, the spread of one octave on a logarithmic scale is *log 2*, and it is the same number for any one octave.

In showing graphs of frequency response where the independent variable is frequency, the frequency is plotted as the logarithm to the base 10 of the frequency. Since this is the standard conventional practice, graph paper is available called *semilogarithmic paper* in which one axis is logarithmic and the other axis is linear. If we were to use ordinary graph paper for frequency-response curves, it would be necessary to calculate the logarithms of the different frequencies used. In semilogarithmic graph paper, the engraved printing plate is designed so that it is laid out proportionally to the logarithms on one axis scale. When this paper is used, there is no need to calculate logarithms; that work was done in the original design of the graph paper. If it is desired to represent 20 to 20,000 Hz, the required logarithmic axis would be 10 to 100 to 1000 to 10,000 to 100,000 or four-cycle semilogarithmic paper. To represent 20 to 8000 Hz, one would need a logarithmic axis of 10 to 100 to 1000 to 10,000 or three-cycle semilogarithmic paper.

Section 11-2 The Decibel In honor of Alexander Graham Bell, the logarithm to the base 10 of the ratio of two powers is defined as a *bel*:

$$\text{Number of bels} \equiv \log_{10} \frac{P_2}{P_1}$$

where P_2 and P_1 represent the two powers that are being compared. For example, P_2 could be the output power and P_1 could be the input power.

The bel as a unit is cumbersome for general use. In order to have convenient numerical results for problems and applications, we define the *decibel dB* as one-tenth of bel:

$$\boxed{dB \equiv 10 \log_{10} \frac{P_2}{P_1} \text{ dB}} \tag{11-1}$$

The symbol dB is used as both the quantity symbol and the unit symbol for the decibel.

In audio work, a change in power level of one decibel is barely perceptible to the ear. A change of two decibels is slightly apparent.

Section 11-3 Logarithms Since the decibel is defined as a logarithm, the technique of the mathematical process of taking logarithms must be studied. A formal mathematical definition may be given to a *logarithm.* If a number N is expressed in the form of the x power of 10, the logarithm of N to the base 10 is x.

If

$$N = 10^x \tag{11-2a}$$

then

$$\log_{10} N = x \tag{11-2b}$$

Using Eq. 11-2a and Eq. 11-2b, we may write

$$\begin{array}{rcr} \text{Since } 10{,}000 = 10^4 & \text{then} & \log_{10} 10{,}000 = 4 \\ 1000 = 10^3 & & \log_{10} 1000 = 3 \\ 100 = 10^2 & & \log_{10} 100 = 2 \\ 10 = 10^1 & & \log_{10} 10 = 1 \\ 1 = 10^0 & & \log_{10} 1 = 0 \end{array}$$

$$0.1 = \frac{1}{10} = 10^{-1} \qquad \log_{10} 0.1 = -1$$

$$0.01 = \frac{1}{100} = 10^{-2} \qquad \log_{10} 0.01 = -2$$

$$0.001 = \frac{1}{1000} = 10^{-3} \qquad \log_{10} 0.001 = -3$$

$$0.0001 = \frac{1}{10{,}000} = 10^{-4} \qquad \log_{10} 0.0001 = -4$$

Thus, as the number N becomes smaller and smaller approaching zero, the value of $\log_{10} N$ becomes a larger and larger negative number approaching negative infinity.

The numbers 4, 3, 2, 1, 0, −1, −2, −3, and −4 are known as the *characteristic*. The characteristic numerically is one less than the number of digits in the number to the left of the decimal point. If the number were 834.24, the characteristic is 2. This means that the logarithm of the number lies between 2 and 3. If the number were 8342.4, the logarithm would have the characteristic 3 and lie between 3 and 4. The exact decimal of the logarithm is called the *mantissa*. The mantissa for 834.24 is the same as the mantissa for 8342.4. It is also the same for 8,342,400 or 8.3424. The mantissa is determined by the sequence of the digits and not by the decimal point. The placement of the decimal point in the original number determines the characteristic.

When using a scientific calculator, we place the number N into the calculator in the conventional manner. Then the key (or keys) is depressed to give the value of $\log_{10}$ into the display. Some examples are:

$$\log_{10} 834240 = 5.921$$
$$\log_{10} 74 \times 10^5 = 6.869$$
$$\log_{10} 231 = 2.364$$
$$\log_{10} 3.85 = 0.585$$
$$\log_{10} 1.005 = 0.002166 = 2.166 \times 10^{-3}$$
$$\log_{10} 0.020 = -1.699$$
$$\log_{10} 0.20 = -0.699$$
$$\log_{10} 0.375 = -0.426$$
$$\log_{10} 0.000674 = -3.171$$
$$\log_{10} 4.23 \times 10^{-8} = -7.374$$

Equation 11-2*b* relates *N* and *x*:

$$\log_{10} N = x \tag{11-2b}$$

If this equation is taken as the initial statement, we then write

$$N = 10^x \tag{11-2a}$$

This inverse procedure is used to determine the *inverse of the logarithm* or the *antilogarithm.* If we have *x* as the value of the $\log_{10}$ of the unknown number *N*, *N* is determined by finding the *x* power of 10 (Eq. 11-2*a*). In the scientific calculator we place the number *x* into the calculator and press the key (or keys) to determine 10^x. The numerical value of *X* may be either a positive number or a negative number. Some examples are:

If $\log_{10} N = 4$ then $N = 10{,}000$ or $N = 10^4$
$\log_{10} N = 2$ $N = 100$ or $N = 10^2$
$\log_{10} N = 0.254$ $N = 1.795$
$\log_{10} N = 3.621$ $N = 4178$ or $N = 4.178 \times 10^3$
$\log_{10} N = -2.00$ $N = 0.01$ or $N = 10^{-2}$
$\log_{10} N = -4.84$ $N = 0.00001445$ or $N = 1.445 \times 10^{-5}$

Problems

11-3.1 Determine the logarithms of the following numbers: (*a*) 2650, (*b*) 132, (*c*) 756,000, (*d*) 1.46, (*e*) 294×10^{16}, (*f*) 0.0023, (*g*) 0.874, (*h*) $\frac{1}{16}$, (*i*) $\frac{3}{64}$, (*j*) 84×10^{-6}.

11-3.2 Determine the numbers for which the logarithms are: (*a*) 2.46, (*b*) 6.92, (*c*) 14.20, (*d*) 23.3, (*e*) 0.024, (*f*) −5.78, (*g*) 0, (*h*) −27.4, (*i*) $\frac{1}{16}$, (*j*) 7.23.

Section 11-4 Decibel Calculations

In Section 11-2, we defined the decibel as

$$dB \equiv 10 \log_{10} \frac{P_2}{P_1} \tag{11-1}$$

Properly speaking, a decibel is a measure of a power ratio, but very often the measurements are taken in terms of voltage, current, or impedance. In most applications, the impedance is purely resistance. If we let

$$P_2 = V_2^2/R_2 \quad \text{and} \quad P_1 = V_1^2/R_1$$

Substitution into Eq. 11-1 yields

$$dB = 10 \log_{10} \frac{V_2^2/R_2}{V_1^2/R_1} = 10 \log_{10} \frac{V_2^2 R_1}{V_1^2 R_2} = 10 \log_{10} \frac{V_2^2}{V_1^2} + 10 \log_{10} \frac{R_1}{R_2}$$

$$\boxed{dB = 20 \log_{10} \frac{V_2}{V_1} + 10 \log_{10} \frac{R_1}{R_2}} \tag{11-3}$$

When R_1 and R_2 have the same value, Eq. 11-3 reduces to

$$\boxed{dB = 20 \log_{10} \frac{V_2}{V_1}} \tag{11-4}$$

If this decibel relation is evaluated in terms of currents instead of voltages, we have

$$P_1 = I_1^2 R_1 \quad \text{and} \quad P_2 = I_2^2 R_2$$

Then

$$dB = 10 \log_{10} \frac{I_2^2 R_2}{I_1^2 R_1}$$

and

$$\boxed{dB = 20 \log_{10} \frac{I_2}{I_1} + 10 \log_{10} \frac{R_2}{R_1}} \tag{11-5}$$

Example 11-1
The voltage across a loudspeaker is 2.3 V and, when the volume control is advanced, the speaker voltage becomes 4.8 V. Determine the increase in output in decibels.

Solution
We do not use the correction factor, $10 \log_{10} R_1/R_2$, because both measurements are taken across the same resistance value. Thus, Eq.

11-4 is used directly.

$$dB = 20 \log_{10} \frac{V_2}{V_1} \tag{11-4}$$

$$dB = 20 \log_{10} \frac{4.8}{2.3} = \mathbf{+6.4\ dB}$$

Example 11-2
The input voltage to a transmission line is 64 V and the output voltage is 18 V. Determine the loss in the transmission line in decibels.

Solution
Since the output voltage is less than the input voltage, the transmission line must show a loss in gain, that is, a negative number for the decibels. Using Eq. 11-4, we find that

$$dB = 20 \log_{10} \frac{V_2}{V_1} \tag{11-4}$$

$$dB = 20 \log_{10} \frac{18}{64} = \mathbf{-11.0\ dB}$$

Example 11-3
A microphone delivers 36 mV to the 300-Ω input of an amplifier. The maximum ac power in a 16-Ω speaker system is 15 W. Determine the gain of the amplifier in dB.

Solution No. 1
The gain may be determined by using the power relation

$$dB = 10 \log_{10} \frac{P_2}{P_1} \tag{11-1}$$

We have the output power (P_2) given as 15 W. We can determine the input power from

$$P_1 = \frac{V_1^2}{R_1} = \frac{0.036^2}{300} = 4.32 \times 10^{-6}\ \text{W}$$

Substituting into Eq. 11-1, we have

$$dB = 10 \log_{10} \frac{P_2}{P_1} = 10 \log_{10} \frac{15}{4.32 \times 10^{-6}} = \mathbf{+65.4\ dB}$$

Solution No. 2

Alternatively, the gain may be determined from the derived equation

$$dB = 20 \log_{10} \frac{V_2}{V_1} + 10 \log_{10} \frac{R_1}{R_2} \qquad (11\text{-}3)$$

The voltage at the speaker is found from

$$P_2 = \frac{V_2^2}{R_2}$$

$$15 = \frac{V_2^2}{16}$$

$$V_2 = 15.48 \text{ V}$$

Substituting into Eq. 11-3, we have

$$dB = 20 \log_{10} \frac{15.48 \text{ V}}{0.036 \text{ V}} + 10 \log_{10} \frac{300 \, \Omega}{16 \, \Omega} = 52.7 + 12.7 = \mathbf{+65.4 \, dB}$$

When the impedance is not specified, it must be assumed that the two values are the same. Then the correction term $\log_{10} R_1/R_2$ is zero. It is standard practice in decibel calculations to insist that the sign + or − be associated with the numerical value. A +7 dB means a gain or increase in level of 7 decibels whereas −4 dB means a decrease in level or a loss of 4 decibels. Sometimes these figures are expressed as *7 dB up* and *4 dB down.*

The decibel values listed in Table 11-1 are very convenient numbers. The power ratios are ratios of whole numbers ($\frac{1}{4}$, $\frac{1}{2}$, 1, 2, and 4) that are often used to specify the properties of electronic devices. The voltage ratio of $1/\sqrt{2}$ is used to define

Table 11-1
Convenient Decibel Values

Decibels	Voltage Ratio	Power Ratio
−6	$\frac{1}{2}$ or 0.500	$\frac{1}{4}$ or 0.250
−3	$1/\sqrt{2}$ or 0.707	$\frac{1}{2}$ or 0.500
0	1	1
+3	$\sqrt{2}$ or 1.414	2
+6	2	4

the bandwidth and Q in ac circuits. In terms of decibels, then, the bandwidth and the Q are determined by the −3 dB values.

Zero dB References Very often it is useful to have a meter that is calibrated to read in decibels. Since the definition of the term *decibel* stated that the decibel is derived from a power ratio, a wattmeter with a new scale can be used. These special wattmeters are used especially at high radio frequencies but they are expensive. An ac voltmeter ordinarily serves as a decibel meter subject to certain restrictions. As 12 V across 30 Ω is not the same power as 12 V across 4000 Ω, the decibel meter needs the additional specification that its scale is accurate only when the meter is used on the specified impedance for which the instrument was calibrated.

There are many different standards for zero dB. Five of these that are commonly used are:

1. Zero dB refers to 6 mW dissipated in 500 Ω.
The reference voltage value corresponding to 0 dB is

$$P = \frac{V^2}{R}$$

$$0.006\ \text{W} = \frac{V^2}{500\ \Omega}$$

Solving for V, we have

$$V = 1.73\ \text{V}$$

2. Zero dB refers to 1 mW dissipated in 600 Ω.
The reference voltage value corresponding to 0 dB is

$$P = \frac{V^2}{R}$$

$$0.001\ \text{W} = \frac{V^2}{600\ \Omega}$$

Solving for V, we have

$$V = 0.774\ \text{V}$$

3. Zero dB refers to a 1-mW dissipation. This reference is given the unit symbol *dBm*. This reference is not dependent upon any particular load impedance value. Calculations are performed by using

$$dBm = 10 \log_{10} \frac{P_2}{0.001\ \text{W}}\ \text{dBm} \tag{11-6}$$

4. Zero dB refers to a 1.0-V level. This reference is given the unit symbol *dBV*. This reference is not dependent upon any particular load resistance value. Calculations are performed by using

$$dBV = 20 \log_{10} V_2\ \text{dBV} \tag{11-7}$$

The *dBV* reference is ideally suited to forming ratings under open-circuit conditions.

5. *The Volume Unit* (*VU*). Zero VU refers to a 1-mW dissipation in 600 Ω.

The VU is used principally in the radio broadcasting field. The VU is used only to read power levels in complex waves such as program lines carrying speech or music. A 0 VU means that the 0-VU complex wave has the same average power content that a 1-mW sinusoidal waveform has at a frequency of 100 Hz.

Problems

11-4.1 The gain of an amplifier is +46 dB. The amplifier delivers 3 W into a 4-Ω load. The amplifier input resistance is 150,000 Ω. What input voltage is necessary to produce full output power?

11-4.2 The input resistance to an amplifier is 175 Ω, and the output resistance is 3000 Ω. The amplifier gain is +28 dB. What is the voltage gain of the amplifier?

11-4.3 An amplifier drives a 16-Ω load. The hum-level rating of the amplifier is 90 dB below the full power-output rating, which is 80 W. What is the hum level in the load, and what voltage does the hum produce across the load?

11-4.4 The input resistance of an amplifier is 75 Ω, and the input current is 6 mA. The output resistance is 1500 Ω, and the output voltage is 16 V. What is the amplifier voltage gain, and what is the power gain? Express both in decibels.

11-4.5 The input to a 1400-foot, 50-Ω transmission line is 64 V. The output is 12 V when the load is matched. What is the loss of the transmission line expressed in decibels per hundred feet? Per hundred meters?

11-4.6 A phonograph pickup develops 15 mV across a 35-Ω input. A 60-W speaker system has an impedance of 16 Ω. What minimum amplifier gain in dB is necessary to produce full power output?

The general equation that is used for transmission of radio signals through free space is

$$P_r = P_t G_t G_r \left(\frac{\lambda}{4\pi R}\right)^2$$

where

G_t is the gain of the transmitting antenna
G_r is the gain of the receiving antenna
R is the distance in meters of the spacecraft from earth
P_t is the transmitter power
P_r is the receiver power
and λ is the wavelength of the radio signal in meters given by $(300/f)$ where f is the frequency in MHz.

$\left(\frac{4\pi R}{\lambda}\right)^2$ is the "space loss" of the signal through space.

11-4.7 The Mariner spacecraft carried a 20-W transmitter operating at 480 MHz. The spacecraft antenna had a gain of +12 dB. The receiving antenna used to track the spacecraft had a gain of +80 dB and an impedance of 50 Ω. What was the available signal voltage from the tracking antenna from the spacecraft signals? The closest approach distance of Venus was 40,000,000 kilometers.

11-4.8 The moon is approximately 368,000 kilometers from earth. If a radio wave is sent to the moon, 15% of the signal is reflected back. On earth an antenna having a gain of +80 dB is available with a receiver that can respond to a signal at a −110 dBm level. Using an identical antenna for transmission at 100 MHz, what transmitter power do we require on earth to obtain a reflected signal from the moon?

11-4.9 Three diodes are connected in series to serve as a variable attenuator. Assume that the dc forward voltage drop across each diode is 0.5 V and that the ac resistance of each diode is given by 25 mV/I. R is a potentiometer. Determine V_{out} when R is set to 0 Ω, to 80 kΩ, to 230 kΩ, to 500 kΩ, and to 2 MΩ. Neglecting the impedance-level correction terms, what is the loss of the attenuator in decibels for each given setting of the potentiometer?

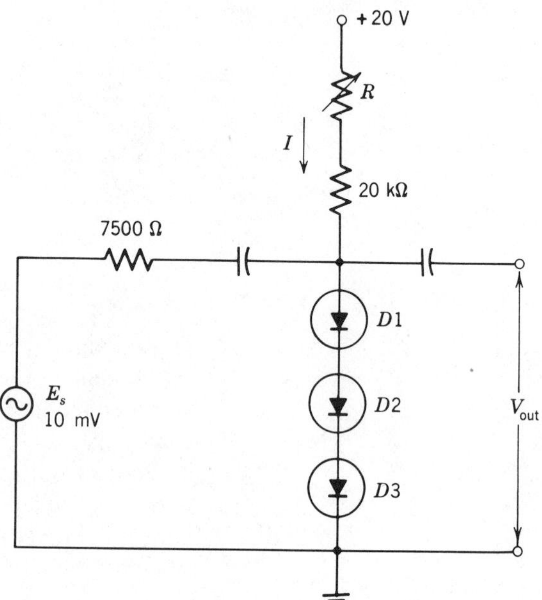

Circuit for Problem 11-4.9.

Chapter 12 Special Amplifiers

A two-stage direct-coupled amplifier can use the principle of complementary symmetry. The circuit can be arranged so that the dc level of the input is identical to the dc level at the output (Section 12-1). The Darlington pair, a form of the emitter follower, is used to obtain a high current gain and a high input resistance (Section 12-2). A differential amplifier can be arranged to provide a balanced output (Section 12-3) or an unbalanced (a single-ended) output (Section 12-4). The effects of an unbalance within the differential amplifier are evaluated by defining the common-mode rejection ratio (Section 12-5). In order to improve the common-mode rejection ratio, constant-current stabilization is often used in a differential-amplifier circuit (Section 12-6). The differential amplifier is the building block used in the operational amplifier (Section 12-7).

Section 12-1 Complementary Symmetry

Complementary symmetry is the circuit connection concept that is used to obtain:

1. Multistage amplifiers without the use of coupling capacitors.
2. Push-pull power amplifiers without the use of transformers (Section 14-6).

Manufacturers provide *complementary symmetry pairs.* For example RCA Corporation makes the pair RCA1C10 and RCA1C11 for use in a 12-W amplifier. The RCA1C10 is an *NPN* transistor and the RCA1C11 is a *PNP* transistor, each having identical magnitude ratings for maximum values of P_C, I_B, I_C, V_{CE} (V_{CEO}), and V_{CB} (V_{CBO}).

The circuit (Fig. 12-1) uses two transistors. In this example, $Q1$ is a *PNP* transistor and $Q2$ is an *NPN* transistor. In order to understand the circuit operation, the dc operating voltages are given on the diagram. Since the base voltage on $Q1$ is

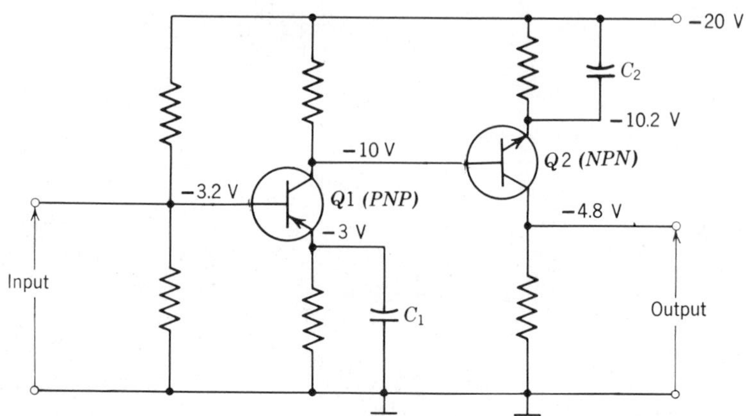

Figure 12-1 Circuit showing complementary symmetry.

−3.2 V and the emitter voltage is −3 V, the emitter is 0.2 V positive with respect to the base, which is the correct bias polarity for the *PNP* transistor. The collector, −10 V, is connected to the base of *Q*2. The emitter voltage of *Q*2 is −10.2 V. The emitter of *Q*2, the *NPN* unit, is negative with respect to the base. The collector voltage of *Q*2 is −4.8 V, making the collector less negative or positive with respect to the emitter. The voltage distribution of this circuit is sketched in Fig. 12-2, showing the relative voltages of the electrodes of the transistors. By adjusting the values of the resistors, the output dc potential can be made the same as the input dc potential.

This two-stage amplifier has two common-emitter stages connected in cascade without coupling capacitors. Each stage of the amplifier gives a 180° phase inversion with the result that the input signal and the output signal are in phase.

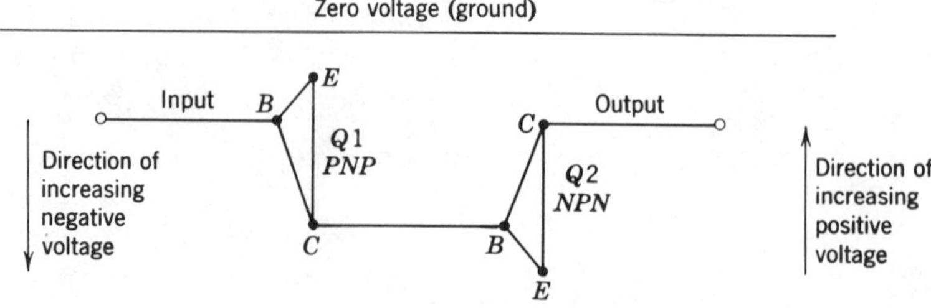

Figure 12-2 Operating voltage levels in complementary symmetry.

Problems **12-1.1** The active elements of the two-stage direct-coupled *IC* amplifier have the following values

$$\begin{array}{llll} Q1: & V_{BE} = 150\text{ mV}, & r'_e = 50\text{ mV}/I_E, & \beta = 40 \\ Q2: & V_{BE} = 180\text{ mV}, & r'_e = 50\text{ mV}/I_E, & I_C = 1.4\text{ mA}, \beta = 35 \\ D1: & r'_e = 50\text{ mV}/I, & V_F = 700\text{ mV} & \end{array}$$

Determine the value of R and the dc levels at the input and the output terminals. What is the circuit voltage gain and what is the maximum peak-to-peak output voltage without clipping?

12-1.2 An *IC* made to the specifications of Problem 12-1.1 now has a

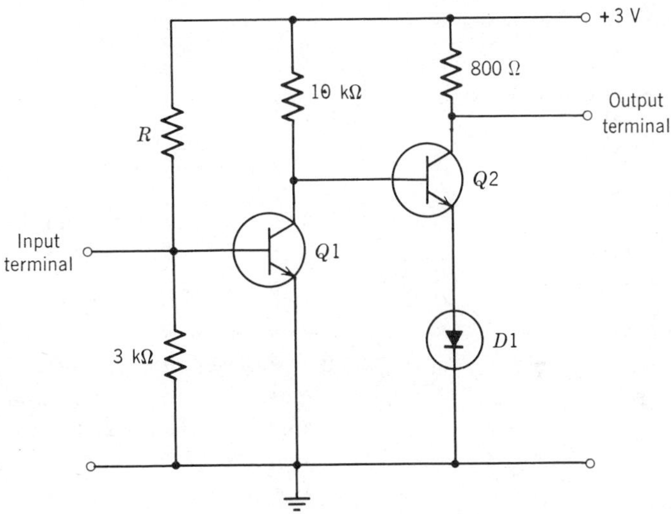

Circuit for Problems 12-1.1 and 12-1.2.

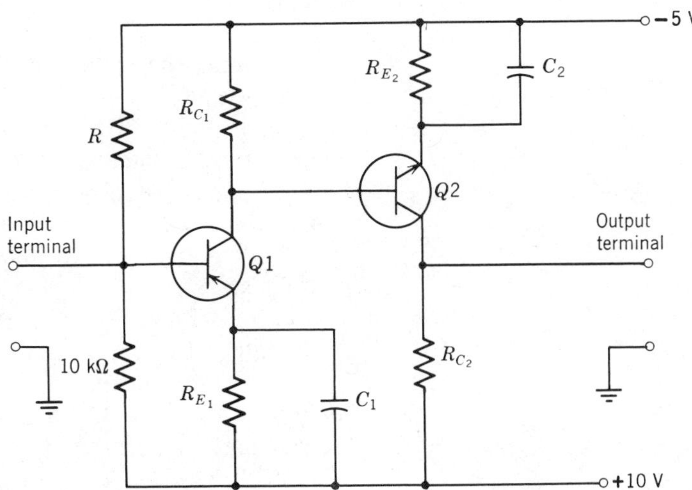

Circuit for Problems 12-1.3. and 12-1.4.

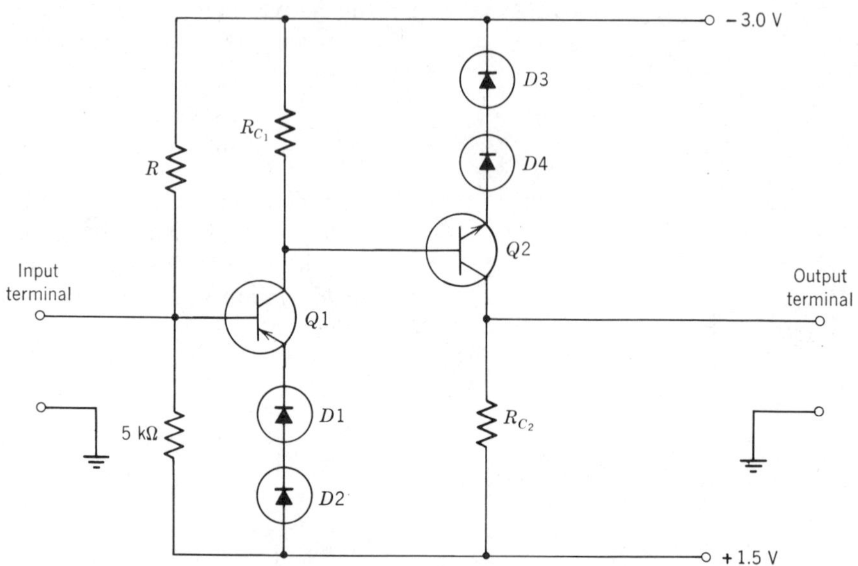

Circuit for Problems 12-1.5 and 12-1.6.

value of 50 for β for each transistor. What is the effect of the increase in β on the performance of the *IC*?

12-1.3 The dc levels at the input terminal and at the output terminal must be zero without signal in the complementary symmetry amplifier. Both transistors have the characteristics:

$$I_C = 5 \text{ mA}, \quad r_e' = 50 \text{ mV}/I_E, \quad V_{BE} = 0.25 \text{ V}, \quad V_{CE} = 3 \text{ V},$$
$$\text{and} \quad \beta = 50$$

Determine the values of the various resistors and the circuit voltage gain.

12-1.4 If C_2 is omitted from the amplifier of Problem 12-1.3, what is the circuit voltage gain?

12-1.5 The complementary symmetry amplifier uses diodes in the emitter circuit. The diodes have a forward voltage drop of 600 mV and an ac resistance given by $r_e' = 50$ mV/I. For the transistors, β is 50, V_{BE} is 300 mV, and $r_e' = 50$ mV/I_E. I_{C_1} is 1 mA and I_{C_2} is 5 mA. The dc input and output levels are at ground potential. Determine the required resistance values and the circuit voltage gain.

12-1.6 Repeat Problem 12-1.5 if three diodes are used in series in each emitter circuit and if the positive voltage supply is raised to 2.1 V.

Section 12-2 The Darlington Pair The *Darlington pair* is the name given to a circuit (Fig. 12-3*a*) in which the emitter of one transistor is connected directly to the base of a second transistor. The emitter current of the first transistor is the base current of the second transistor. Darlington pairs are available commercially mounted in a single case that has only three leads: the collector lead, the base input lead to the first transistor, and the emitter output lead from the second transistor. The Darlington pair connection is readily formed from two adjacent transistors in microcircuits.

In this section and in the sections on the differential amplifiers, we will summarize the properties of the circuit and then proceed to show how they are obtained.

The essential properties of the Darlington pair as compared to an emitter follower are:

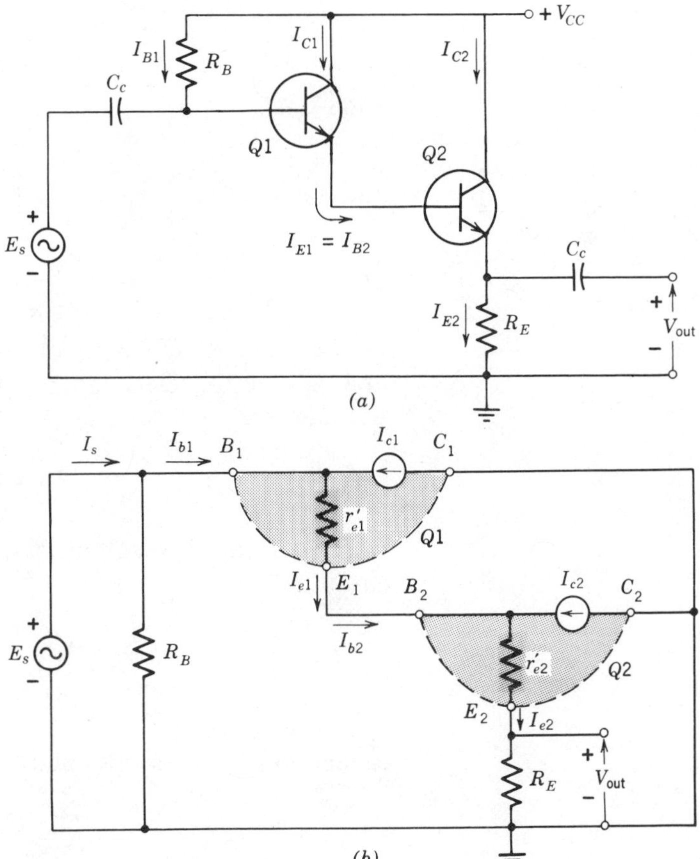

Figure 12-3 The Darlington pair. (*a*) Circuit. (*b*) ac model.

1. The input resistance $(1+\beta)^2(r'_e+R_E)$ to the Darlington pair is higher than the input resistance to the emitter follower $(1+\beta)(r'_e+R_E)$.
2. The current gain of the Darlington pair $(1+\beta)^2$ is higher than the current gain of the emitter follower $(1+\beta)$.
3. The voltage gain of the Darlington pair is identical to the voltage gain of the emitter follower $R_E/(r'_e+R_E)$.

A Darlington pair is usually formed by connecting two adjacent transistors in an IC. The parameters of adjacent transistors are so close that we can assume

$$\beta_1 = \beta_2 = \beta$$

and

$$V_{BE_1} = V_{BE_2} = V_{BE}$$

The dc analysis of the circuit requires a Kirchhoff's voltage loop equation through the base circuits of Fig. 12-3*a*.

$$V_{CC} = R_B I_{B_1} + 2V_{BE} + R_E I_{E_2} \qquad (12\text{-}1)$$

where

$$I_{B_2} = \frac{I_{E_2}}{1+\beta} = I_{E_1}$$

and

$$I_{B_1} = \frac{I_{E_1}}{1+\beta} = \frac{I_{E_2}}{(1+\beta)^2}$$

or

$$I_{E_2} = (1+\beta)^2 I_{B_1}$$

The exact ac signal model (Fig. 12-3*b*) is used for the ac signal analysis of the Darlington pair. In this analysis we will also assume that

$$r'_{e_1} = r'_{e_2} = r'_e$$

The Kirchhoff's voltage loop equation for the input to the circuit is

$$\begin{aligned} E_s &= r'_e I_{e_1} + r'_e I_{e_2} + R_E I_{e_2} \\ &= r'_e I_{e_1} + (r'_e + R_E) I_{e_2} \qquad (12\text{-}2) \end{aligned}$$

Inspection of the model shows that

$$(1+\beta) I_{b_2} = I_{e_2} \qquad (12\text{-}3a)$$

$$I_{b_2} = I_{e_1}$$

and
$$(1+\beta)I_{b_1} = I_{e_1} \tag{12-3b}$$

We can write

$$I_{e_2} = (1+\beta)I_{b_2} = (1+\beta)I_{e_1} = (1+\beta)^2 I_{b_1} \tag{12-3c}$$

Substituting Eq. 12-3*b* and Eq. 12-3*c* into Eq. 12-2, we have

$$E_s = [(1+\beta)r_e' + (1+\beta)^2(r_e' + R_E)]I_{b_1} \tag{12-4}$$

If both sides of this equation are divided by I_{b_1}, we have the input resistance r_{in} to the Darlington amplifier

$$r_{\text{in}} = (1+\beta)r_e' + (1+\beta)^2(r_e' + R_E)$$

Since

$$(1+\beta)^2(r_e' + R_E) \gg (1+\beta)r_e'$$

$$\boxed{r_{\text{in}} \approx (1+\beta)^2(r_e' + R_E)} \tag{12-5}$$

If both sides of Eq. 12-3*c* are divided by I_{b_1}, we have the current gain A_i of the circuit.

$$A_i = \frac{I_{e_2}}{I_{b_1}}$$

$$\boxed{A_i = (1+\beta)^2} \tag{12-6}$$

The output voltage of the circuit is

$$V_{\text{out}} = R_E I_{e_2}$$

Substituting Eq. 12-3*c* for I_{e_2}, we have

$$V_{\text{out}} = (1+\beta)^2 R_E I_{b_1}$$

and dividing this equation by Eq. 12-4, we have the voltage gain A_v across the Darlington pair.

$$A_v = \frac{V_{\text{out}}}{E_s} = \frac{(1+\beta)^2 R_E}{(1+\beta)r_e' + (1+\beta)^2(r_e' + R_E)}$$

Dividing each term by $(1+\beta)^2$, we have

$$A_v = \frac{R_E}{\dfrac{r'_e}{1+\beta} + r'_e + R_E}$$

But $r'_e/(1+\beta)$ is very small. Then

$$A_v = \frac{R_E}{r'_e + R_E} \leq 1 \qquad (12\text{-}7)$$

Problems For all problems, r'_e is $50\text{ mV}/I_E$.

12-2.1 R_E is $1000\ \Omega$, R_2 is infinite, R_s is $10\text{ k}\Omega$, and V_{CE} for $Q2$ is 5 V. V_{BE} is 700 mV and β is 50 for $Q1$ and $Q2$. Find R_1. Find E_s and I_s that produce the largest output signal without clipping.

12-2.2 R_E is $10\ \Omega$, R_s is $50\text{ k}\Omega$, R_2 is $1\text{ M}\Omega$, V_{BE} is 600 mV, and β is 60. V_{CE} is 5 V for $Q2$. Find the current gain, the voltage gain, and the input impedance.

12-2.3 Repeat Problem 12-2.2 if transistors that have a β of 120 are used. Compare the results.

12-2.4 The load R_E is a control winding of a dc generator. The transistors have a β of 100 and a value of 500 mV for V_{BE}. R_1

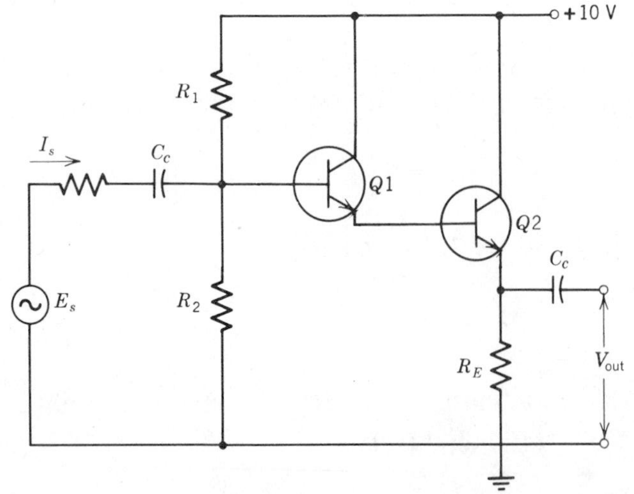

Circuit for Problems 12-2.1 through 12-2.3.

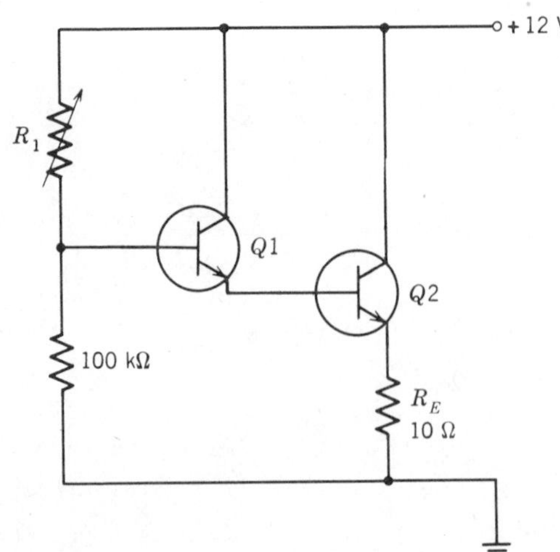

Circuit for Problems 12-2.4 and 12-2.5.

is a control resistor that is used to vary the current in the generator control winding from 0.3 A to 0.9 A. What is the required range of variation in R_1?

12-2.5 Repeat Problem 12-2.4 if the transistors with a β of 140 are used.

Section 12-3 The Differential Amplifier with Balanced Output

The circuit for the *differential amplifier* or *difference amplifier* having a *balanced output* is shown in Fig. 12-4*a*. There are two input signals to the differential amplifier: V_{in_1} to transistor $Q1$ and V_{in_2} to transistor $Q2$. The output from the circuit V_{out} is taken between the collectors of $Q1$ and $Q2$.

The differential amplifier is not intended to amplify each of V_{in_1} and V_{in_2}. It is intended only to amplify the difference between V_{in_1} and V_{in_2}. In our analysis of this circuit and the circuit in the next section, we will assume that V_{in_1} is greater than V_{in_2}.

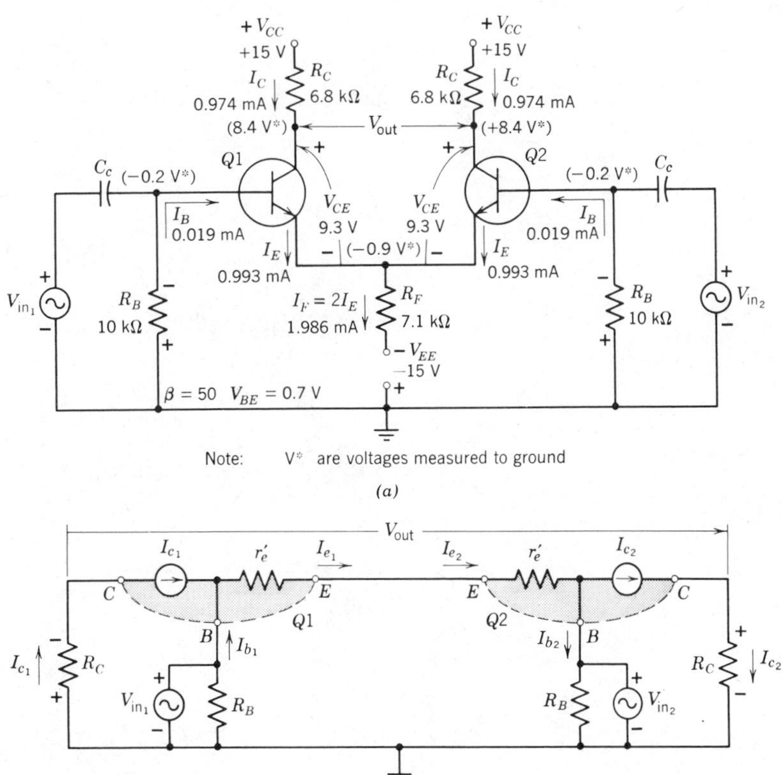

Figure 12-4 Differential amplifier with balanced output. (*a*) Circuit. (*b*) ac model.

$$V_{\text{in}_1} > V_{\text{in}_2} \tag{12-8a}$$

Then the circuit amplifies V_{in} where

$$V_{\text{in}} = V_{\text{in}_1} - V_{\text{in}_2} \tag{12-8b}$$

Transistor $Q1$ is an amplifier in the common-emitter circuit arrangement. Also transistor $Q2$ is an amplifier in the common-emitter circuit arrangement. However, there is a major difference between the differential amplifier and the conventional amplifier.

The emitters of $Q1$ and $Q2$ are tied together to a common resistor R_F that goes to the emitter supply $-V_{EE}$.

Before we examine the ac model, let us verify the dc values of current and voltage shown on the circuit diagram (Fig. 12-4a). The Kirchhoff's voltage loop equation through the base circuit is

$$|-V_{EE}| = R_B I_B + V_{BE} + R_F(2I_E)$$

Recalling from Table 4-2 that

$$I_E = (1+\beta)I_B$$

We see that

$$|-V_{EE}| = R_B I_B + V_{BE} + R_F[2(1+\beta)I_B]$$

Substituting values, we have

$$15\text{ V} = 10\text{ k}\Omega \times I_B + 0.7\text{ V} + 7.1\text{ k}\Omega \times 2 \times (1+50) \times I_B$$

Solving for I_B, we have

$$I_B = 0.019\text{ mA}$$

Then

$$I_C = \beta I_B = 50 \times 0.019\text{ mA} = 0.974\text{ mA}$$

and $$I_E = (1+\beta)I_B = 51 \times 0.019\text{ mA} = 0.993\text{ mA}$$

and $$I_F = 2I_E = 2 \times 0.993\text{ mA} = 1.986\text{ mA}$$

The Kirchhoff's voltage loop equation through the collector is

$$V_{CC} - (-V_{EE}) = I_C R_C + V_{CE} + 2I_E R_F$$

Substituting values, we have

$$15\text{ V} - (-15\text{ V}) = 0.974\text{ mA} \times 6.8\text{ k}\Omega + V_{CE} + 2 \times 0.993\text{ mA} \times 7.1\text{ k}\Omega$$

Solving for V_{CE}, we find

$$V_{CE} = 9.3\text{ V}$$

The voltage measured from the collector to ground is

$$V_{CC} - I_C R_C = 15\text{ V} - 0.974\text{ mA} \times 6.8\text{ k}\Omega = +8.4\text{ V}$$

The voltage measured from the emitter to ground is

$$-V_{EE} + 2I_E R_F = -15\text{ V} + 2 \times 0.993\text{ mA} \times 7.1\text{ k}\Omega = -0.9\text{ V}$$

The voltage measured from the base to ground is

$$-I_B R_B = -0.019\text{ mA} \times 10\text{ k}\Omega \approx -0.2\text{ V}$$

These voltages measured to ground are those values we would obtain using a voltmeter when servicing or testing the differential amplifier.

The emitter current is approximately 1 mA. The value of r'_e is determined from

$$\frac{25\text{ mV}}{I_E} \leq r'_e \leq \frac{50\text{ mV}}{I_E} \tag{7-5}$$

Thus, for this circuit, r'_e can range between 25 Ω and 50 Ω. Since R_F is 7.1 kΩ, it is obvious that

$$R_F \gg r'_e \tag{12-8c}$$

The resistor R_F that is common to the two emitters serves as a means of obtaining a *constant current* in R_F. For example, if I_E in $Q1$ increases by 20% from 0.993 mA to 1.192 mA, the current in R_F becomes 1.192 mA from $Q1$ plus 0.993 mA from $Q2$ or a total of 2.185 mA. The voltage drop across R_F is now

$$I_F R_F = 2.185 \text{ mA} \times 7.1 \text{ k}\Omega = 15.51 \text{ V}$$

The voltage measured from both emitters to ground is

$$-V_{EE} + I_F R_F = -15 \text{ V} + 15.51 \text{ V} = +0.51 \text{ V}$$

If the voltage from the emitter to ground were +0.51 V, the base-to-emitter junctions of both transistors would have a reverse bias and all currents in $Q1$ and $Q2$ would be zero. Consequently, the condition we assumed, that I_E in $Q1$ can increase while I_E in $Q2$ does not change, is impossible.

Therefore, when the emitter current in $Q1$ rises 20% from 0.993 mA to 1.192 mA, the emitter current in $Q2$ must decrease by a like amount from 0.993 mA to 0.794 mA in order that I_F $(I_{E_1} + I_{E_2})$ keep constant at 1.986 mA.

From the viewpoint of the ac model, this constant current feature of R_F means that R_F does not appear in the ac model at all. By Norton's theorem, the ac resistance associated with a constant-current source is infinite (an open circuit).

In order to care for the constant-current concept in the ac model, we must show the direction of I_{e_2} to be the same as the direction of I_{e_1}. The ac model is drawn in Fig. 12-4*b*. Inspection of the ac model shows that

$$V_{\text{out}} = I_{c_1} R_C + I_{c_2} R_C$$

and

$$V_{\text{in}_1} - V_{\text{in}_2} = I_{e_1} r'_e + I_{e_2} r'_e$$

Since I_{c_1} must equal I_{c_2} and I_{e_1} must equal I_{e_2}, these equations become

$$V_{\text{out}} = I_c R_C + I_c R_C = 2 I_c R_C$$

and

$$V_{\text{in}_1} - V_{\text{in}_2} = I_e r'_e + I_e r'_e = 2 I_e r'_e$$

We define the voltage gain of the differential amplifier as

$$A_v \equiv \frac{V_{out}}{V_{in_1} - V_{in_2}} \tag{12-9}$$

Substituting, we have

$$A_v = \frac{V_{out}}{V_{in_1} - V_{in_2}} = \frac{2I_cR_C}{2I_er'_e} = \frac{I_cR_C}{I_er'_e}$$

Using the conversion factors from Table 4-2, we find

$$A_v = \frac{\beta I_bR_C}{(1+\beta)I_br'_e} = \frac{\beta R_C}{(1+\beta)r'_e}$$

Since the ratio $\beta/(1+\beta)$ is approximately 1, we have

$$A_v = \frac{R_C}{r'_e} \tag{12-10}$$

When external resistors R_E are placed in each emitter lead, Eq. 12-10 becomes

$$A_v = \frac{R_C}{R_E + r'_e} \tag{12-11}$$

We specified for this analysis that

$$V_{in_1} > V_{in_2} \tag{12-8a}$$

Therefore, the base of $Q1$ is + with respect to the base of $Q2$. In the ac model (Fig. 12-4*b*) the collector of $Q2$ is + with respect to the collector of $Q1$. Thus there is the expected phase inversion through the differential amplifier circuit.

Problems For these problems use the circuit diagram of Fig. 12-4. All transistors have a β of 80 and are silicon. Use $r'_e = 25\ \text{mV}/I_E$; $V_{BE} = 0.7\ \text{V}$.

12-3.1 V_{CC} and V_{EE} are each 15 V. R_C is 20 kΩ, R_F is 20 kΩ, and R_B is 47 kΩ. Determine the dc operating levels in the circuit. Determine the circuit voltage gain, and the maximum available peak-to-peak output voltage without clipping.

12-3.2 Repeat Problem 12-3.1 if 200-Ω resistors R_E are placed externally in series with each emitter.

12-3.3 V_{CC} is +10 V, V_{EE} is −4 V, R_C is 10 kΩ, R_F is 3.9 kΩ, and R_B is 10 kΩ. Determine the dc operating levels of the transistors. Determine the circuit voltage gain and the maximum available peak-to-peak output voltage without clipping.

12-3.4 Repeat Problem 12-3.3 if 120-Ω resistors R_E are placed externally in series with each emitter.

Section 12-4 The Differential Amplifier with Unbalanced Output

The circuit for the *differential amplifier with unbalanced output* is shown in Fig. 12-5*a*. The modifications made to the circuit of the differential amplifier with balanced output are:

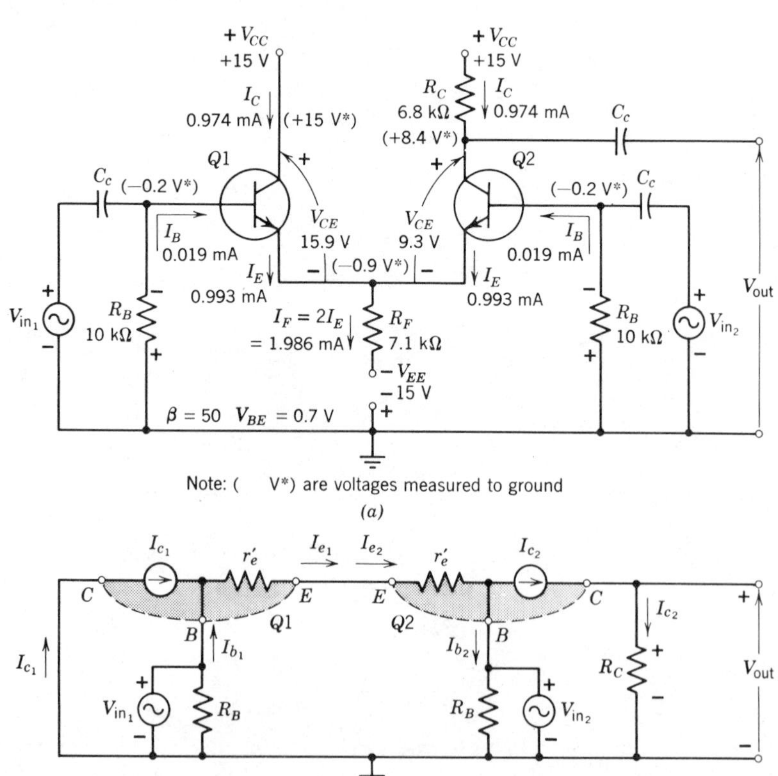

Figure 12-5 Differential amplifier with unbalanced output. (*a*) Circuit. (*b*) ac model.

1. The collector load resistor in $Q1$ is omitted ($R_C = 0\ \Omega$).
2. The output voltage is taken from the collector of $Q2$ to ground.

All the other circuit component values remain the same. In the previous section, we determined the dc currents in each transistor from a Kirchhoff's voltage loop equation through the base and emitter. This equation is identical for this circuit. Thus all the current values remain the same. The only change in the dc analysis is the value for V_{CE} for $Q1$. Since the collector of $Q1$ is connected directly to $+V_{CC}$, the value of V_{CE} is

$$V_{CE} = V_{CC} - (-0.9\text{ V}) = 15\text{ V} + 0.9\text{ V} = +15.9\text{ V}$$

This value is placed on Fig. 12-5*a*. All other dc values are carried over from Fig. 12-4*a*. R_F continues to serve as a constant current source and

$$R_F \gg r'_e \tag{12-8c}$$

In order to establish the current directions in the ac model (Fig. 12-5*b*) we also require for this analysis that the input voltage V_{in} is

$$V_{\text{in}} = V_{\text{in}_1} - V_{\text{in}_2} \tag{12-8b}$$

We also show in the ac model that the output voltage V_{out} is taken across R_C in the collector of $Q2$. The magnitude of the output voltage is

$$V_{\text{out}} = I_{c_2} R_C$$

The Kirchhoff's voltage loop equation through the base circuit is

$$V_{\text{in}_1} - V_{\text{in}_2} = I_{e_1} r'_e + I_{e_2} r'_e$$

Since I_{e_1} and I_{e_2} are equal

$$V_{\text{in}_1} - V_{\text{in}_2} = I_{e_2} r'_e + I_{e_2} r'_e = 2 I_{e_2} r'_e$$

The voltage gain is defined as

$$A_v \equiv \frac{V_{out}}{V_{in_1} - V_{in_2}} \tag{12-9}$$

Substituting, we have

$$A_v = \frac{I_{c_2} R_C}{2 I_{e_2} r'_e}$$

Using the conversion factors of Table 4-2, we have

$$A_v = \frac{\beta I_{b_2} R_C}{2(1+\beta) I_{b_2} r'_e} = \frac{\beta R_C}{2(1+\beta) r'_e}$$

Since the ratio $\beta/(1+\beta)$ is approximately 1, we have

$$A_v = \frac{R_C}{2r'_e} \tag{12-12}$$

When external resistors R_E are added to each emitter lead, the circuit gain becomes

$$A_v = \frac{R_C}{2(r'_e + R_E)} \tag{12-13}$$

This circuit also has the same phase relation between the output and the input as the previous circuit. We can summarize the phase relations by stating that, if we use the instantaneous polarity of the output voltage V_{out} as the reference, the input to the base of $Q1$ is the *noninverting input* and the input to the base of $Q2$ is the *inverting input.*

Problems For these problems use the circuit diagram of Fig. 12-5.
All transistors have a β of 100 and are silicon.
Use $r'_e = 25\ \text{mV}/I_E$; $V_{BE} = 0.7$ V.

12-4.1 V_{CC} and V_{EE} are each 20 V. R_C is 6.8 kΩ, R_F is 6.8 kΩ, and R_B is 24 kΩ. Determine the dc operating levels of the transistors.

Determine the circuit voltage gain and the maximum available peak-to-peak output voltage without clipping.

12-4.2 Repeat Problem 12-4.1 if 130-Ω external resistors R_E are placed in each emitter lead.

12-4.3 V_{CC} is +10 V, V_{EE} is −4V, R_C is 20 kΩ, R_F is 7.5 kΩ, and R_B is 20 kΩ. Determine the dc operating levels of the transistors. Determine the circuit voltage gain and the maximum available peak-to-peak output voltage without clipping.

12-4.4 Repeat Problem 12-4.3 if 270-Ω resistors R_E are placed in each emitter lead.

Section 12-5 Common-Mode Rejection Ratio

The two signals into a differential amplifier are V_{in_1} and V_{in_2} (Fig. 12-6*a*). The triangle is used to serve as a block diagram for the complete internal circuit. The gain of the amplifier is

$$A_v = \frac{V_{\text{out}}}{V_{\text{in}_1} - V_{\text{in}_2}}$$

or

$$V_{\text{out}} = A_v(V_{\text{in}_1} - V_{\text{in}_2}) \qquad (12\text{-}14)$$

Assume the gain of the amplifier is 100. Also, assume the peak values of two in-phase signals are 4.22 V (V_{in_1}) and 4.10 V (V_{in_2}). The net input signal is

$$(V_{\text{in}_1} - V_{\text{in}_2}) = 4.22 - 4.10 = 0.12 \text{ V}$$

and

$$V_{\text{out}} = A_v(V_{\text{in}_1} - V_{\text{in}_2}) = 100 \times 0.12 \text{ V} = 12 \text{ V}$$

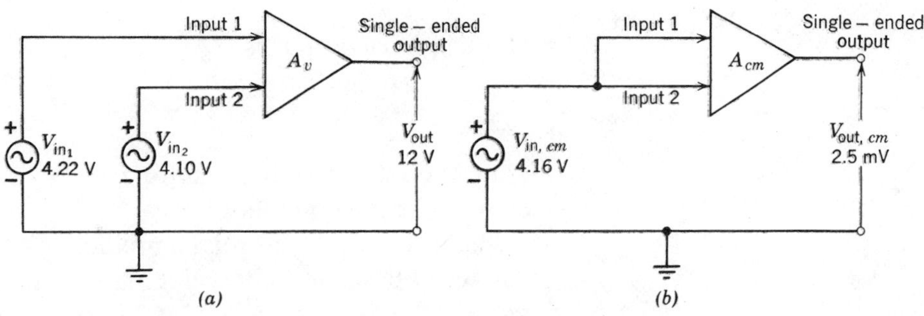

Figure 12-6 The differential amplifier. (*a*) Normal operation. (*b*) Test arrangement for common-mode output.

This 12-V output signal is the amplified value of the difference between V_{in_1} and V_{in_2}.

The mean value of V_{in_1} and V_{in_2} is

$$\frac{V_{in_1} + V_{in_2}}{2} = \frac{4.22 + 4.10}{2} = 4.16 \text{ V}$$

The net input signal to V_{in_1} is equivalent to $(4.22 - 4.16)$ or $+0.06$ V. The net input signal to V_{in_2} is equivalent to $(4.10 - 4.16)$ or -0.06 V. If we use these values

$$V_{in_1} - V_{in_2} = +0.06 - (-0.06) = 0.12 \text{ V}$$

and $$V_{out} = A_v(V_{in_1} - V_{in_2}) = 100 \times 0.12 \text{ V} = 12 \text{ V} \qquad (12\text{-}14)$$

which is the same output as obtained for input signals of 4.22 V and 4.10 V.

The average or mean value, 4.16 V, is *not* amplified; only the difference is amplified. The average value that is not amplified is the *common-mode signal* $V_{in,cm}$. The purpose of the differential amplifier is to *reject* the common-mode signal and to amplify only the difference signal. Obviously, there is a limit to the amount of the common-mode signal that can be applied to an amplifier without saturating it or even destroying it. This limiting common-mode value is specified by the manufacturer for commercial units.

Now, let us set one signal source to the common-mode value, 4.16 V, and connect this signal to *both* inputs (Fig. 12-6*b*). The difference signal is

$$V_{in_1} - V_{in_2} = 4.16 \text{ V} - 4.16 \text{ V} = 0 \text{ V}$$

The amplifier output is

$$V_{out} = A_v(V_{in_1} - V_{in_2}) = 100 \times 0 = 0 \text{ V} \qquad (12\text{-}14)$$

Thus the output signal resulting from common-mode input is zero in the ideal amplifier.

In a practical differential amplifier, we find that the output is not zero but some finite value. Let us assume that a *common-mode output voltage* $V_{out,cm}$ of 25 mV is produced by the 4.16-V common-mode input voltage $V_{in,cm}$ (Fig. 12-6*b*).

The presence of a common-mode output voltage results because the circuit is not exactly balanced. The two transistors are not exactly identical and corresponding component parts are also not exactly identical. In deriving the gain equation for A_v, we assumed

$$R_F \gg r'_e$$

Without this assumption, we could not have reduced the gain equation to

$$A_v = \frac{R_C}{2r'_e} \tag{12-12}$$

The total output voltage V'_{out} is the sum of the output produced by the difference input plus the output produced by the presence of a common-mode input signal.

$$V'_{\text{out}} = V_{\text{out}} + V_{\text{out},cm} \tag{12-15}$$

In the example we are using

$$V'_{\text{out}} = V_{\text{out}} + V_{\text{out},cm} = 12\ \text{V} + 25\ \text{mV} = 12.025\ \text{V}$$

In this case the 25-mV variation from the ideal value of 12 V is negligible.

Now we define the *common-mode* gain A_{cm} as

$$\boxed{A_{cm} \equiv \frac{V_{\text{out},cm}}{V_{\text{in},cm}}} \tag{12-16}$$

Using the numerical values we have assumed

$$A_{cm} = \frac{V_{\text{out},cm}}{V_{\text{in},cm}} = \frac{25\ \text{mV}}{4.16\ \text{V}} = \frac{25\ \text{mV}}{4160\ \text{mV}} = 0.006$$

The common-mode rejection ratio, CMRR, is defined as the ratio of the differential amplifier gain A_v to the common-mode gain A_{cm}.

$$\boxed{\text{CMRR} \equiv \frac{A_v}{A_{cm}}} \qquad (12\text{-}17)$$

Using numerical values, we have

$$\text{CMRR} = \frac{A_v}{A_{cm}} = \frac{100}{0.006} = 16667$$

The common-mode rejection ratio is usually given in decibels

$$\text{CMRR}_{dB} = 20 \log_{10} \text{CMRR dB} \qquad (12\text{-}18)$$

and for the numerical values

$$\text{CMRR}_{dB} = 20 \log_{10} \text{CMRR} = 20 \log_{10} 16667 = 84.4 \text{ dB}$$

In the numerical example we have used, when the common-mode rejection ratio of the differential amplifier is 84.4 dB, the common-mode output voltage $V_{\text{out},cm}$ is 25 mV for a common-mode input of 4.16 V. The effect of $V_{\text{out},cm}$ is negligible compared to the ideal output voltage V_{out} of 12 V.

Now, let us assume we have a less expensive differential amplifier that has a common-mode rejection ratio of 20 dB. Then

$$\text{CMRR}_{dB} = 20 \log_{10} \text{CMRR} \qquad (12\text{-}18)$$

Substituting, we have

$$20 = 20 \log_{10} \text{CMRR}$$

or

$$\text{CMRR} = 10$$

By Eq. 12-17

$$A_{cm} = \frac{A_v}{\text{CMRR}} = \frac{100}{10} = 10 \qquad (12\text{-}17)$$

Then

$$V'_{\text{out},cm} = A_{cm} V_{cm} = 10 \times 4.16 \text{ V} = 41.6 \text{ V} \qquad (12\text{-}16)$$

and

$$V'_{\text{out}} = V_{\text{out}} + V_{\text{out},cm} = 12 + 41.6 = 53.6 \text{ V}$$

Since the desired output signal is only 12 V, the low common-mode rejection ratio causes the desired signal to be masked out completely.

It should be noted very carefully that, under situations where there is no common-mode input voltage ($V_{cm} = 0$), the amplifier that has a CMRR of 20 dB will perform just as satisfactorily as the unit with the CMRR of 84.4 dB. In such an application, the extra cost of the higher CMRR would not be justified.

If Eq. 12-13, Eq. 12-16, and Eq. 12-17 are substituted into Eq. 12-15, we can show that

$$V'_{\text{out}} = \left[1 + \frac{1}{\text{CMRR}} \times \frac{V_{\text{in},cm}}{(V_{\text{in}_1} - V_{\text{in}_2})}\right] V_{\text{out}} \tag{12-19}$$

Substituting numerical values into Eq. 12-19, we have

$$V'_{\text{out}} = \left[1 + \frac{1}{10} \times \frac{4.16\text{ V}}{0.12\text{ V}}\right] 12\text{ V} = 4.47 \times 12\text{ V} = 53.6\text{ V}$$

Problems

12-5.1 A 20-V peak-to-peak signal is connected simultaneously to the two inputs of a differential plug-in to an oscilloscope. In the differential mode the signal on the oscilloscope is 100 μV peak to peak. What is the common-mode rejection ratio specification in decibels?

12-5.2 Another plug-in unit produces a 20-mV peak-to-peak signal on the oscilloscope when the common-mode signal is 5 V peak to peak. What is the common-mode rejection ratio of this plug-in unit in decibels?

12-5.3 A difference amplifier has a difference gain of 80 dB and a common-mode rejection ratio of 86 dB. One signal is 3.000 + 0.001 V and the other signal is 3.000 − 0.001 V. What is the desired output voltage and what is the actual output voltage?

12-5.4 A difference amplifier has a difference gain of 40 dB and a common-mode rejection ratio of 40 dB. One signal is 10.000 + 0.001 V and the other signal is 10.000 − 0.001 V. What is the desired output voltage, and what is the actual output voltage?

Section 12-6 Constant-Current Stabilization

In deriving the gain equation

$$A_v = \frac{R_C}{r'_e} \tag{12-10}$$

for the differential amplifier we assumed that

$$R_F \gg r'_e \tag{12-8c}$$

We will use an oversimplified approach to explain what has to be done to a circuit to get a high common-mode rejection ratio. Figure 12-7*a* shows one transistor of the circuit for a differential amplifier. It is evident that Eq. 12-10 is the gain equation for this circuit.

When there is an unbalance between the two transistors, R_F becomes involved in the gain equation (Fig. 12-7*b*) for the common-mode gain A_{cm}.

$$A_{cm} = \frac{R_C}{r'_e + KR_F} \approx \frac{R_C}{KR_F} \tag{12-20}$$

The symbol K stands for a number that is ideally infinite if the balance is ideal. Then, the common-mode output voltage $V_{\text{out},cm}$ is zero. When K is a finite number, A_{cm} is not zero and there is a common-mode output voltage $V_{\text{out},cm}$.

We will replace R_F with an ideal constant-current source (Fig. 12-7*c*). The constant current is equal to I_E of $Q1$ plus I_E of $Q2$. By Norton's theorem, an ideal constant-current source is in parallel with an infinite resistance. Consequently, KR_F in Eq. 12-20 becomes infinite.

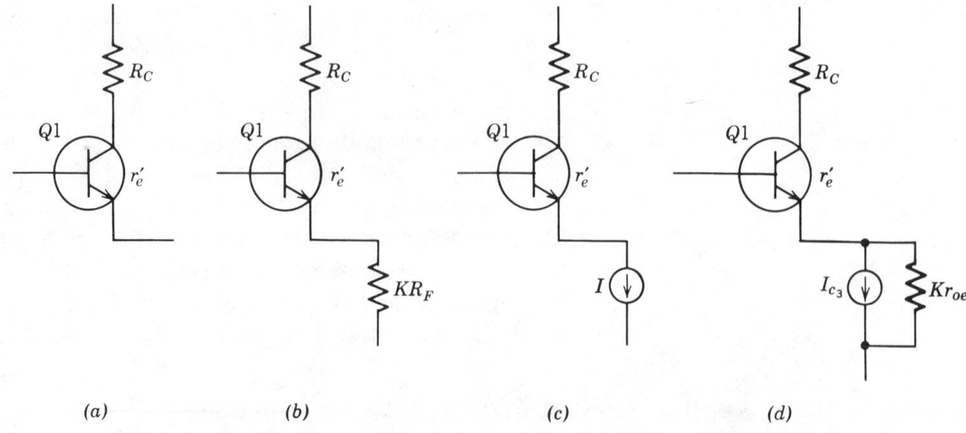

Figure 12-7 Circuits used to show common-mode gain. (*a*) Differential gain. (*b*) Use of R_F. (*c*) Ideal current source. (*d*) Transistor as a current source.

$$A_{cm} \approx \frac{R_C}{KR_F} = \frac{R_C}{\infty} = 0 \qquad (12\text{-}20)$$

Now the common-mode rejection ratio is infinite and the common-mode output voltage $V_{\text{out},cm}$ is zero.

We can only approximate an ideal current source in practice. We replace R_F with a transistor, $Q3$, making use of the fact that the collector current in a transistor is a constant current. The model for $Q3$ is shown in Fig. 12-7*d*. The transistor $Q3$ is not ideal but has an output resistance r_{oe}* in parallel with the constant-current source. Now the common-mode gain is

$$A_{cm} \approx \frac{R_C}{Kr_{oe}} \qquad (12\text{-}21)$$

The usual value of R_F is of the order of $5\,\text{k}\Omega$ and the expected value of r_{oe} is $100\,\text{k}\Omega$. With the same unbalance, there is an improvement of $100\,\text{k}\Omega/5\,\text{k}\Omega$ or 20 to 1. In decibels, this represents an improvement in the common-mode rejection ratio of $20 \log_{10} 20$ or 26 dB. As a result, most practical circuits use a transistor $Q3$ in place of R_F. A typical circuit using $Q3$ is shown in Fig. 12-8. In practice, a resistor R_E is usually added in series with each of the emitters of $Q1$ and $Q2$.

The constant-current stabilizing circuit shown in Fig. 12-9 uses a voltage-dividing circuit of R_1, R_2, R_3, $D1$, and $D2$. The use of two or more diodes provides *diode compensation* that compensates $Q3$ for changes in a variation of V_{BE} with temperature. As a result, an increase in temperature prevents any significant change in I_{C_3} or in $(I_{E_1} + I_{E_2})$.

In linear *IC*s, we often find that a *current mirror* (Fig. 12-10), is used to duplicate constant-current values. When two transistors, $Q1$ and $Q2$, are adjacent on a monolithic chip, their characteristics are very close to each other. Examination

*Transistor specification sheets usually list a value for h_{ob}, the conductance between collector and base in siemens. The reciprocal of h_{ob} is r_{ob} in ohms. However, when we use a common-emitter circuit, we must convert r_{ob} to r_{oe} by dividing by $(1+\beta)$ much with the same reasoning that we multiplied I_{CBO} by $(1+\beta)$ to obtain I_{CEO}.

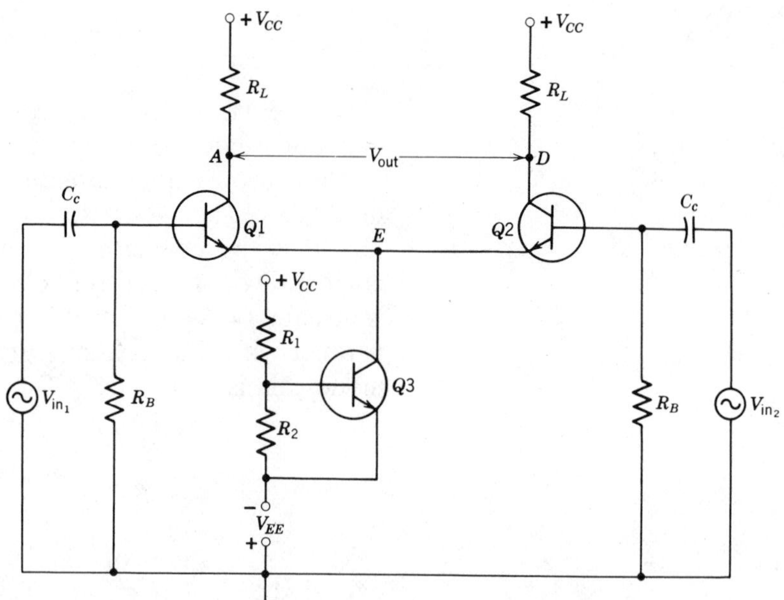

Figure 12-8 Differential amplifier with balanced output using constant-current stabilization.

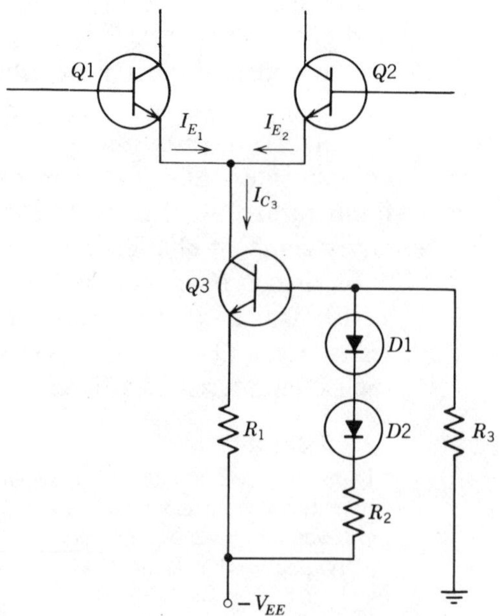

Figure 12-9 Modified constant-current circuit.

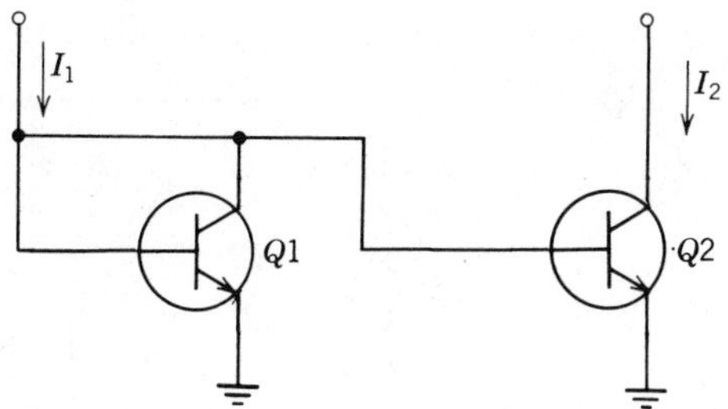

Figure 12-10 The current mirror.

of the circuit shows that

$$I_1 = I_{B_1} + I_{C_1} + I_{B_2}$$

and

$$I_2 = \beta_2 I_{B_2}$$

If the β values of $Q1$ and $Q2$ are equal and high, I_{B_1} is small with respect to I_{C_1}. Then

$$I_1 = I_{C_1}$$

and

$$I_2 = I_{C_2}$$

In the current mirror, if I_1 is a *forced current*, the *driven current* I_2 in $Q2$ is identical to I_1. Thus, by the use of this circuit, we can maintain two different constant currents at the same value. If the forced current is the current in Circuit A, the current mirror causes the current in Circuit B, the driven current, to be the same value at all times. If the current in Circuit A changes, so will the current in Circuit B maintain the same change.

Section 12-7 The Basic Operational-Amplifier Circuit

The basic circuit for a typical IC operational amplifier is shown in Fig. 12-11. The transistor pair $Q1$–$Q2$ forms a differential amplifier having a balanced output as discussed in Section 12-3. The outputs from the collectors of $Q1$ and $Q2$ drive the bases of transistors $Q3$ and $Q4$. The transistor pair $Q3$–$Q4$ forms a differential amplifier with an unbalanced output as dicusssed in Section 12-4. The single output from the collector of $Q4$ drives the base of transistor $Q5$. Transistors $Q5$ and $Q6$ form a Darlington pair as discussed in

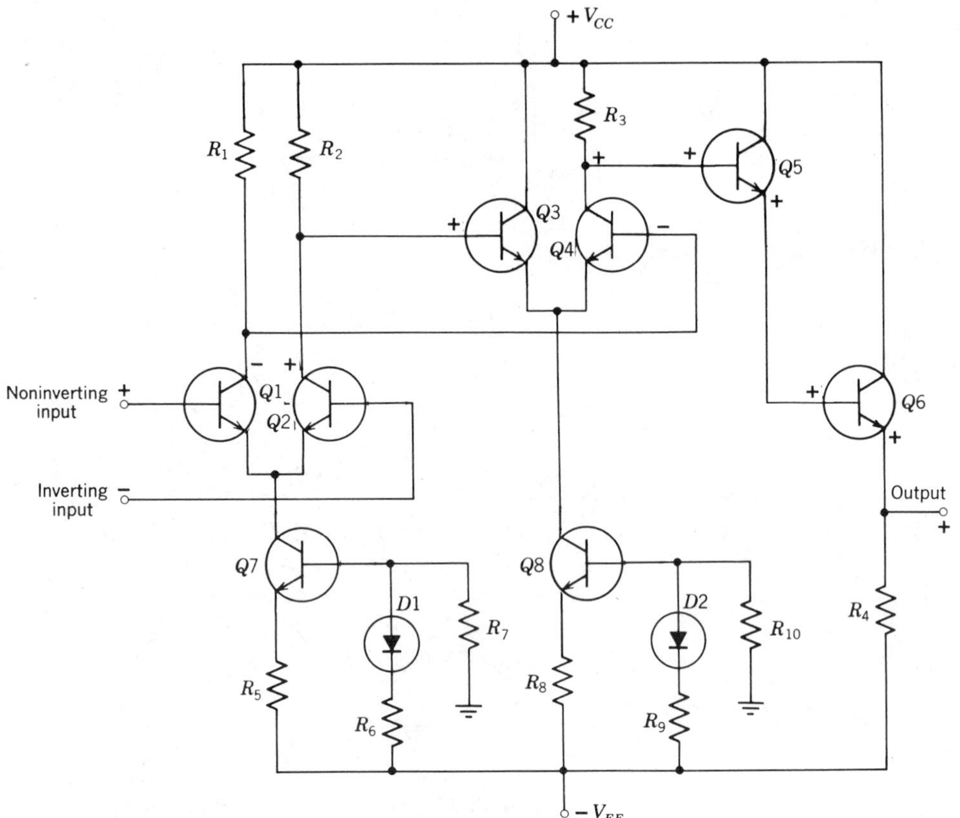

Figure 12-11 The basic circuit of the operational amplifier.

Section 12-2. The output voltage is taken from R_4, which is in the emitter of *Q*6.

Transistor *Q*7 is a constant-current stabilizing circuit (Section 12-6) for the differential amplifier pair *Q*1–*Q*2. Then R_3, R_5, R_6, and *D*1 form the temperature-compensating circuit. Similarly, transistor *Q*8 forms the constant-current stabilizing circuit for the differential-amplifier pair *Q*3 and *Q*4. The purpose of these stabilizing circuits is to improve the common-mode rejection ratio.

The noninverting signal terminal is marked +. The *Q*1 inverts the phase and a label – is placed on the collector. The signal fed into the base of *Q*4 is –. Again *Q*4 inverts the phase and a label + is placed on the collector. This + signal drives the base of *Q*5. Since *Q*5 and *Q*6 form a Darlington pair, the label + is placed on the emitter of $Q5$, on the base of $Q6$, and on the emitter of $Q6$. Thus the output terminal has the label +, that is, the output has the same phase as the noninverting input. Accordingly, the inverting input terminal is marked −. Transistor $Q2$ inverts this signal. The collector of $Q2$ and the base of $Q3$ are marked +.

In a practical IC operational amplifier circuit, we find two or three additional transistors connected into the circuit. The function of these additional transistors is to provide feedback circuits that further improve the balance of the differential amplifier and consequently improve the common-mode rejection ratio. These additional circuits also minimize dc level shifting.

All of the circuits used for differential amplifiers and for constant-current stabilizing circuits can be constructed from FETs in place of the bipolar transistors we have used in this discussion. The FETs have the advantage of having a higher input impedance than transistor circuits. Also the FETs are simpler to fabricate.

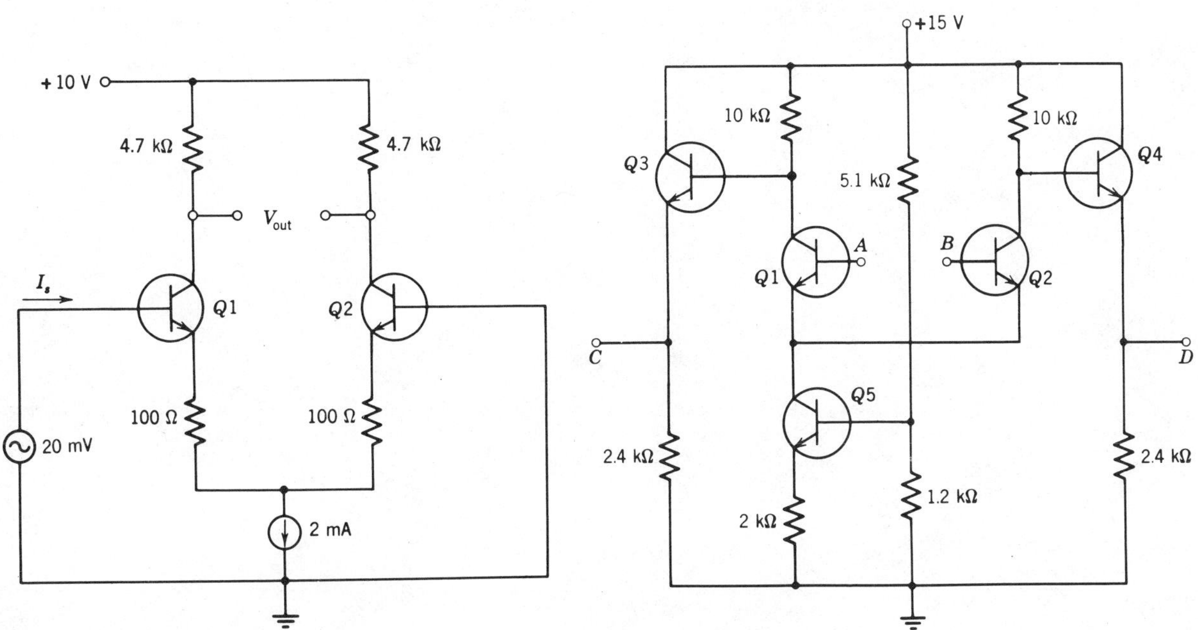

Circuit for Problem 12-7.1.

Circuit for Problem 12-7.2.

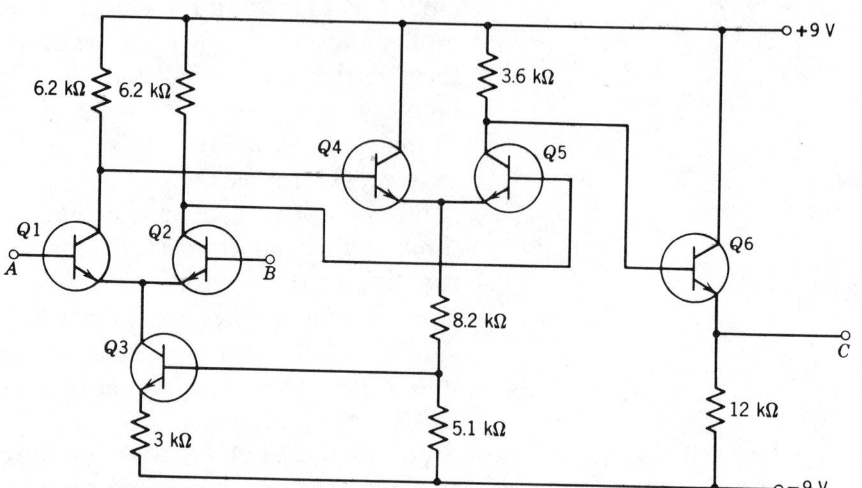

Circuit for Problem 12-7.3.

Problems All transistors are silicon with a β of 80; $r'_e = 25\ \text{mV}/I_E$; $V_{BE} = 0.7$ V.

12-7.1 Determine V_{out} and I_s.

12-7.2 For the given circuit of the IC chip, determine the differential gain and output voltage swing.

12-7.3 Repeat the requirements of Problem 12-7.2 for this IC circuit.

Chapter 13 Single-Ended Power Amplifiers

Constant power dissipation curves plotted on a collector characteristic are hyperbolas. This property enables us to determine the properties for maximum power dissipation in a transistor (Section 13-1). The technician should be able to select a suitable heat sink for a transistor (Section 13-2). Transformer coupling is used in power amplifiers. The method of showing a load line on the collector characteristic is shown for an amplifier using a transformer (Section 13-3). The ac load power in a class-A amplifier is examined. Maximum output power conditions are established together with values for collector efficiency and collector power dissipation. Also the method of determining the required input power to the stage is explained (Section 13-4).

Section 13-1 Power Dissipation

If V_{CE} is the collector voltage and if I_C is the collector current in a common-emitter amplifier, the collector dissipation P_C in watts is the product of V_{CE} and I_C.

$$\boxed{P_C = V_{CE} I_C} \qquad (13\text{-}1)$$

The collector dissipation produces heating within the transistor at the collector-to-base junction. This heating causes the *junction temperature* T_j to increase. By destructive testing, the manufacturer determines the maximum permissible value of T_j and accordingly gives the transistor a maximum allowable value for P_C.

When Eq. 13-1 is plotted for a specific value of P_C on a collector characteristic (Fig. 13-1), the resulting curve is a hyperbola. Now, at a point Q, a tangent is drawn to the constant power-dissipation curve. This tangent intersects the

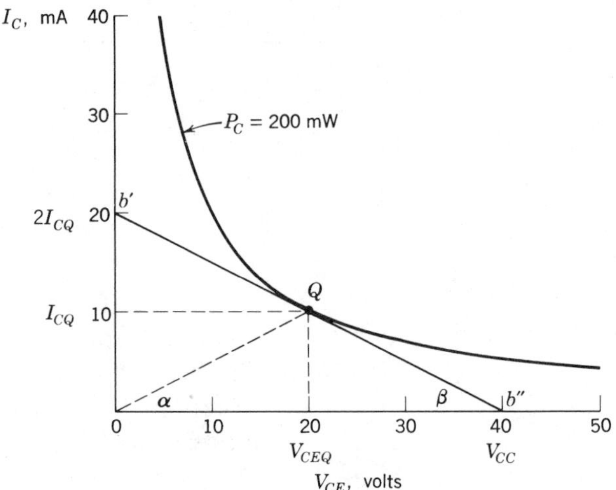

Figure 13-1 Tangent drawn to a power-dissipation curve.

axes at b' and at b'', respectively. If we call Q the operating point of a common-emitter amplifier, point b'' is the supply voltage V_{CC} to the circuit and point b' is the value of collector current determined by V_{CC}/R_C. The coordinates of the operating point are I_{CQ} and V_{CEQ}. We wish to show that point Q must be the midpoint of the tangent line between b' and b''.

If we solve Eq. 13-1 for I_C and differentiate the result using methods of calculus, we obtain the value of the slope m_1 of the power dissipation curve at point Q. The collector current is

$$I_C = \frac{P_C}{V_{CE}}$$

Differentiating with respect to V_{CE}, we find

$$m_1 = \frac{dI_C}{dV_{CE}} = -\frac{P_C}{V_{CE}^2}$$

Substituting Eq. 13-1 into this result, we have

$$m_1 = -\frac{P_C}{V_{CE}^2} = -\frac{V_{CE}I_C}{V_{CE}^2} = -\frac{I_C}{V_{CE}}$$

At the operating point

$$I_C = I_{CQ} \quad \text{and} \quad V_{CE} = V_{CEQ}$$

Then, at the operating point,

$$m_1 = -\frac{I_{CQ}}{V_{CEQ}}$$

If we draw a line from the origin to the operating point, this line has a slope m_2 given by

$$m_2 = \frac{I_{CQ} - 0}{V_{CEQ} - 0} = \frac{I_{CQ}}{V_{CEQ}}$$

A comparison of m_1 and m_2 shows that

$$m_2 = -m_1$$

Thus, the "going-up" slope of 0–Q is identical to the "going-down" slope of Q–b''. Angle α equals angle β. Triangle 0–Q–b'' is isosceles. Then, by congruency, V_{CEQ} must be the midpoint between 0 and V_{CC}. Similarly, I_{CQ} must be the midpoint of the line between 0 and b'. Thus, the value of current at b' is $2I_{CQ}$. We can summarize by stating that:

Any load line that is tangent to a curve of constant power dissipation P_C is bisected at the point of tangency.

In Fig. 13-2, we have the same dissipation curve and the same load line that were used in Fig. 13-1. Additionally, we

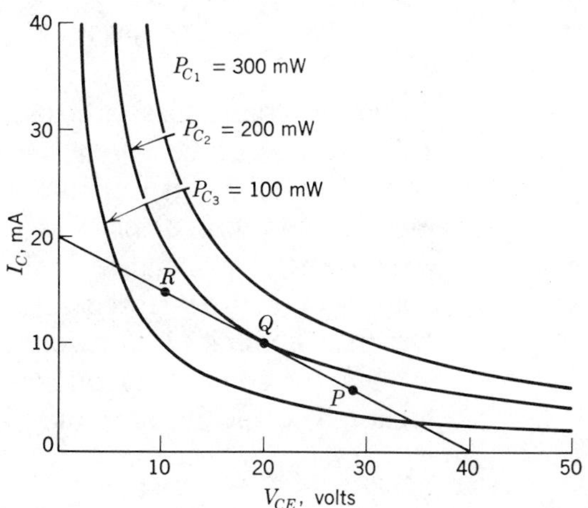

Figure 13-2 Collector characteristics showing different dissipation curves.

have shown two other power dissipation curves where P_{C_1} is greater than P_{C_2} and where P_{C_3} is less than P_{C_2}. It is obvious from this graph that the load line cannot intersect P_{C_1}. On the other hand, the load line intersects P_{C_3} twice.

We can draw a very important conclusion from these curves:

> *If we design a load line to have the operating point at the midpoint of the load line, any shift in the operating point, from Q toward R or from Q toward P, must cause the collector dissipation to decrease. Accordingly, the midpoint of the load line, point Q, is the worst case. If we design the heat dissipating arrangement for the worst case, we are unconditionally safe.*

Section 13-2 Heat Sinks

All semiconductors have a collector dissipation rating that is stated as a function of temperature. These dissipation ratings are established by the manufacturer as the result of extensive destructive testing. As may be recalled, at a certain temperature, the crystalline structure is destroyed, and there is no recovery or second chance once this has happened. The critical point in the transistor is the junction between the collector and the base. The maximum allowable junction temperature is given as T_J in degrees centigrade. The lower limit of a semiconductor is taken as −65°C. Therefore the restriction on the operating temperature range of a semiconductor is

$$-65°\text{C} \leq T_J \leq T_{J,\text{MAX}}$$

Before transistors are used, they are stored and consideration must be given to storage temperature, T_{STG}. The storage concept also applies to a completed piece of equipment that is turned off or that is serving as a spare. Military equipment, for instance, can be placed in a warehouse in Alaska or can be stored in a metal container in the Sahara Desert. The limit of storage temperature is usually the limit of T_J, but for some semiconductors the limit can be somewhat higher. The specification may read, for example,

$$-65°\text{C} \leq T_J = T_{STG} \leq 150°\text{C}$$

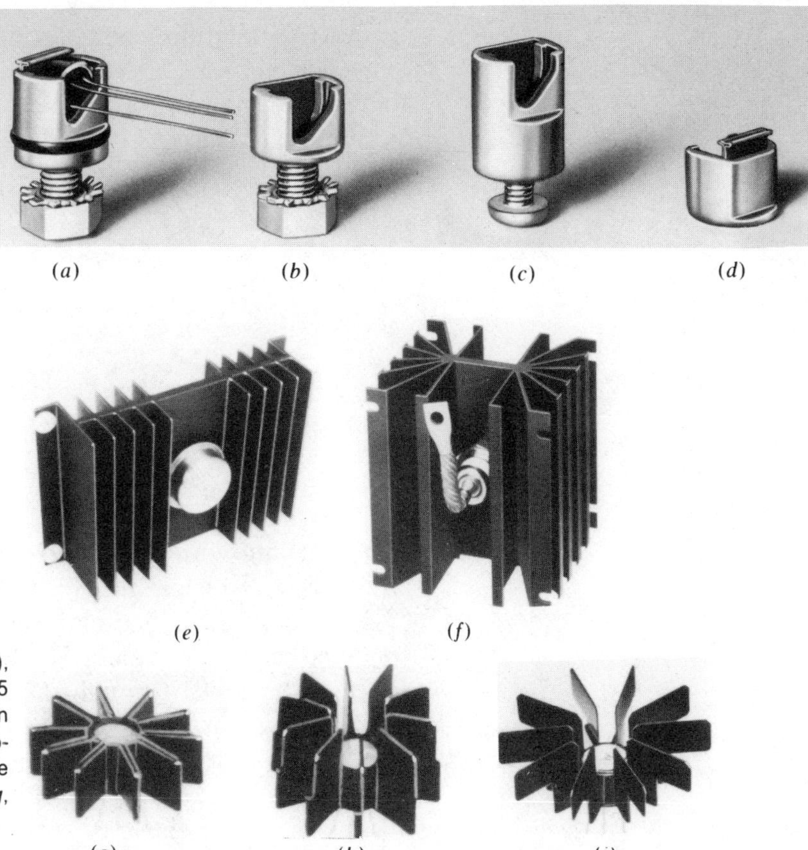

Figure 13-3 Typical heat sinks. (*a*), (*b*), (*c*), and (*d*) Thermal retainers for TO-5 case. (*e*) and (*f*) Natural convection coolers. (*g*), (*h*), and (*i*) Typical snap-on heat dissipators for various case sizes. (*Courtesy Wakefield Engineering, Inc.*)

A heat sink (Fig. 13-3) is a mechanical device connected to the case of the semiconductor that provides a path for the developed heat. The heat flows through the heat sink and is carried off to the surrounding air. If a heat sink is not used, all the heat must transfer from the case to the surrounding air. The heat sink causes the temperature of the case to be lowered.

If all the heat that is generated at a collector junction could be transferred out of the transistor instantaneously, the allowable collector dissipation would be infinite. There is, however, a finite *thermal lag*, and heat can only flow in a path where there is a temperature difference.

This heat problem is very similar to a simple electric circuit and Ohm's law. The heat unit that corresponds to voltage V is the difference in temperature $(T_2 - T_1)$ across the element. The quantity of heat flow corresponding to electric current I is the heat flow P_C. Then by the Ohm's law concept, we obtain thermal resistance θ as

$$\boxed{\theta = \frac{T_2 - T_1}{P_C} \ ^\circ\text{C/W or } ^\circ\text{C/mW}} \tag{13-2a}$$

As with Ohm's law, other forms are very useful.

$$T_2 - T_1 = \theta P_C \ ^\circ\text{C} \tag{13-2b}$$

and

$$P_C = \frac{T_2 - T_1}{\theta} \ \text{W or mW} \tag{13-2c}$$

In working with heat-sink problems, we find that the usual circuit is a series circuit (Fig. 13-4). The junction temperature (T_J) at the base-collector junction (J) of a transistor is the specified maximum value. The heat of the collector dissipation P_C in watts flows through the transistor to the case (C) and establishes a case temperature (T_C). There is some separation between the transistor case and the heat sink that creates a thermal resistance. In many units, a washer often is used for insulation, since the case of the transistor itself frequently serves as the electrical connection to the collector. A special silicone grease often is used to establish a good heat-conducting path between the case and the heat sink. Consequently, the sink temperature (T_S) differs from the case temperature.

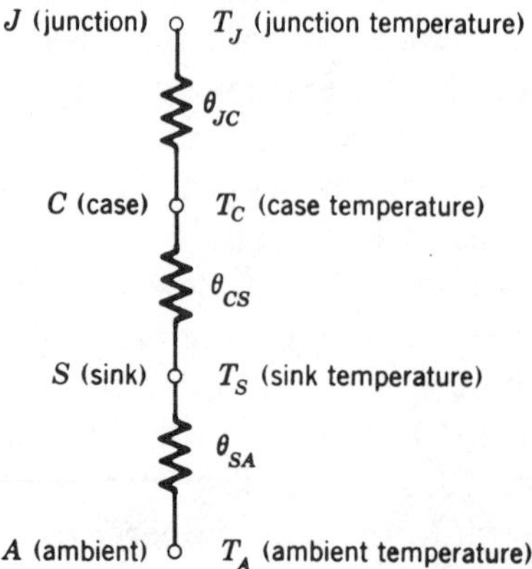

Figure 13-4 Series thermal circuit.

The sink is provided with fins designed to transfer the heat to the ambient (A) or to the surrounding air that is at the ambient temperature (T_A).

As in electric circuits, the property of a series circuit is that the total resistance is the sum of the individual resistances

$$\theta_{JA} = \theta_{JC} + \theta_{CS} + \theta_{SA} \quad °C/W \text{ or } °C/mW \tag{13-3a}$$

and an overall equation can be written from Eq. 13-2b as

$$T_J = T_A + \theta_{JA} P_C \quad °C \tag{13-3b}$$

where P_C is the power dissipated in the collector.

Example 13-1

A transistor may be used from −65°C to +100°C. In a 25°C ambient, the transistor can dissipate 1 W without a heat sink and 5 W with a proper heat sink. Determine the allowable dissipation in each case at 55°C.

Solution

The transistor is rated 1 W without a heat sink and 5 W with a heat sink from −55°C to +25°C. These ratings are horizontal lines on the graphs of Fig. 13-5. The ratings decrease linearly from maximum at 25°C to zero at 100°C as shown by the straight lines. The desired ratings are shown on the graphs by vertical dashed lines at 55°C. The resulting triangles are similar triangles and we can form ratios

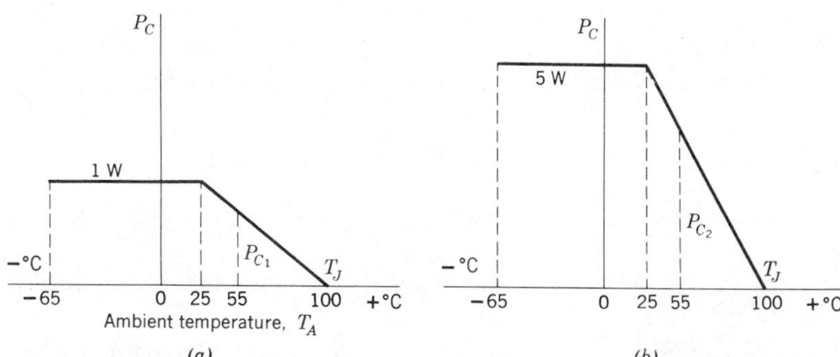

Figure 13-5 Derating curves. (*a*) Without heat sink. (*b*) With heat sink.

from corresponding sides. For the case without a heat sink

$$\frac{100-55}{100-25}=\frac{P_{C_1}}{1\ \text{W}}$$

Solving $$P_{C_1} = \mathbf{0.6\ W}$$

and for the case with a heat sink

$$\frac{100-55}{100-25}=\frac{P_{C_2}}{5\ \text{W}}$$

Solving, we have

$$P_{C_2} = \mathbf{3.0\ W}$$

Example 13-2

Assume that, when the transistor of Example 13-1 is operated in free air (25°C), the actual case temperature of the transistor is 90°C. Determine the values of θ_{JC} and θ_{CA} and the minimum value of θ_{SA} for a heat sink that can be used in a 25°C ambient to dissipate 5 W.

Solution

The value of the thermal resistance from junction to case (θ_{JC}) is

$$\theta_{JC}=\frac{T_2-T_1}{P_C}=\frac{100-90}{1}=\mathbf{10\,°C/W}$$

The effective thermal resistance from case to ambient must be

$$\theta_{CA}=\frac{90-25}{1}=\mathbf{65\,°C/W}$$

We omit any washer between the transistor and the heat sink to simplify the calculation; that is, in the series circuit of Fig. 13-4

$$\theta_{CS}=0 \quad \text{and} \quad T_C=T_S$$

If P_C is to be 5 W, we have from Eq. 13-2*b*

$$T_J-T_C=T_J-T_S=\theta_{JC}P_C=10\times 5=50°\text{C} \qquad (13\text{-}2b)$$

T_J has a maximum value of 100°C. Then by rearranging Eq. 13-2*b*, we have

$$T_C = T_S = T_J - \theta_{JC}P_C = 100 - 50 = 50°\text{C}$$

We can obtain the minimum required heat-sink rating from Eq. 13-2*a*

$$\theta_{SA} = \frac{T_S - T_A}{P_C} = \frac{50 - 25}{5} = \mathbf{5\ °C/W} \qquad (13\text{-}2a)$$

The sense of this last rating is this: We purchase a heat sink that has a rating of 5°C/W. If we try to dissipate 100 W through this heat sink, the temperature drop across the heat sink is $\theta_{SA}P_C$ or 5×100 or 500°C. If we use this heat sink in an application where we only have to dissipate 100 mW (0.1 W), the temperature drop across the heat sink is only 5×0.1 or 0.5°C. Obviously, in the first case the heat sink is too small, and in the second case it is unnecessarily large.

Example 13-3

Use the transistor and the heat-sink data of Example 13-1 and Example 13-2. The transistor and the heat sink are in an ambient of 40°C. What is the maximum allowable value of P_C for the combination?

Solution

The total circuit thermal resistance is

$$\theta_{JA} = \theta_{JC} + \theta_{CS} + \theta_{SA} = 10 + 0 + 5 = 15°\text{C/W} \qquad (13\text{-}3a)$$

Substituting this value into Eq. 13-3*b*, we have

$$T_J = T_A + \theta_{JA}P_C \qquad (13\text{-}3b)$$

$$100 = 40 + 15P_C$$

Solving for P_C, we have

$$P_C = \mathbf{4\ W\ maximum}$$

In practice the thermal resistance θ_{CS} cannot be neglected unless the value of θ for the heat sink is for a device that is not used with a washer but one that fits on or clips on the

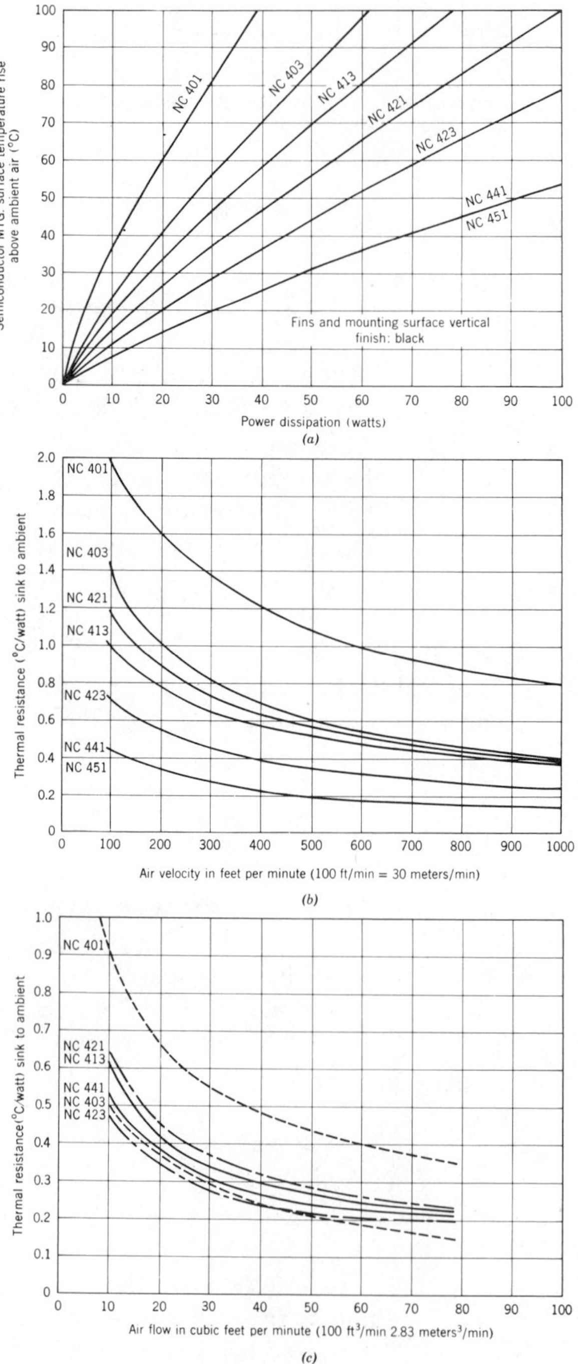

Figure 13-6 Convection characteristics. Data for typical large heat sinks listed in Table 13-2 and illustrated in Figure 13-3 (*e*) and (*f*). (*a*) Natural convection. (*b*) Forced convection for air velocity. (*c*) Forced convection for air flow. (*Courtesy Wakefield Engineering, Inc.*)

Table 13-1 **Specific Thermal Resistance of Interface Materials**

Material	ρ(°C × inches/watt)	ρ_1(°C × meters/watt)
Still air	1200	30.48
Silicone grease	204	5.182
Mylar film	236	5.994
Mica	66	1.676
Wakefield Type 120 compound	56	1.422
Wakefield Delta Bond 152	47	1.194
Anodize	5.6	0.1422
Aluminum	0.19	4.826×10^{-3}
Copper	0.10	2.540×10^{-3}

$\theta = \rho t/A$°C/W where t is thickness in inches and A is in square inches.
$\theta = \rho_1 t_1/A_1$°C/W where t_1 is thickness in meters and A_1 is in square meters.
***Source*:** Courtesy of Wakefield Engineering Inc.

transistor case (Fig. 13-3*g*, *h*, and *i*.). A list of typical materials, with their thermal resistivities, used to separate a case from a heat sink is given in Table 13-1. By using this table, we find that the value of θ for a mica washer (area = 0.75 in.^{2} = 484×10^{-6} m^2 and thickness = 0.003 in. = 76×10^{-6} m) is

$$\theta = \frac{\rho t}{A} = \frac{66(°\text{C} \times \text{in/W}) \times 0.003 \text{ (in.)}}{0.75 \text{ (in.}^2)} = 0.264°\text{C/W}$$

or

$$\theta = \frac{\rho_1 t_1}{A_1} = \frac{1.676 \, (°\text{C} \times \text{m/W}) \times 76 \times 10^{-6} \text{ (m)}}{484 \times 10^{-6} \text{ (m}^2)} = 0.264°\text{C/W}$$

If this washer is used on a semiconductor that dissipates 100 W, the temperature drop across the washer alone is over 26°C, which certainly is not negligible.

A silicone grease, by filling in scratches and airgaps, can reduce the thermal temperature drop from the stud mounting of a 200-W semiconductor to its heat sink by about 15°C.

A fan can be used to improve heat transfer from the heat sinks of large semiconductors. A fan with a guaranteed life of 5 years costs approximately 5 dollars. When the total power dissipation is of the order of 300 W or more, the use of the cooling fan becomes practical.

Table 13-2 **Typical Heat-Sink Data**

Figure[a]	Degrees Centigrade per watt (Case to Chassis)	Approximate Cost
a	6.0	25 ¢
b	4.1	18 ¢
c	5.3	20 ¢
d	4.0	13 ¢

Figure[a]	Type	Size HWD (Inches)	Size HWD (cm)	Approximate Cost
e	NC 401	1.50 × 4.81 × 1.25	3.81 × 12.22 × 3.18	$1.50
e	NC 403	3.00 × 4.81 × 1.25	7.62 × 12.22 × 3.18	$1.75
e	NC 413	3.00 × 4.81 × 1.87	7.62 × 12.22 × 4.75	$2.40
e	NC 421	3.00 × 4.81 × 2.63	7.62 × 12.22 × 6.68	$2.50
e	NC 423	5.50 × 4.81 × 2.63	13.97 × 12.22 × 6.68	$3.60
f	NC 441	5.50 × 4.75 × 4.50	13.97 × 12.04 × 11.43	$6.40

[a]Illustrated in Fig. 13-3.

Source: Courtesy of Wakefield Engineering, Inc.

Example 13-4

A small transistor has the following ratings:

$$T_A \text{ to } 25°\text{C} \qquad P_C = 0.8 \text{ W}$$

$$T_C \text{ to } 25°\text{C} \qquad P_C = 3.0 \text{ W}$$

$$-65°\text{C} < T_J = T_{STG} < 200°\text{C}$$

$$\theta_{JC} = 58.3°\text{C/W} \qquad \theta_{JA} = 219°\text{C/W}$$

Heat sink *b* from Table 13-2 is used to clamp the assembly to a chassis that is held to 30°C. What is the maximum value of P_C with this arrangement?

Solution

Heat sink *b* has a value for θ_{SA} of 4.1°C/W. Since we are using a heat sink, the total thermal resistance is

$$\theta_{JA} = \theta_{JC} + \theta_{SA} = 58.3 + 4.1 = 62.4°\text{C/W} \qquad (13\text{-}3a)$$

Then, using Eq. 13-3*b*, we find

$$T_J = T_A + \theta_{JA}P_C \qquad (13\text{-}3b)$$

$$200 = 30 + 62.4P_C$$

$$P_C = \mathbf{2.73\ W\ maximum}$$

***Note*:** In heat sink problems, we always take a worst-case condition to protect the transistor.

Example 13-5

A large power transistor has the ratings

$$-65°\text{C} \leq T_J = T_{STG} \leq 100°\text{C}$$

$$P_C = 30\text{ W for } T_{MF} \leq 55°\text{C}$$

T_{MF} is the temperature of the mounting frame (the heat sink). What heat sink can be used to handle the 30 W if the ambient is 25°C?

Solution

The allowable temperature drop across the heat sink is

$$T_{SA} = T_{MF} - T_A = 55 - 25 = 30°\text{C}$$

Locate a point on the curves of Fig. 13-6*a* corresponding to 30 W and a 30°C rise. Only three of the heat sinks are capable of handling the dissipation:

NC 423 giving a 28°C rise

NC 441 and NC 451 giving a 20°C rise

The NC 423 heat sink provides a 2°C margin whereas the NC 441 or the NC 451 provides a 10°C margin in case the ambient rises above 25°C. A few degrees rise in ambient could easily occur if the circulation of air were inadvertently blocked, say with a coat or jacket being placed on the equipment. The small reserve margin of the NC 423 heat sink is not enough to prevent the transistor from being destroyed in that case.

When heat is generated at a collector junction, there is a finite time lag for it to flow away from the junction. Transient pulses can destroy a semiconductor because of this thermal lag. If the heat is not drawn off, the temperature increases and more current is produced. Therefore, the dissipation increases. This upward spiraling of heat at a junction is called *thermal runaway.* If an ammeter is in the collector circuit, the meter reading steadily increases to the destruction of the transistor.

According to the analysis developed from Fig. 13-1, when the operating point is in the center of the load line at $V_{CE} = \frac{1}{2}V_{CC}$, any change in the operating point from this tangent point to the constant collector-dissipation line results in a reduced collector dissipation. If R is the resistance value of the load line, and if the operating point is at $\frac{1}{2}V_{CC}$, the collector dissipation P_C is

$$P_C = (\tfrac{1}{2}V_{CC})^2/R$$

Solving for R, we have

$$R = \frac{V_{CC}^2}{4P_C}$$

If P_C is the maximum allowable collector dissipation under a given set of operating conditions for a particular ambient temperature with a particular heat sink, R becomes the least value of dc resistance required in the collector circuit to protect the semiconductor from a thermal runaway.

$$R_{\min} = \frac{V_{CC}^2}{4P_C} \tag{13-4}$$

Problems **13-2.1** A transistor has the following ratings:

$$-65°\text{C} \leqslant T_J = T_{STG} \leqslant 100°\text{C}$$

$$\text{to } 25°\text{C } T_A, \qquad P_C = 1.0 \text{ W}$$

$$\text{to } 25°\text{C } T_C, \qquad P_C = 7.5 \text{ W}$$

Find θ_{JA} and θ_{JC}.

13-2.2 Using the data given for the transistor in Problem 13-2.1, find the power ratings for an ambient temperature of 60°C and also for a case temperature of 60°C.

13-2.3 A 2N404 transistor is rated in free air at 150 mW and has a maximum value for T_J of 85°C. What is the derating value in mW/°C for elevated temperatures?

13-2.4 A transistor in free air has a value for θ_{JA} of 0.25°C/mW, and for an infinite heat sink θ_{JC} is 0.11°C/mW. The maximum junction temperature T_J is 85°C. The heat sink *c* listed in Table 13-2 is used in an ambient of 35°C. What is the maximum allowable collector dissipation? What is the maximum allowable collector dissipation in a 35°C ambient without a heat sink?

13-2.5 The operating junction temperature of a transistor is 125°C. The total dissipation at a 25°C case temperature is 0.5 W and at a 25°C ambient the total dissipation is 0.2 W. What is the value of θ_{CA}?

13-2.6 A transistor that operates at a junction temperature of 170°C has a power dissipation capability of 12 W when the case is at 25°C. The transistor is used with heat sink *b* listed in Table 13-2. The ambient temperature is 60°C. What is the maximum allowable collector dissipation for the transistor with the heat sink?

13-2.7 The allowable power dissipation of a transistor at or below a mounting frame temperature of 70°C is 60 W. Select a heat sink from Fig. 13-6 that will dissipate this power by natural convection cooling in an ambient of 25°C. What is the reserve for this heat sink in ambient rise?

13-2.8 Repeat Problem 13-2.7 if the ambient is 50°C. Forced air cooling is required. Determine the air flow required. The air temperature is also 50°C.

13-2.9 The surface area of a transistor in contact with a heat sink is 1.20 in^2. The transistor dissipates 40 W. Irregularities in the surfaces of the transistor and its heat sink create an effective air gap of 0.0002 in. What is the temperature drop across the gap? What is the temperature drop across the gap if silicone grease is used? If Delta Bond 152 is used? Thermal resistances are given in Table 13-1.

13-2.10 A transistor can be operated with a junction temperature of 200°C. The transistor can dissipate 1 W without a heat sink in a 25°C ambient. The transistor can dissipate 10 W when used with an infinite heat sink in a 25°C ambient. The transistor is used with a heat sink that is rated at 10°C/W to a 25°C ambient. What dissipation can be tolerated for this transistor with this heat sink?

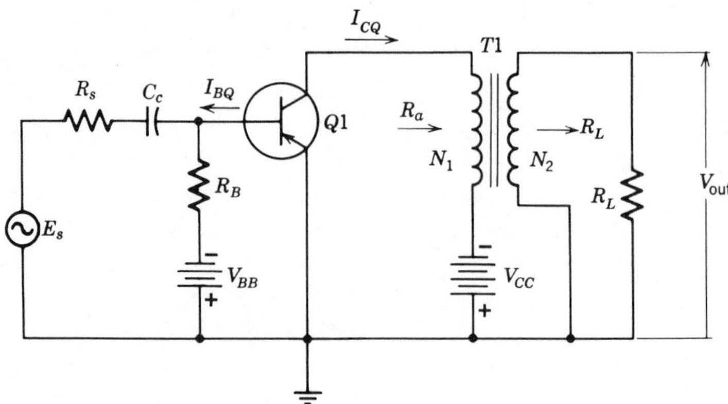

Figure 13-7 A transformer-coupled power amplifier.

Section 13-3 The Transformer-Coupled Amplifier

A *transformer-coupled amplifier* circuit is shown in Fig. 13-7. A transformer $T1$ is used to *couple* the transistor $Q1$ to the load R_L. The turns ratio α of the transformer converts the load resistance R_L into a new *reflected value* R_a. The transformer is used to *match* the load to the transistor. As we recall from ac circuit theory, we define the turns ratio α of the transformer as

$$\alpha \equiv \frac{N_2}{N_1} = \frac{V_2}{V_1} = \frac{V_{out}}{V_{ce}} \tag{13-5a}$$

Then

$$R_a = \frac{1}{\alpha^2} R_L \tag{13-5b}$$

In this section, we show how a load line for R_a is placed on the collector characteristics of the transistor $Q1$.

When the dc resistance of the primary is taken into consideration, a dc load line AB is drawn on the characteristic curves for the transistor (Fig. 13-8) in the same manner as the load lines in Chapter 6. Usually, this dc resistance is neglected, and the load line for this value of zero dc resistance is the vertical load line AC. The operating point must lie on the dc load line. For the dc load line AC, we notice that the operating

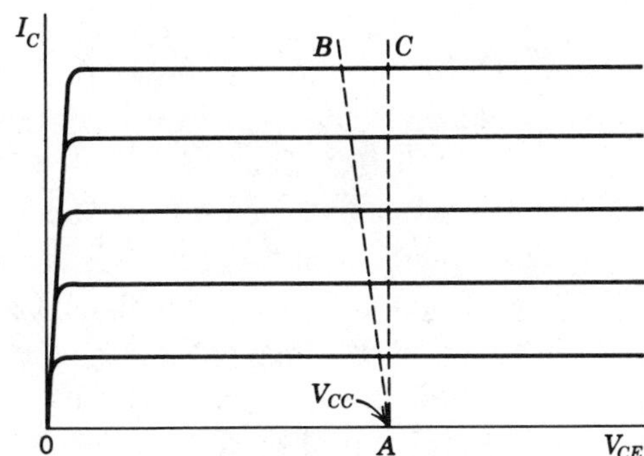

Figure 13-8 Determination of the operating point.

collector voltage, V_{CEQ}, is identical with the supply voltage V_{CC}. The operating point is then determined by the value of the bias current I_{BQ} (Fig. 13-7). An ac load line equal to the value of the reflected resistance R_a is drawn through the operating point.

Let us assume the following operating point values for $Q1$ in Fig. 13-7.

$$V_{CC} = V_{CEQ} = -20\text{ V} \qquad I_{CQ} = 372\text{ mA}$$

$$R_a = 60\ \Omega \qquad I_{BQ} = 3.72\text{ mA}$$

Assume the dc resistance of the transformer primary is negligible.

At the collector supply voltage, 20 V, a vertical line is drawn (Fig. 13-9). The operating point Q must lie on this vertical line. From the bias calculation, it must be the intersection of this vertical line with the $I_{BQ} = 3.72$ mA curve. A simple method used to draw the ac load line is to assume a

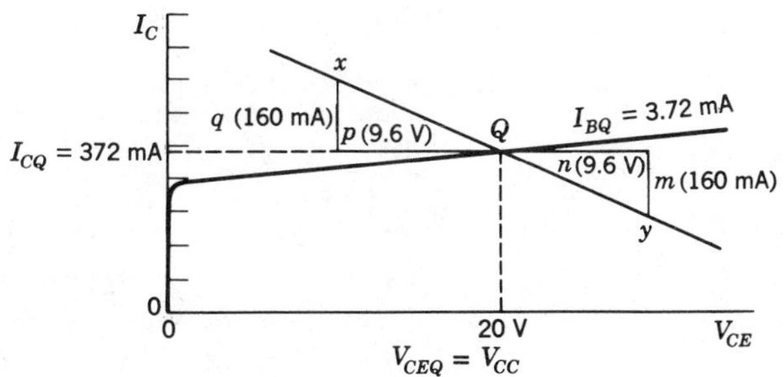

Figure 13-9 Load line for a transformer-coupled amplifier.

small convenient current change and, by Ohm's law, to determine the corresponding voltage change. For 60 Ω, an assumed current change of 160 mA gives a voltage change of 60 Ω × 0.160 A or 9.6 V. On Fig. 13-9, if we shift 9.6 V to the left along path *p* and up 160 mA on the I_C scale along path *q*, we locate point *x*. Likewise, a shift of 9.6 V to the right along path *n* and down 160 mA along path *m* locates point *y*. The line that is drawn through points *Q*, *x*, and *y* and extended beyond *x* and *y* is the ac load line. The load line is extended beyond *x* and beyond *y* to give the length of the load line that is required to handle the signal swing.

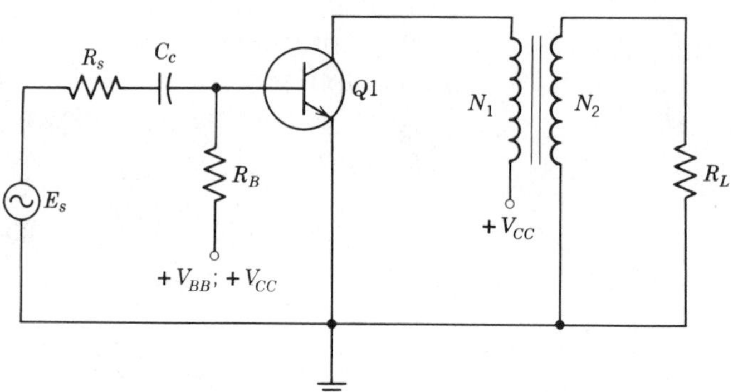

Circuit for Problems 13-3.1 and 13-3.2.

Problems Assume the silicon transistor has an ideal collector characteristic and its parameters are

$$\beta = 60 \quad \text{and} \quad r'_e = \frac{50\ \text{mV}}{I_E}$$

13-3.1 The supply voltage is +12 V. The turns ratio of the transformer N_2/N_1 is 4. The value of the load resistance R_L is 2500 Ω. Determine the value of R_B required to set I_C to 100 mA. Draw the load line on the ideal collector characteristic.

13-3.2 The supply voltage is +20 V. The turns ratio of the transformer N_2/N_1 is 1/4. The value of the load R_L is 30 Ω. Determine the bias resistor R_B required to set I_C to 25 mA. Draw the load line on the ideal collector characteristic.

Section 13-4 The Class-A Power Amplifier

A *class-A amplifier* is defined as an amplifier in which the signal current in the output is not limited by clipping caused by either saturation or cutoff. All the amplifiers we have considered so far in this text are class-A amplifiers.

The load line for the circuit of Fig. 13-10 is shown on the collector characteristic given in Fig. 13-11. The bias sets the operating point at point Q on the load line. A signal applied to the base swings the dynamic operating point between points 1

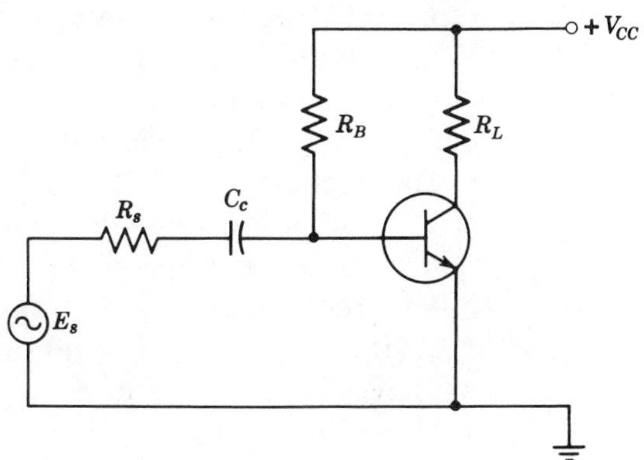

Figure 13-10 Power amplifier with resistive load.

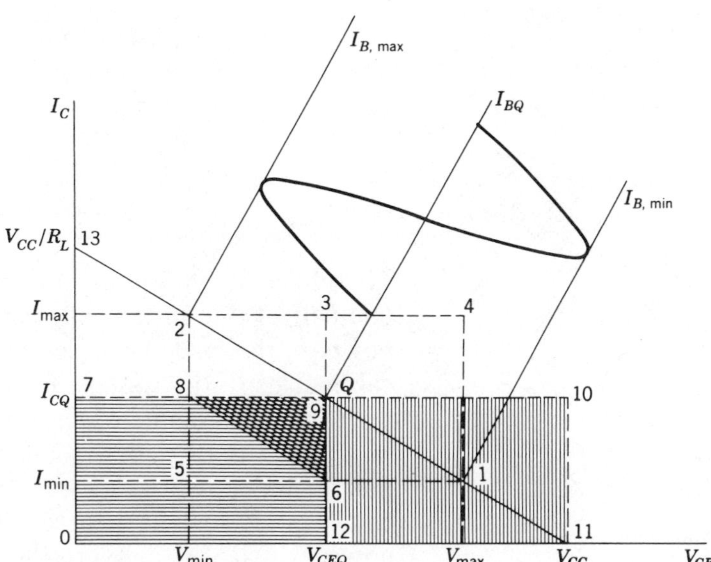

Figure 13-11 Representations of powers on a collector characteristic.

and 2. Let us construct the rectangles

0-7-10-11-0

and 2-4-1-5-2

and extend the vertical line from point 12 through the operating point Q to point 3.

The large rectangle 0-7-10-11-0 has the area $I_{CQ}V_{CC}$. This area gives the total power supplied by the power supply to the collector circuit.

The rectangle 0-7-9-12-0 has the area $I_{CQ}V_{CE}$. This area represents the total power delivered to the transistor itself. When there is no signal, this power is the value of the collector dissipation P_C. The heat sink used for the transistor must be capable of dissipating P_C.

The rectangle 12-9-10-11-12 is the difference between the total input power and the collector dissipation. Accordingly, this is the dc heat loss in R_L.

The rectangle 2-4-1-5-2 has the area $(V_{max} - V_{min})(I_{max} - I_{min})$. If we divided the peak-to-peak value (max − min) by 2, we have the peak value. If we divide the peak value by $\sqrt{2}$, we have the effective value. The product of effective voltage times the effective current is the average ac power in the load.

$$P_L = \frac{V_{max} - V_{min}}{2\sqrt{2}} \times \frac{I_{max} - I_{min}}{2\sqrt{2}}$$

$$\boxed{P_L = \frac{(V_{max} - V_{min})(I_{max} - I_{min})}{8}} \qquad (13\text{-}6)$$

Therefore, if we divide this rectangle into four equal parts, one-half of one of the four parts graphically represents the ac power in the load. Then, let us take the rectangle 5-8-9-6-5 and call half of it, the triangle 8-9-6, the *ac power in the load.*

From this graphical picture, we see that the circuit takes $V_{CC}I_{CQ}$ watts (the total shaded area) from the power supply. Part of this power $(V_{CC} - V_{CEQ})I_{CQ}$ (the area shaded by vertical lines) is lost as heat dissipation in R_L. The rest of the power $V_{CEQ}I_{CQ}$ is delivered to the transistor. Any part of this power that is not converted into ac output power is the heat dissipation P_C in the transistor. Figure 13-11 shows this heat dissipation as the area with horizontal shaded lines. If the signal is reduced to zero, the output power is zero and the collector dissipation is $V_{CEQ}I_{CQ}$.

When the Q-point is located at the optimum value, it is at the midpoint of the load line, point Q, Fig. 13-12. When the signal input to the amplifier is increased from zero, the maximum and minimum values move to the ends of the load line simultaneously. Now the amplifier is delivering the maximum possible undistorted power to the load. It is apparent from the diagram that the area of triangle B equals the area of triangle C. Also the rectangle formed by triangles B and C equals the area of rectangle A. Therefore, when we define the overall efficiency of the amplifier as

$$\eta_{\text{overall}} \equiv \frac{P_L}{V_{CC}I_{CQ}} \times 100\% \tag{13-7a}$$

we see that

The maximum possible overall efficiency of a class-A amplifier with resistive load is 25%.

Naturally, if the input signal is reduced to zero, this efficiency is zero.

Similarly, if we define the collector efficiency of the transistor as

$$\eta_{\text{coll}} \equiv \frac{P_L}{V_{CEQ}I_{CQ}} \times 100\% \tag{13-7b}$$

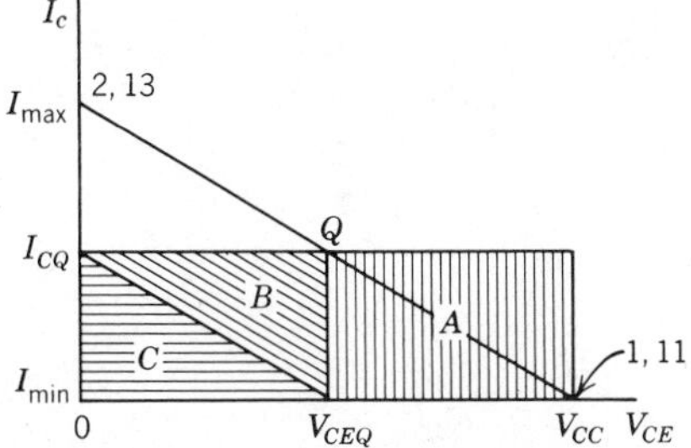

Figure 13-12 Power distribution in a class-A amplifier (resistive load) with optimum bias and maximum signal.

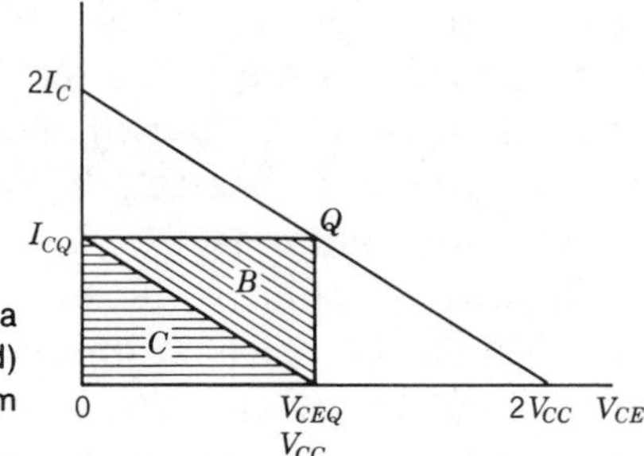

Figure 13-13 Power distribution in a class-A amplifier (transformer load) with optimum bias and maximum signal.

we see that

The maximum possible collector efficiency of a class-A amplifier with resistive load is 50%.

If we use a transformer to drive the load (Fig. 13-13), V_{CEQ} and V_{CC} are the same value. The load line of Fig. 13-13 becomes the ac load line drawn through the operating point (Fig. 13-13). The shaded rectangle marked *A* that we had in Fig. 13-12 cannot exist. All the power supplied by the power supply is delivered to the transistor. Now, the overall efficiency and the collector efficiency are identical

$$\eta_{\text{overall}} = \eta_{\text{coll}} = \frac{P_L}{V_{CC}I_{CQ}} \times 100\% = \frac{P_L}{V_{CEQ}I_{CQ}} \times 100\% \quad (13\text{-}7c)$$

We see that

The maximum possible overall efficiency and the maximum possible collector efficiency for a class-A amplifier using an output transformer are both 50%.

Example 13-6
Determine the turns ratio of the transformer, the collector current, and the collector dissipation for the amplifier shown in Fig. 13-14*a*.

Solution
The equation for the ac power in the load is

$$P_L = \frac{V_L^2}{R_L} = \frac{(V_m/\sqrt{2})^2}{R_L} = \frac{V_m^2}{2R_L}$$

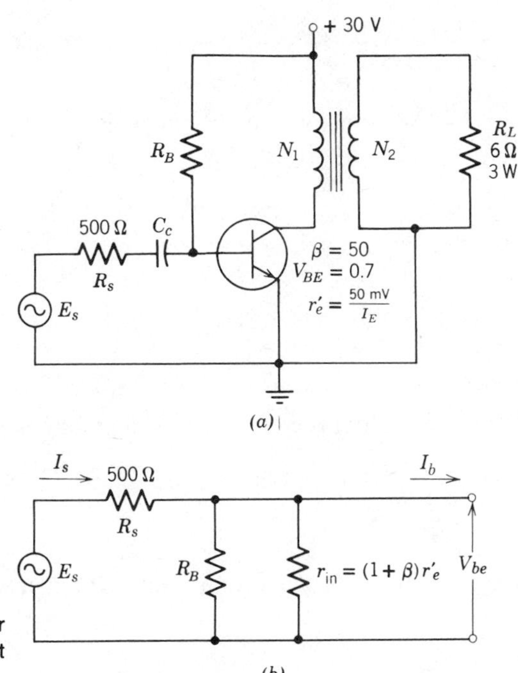

Figure 13-14 Class-A power amplifier calculation. (*a*) Circuit. (*b*) Input equivalent circuit.

Solving for V_m, we find

$$V_m = \sqrt{2P_LR_L} \tag{13-8}$$

Eq. 13-8 is very useful because, if we have the value of R_L and the value of P_L, we obtain the peak value of V_L. When this peak value is reflected through the transformer, we have the supply voltage V_{CC} for the circuit.

Using Eq. 13-8, we find that the peak value of the load voltage is

$$V_m = \sqrt{2P_LR_L} = \sqrt{2 \times 3\text{ W} \times 6\ \Omega} = 6\text{ V} \tag{13-8}$$

The peak-to-peak collector signal is twice the supply voltage if the circuit is optimally biased. In this example, the peak value of the signal at the collector is

$$V_{c,\max} = V_{CC} = 30\text{ V}$$

The turns ratio required for the transformer is

$$\alpha = \frac{V_2}{V_1} = \frac{V_m}{V_{CC}} = \frac{6\text{ V}}{30\text{ V}} = \frac{1}{5} = \frac{N_2}{N_1} \tag{13-5a}$$

If the required load power is 3 W, the total supply power is 6 W since the overall efficiency of the circuit is 50%. The collector current is determined by

$$V_{CC}I_C = P_{dc} = 6\text{ W}$$
$$30\text{ V} \times I_C = 6\text{ W}$$
$$I_C = \frac{6\text{ W}}{30\text{ V}} = 0.2\text{ A} = \mathbf{200\ mA}$$

When full output power is obtained from the circuit, the efficiency is 50%.

$$P_C = P_{dc} - P_L = 6 - 3 = 3\text{ W}$$

However, if E_s becomes smaller than its maximum value, P_C goes up. When E_s is reduced to zero,

$$P_C = P_{dc} = \mathbf{6\ W}$$

Now the required transistor must have the specifications

$$I_C = 0.2\text{ A} \qquad P_C = 6\text{ W} \quad \text{and} \quad BV_{CE} = 60\text{ V}$$

Also we must provide a heat-sink arrangement that can dissipate 6 W.

Example 13-7
Determine the values of V_{be}, E_s, and the power from the source required to obtain 3 W in the load of the circuit of Fig. 13-14*a*. Find the power gain of the circuit in decibels.

Solution
The model for the input circuit is shown in Fig. 13-14*b*. The unknown quantities are labeled on this model.
If the value of β is 50, the peak base current is

$$I_B = \frac{I_C}{\beta} = \frac{200\text{ mA}}{50} = 4\text{ mA}$$

The value of I_C is 200 mA and the value of I_E is

$$I_E = I_C + I_B = 200 + 4 = 204\ \text{mA} \tag{4-1}$$

The value of r'_e is

$$r'_e = \frac{50\ \text{mV}}{I_E} = \frac{50\ \text{mV}}{204\ \text{mA}} = 0.245\ \Omega \tag{7-5}$$

The ac input resistance r_{in} to the base is

$$r_{\text{in}} = (1+\beta)r'_e = (1+50)0.245 = 12.5\ \Omega \tag{7-7}$$

The Kirchhoff's voltage loop equation through the base is

$$\begin{aligned} V_{CC} = V_{BB} &= I_B R_B + V_{BE} \\ 30\ \text{V} &= 0.004\ \text{A} \times R_B + 0.7\ \text{V} \\ 0.004\ \text{A} \times R_B &= 29.3\ \text{V} \\ R_B &= 7325\ \Omega \end{aligned}$$

The ac load on the collector r_c is given by Ohm's law as

$$r_c = \frac{V_{c,\max}}{I_{c,\max}} = \frac{30\ \text{V}}{0.2\ \text{A}} = 150\ \Omega$$

The voltage gain A_v across the transistor is

$$A_v = \frac{r_c}{r'_e} = \frac{150\ \Omega}{0.245\ \Omega} = 612 \tag{7-8a}$$

The peak value of V_{be} is

$$V_{be,\max} = \frac{V_{c,\max}}{A_v} = \frac{30\ \text{V}}{612} = 0.049\ \text{V} = \mathbf{49\ mV}$$

The peak value of E_s is found from the voltage-divider rule.

$$V_{be} = \frac{r_{\text{in}}}{r_{\text{in}} + R_s} E_s \tag{7-2}$$

Substituting values, we have

$$49\ \text{mV} = \frac{12.5\ \Omega}{12.5\ \Omega + 500\ \Omega} E_s$$

Solving for E_s, we find

$$E_s = 2000\text{ mV} = \mathbf{2\ V\ peak}$$

The load on E_s is $(R_s + r_{\text{in}})$ and the current I_s in E_s is

$$I_s = \frac{E_s}{R_s + r_{\text{in}}} = \frac{2\text{ V}}{500\ \Omega + 12.5\ \Omega} \approx 0.004\text{ A} = 4\text{ mA peak}$$

The peak driving power is

$$E_{s,\max} I_{s,\max} = 2\text{ V} \times 4\text{ mA} = 8\text{ mW peak}$$

and the average power is half the peak power.

$$P_s = \mathbf{4\ mW}$$

The power gain is

$$dB = 10\log_{10}\frac{P_{\text{out}}}{P_s} = 10\log_{10}\frac{3\text{ W}}{0.004\text{ W}} = \mathbf{+29\ dB}$$

We must recognize that these numerical values are ideal values predicated upon obtaining the theoretical value of 50% efficiency and upon the use of an ideal transformer that has an efficiency of 100%. Practically, we can obtain a collector efficiency of the order of 45 to 48%. We also find that the transformer efficiency is in the range of 90 to 95%. Accordingly, the numerical values we have obtained are used as a guide to the final design. The supply voltage, E_s and the collector dissipation must have somewhat larger values in order to obtain an actual 3-W power in the load.

The effects of variation of the input signal E_s on the circuit of Fig. 13-14*a* are shown in Fig. 13-15. The circuit is a linear circuit. Therefore, the plot of peak collector signal voltage against E_s must be a straight line (Fig. 13-15*a*). Since the output power is proportional to the square of the voltage, the output power (Fig. 13-15*b*) is a squared curve. The input power $V_{CC}I_{CQ}$ is a constant value, a horizontal line. The distance between $V_{CC}I_{CQ}$ and P_L at any value of E_s is the collector dissipation P_C at that value of E_s. The variation of

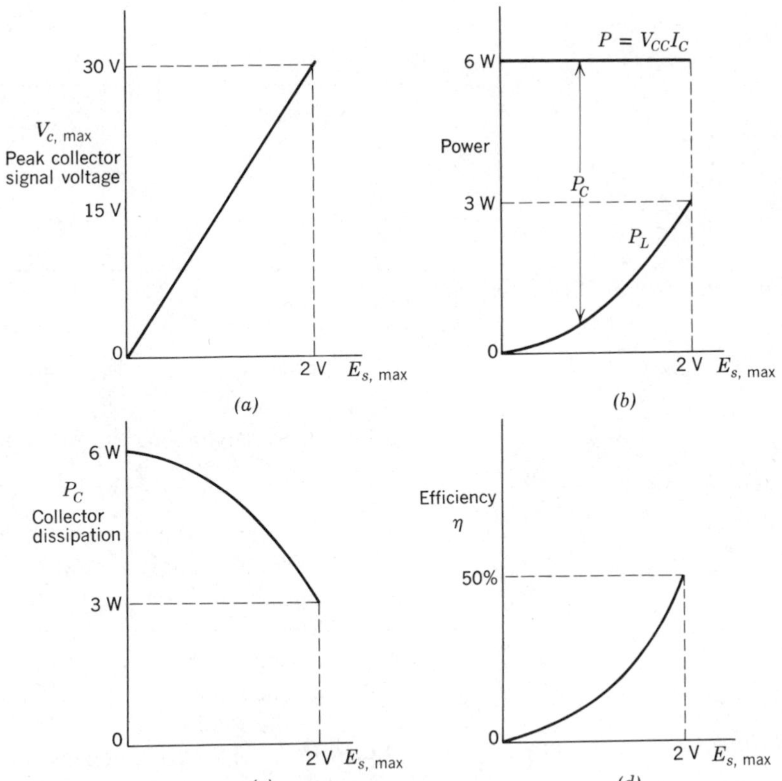

Figure 13-15 Effect of input-signal level variation. (*a*) Variation of peak collector signal. (*b*) Variation of load power. (*c*) Variation of collector dissipation. (*d*) Variation of efficiency.

the collector dissipation is shown in Fig. 13-15*c*. The efficiency is

$$\eta_{\text{overall}} = \eta_{\text{coll}} = \frac{P_L}{V_{CC}I_{CQ}} \times 100\% \qquad (13\text{-}7c)$$

Since $V_{CC}I_{CQ}$ is a constant value, the efficiency curve (Fig. 13-15*d*) must have the same shape as the load power curve. The efficiency curve rises to a maximum value of 50%.

Problems In Problems 13-4.1 through 13-4.8 determine the required transformer turns ratio and the values for I_{CQ}, P_C, and BV_{CE} for the required power transistor.

13-4.1 $P_L = 2$ W $\quad R_L = 8\ \Omega \quad V_{CC} = +12$ V

13-4.2 $P_L = 200$ mW $\quad R_L = 200\ \Omega \quad V_{CC} = +30$ V

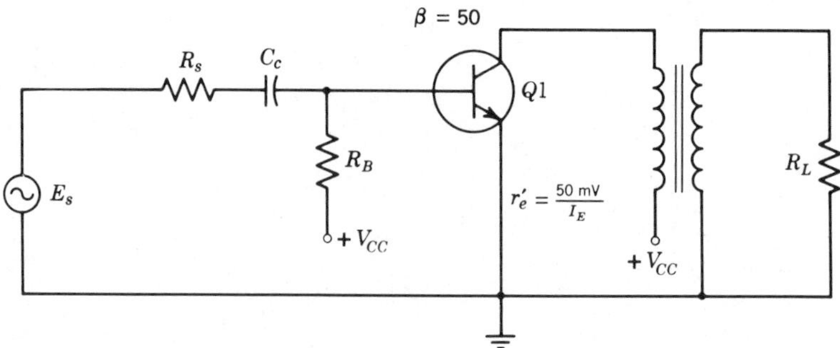

Circuit for Problems 13-4.1 through 13-4.8.

13-4.3 $P_L = 10$ W $\quad R_L = 20\ \Omega \quad V_{CC} = +20$ V
13-4.4 $P_L = 1$ W $\quad R_L = 16\ \Omega \quad V_{CC} = +24$ V
13-4.5 A transistor has the maximum ratings:

$$P_C = 15 \text{ W},\ I_{CQ} = 2 \text{ A, and } BV_{CE} = 30 \text{ V}$$

Determine the transformer turns ratio for the greatest possible collector supply voltage that can be used to deliver maximum possible power to an 8-Ω load.

13-4.6 Using the data given in Problem 13-4.5, determine the transformer turns ratio for the least possible collector supply voltage that can be used to deliver maximum possible power to the load.

13-4.7 If R_s is 200 Ω, determine the value of E_s required to deliver the full-load power in Problem 13-4.1. What is the gain in dB?

13-4.8 If R_s is 500 Ω, determine the value of E_s required to deliver the full-load power in Problem 13-4.4. What is the gain in dB?

Supplementary Problems

13-1 A transistor used with a voltage regulator can dissipate 100 W with a heat sink that keeps the case temperature to 100°C. The free-air (25°C) rating of this transistor is 2 W. Its junction temperature limit is 200°C. We wish to use this transistor in an application that causes the transistor to dissipate 60 W in an ambient temperature of 35°C. What heat sink (Fig. 13-6) should be used?

13-2 A 2N1491 transistor has a maximum allowable junction temperature of 175°C. In free air (25°C) the transistor can dissipate 0.5 W. What is its allowable dissipation if the ambient rises to 100°C?

13-3 The transistor of Problem 13-2 dissipates 0.3 W in an ambient temperature of 35°C. What is the junction temperature?

13-4 The 2N3119 transistor has a maximum allowable junction temperature of 200°C. In free air (25°C) the transistor can dissipate

1 W without a heat sink. With an infinite heat sink that keeps the case at 25°C, the transistor can dissipate 4 W. If we use heat sink *a* (Fig. 13-3, Table 13-2), how much power can be dissipated in an ambient of 40°C?

13-5 The T_J for a transistor is 150°C. When it is connected to an infinite heat sink at 25°C, the transistor can dissipate 10 W. This transistor is used in a 45°C ambient with a heat sink that has a rating of 8°C/W for θ. What is maximum dissipation of the combination?

13-6 A class-A amplifier is required to supply 5 W to an automobile high-fidelity system that uses an 8-Ω speaker. Specify the transistor and the output transformer. The supply is 13 V.

13-7 The supply to a class-A amplifier that uses a transformer is 20 V. The maximum allowable collector current is 2 A. Specify the transformer that must be used to obtain maximum power in a 16-Ω speaker. What is the audio power?

13-8 The operating point of a class-A power amplifier using a transformer is 50 mA (I_{CQ}) and 10 V (V_{CEQ}). Sketch a graph showing load lines for reflected values of resistance into the primary of 50 Ω, 100 Ω, 200 Ω, 500 Ω, 1000 Ω, and 2000 Ω. What are the maximum peak-to-peak values of collector current and collector voltage for each case without clipping? What is the maximum possible output power for each without clipping?

Chapter 14 Push-Pull Amplifiers

The principles of the operation of the push-pull amplifier are examined (Section 14-1). Phase inverters are required to provide the balanced driving signals to the push-pull amplifier (Section 14-2). Class-A operation, class-AB operation, and class-B operation are defined (Section 14-3). Methods of calculating the input and the output conditions for a class-A power amplifier using an output transformer (Section 14-4) and a class-B power amplifier using an output transformer (Section 14-5) are detailed. The class-A and the class-B circuits are compared in Table 14-1. Complementary symmetry (Section 14-6) can be used to avoid the use of a transformer. The transformer-type circuits and the complementary symmetry circuits are compared in Table 14-2. Representative commercial audio amplifier circuits are given in Section 14-7.

Section 14-1 The Basic Circuit

The signal voltage is the primary voltage of the input transformer $T1$ in Fig. 14-1. The secondary winding of the input transformer is grounded at the center tap. When the center tap on the winding is made the reference point (in this case, ground), the voltage from the center tap to the top of the winding is 180° out of phase with the voltage from the center tap to the bottom of the winding. By use of a center tap, the number of turns in the top half of the winding equals the number of turns in the bottom half of the winding, and V_1 is exactly equal in magnitude to V_2. Thus, if we consider the input voltage to $Q1$ to be instantaneously $+1$ V, the input voltage to $Q2$ must be -1 V at that instant. Accordingly, when V_1 is positive, the forward bias on transistor $Q1$ decreases and its collector current I_{C_1} decreases. Simultaneously, V_2 is negative. The increasing forward bias on $Q2$ causes the collector current I_{C_2} in $Q2$ to increase in magnitude. If we assume the circuit is ideally linear, the decrease in I_{C_1} equals in magnitude

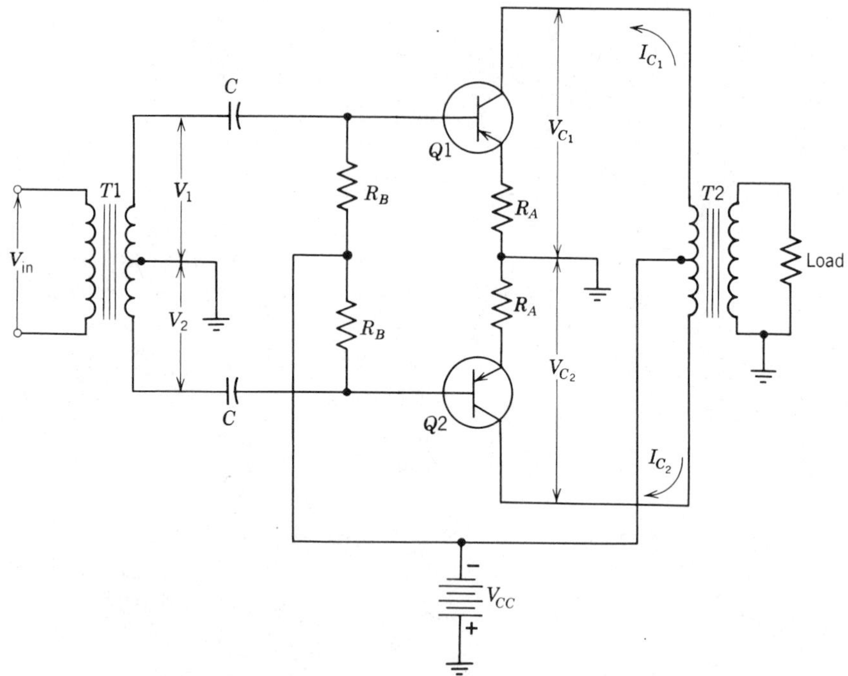

Figure 14-1 The basic push-pull circuit.

the increase in I_{C_2}. Correspondingly, V_{C_1} and V_{C_2} are out of phase with each other. Since the action of a transistor as a common-emitter amplifier introduces a 180° phase inversion, V_{C_1} is in phase with V_2 and V_{C_2} is in phase with V_1. Since I_{C_1} decreases as I_{C_2} increases, the sum of I_{C_1} and I_{C_2} is a constant and does not vary with signal.

Let us assume that the flux in the primary of $T2$ caused by I_{C_1} acts upward and that the flux caused by I_{C_2} acts downward. Without a signal, I_{C_1} and I_{C_2} are equal. The two fluxes are equal and cancel with the result that the net flux in the transformer is zero. With a signal, I_{C_1} and I_{C_2} differ. Then $(I_{C_1} - I_{C_2})$ produces the net primary flux, which develops the load voltage and the load power in the secondary winding of $T2$.

In checking the operation of a push-pull circuit with a test signal, we find that the magnitude of V_1 should equal the magnitude of V_2, using either an oscilloscope or an ac meter. The observed ac collector voltages on $Q1$ and $Q2$ should also be equal in magnitude.

Small resistors R_A are placed in the emitter of $Q1$ and $Q2$ to provide stability and to prevent thermal runaway. The proper operating bias on the transistors is obtained by means of a base resistor R_B for each transistor. To prevent the bias currents from being short-circuited through the transformer, blocking capacitors C are required.

The circuits shown in Fig. 14-2 illustrate other basic push-pull circuits. The bias can be obtained from a voltage-dividing network (Fig. 14-2*a*) that is applied simultaneously to both bases. In the circuit of Fig. 14-2*b*, the bias circuit is modified to provide a temperature compensation. If the ambient temperature rises, the collector characteristics shift in the direction of an increased collector current. In order to keep the

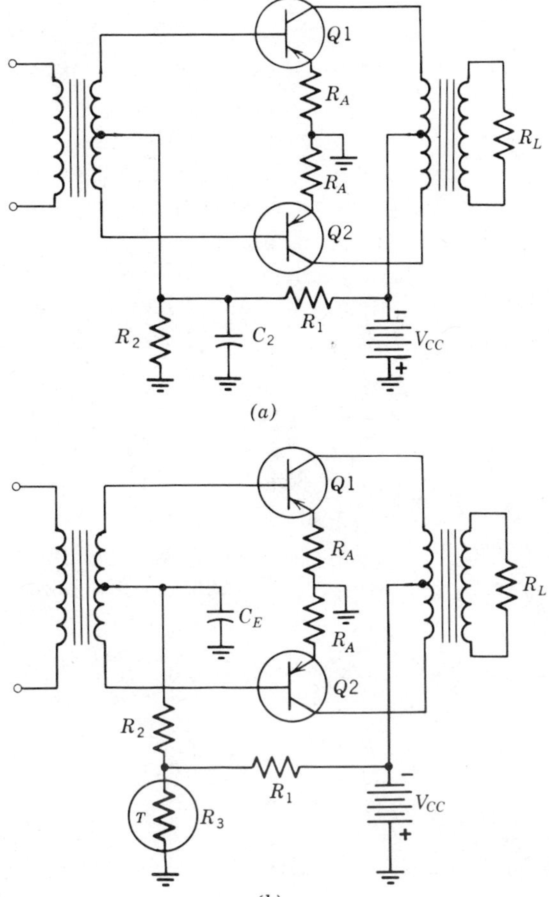

Figure 14-2 Typical push-pull circuits. (*a*) Bias obtained from voltage divider. (*b*) Temperature-compensated bias.

operating point at the center of the load line, the bias current must decrease when the curves rise. The resistor R_3 is a *thermistor* used for the compensation. As the ambient temperature increases, the resistance of the thermistor decreases, causing more current to be shunted to ground and less bias current to enter into the transistor. Proper design of this circuit keeps the operating point at the center of the load line at different temperatures.

Section 14-2 Phase Inverters

A circuit arrangement to produce balanced voltages which are 180° out of phase for the inputs of the push-pull stage is termed a *phase inverter* or *driver*. Many circuit variations have been developed for this purpose. We will consider some of the fundamental designs that are in common use.

The circuit of Fig. 14-3 provides a simple and effective means of obtaining the balanced driving voltages. The balance in this circuit is determined by the exactness of the location of the center tap. The simplicity of the circuit is often outweighed by the expense, size, and weight of the driver transformer. When there is a large current requirement for the input of push-pull stage, the use of the driver transformer cannot be avoided without producing a very large distortion caused by an *IZ* drop in the driving circuit and a shift in the operating point.

In the cascade phase inverter (Fig. 14-4) two identical amplifier stages are used. In order to understand the operation of this circuit, assume that the input signal at *a* has a positive phase direction. *Q*1 amplifies this signal with a 180° phase reversal. Thus the signal at *b* has a negative phase direction and $V_{out,1}$ has a negative phase direction. The capacitor *C* is large and functions as a blocking capacitor feeding the signal

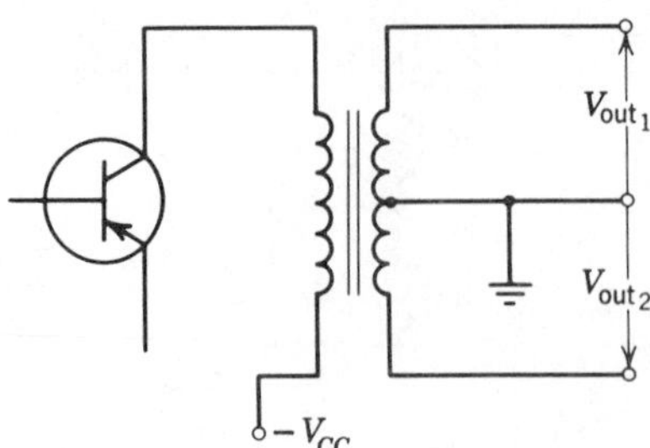

Figure 14-3 Transformer phase inverter.

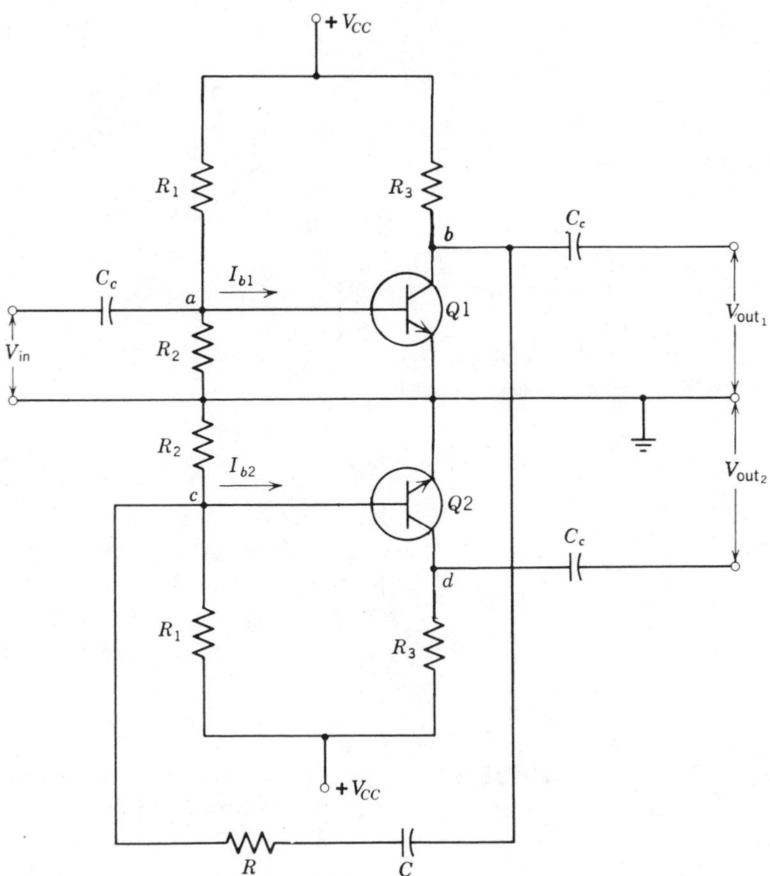

Figure 14-4 Cascade phase inverter.

from b to c without phase change. The signal at c has a negative phase direction and the amplifier $Q2$ amplifies this signal with a phase inversion to d. Now the signal at d has a positive phase direction and the proper phase relations for a driving signal for a push-pull amplifier are obtained.

The input signal V_{in} produces an ac signal at the base I_{b_1}. The value of R is selected, so that the base current I_{b_2} in $Q2$ is identical to I_{b_1}. Therefore, if $Q1$ and $Q2$ are identical in characteristics, the outputs V_{out_1} and V_{out_2} are balanced. An experimental adjustment of R can provide an exact balance.

The split-load phase inverter (Fig. 14-5) uses the concept of simultaneous collector and emitter outputs. The signal from the collector V_{out_1} is 180° out of phase with V_{in}, and the signal from the emitter V_{out_2} is in phase with V_{in}. Since the emitter

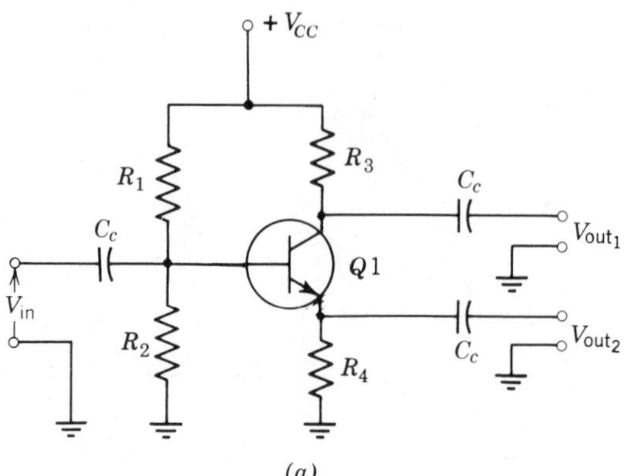

Figure 14-5 Split-load phase inverter.

follower amplifier cannot have a voltage gain exceeding unity, the value of R_3 must be decreased (or R_4 increased) to match V_{out_1} to V_{out_2} in amplitude to provide the requisite balance of the two outputs. If R_3 and R_4 are equal, the outputs are almost exactly balanced. An exact balance can be obtained by adjusting or by selecting either R_3 or R_4.

Problems Use 50 mV/I_E for r'_e for the transistors.

14-2.1 In the circuit of Fig. 14-4, the silicon transistors have a β of 100. The supply voltages are 10 V each. R_3 is 10 kΩ, R_2 is 10 kΩ, and V_{CE} is 5 V. Find R_1 and R. What is the gain of the circuit?

14-2.2 In the circuit of Fig. 14-5, the silicon transistor has a β of 60. The supply voltage is 15 V, and V_{CE} is 5 V. If R_2, R_3, and R_5 are 10-kΩ precision resistors, what are the exact values of the output voltages when the input signal is 2 V? What is the value of R_1?

Section 14-3 Class-A, Class-AB, and Class-B Amplifiers

The term *class* describes the operation of an amplifier by specifying the conditions of collector current flow for an ac cycle of the signal (Fig. 14-6). In a *class-A* amplifier, the collector current flows for the full ac cycle (360°). In a *class-AB* amplifier, collector current flows for more than half the ac cycle but less than a full ac cycle. In a *class-B* amplifier, the collector current flows for exactly 180° of the full ac cycle. In *class-C* operation, the collector current flows for less than 180° of the full ac cycle.

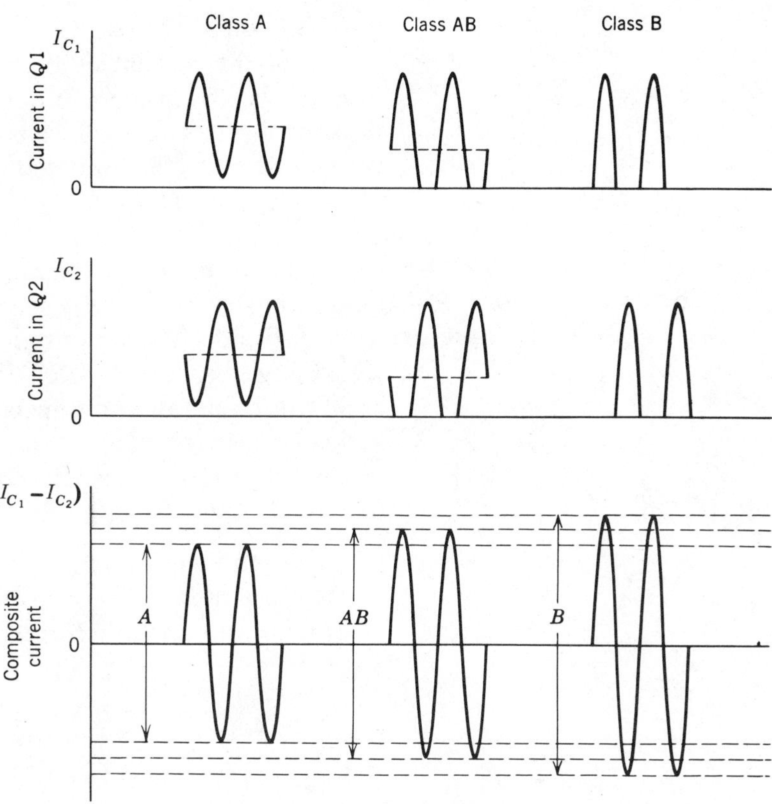

Figure 14-6 Currents in a push-pull circuit.

All of the amplifiers that we have considered to this point have been class-A amplifiers. Particular efforts were taken to insure that the amplifiers were linear over the full ac signal cycle.

The power amplifier we considered in detail in Section 13-4 is properly called *a single-ended class-A amplifier.* In the next section, Section 14-5, we will treat class-A push-pull amplifiers. In Section 14-6, we will examine class-B push-pull amplifiers.

When we discussed the theory of operation of a push-pull amplifier, we showed that the ac flux in the primary of the output transformer is produced by the difference of I_{C_1} and I_{C_2}. Accordingly, in Fig. 14-6, the composite current $(I_{C_1} - I_{C_2})$ is this difference. The individual collector currents for $Q1$ and for $Q2$ have identical peak values for the three classes. However, the peak-to-peak values of the composite currents differ for the three classes. Since power is proportional to the square of the current, it should be evident that we can obtain more power from class-B operation than from class-A operation for a given pair of transistors. In the next two sections, we will develop this in detail.

Section 14-4 The Class-A Push-Pull Amplifier

The class-A push-pull amplifier shown in Fig. 14-7 uses an input transformer $T1$ to supply the required 180° signals to drive $Q1$ and $Q2$. The output transformer $T2$ supplies the ac signal power to the load R_L. The bias resistors R_{B_1} and R_{B_2} are adjusted to operate the transistors at the optimum point on the ac load line.

When these operating conditions are obtained, we may use the conclusions drawn in Section 13-4.

The maximum possible overall efficiency and the maximum possible collector efficiency for a class-A amplifier using an output transformer are both 50%.

$$\boxed{\eta_{\text{overall}} = \eta_{\text{collector}} = 50\%} \tag{14-1a}$$

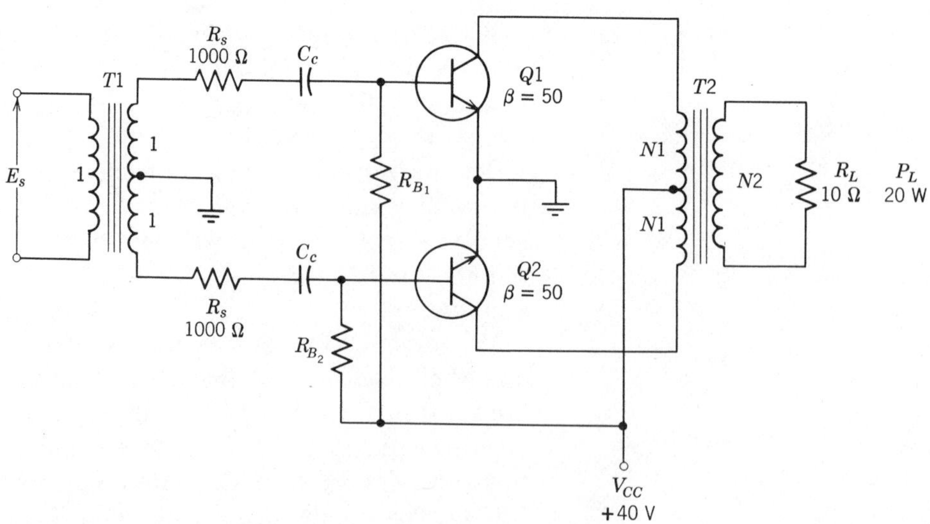

Figure 14-7 Circuit for class-A push-pull amplifier.

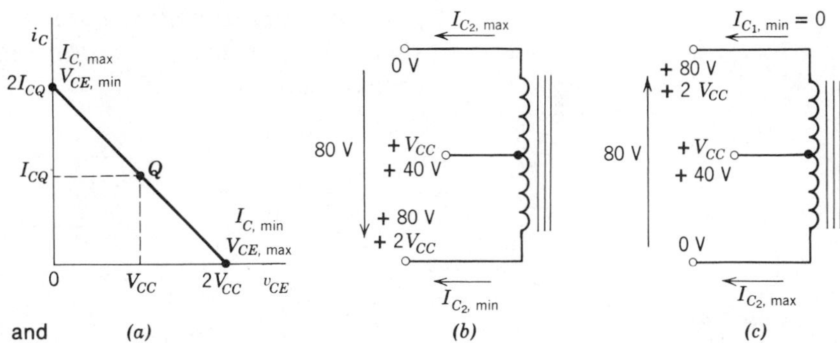

Figure 14-8 Instantaneous peak and minimum conditions. (*a*) AC load line. (*b*) Negative peak condition. (*c*) Positive peak condition.

The waveforms for collector currents, in Fig. 14-6, show that, when the collector current in $Q1$ is a maximum, the collector current in $Q2$ is a minimum. When the collector current in $Q1$ is a minimum, the collector current in $Q2$ is a maximum. When the load line and the operating point meet optimum conditions (Fig. 14-8*a*) the maximum collector current in one transistor is $2I_{CQ}$ when V_{CE} is zero. Also the minimum collector current in one transistor is zero when V_{CE} is $2V_{CC}$. These endpoint values are used to show the instantaneous currents and voltages existing in the primary of the output transformer at the instants these endpoint conditions occur (Fig. 14-8*b* and Fig. 14-8*c*).

The peak voltage across the full primary in Fig. 14-8*b* is $2V_{CC}$ (80 V) directed downward. The peak voltage across the full primary in Fig. 14-8*c* is $2V_{CC}$ directed upward. Then the *peak-to-peak ac voltage* across the full primary winding is $4V_{CC}$(160 V).

The load power is P_L watts. Since this circuit has an efficiency of 50%, the supply power $V_{CC}I_{dc}$ must be $2P_L$.

$$2P_L = V_{CC}I_{dc} = V_{CC}(I_{CQ_1} + I_{CQ_2}) = 2\ V_{CC}I_{CQ}$$

If the operating or quiescent current in $Q1$ is I_{CQ}, the collector current in $Q1$ has a peak-to-peak value of $2I_{CQ}$. If the input signal is turned off, the combined collector dissipation of $Q1$ and $Q2$ is $2P_L$. Therefore:

Each transistor must have a heat sink capable of dissipating P_L when P_L is the maximum available power output from the stage.

$$P_C = P_L \tag{14-1b}$$

Example 14-1
For the circuit of Fig. 14-7, find the transformer turns ratio. What is the dc current value I_C, and the power dissipation rating, P_C, for each transistor? What is the peak reverse voltage BV_{CE} on the transistors?

Solution
The peak load voltage is given by Eq. 13-8 as

$$V_{L,\max} = \sqrt{2P_LR_L} = \sqrt{2 \times 20\ \text{W} \times 10\ \Omega} = 20\ \text{V} \tag{13-8}$$

The peak-to-peak primary ac voltage is $4V_{CC}$ or 160 V. Using a peak value of 80 V, we find that the turns ratio of the required output transformer is

$$\frac{N_2}{(N_1 + N_1)} = \frac{20\ \text{V}}{80\ \text{V}} = \frac{1}{4} \quad \text{or} \quad \frac{N_2}{N_1} = \frac{1}{2} \tag{13-5a}$$

Since the load power P_L is 20 W, the value of P_C for each transistor is 20 W and each heat sink must dissipate 20 W.

$$P_C = P_L = 20\ \text{W} \tag{14-1b}$$

Since the ideal efficiency of the class-A amplifier is 50%, the supply power is 2×20 or 40 W.

$$V_{CC}(I_{CQ_1} + I_{CQ_2}) = 2P_L$$
$$40\ \text{V} \times (I_{CQ_1} + I_{CQ_2}) = 40\ \text{W}$$
$$I_{CQ_1} = I_{CQ_2} = 0.5\ A$$

The peak value of reverse voltage at the collector is twice the supply voltage.

$$BV_{CE} = 2V_{CC} = 2 \times 40 = 80\ \text{V}$$

Then, the minimum specifications for $Q1$ and $Q2$ are

$$I_{CQ} = \mathbf{0.5\ A} \quad P_C = \mathbf{20\ W} \quad \text{and} \quad BV_{CE} = \mathbf{80\ V}$$

Example 14-2

Find the drive power to the bases of the transistors of Example 14-1. What is the gain in decibels? What is the value of E_s and what is the overall power gain?

Solution

The input equivalent circuit is drawn in Fig. 14-9. The input resistances to the transistors are designated as r_{in}. Note that the transformer turns ratio is 1 : 1 : 1. In order to determine r'_e we must find I_{BQ} from the values given in the original circuit, Fig. 14-7. The base current I_{BQ} is

$$I_{BQ} = \frac{I_{CQ}}{\beta} = \frac{0.200}{50} = 0.004\ \text{A} = 4\ \text{mA}$$

The Kirchhoff's voltage loop equation for the dc circuit through the base is

$$V_{CC} = V_{BB} = I_B R_B + V_{BE}$$
$$40 = 0.004 R_B + 0.7$$
$$R_B = 9825\ \Omega$$

The input resistance to each transistor is

$$r_{\text{in}} = (1+\beta) r'_e = (1+\beta)\frac{50\ \text{mV}}{I_E} = (1+\beta)\frac{50\ \text{mV}}{I_C + I_B}$$
$$= (1+50)\frac{50\ \text{mV}}{200\ \text{mA} + 4\ \text{mA}} = 12.5\ \Omega \qquad (7\text{-}7)$$

In the equivalent circuit for the input to the circuit, R_B is in parallel with r_{in}. Since R_B is very much larger than r_{in}, we may neglect R_B in

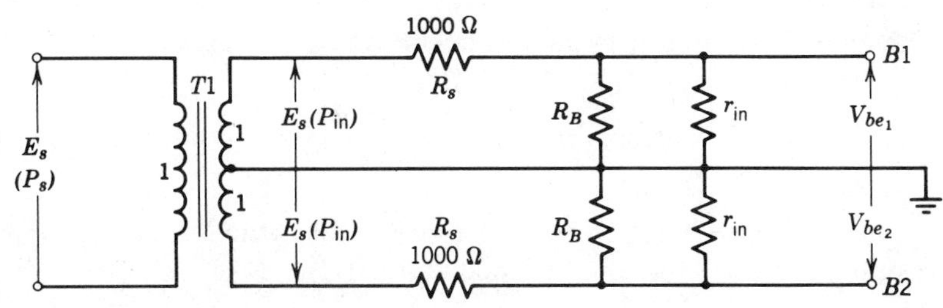

Figure 14-9 Input equivalent circuit.

Fig. 14-9. Since I_{BQ} is 4 mA, the peak value of the base signal current is 4 mA. Then the peak value of the base signal voltage is

$$V_{be,\max} = I_{b,\max} r_{\text{in}} = 4\text{ mA} \times 12.5\ \Omega = 50\text{ mV}$$

The ac driving power to each of $Q1$ and $Q2$ is

$$P_{\text{in}} = \frac{V_{be,\max} I_{b,\max}}{2} = \frac{50\text{ mA} \times 4\text{ mA}}{2} = \mathbf{100\ \mu W}$$

The ac driving power required for the two transistors is 200 μW or 0.2 mW. Then the overall power gain in decibels is

$$dB = 10 \log_{10} \frac{P_L}{P_{\text{in}}} = 10 \log_{10} \frac{20\text{ W}}{0.0002\text{ W}} = \mathbf{+50\ dB} \qquad (11\text{-}1)$$

The input circuit (Fig. 14-9) is a simple voltage divider in which

$$V_{be,\max} = \frac{r_{\text{in}}}{r_{\text{in}} + R_s} E_{s,\max}$$

$$0.05\text{ V} = \frac{12.5\ \Omega}{12.5\ \Omega + 1000\ \Omega} E_{s,\max}$$

Solving for $E_{s,\max}$, we have

$$E_{s,\max} = \mathbf{4.0\ V}$$

Each half of the secondary of the driving transformer $T1$ must provide

$$P'_{\text{in}} = \frac{E_{s,\max} I_{b,\max}}{2} = \frac{4.0\text{ V} \times 4\text{ mA}}{2} = 8.0\text{ mW}$$

The total power P_s required from the source E_s is

$$P_s = 2P'_{\text{in}} = 2 \times 8 = \mathbf{16\ mW} \text{ or } 0.016\text{ W}$$

The overall circuit power gain in decibels is

$$dB = 10 \log_{10} \frac{P_L}{P_s} = 10 \log_{10} \frac{20\text{ W}}{0.016\text{ W}} = \mathbf{+31\ dB} \qquad (11\text{-}1)$$

The difference between +31 dB and +50 dB is the loss, 19 dB, that results from the source impedance of 1000 Ω.

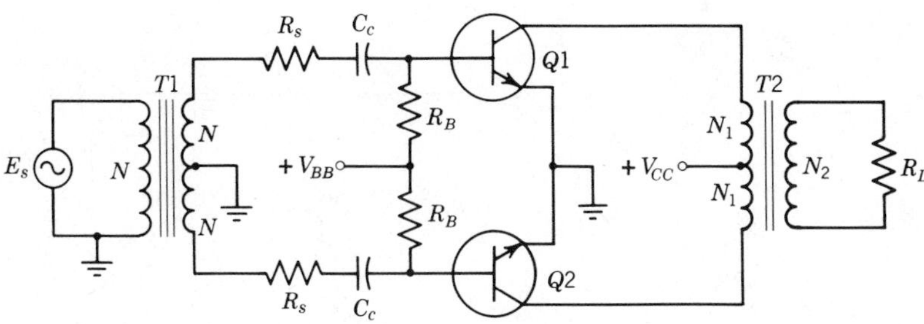

Circuit for problems 14-4.1 through 14-4.6.

For silicon transistors, $\beta = 60$ and $r_e' = \frac{50 \text{ mV}}{I_E}$

Problems

14-4.1 The maximum power in the 20-Ω load is 20 W. If the supply voltage V_{CC} is +60 V, determine the turns ratio required for $T2$. Determine R_B. Determine I_C, P_C, and BV_{CE} for each transistor.

14-4.2 Repeat Problem 14-4.1 for a supply voltage of +80 V and a load power of 50 W into 16 Ω.

14-4.3 Repeat Problem 14-4.1 for a supply voltage of +12 V and a load power of 4 W into 8 Ω.

14-4.4 Repeat Problem 14-4.1 for a supply voltage of +30 V and a load power of 10 W into 600 Ω.

14-4.5 Determine the value of E_s required to obtain the full power output using the data given in Problem 14-4.1. R_s is 600 Ω. What is the overall power gain in decibels?

14-4.6 Determine the value of E_s required to obtain the full power output using the data given in Problem 14-4.2. R is 100 Ω. What is the overall power gain in decibels?

Section 14-5 The Class-B Push-Pull Amplifier

A typical class-B amplifier circuit is shown in Fig. 14-10. The class-B amplifier is biased at cutoff. For transistors, this is very simple. Zero bias is the required cutoff bias. Thus the base leads are returned directly to the return path, ground. When the bias is zero, one-half of the signal cycle is a forward bias causing collector current, while the other half of the cycle is a reverse bias preventing collector current. On the other hand, a separate fixed bias equal to V_p, the pinch-off voltage, is required for class-B operation for the FET.

The waveforms for two cycles of a class-B amplifier are shown in Fig. 14-11. The waveform of current in each transistor is the waveform of the current in a half-wave rectifier with resistive load. Consequently, the average or dc collector

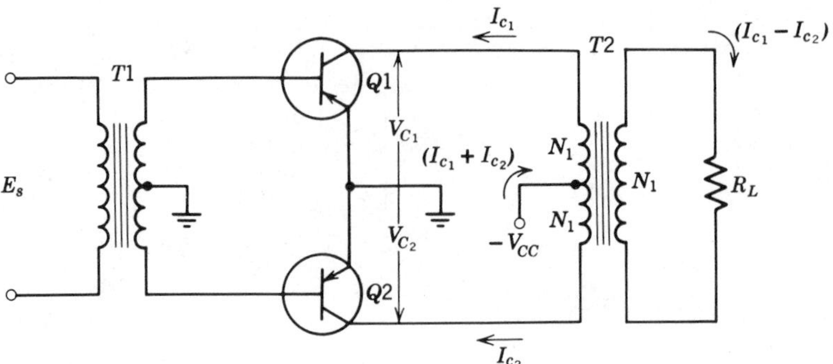

Figure 14-10 The class-B amplifier.

current in each transistor is

$$I_C = \frac{I_m}{\pi} \tag{14-2a}$$

Then for two transistors, the dc supply current I_{dc} is $2I_C$

$$I_{dc} = 2I_C = \frac{2I_m}{\pi} \tag{14-2b}$$

and the power supply delivers

$$P_{dc} = V_{CC} \times I_{dc} = \frac{2I_m}{\pi} V_{CC} \text{ watts} \tag{14-3a}$$

to the collector circuit.

For simplification of the analysis, let us assume the transformer turns ratio is $(N_1 + N_1)$: N_1 as shown in Fig. 14-10. Then, if V_m is the peak voltage at either collector, the peak load voltage is also V_m. The peak current in the load is also I_m. Then the ac load power P_L is

$$P_L = \frac{V_m}{\sqrt{2}} \times \frac{I_m}{\sqrt{2}} = \frac{V_m I_m}{2} \tag{14-3b}$$

The total collector dissipation for the two transistors is

$$2P_C = P_{dc} - P_L = \frac{2I_m V_{CC}}{\pi} - \frac{V_m I_m}{2} = 2I_m \left(\frac{V_{CC}}{\pi} - \frac{V_m}{4}\right) \tag{14-3c}$$

where P_C is the dissipation for one transistor.

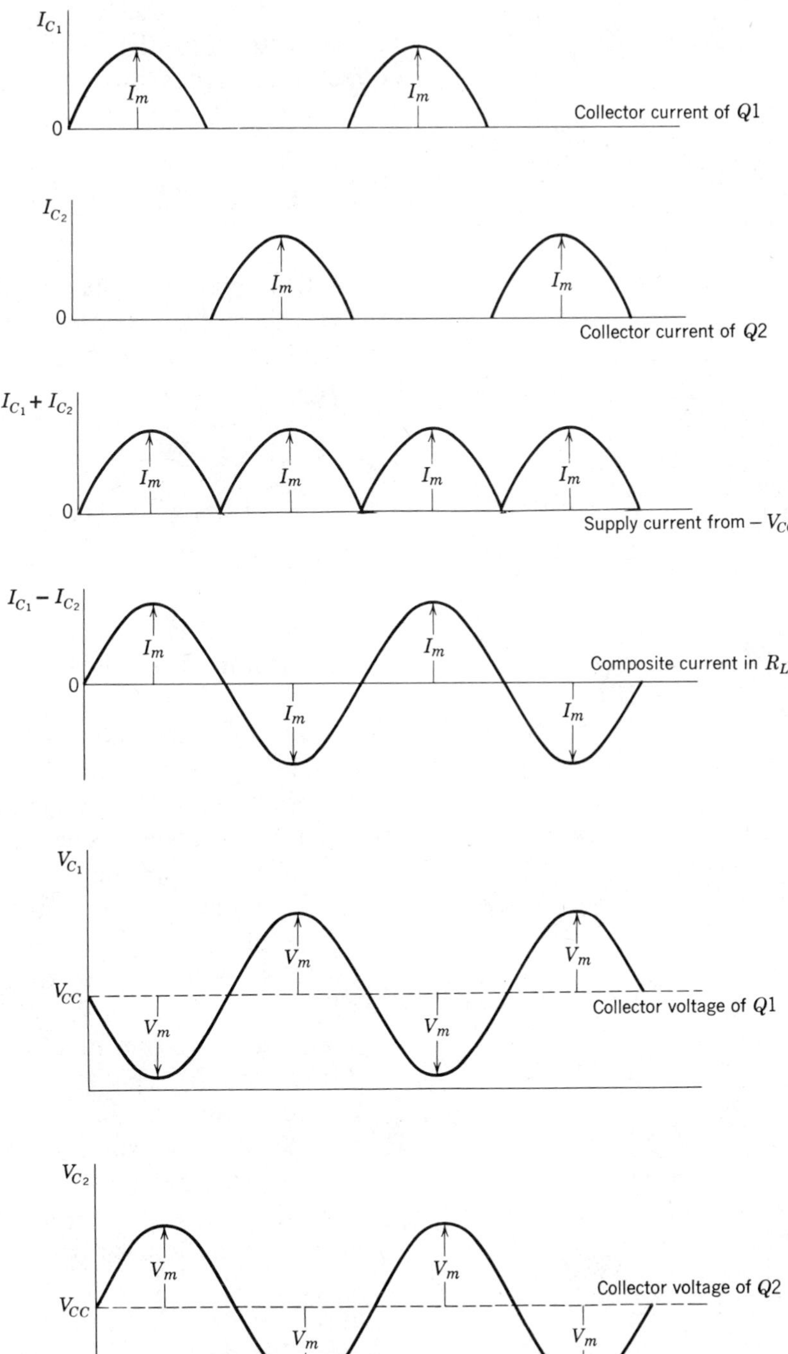

Figure 14-11 Waveforms of a class-B amplifier.

Since a transformer is used to transfer the ac power from the collectors to the load, the overall efficiency and the collector efficiency are identical. This efficiency is found by dividing Eq. 14-3*b* by Eq. 14-2*a*:

$$\eta_{\text{overall}} = \eta_{\text{coll}} = \frac{V_m I_m/2}{2I_m V_{CC}/\pi} \times 100 = \frac{\pi}{4}\frac{V_m}{V_{CC}} \times 100\% \quad (14\text{-}4)$$

When the operation of the circuit is ideal for maximum power in the load, the collector voltage is zero at the instant the peak of the alternating collector voltage V_m equals V_{CC}. Then Eq. 14-4 becomes

$$\boxed{\eta_{\text{overall}} = \eta_{\text{coll}} = \frac{\pi}{4} \times 100\% = 78.5\%} \quad (14\text{-}5)$$

Under maximum signal conditions the efficiency of a class-B amplifier is $\pi/4$ or 78.5%.

This value of 78.5% contrasts with the 50% value obtained from a class-A power amplifier. Obviously, the efficiency of a class-AB amplifier must lie between 50% and 78.5% depending upon the exact angle of current flow.

Let us assume that the dc input power to a class-B amplifier is 100 W. From Eq. 14-5, the value of P_L is 78.5 W. The class-B amplifier is a linear circuit; that is, V_m rises linearly to V_{CC} with the input signal level E_s. The peak current I_m also rises linearly with E_s as shown in Fig. 14-12*a*. The load power is a squared curve that must equal 78.5 W at maximum signal level (Fig. 14-12*b*). The dc supply power given by Eq. 14-3*a* is $2I_m V_{CC}/\pi$. Therefore, the dc supply power rises *linearly* with E_s to 100 W at maximum input signal (Fig. 14-12*b*). The distance between the two curves represents the power dissipation of the two transistors together, $2P_C$.

The power dissipation is now plotted in Fig. 14-12*c*.

It is obvious from this curve that maximum power dissipation does not occur in a class-B amplifier at maximum signal level but at some lower signal level.

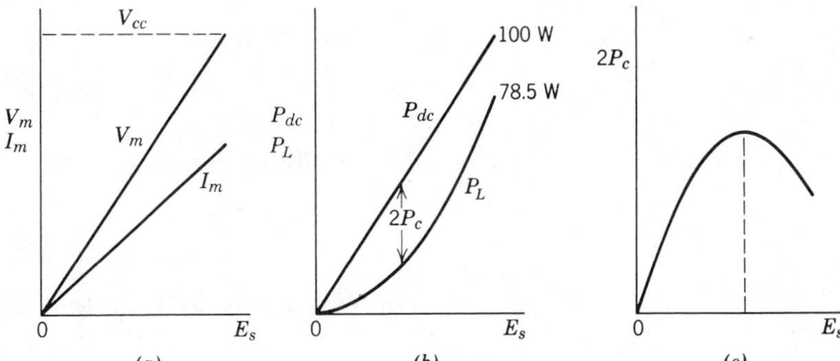

Figure 14-12 Power relations in a class-B amplifier. (*a*) Load voltage and current. (*b*) Input and load powers. (*c*) Dissipation.

The total transistor dissipation is given by Eq. 14-3*c*.

$$P_{\text{diss}} = 2P_C = 2I_m\left(\frac{V_{CC}}{\pi} - \frac{V_m}{4}\right) \tag{14-3c}$$

Using the simplified unity turns ratio used in Fig. 14-10, we have

$$I_m = \frac{V_m}{R_L}$$

and substituting into Eq. 14-3*c*, we have

$$P_{\text{diss}} = 2\frac{V_m}{R_L}\left(\frac{V_{CC}}{\pi} - \frac{V_m}{4}\right) = \frac{2V_{CC}}{\pi R_L}V_m - \frac{2}{4R_L}V_m^2 \tag{14-6}$$

To find the maximum dissipation, we take the first derivative of Eq. 14-6 and set the result to zero. The resulting solution for V_{m_2} gives the value of V_m at which maximum dissipation occurs.

$$\frac{dP_{\text{diss}}}{dV_m} = \frac{2V_{CC}}{\pi R_L} - \frac{4}{4R_L}V_{m_2} = 0$$

Solving for V_{m_2}, we find

$$\frac{4}{4R_L}V_{m_2} = \frac{2V_{CC}}{\pi R_L}$$

or

$$V_{m_2} = \frac{2}{\pi} V_{CC} = 0.636\ V_{CC} \tag{14-7}$$

At maximum output power

$$V_{m_1} = V_{CC}$$

The maximum possible load power is

$$P_{L_1} = \frac{V_{m_1}^2}{2R_L} = \frac{V_{CC}^2}{2R_L} \tag{14-8}$$

At maximum dissipation, the load power is

$$P_{L_2} = \frac{V_{m_2}^2}{2R_L} = \frac{\left(\frac{2}{\pi} V_{CC}\right)^2}{2R_L} = \frac{2V_{CC}^2}{\pi^2 R_L} \tag{14-9}$$

If we divide Eq. 14-9 by Eq. 14-8, we have

$$\frac{P_{L_2}}{P_{L_1}} = \frac{\dfrac{2V_{CC}^2}{\pi^2 R_L}}{\dfrac{V_{CC}^2}{2R_L}} = \frac{4}{\pi^2} = 0.405 \tag{14-10}$$

Eq. 14-10 states that maximum collector dissipation occurs when the load power is 40.5% of the maximum possible load power. Since this maximum dissipation is for two transistors, the dissipation for each transistor is

$$\boxed{P_{C,\max} = 0.203\ P_{L,\max} \approx 0.20\ P_{L,\max}} \tag{14-11}$$

These lengthy derivations may be summarized very simply.

A class-B amplifier has an efficiency of $\pi/4$ or 78.5% at maximum output power $P_{L,\max}$. Maximum collector dissipation occurs at 40% of $P_{L,\max}$ and at that point each transistor dissipates 20% of $P_{L,\max}$.

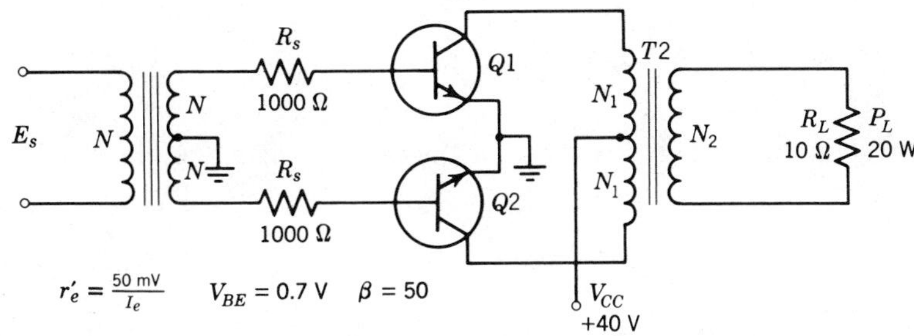

Figure 14-13 Class-B amplifier.

Example 14-3

For the circuit of Fig. 14-13, find the transformer turns ratio. What is the dc current value and the power dissipation rating for each transistor? What is the peak reverse voltage on each transistor?

Solution

This amplifier has the same supply voltage and load requirement as those we used for the class-A amplifier in Example 14-1 and in Example 14-2.

The peak load voltage is the same value as before.

$$V_{L,\text{max}} = \sqrt{2P_L R_L} = \sqrt{2 \times 20 \text{ W} \times 10 \ \Omega} = 20 \text{ V}$$

The peak-to-peak primary ac voltage is 4 V_{CC} or 160 V. The peak value is 80 V. The turns ratio of the output transformer is the same as before.

$$\frac{N_2}{(N_1 + N_1)} = \frac{20 \text{ V}}{80 \text{ V}} = \frac{1}{4} \quad \text{or} \quad \frac{N_2}{N_1} = \frac{1}{2} \tag{13-4a}$$

Since the efficiency is 78.5% and the required load power is 20 W, the dc supply power is

$$P_{\text{dc}} = V_{CC}(2I_C) = \frac{P_L}{\eta}$$

$$P_{\text{dc}} = 40 \text{ V}(2I_C) = \frac{20 \text{ W}}{0.78}$$

Solving for I_C, we find

$$I_C = 0.32 \text{ A} = 320 \text{ mA}$$

The dissipation requirement for each transistor and for each heat sink is

$$P_C = 0.20\ P_L = 0.20 \times 20 = 4\ \text{W} \qquad (14\text{-}11)$$

The peak value of reverse voltage at the collector is twice the supply voltage.

$$BV_{CE} = 2V_{CC} = 2 \times 40 = 80\ \text{V}$$

Then, the minimum specifications for $Q1$ and $Q2$ are

$$I_{CQ} = \mathbf{320\ mA} \qquad P_C = \mathbf{4\ W} \quad \text{and} \quad BV_{CE} = \mathbf{80\ V}$$

Example 14-4
Find the drive power to the bases of the transistors of Example 14-3. What is the gain in decibels? What is the value of E_s and what is the overall power gain?

Solution
The peak collector current is determined from

$$I_C = \frac{I_m}{\pi} \qquad (14\text{-}2a)$$

$$320\ \text{mA} = \frac{I_m}{\pi}$$

$$I_m = 1005\ \text{mA}$$

Then the peak current in the base is

$$I_{b,\max} = \frac{I_m}{\beta} = \frac{1005\ \text{mA}}{50} = 20\ \text{mA}$$

The peak current in the emitter is

$$I_{e,\max} = I_m + I_{b,\max} = 1005 + 20 = 1025\ \text{mA} \qquad (4\text{-}1)$$

The value of r'_e at the instant the emitter current is $I_{e,\max}$ is

$$r'_e = \frac{50\ \text{mV}}{I_{e,\max}} = \frac{50}{1025} = 0.049\ \Omega \qquad (7\text{-}5)$$

Then

$$r_{\text{in}} = (1 + \beta)r'_e = (1 + 50)(0.049) = 2.5\ \Omega \qquad (7\text{-}7)$$

Since $I_{b,\max}$ is 20 mA, then, by Ohm's law

$$V_{be,\max} = r_{\text{in}} I_{b,\max} = 2.5\ \Omega \times 20\ \text{mA} = 50\ \text{mV}$$

The input signal power to the two bases is

$$P_{\text{in}} = \frac{V_{be,\max} I_{b,\max}}{2} = \frac{50\ \text{mV} \times 20\ \text{mA}}{2} = \mathbf{500\ \mu W}$$

The power gain is

$$dB = 10 \log_{10} \frac{P_L}{P_{\text{in}}} = 10 \log_{10} \frac{20\ \text{W}}{0.0005\ \text{W}} = \mathbf{+46\ dB} \qquad (11\text{-}1)$$

The source resistance, 1000 Ω, is much greater than r_{in} and, as a result, the peak value of E_s is

$$E_{s,\max} = I_{b,\max} R_s = 0.020\ \text{A} \times 1000\ \Omega = \mathbf{20\ V}$$

The power required from E_s to drive the amplifier to full power output is

$$P_s = \frac{E_{s,\max} I_{b,\max}}{2} = \frac{20\ \text{V} \times 0.020\ \text{A}}{2} = \mathbf{0.20\ W}$$

Now the overall circuit gain is

$$dB = 10 \log_{10} \frac{P_L}{P_s} = 10 \log_{10} \frac{20\ \text{W}}{0.20\ \text{W}} = \mathbf{+20\ dB} \qquad (11\text{-}1)$$

The difference between +20 dB and +46 dB is the loss, 26 dB, that results from the source impedance of 1000 Ω.

The results of both sets of calculations are shown in Table 14-1. The big difference between the two circuits is that the "size," that is, the heat dissipation requirement, is reduced by a factor of five for the class-B connection. Also, it should be noted that, in order to obtain this saving in "size," the driving voltage and power must be greater for the class-B amplifier.

When the signal in the class-A amplifier is reduced to zero, there is a heat dissipation in each transistor of 20 W. When the signal in the class-B amplifier is reduced to zero, the heat dissipation in each transistor is zero. As a result of this

Table 14-1

	Class A Push-Pull	Class B Push-Pull
P_L, load power	20 W	20 W
V_{CC}, supply voltage	40 V	40 V
I_C, transistor current (each)	500 mA	320 mA
BV_{CE}, breakdown voltage	80 V	80 V
P_C, maximum dissipation per transistor	20 W	4 W
E_s, maximum peak required	4 V	20 V
Transistor gain	+50 dB	+46 dB
Overall gain	+31 dB	+20 dB
Driving circuit loss	19 dB	26 dB

Note**:** Identical input and output transformers are used.

transistor heating difference, it is essential that the power output stage in a portable radio is class B in order to minimize the drain on the batteries.

At the 0°, 180°, and 360° points in the signal input cycle, the v_{be} falls to zero in the class-B amplifier. At many points in this text, we have required the input signal to be greater than 0.7 V

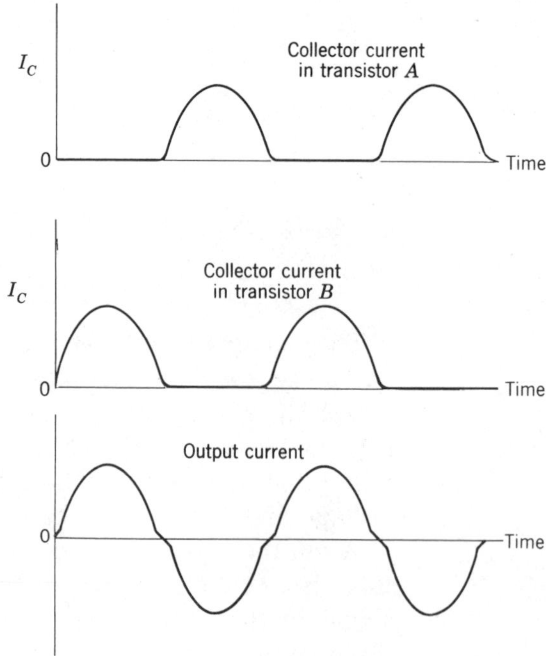

Figure 14-14 Crossover distortion.

for a silicon transistor (0.3 V for germanium). Actually, some current does flow in the base for values of v_{be} less than 0.7 V (0.3 V). Therefore there is insufficient current in the collector to follow a pure sinusoid and we observe a slight deviation from a pure sinusoid, Fig. 14-14. This deviation, called *cross-over distortion*, shows in both the positive and negative parts of the output waveform. This distortion, an inherent characteristic of the class-B transistor circuit, can be avoided only by keeping the operating point slightly off class B and into class AB.

Problems

14-5.1 The maximum power in the 4-Ω load is 200 mW. If the supply voltage V_{CC} is +9 V, determine the transformer turns ratio for $T2$. Determine I_C, P_C, and BV_{CE} for each transistor.

14-5.2 Repeat Problem 14-5.1 if the supply is +80 V and the load is 50 W at 16 Ω.

14-5.3 Repeat Problem 14-5.1 if the supply is +12 V and the load is 4 W at 8 Ω.

14-5.4 Repeat Problem 14-5.1 if the supply is +30 V and the load is 10 W at 600 Ω.

14-5.5 Determine the value of E_s required to obtain the full output power given in Problem 14-5.1. $N_1 = N_2$, and R_s is 100 Ω. What is the overall power gain in decibels?

14-5.6 Determine the value of E_s required to obtain the full output power given in Problem 14-5.2. $N_1 = 2N_2$, and R_s is 30 Ω. What is the overall power gain in decibels?

14-5.7 The supply voltage is 80 V and the load resistance is 8 Ω. $N_3 = 2N_4$ and $N_1 = 4N_2$, R_s is 50 Ω. Determine E_s to develop maximum power in the load. What is the load power?

14-5.8 Solve Problem 14-5.7 if the supply is 50 V and the load is 50 Ω. $N_1 = 5N_2$ and $N_3 = 2N_4$; R_s is 10 Ω.

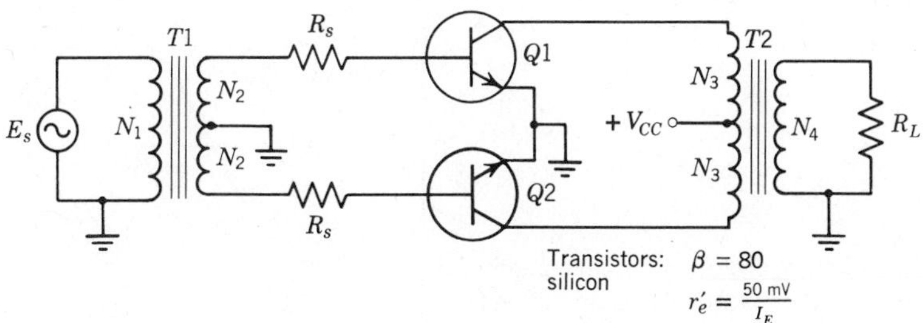

Circuit for Problem 14-5.1 through Problem 14-5.8.

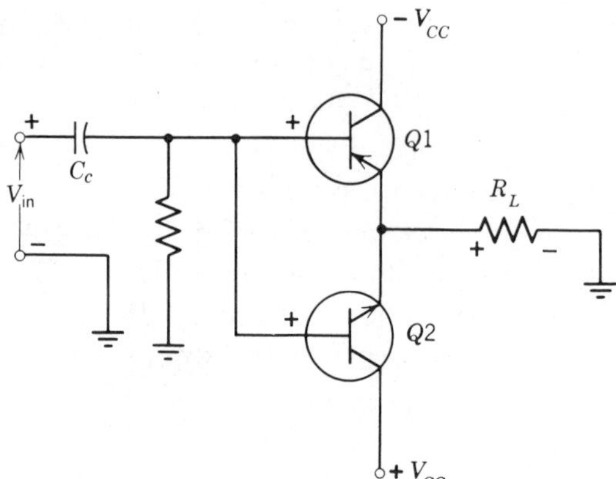

Figure 14-15 Class-B push-pull using complementary symmetry.

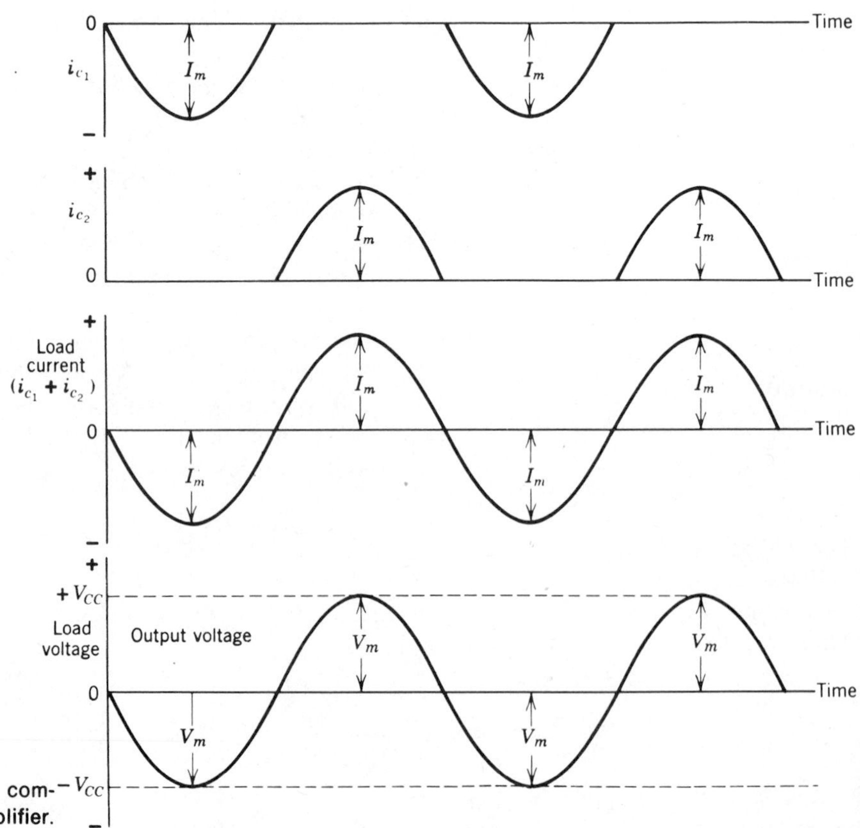

Figure 14-16 Waveforms for the complementary symmetry class-B amplifier.

Section 14-6 Complementary Symmetry in Push-Pull

The principles of complementary symmetry described in Section 12-1 may be applied to the push-pull amplifier, Fig. 14-15. This circuit uses neither an input nor an output transformer. If V_{in} is zero, the bias on both transistors, $Q1$ and $Q2$, is zero. When the bias on a transistor is zero, all the currents in the transistor are zero.

If V_{in} is turned on and instantaneously has the polarity shown in Fig. 14-15, transistor $Q2$ has a forward bias but transistor $Q1$ is driven deeper into cutoff. A collector current I_{c_2} flows in $Q2$. This must flow through R_L producing a voltage drop with the indicated polarity across R_L. When the polarity of V_{in} changes, $Q1$ is "on" and $Q2$ is cut off. Now a current I_{c_1} flows in $Q1$ and the polarity of the voltage drop across R_L is reversed. Thus, each stage, $Q1$ and $Q2$, serves as an emitter follower for its half of the incoming ac cycle. The whole circuit, then, is a class-B amplifier. The waveforms detailing the above explanation are given in Fig. 14-16. We also have the problem of crossover distortion in this circuit. For simplicity, the waveforms of Fig. 14-16 do not show this crossover distortion.

Example 14-5

The required output for the class-B circuit of Fig. 14-15 is 20 W into a 10-Ω resistive load. Determine the values of $+V_{CC}$ and $-V_{CC}$. Determine the values of I_C and P for each transistor. What is the peak reverse voltage rating for the transistors? What is the peak-to-peak voltage required for the input signal?

Solution

By Eq. 13-8, the peak value of voltage across the load is

$$V_m = \sqrt{2P_LR_L} = \sqrt{2 \times 20\ \text{W} \times 10\ \Omega} = 20\ \text{V} \qquad (13\text{-}8)$$

The waveform for the load voltage in Fig. 14-16 shows that the load voltage equals $+V_{CC}$ or $-V_{CC}$ at the instant the load voltage is at its peak value, V_m. We are considering an ideal case where $V_{CE,\text{sat}}$ is 0 V. Therefore, the value of V_m determines the required supply voltages.

$$V_{CC} = +V_m = \mathbf{+20\ V}$$

and

$$-V_{CC} = -V_m = \mathbf{-20\ V}$$

Since the efficiency of the class-B amplifier is $\pi/4$ or 78.5%, we have

$$\eta = \frac{P_L}{2V_{CC}I_C} 100\%$$

$$78.5 = \frac{20 \text{ W}}{2 \times 20 \text{ V} \times I_C} 100$$

$$I_C = 0.637\text{A} = \mathbf{637\ mA}$$

We may obtain I_C by a different approach. The peak value of the load current is

$$I_m = \frac{V_m}{R_L} = \frac{20 \text{ V}}{10\ \Omega} = 2 \text{ A}$$

Therefore

$$I_C = \frac{I_m}{\pi} = \frac{2}{\pi} = 0.637 \text{ A} = \mathbf{637\ mA} \qquad (14\text{-}2a)$$

The maximum collector dissipation for each transistor is

$$P_C = 0.20 P_L = 0.20 \times 20 = \mathbf{4\ W} \qquad (14\text{-}11)$$

The maximum reverse voltage on each transistor is twice V_{CC}

$$BV_{CE} = 2V_{CC} = 2 \times 20 = \mathbf{40\ V}$$

Since the transistors in the complementary symmetry power amplifiers are emitter-follower circuits, the value of A_v is unity. Then

$$V_{\text{in}} = 2V_m = 2 \times 20 = \mathbf{40\ V\ peak\text{-}to\text{-}peak}$$

There is a very important difference between the class-B amplifier using complementary symmetry and the class-B amplifier using an output transformer. The value of the load resistance and the value of the required load power determine the values of $+V_{CC}$ and $-V_{CC}$ in complementary symmetry. If a different output power is required, different supply voltages are required. In the circuit using an output transformer, we can control this by using a different turns ratio and not by changing the supply voltage. It should also be noted that the basic complementary symmetry circuit requires two supply voltage sources whereas the transformer circuit requires only one supply voltage.

In a class-A complementary symmetry push-pull amplifier, we require each transistor to be biased at I_{CQ}. The load current varies between $+2I_{CQ}$ and $-2I_{CQ}$ with an efficiency of 50%. When the signal is zero, the quiescent currents are within the transistors; there is zero current in the load requiring us to use an efficiency of 50%.

Example 14-6

The power in a 10-Ω resistive load is 20 W in the class-A circuit shown in Fig. 14-17. Determine the values of $+V_{CC}$ and $-V_{CC}$. Determine the values of I_C and P_C for each transistor. What is the peak-to-peak value required for V_{in} at maximum load power?

Solution

This load condition is the same as those used in the previous examples in this chapter for comparison purposes.

The peak value of voltage across the load is

$$V_m = \sqrt{2P_LR_L} = \sqrt{2 \times 20\text{ W} \times 10\ \Omega} = 20\text{ V}$$

The magnitude of the collector supply voltages each equals V_m. Then,

$$+V_{CC} = +V_m = \mathbf{+20\ V}$$

and

$$-V_{CC} = -V_m = \mathbf{-20\ V}$$

The maximum efficiency of a class-A amplifier is 50%.

$$\eta = \frac{P_L}{P_{dc}}100 = \frac{P_L}{V_{CC}I_C + (-V_{CC})(-I_C)}100 = \frac{P_L}{2V_{CC}I_C}100\% \quad (14\text{-}1a)$$

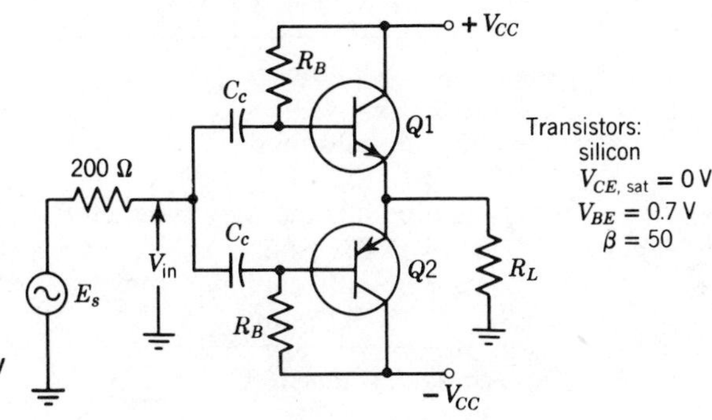

Figure 14-17 Class-A complementary symmetry push-pull amplifier.

Substituting numerical values, we have

$$50\% = \frac{20\ \text{W}}{P_{\text{dc}}}100 = \frac{20\ \text{W}}{2 \times 20\ \text{V} \times (I_C)}100$$

Solving, we have

$$P_{\text{dc}} = 40\ \text{W}$$

and

$$\boldsymbol{I_C = 1\ \text{A}}$$

In a class-A amplifier the maximum possible value of P_C occurs when the input signal is zero. Then

$$P_C = \frac{P_{\text{dc}}}{2} = \frac{40}{2} = \mathbf{20\ W}$$

Transistors $Q1$ and $Q2$ are emitter-follower circuits. The voltage gain A_v from the base to the emitter is unity. Therefore

$$V_{\text{in}} = 2V_m = 2 \times 20 = \textbf{40 V peak-to-peak}$$

Table 14-2 shows a comparison between the four basic circuits that were worked out in the illustrative examples. This comparison is made for the output circuits only. It must be stressed that these results assume that the transistors are ideal

Table 14-2
Comparison of Push-Pull Amplifier Circuits
($P_L = 20$ W and $R_L = 10\ \Omega$ for all circuits)

Circuit Type	V_{CC}	$*I_C$	P_{dc}	$*P_C$
Class-A transformer (Fig. 14-7 Ex. 14-1)	40 V	0.5 A	40 W	20 W
Class-B transformer (Fig. 14-13 Ex. 14-3)	40 V	0.32 A	25.6 W	4 W
Class A complementary symmetry (Fig. 14-17 Ex. 14-6)	±20 V	1.0 A	40 W	20 W
Class B complementary symmetry (Fig. 14-15 Ex. 14-5)	±20 V	0.64 V	25.6 W	4 W

*For each transistor.

and that $V_{CE,\text{sat}}$ is zero. In a practical design, some margin must be added. For instance, if we find our calculation requires 14 V for E_s and 20 W for P_C, the actual circuit could require a slightly larger E_s, slightly larger values for V_{CC}, and perhaps the use of transistors having 25-W dissipation ratings.

Problems

14-6.1 If $+V_{CC}$ is +80 V and $-V_{CC}$ is −80 V and if R_L is 16 Ω, what power can be obtained in the load? What is the value of P_C for each transistor?

14-6.2 If the supply voltages are +12 V and −12 V, what power can be obtained in a 12-Ω load? What is the value of P_C for each transistor?

14-6.3 Determine $+V_{CC}$, $-V_{CC}$, and E_s if the power in the 16-Ω load is to be 10 W. What are the ratings (P_C, I_C, and BV_{CE}) for each transistor? Determine the overall power gain in decibels.

14-6.4 Repeat Problem 14-6.3 if the required output is 500 mW in 40 Ω.

14-6.5 The power required in the 50-Ω load (R_L) is 1 W. Determine $+V_{CC}$, $-V_{CC}$, R, and the value of E_s required to obtain the full output power. Determine the ratings (I_C, P_C, and BV_{CE}) for each transistor. Find the overall power gain in decibels. Use the circuit and data of Fig. 14-17.

14-6.6 Repeat Problem 14-6.5 if the required load power is 4 W and the load resistance is 16 Ω.

14-6.7 If Problem 14-6.1 required the use of a class-A circuit, specify values for R_B, I_C, and P_C for each transistor. What is P_L?

14-6.8 If Problem 14-6.2 required the use of a class-A circuit, specify values for R_B, I_C, and P_C for each transistor. What is P_L?

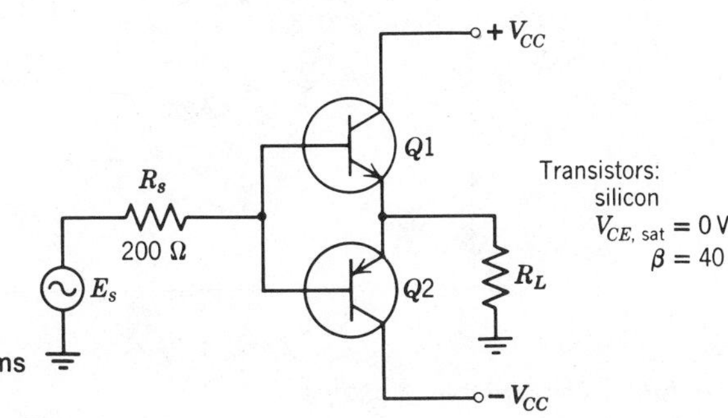

Class-B amplifier circuit for Problems 14-6.1 through 14-6.4.

14-6.9 If Problem 14-6.3 required the use of a class-A circuit, specify values for R_B, I_C, and P_C for each transistor. What is P_L?

14-6.10 If Problem 14.6.4 required the use of a class-A circuit, specify values for R_B, I_C, and P_C for each transistor. What is P_L?

Section 14-7 Commercial Audio Amplifiers

Three commercially used circuits are given to conclude this chapter. The amplifier shown in Fig. 14-18 is the complete electronic circuit used in a portable record player. A driver transformer and an output transformer are used in the push-pull output amplifier. A thermistor is used to temperature-compensate the power stage. The circuit is simple, and the only control is to regulate volume. The expected dc voltage values are given on the circuit for a no-signal condition.

The amplifier shown in Fig. 14-19 is one channel of a two-channel stereo amplifier for a record player. The output is a complementary symmetry push-pull arrangement in which only one power source is required. The complementary symmetry arrangement in the driver stages improves the low-frequency response of the system by avoiding coupling capacitors. The bass control effectively removes and inserts a

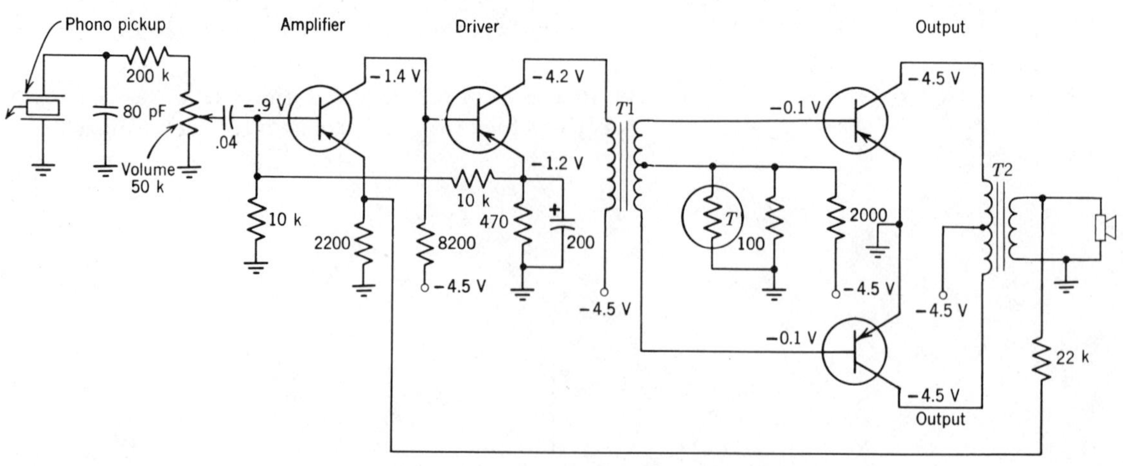

Figure 14-18 Amplifier for a small portable phonograph.

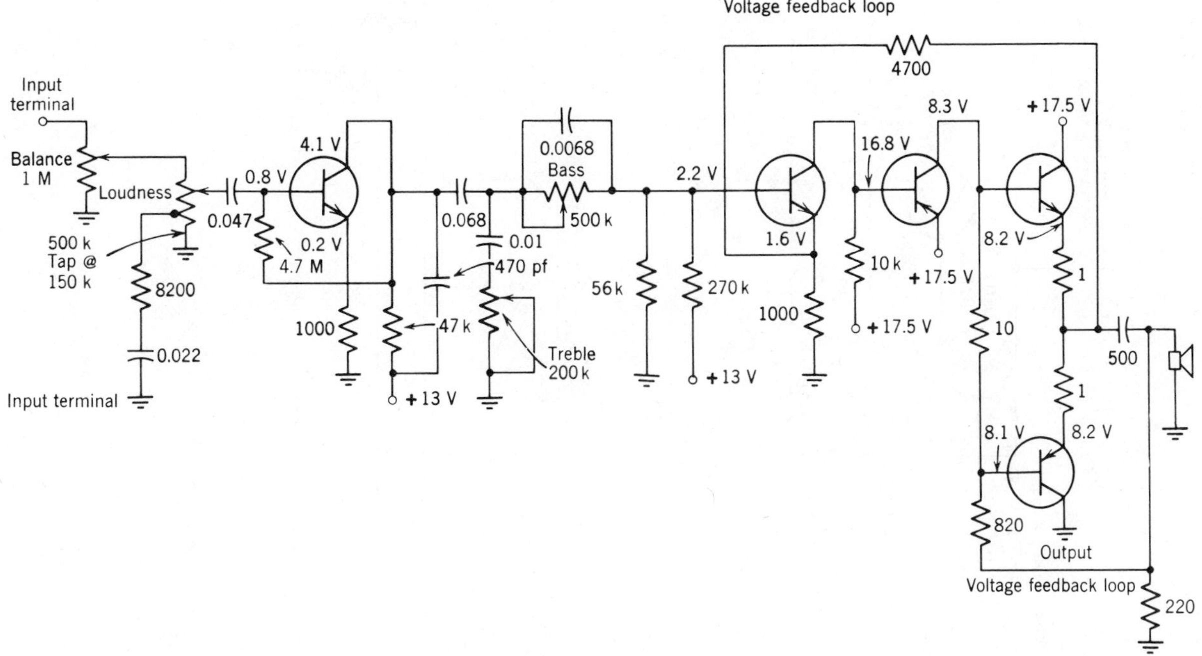

Figure 14-19 One channel of a high fidelity amplifier.

small coupling capacitor. The treble control controls the amount of the high frequencies that is shorted to ground. The balance control sets the level of this channel in relation to the other channel. Like channel controls are ganged together so that the same effect is simultaneously introduced in both channels.

The power amplifier section of a high-fidelity radio receiver is given in Fig. 14-20. This circuit is representative of the maximum quality audio amplifiers that are available. Transistors Q806 and Q808 are used in what is called a *quasi-complementary symmetry* circuit. Without signal the center point of the circuit (the junction between resistors R and R) is carefully adjusted to +31.5 V, which is half the supply voltage. An incoming signal to the amplifier causes this junction potential to vary with the incoming signal.

The feedback loops in these circuits will be considered in Chapter 16.

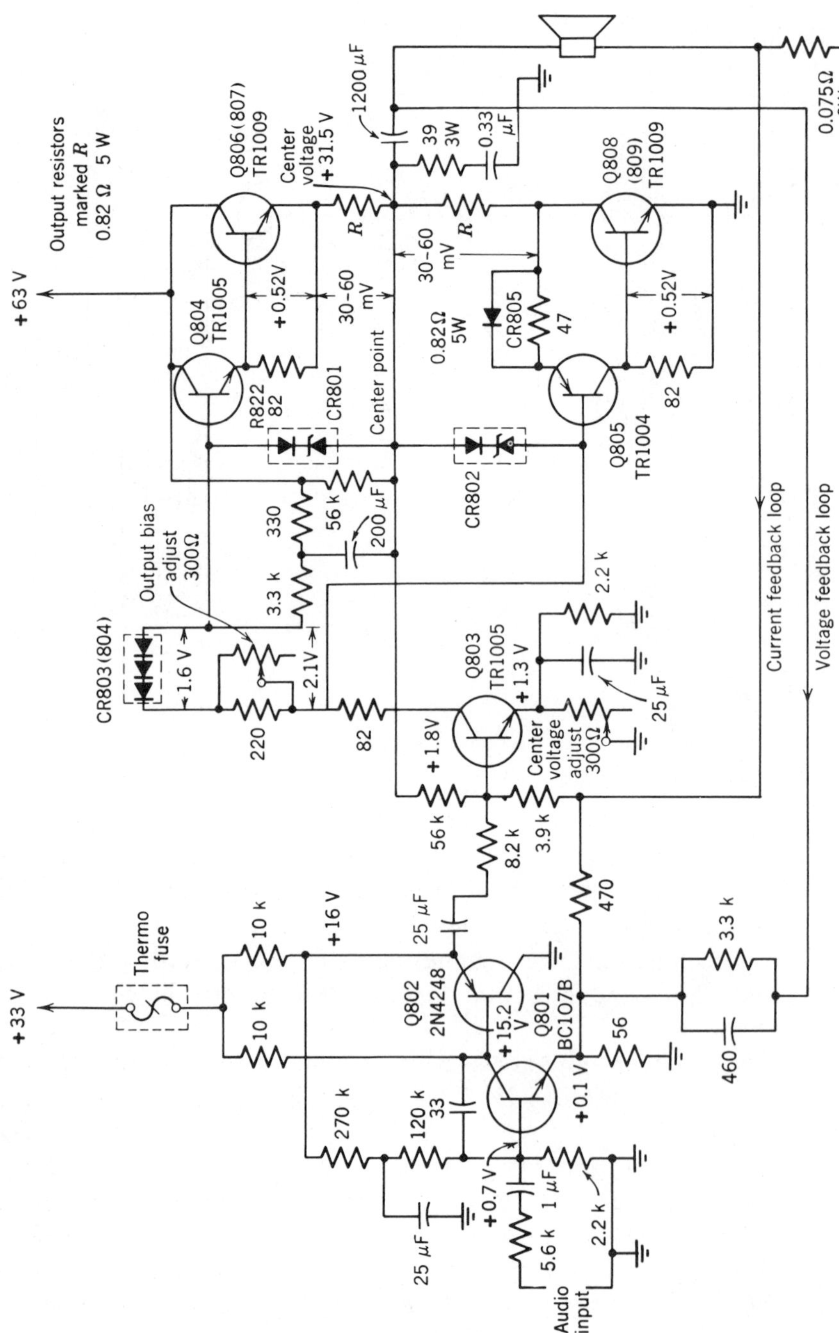

Figure 14-20 Fisher Radio Corporation 45-W audio control amplifier TX-1000 Dual Channel Amplifier. *(Courtesy Fisher Radio Corporation.)*

Supplementary Problems

Load A is 2 W in 8 Ω and Load B is 30 W in 16 Ω.

14-1 The supply to a class-A push-pull amplifier using a transformer is 24 V. Determine the specifications for the transformer and for the transistors if Load A is used.

14-2 Repeat Problem 14-1 if the supply voltage is 40 V and if Load B is used.

14-3 The supply to a class-B push-pull amplifier using a transformer is 24 V. Determine the specifications for the transformer and for the transistors if Load A is used.

14-4 Repeat Problem 14-3 if the supply voltage is 40 V and if Load B is used.

14-5 A complementary symmetry class-A amplifier is used to supply Load A. Determine the specifications for the supply voltages and for the transistors.

14-6 Repeat Problem 14-5 if the load is Load B.

14-7 A complementary symmetry class-B amplifier is used to supply Load A. Determine the specifications for the supply voltages and for the transistors.

14-8 Repeat Problem 14-7 if the load is Load B.

Chapter 15 Frequency Response

A network that has a loss in output voltage together with a phase change at low frequencies (Section 15-1) and one that has a loss in output voltage together with a phase change at high frequencies (Section 15-2) are examined using methods of ac circuit analysis. A simplified approach to the gain and phase properties of the network uses Bode plots for both the low-frequency response (Section 15-3) and the high-frequency response (Section 15-4). A comparison of the simplified approximation of the Bode technique is made with the exact results of the ac circuit analysis (Section 15-5). The frequency and phase response of a two-stage amplifier is obtained using Bode plots (Section 15-6). The input capacitance to a semiconductor is evaluated by using the Miller theorem (Section 15-7).

Section 15-1 Low-Frequency Response

As a result of capacitive effects, the gain of an amplifier (or of a circuit) falls off at low frequencies and at high frequencies. The amplifier gains that we have been working with in this text are *mid-frequency gains* or *mid-band gains.* In the laboratory we usually take measurements at 400 Hz or at 1 kHz to be sure that what we measure is mid-band gain.

Before we examine amplifier circuits, we will show how the output voltage of a passive circuit falls off at low frequencies. Then, in a later section, we will show how the output of a passive circuit falls off at high frequencies.

The circuit shown in Fig. 15-1a is a simple series circuit. Since there is a capacitor C_1 in the circuit, the current I_{in} leads V_{in} by θ degrees as shown on the phasor diagram (Fig. 15-1b). The output voltage V_{out} is the voltage drop across R_2, and therefore V_{out} *leads* V_{in} by the phase angle θ.

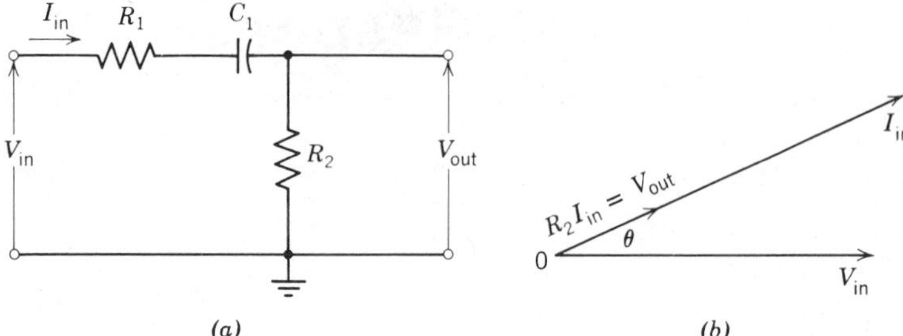

Figure 15-1 Passive circuit showing a drop in V_{out} at low frequencies. (*a*) Circuit diagram. (*b*) Phasor diagram.

Since we have a series circuit,

$$I_{in} = \frac{V_{in}}{R_1 + R_2 - jX_{C_1}}$$

and

$$V_{out} = R_2 I_{in} = \frac{R_2}{R_1 + R_2 - jX_{C_1}} V_{in}$$

If we divide both sides by V_{in}, we *define* the low-frequency gain A_{LF} at any low frequency f as

$$A_{LF} \equiv \frac{V_{out}}{V_{in}} = \frac{R_2}{R_1 + R_2 - jX_{C_1}} \tag{15-1}$$

At mid-frequencies or at mid-band, the capacitor has no effect. At these higher frequencies, the reactance of the capacitor becomes negligibly small.

$$X_{C_1} \leq \frac{1}{10}(R_1 + R_2)$$

At mid-band the output voltage is the result of a simple voltage divider formed by R_1 and R_2.

$$V_{out} = \frac{R_2}{R_1 + R_2} V_{in}$$

If we divide both sides by V_{in}, we *define* A_v, the mid-band gain.

$$A_v \equiv \frac{R_2}{R_1 + R_2} \tag{15-2}$$

If we divide Eq. 15-1 by Eq. 15-2, we *define* the ratio K_{LF} as the ratio of the gain at a low frequency f to the gain at mid-band.

$$K_{LF} \equiv \frac{A_{LF}}{A_v} \tag{15-3}$$

where $$0 \leq K_{LF} \leq 1$$

Substituting, we have

$$K_{LF} = \frac{A_{LF}}{A_v} = \frac{\dfrac{R_2}{R_1 + R_2 - jX_{C_1}}}{\dfrac{R_2}{R_1 + R_2}} = \frac{R_1 + R_2}{R_1 + R_2 - jX_{C_1}}$$

Dividing through by $(R_1 + R_2)$, we find

$$K_{LF} = \frac{1}{1 - j\dfrac{X_{C_1}}{R_1 + R_2}} \tag{15-4}$$

Equation 15-4 can be expressed in the rectangular form

$$K_{LF} = \frac{1}{\sqrt{1 + \left(\dfrac{X_{C_1}}{R_1 + R_2}\right)^2}} \angle + \tan^{-1}\frac{X_{C_1}}{R_1 + R_2} \tag{15-5}$$

Example 15-1
Determine data for a plot of the magnitude and phase angle for K_{LF} for the circuit shown in Fig. 15-2.

Solution
The numerical values are substituted either into Eq. 15-4 or into Eq. 15-5 for frequencies between 1000 Hz and 0.1 Hz. These results are summarized in Table 15-1 and they are plotted in Fig. 15-3.

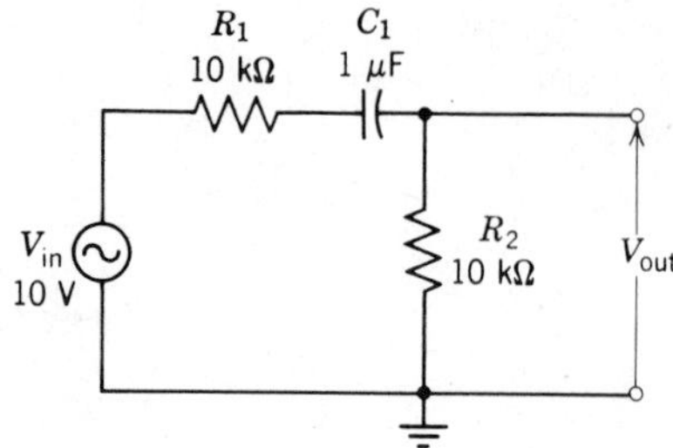

Figure 15-2 Circuit used to show low-frequency response.

Table 15-1
Values for K_{LF} for the Circuit of Fig. 15-2

f (Hz)	θ (degrees)	$\lvert K_{LF} \rvert$
1000	+0.5	1.000
500	+0.9	1.000
100	+4.6	0.997
20	+21.7	0.929
10	+38.5	0.782
7.96	+45.0	0.707
5	+57.9	0.532
2	+75.9	0.244
1	+82.8	0.125
0.5	+86.4	0.063
0.2	+88.6	0.025
0.1	+89.3	0.013

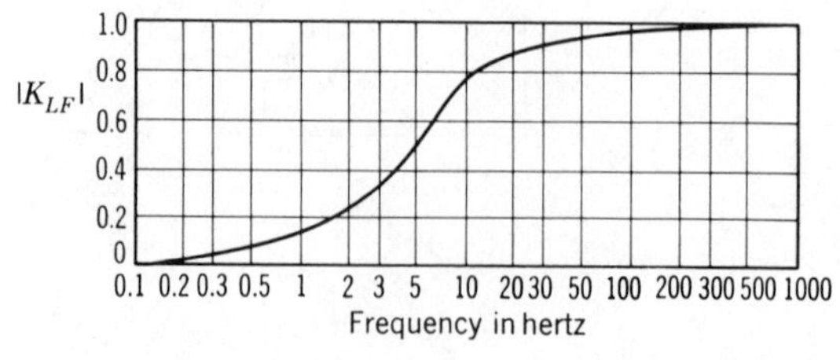

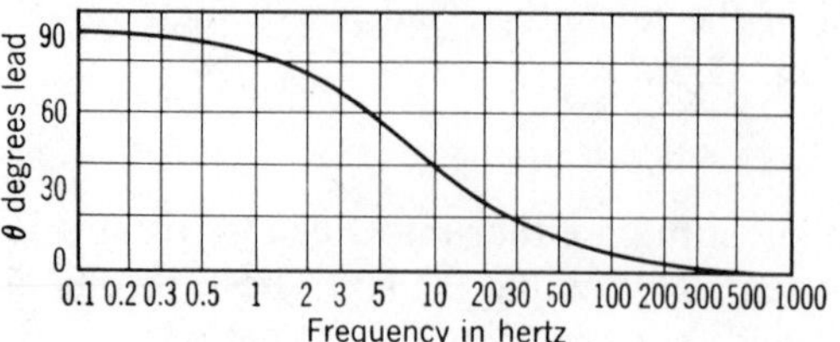

Figure 15-3 Magnitude and phase plots for K_{LF} for the circuit given in Fig. 15-2.

Inspection of the graphs of Fig. 15-3 shows that K_{LF} is 1 at 400 Hz and at 1 kHz and at all higher frequencies. At frequencies below 100 Hz, the value of K_{LF} falls off sharply. At 0.1 Hz, the output voltage is only about 1% of the output at 1000 Hz. The phase angle, which is negligible at 1 kHz, rises to a lead of over 89° at 0.1 Hz. Since V_{out} at midband is

$$V_{out} = \frac{R_2}{R_1 + R_2} V_{in} = \frac{10\ \text{k}\Omega}{10\ \text{k}\Omega + 10\ \text{k}\Omega} 10\ \text{V} = 5\ \text{V}$$

the value of V_{out} can be obtained at any frequency by multiplying the value of K_{LF} at that frequency by 5 V.

Problems For each of the problems, determine sufficient data for V_{out} and for the phase shift to plot the response curves.

15-1.1 C is 10 μF and R is 100 Ω. Determine V_{out} and the phase shift at 100 Hz.

15-1.2 C is 0.02 μF and R is 82 kΩ. Determine V_{out} and the phase shift at 150 Hz.

15-1.3 C is 0.01 μF and R is 500 Ω. Determine V_{out} and the phase shift at 20 kHz.

15-1.4 C_1 is 0.2 μF, R_1 is 2 kΩ, and R_2 is 10 kΩ. Determine V_{out} and the phase shift at 12 Hz.

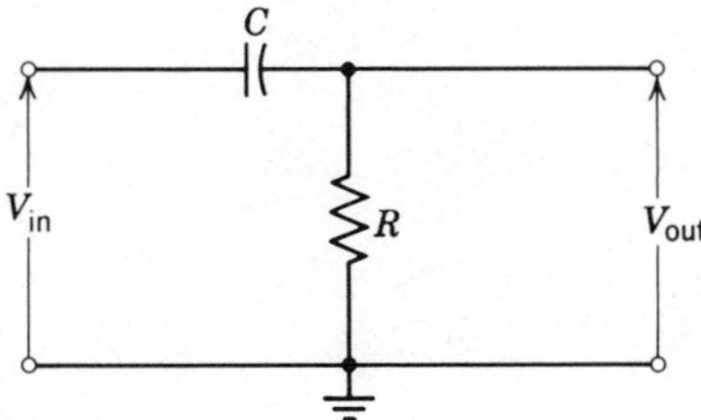

Circuit for Problems 15-1.1 to 15-1.3.

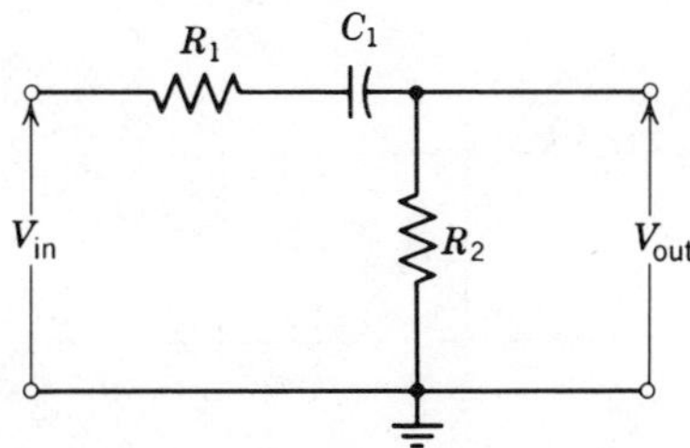

Circuit for Problems 15-1.4 to 15-1.6.

15-1.5 C_1 is 0.33 μF, R_1 is 20 kΩ, and R_2 is 30 kΩ. Determine V_{out} and the phase shift at 4 Hz.

15-1.6 C_1 is 0.05 μF, R_1 is 4.7 kΩ, and R_2 is 4.7 kΩ. Determine V_{out} and the phase at 100 Hz.

Section 15-2 High-Frequency Response

The input current I_{in} in the circuit shown in Fig. 15-4*a* divides into I_1 and I_2. We draw the phasor diagram (Fig. 15-4*b*) by using V_{out} as the reference phasor. Then I_2 is in phase with V_{out} and I_1 leads V_{out} by 90°. The phasor sum of I_1 and I_2 is I_{in}. The voltage drop R_1I_{in} is in phase with I_{in} and it is drawn as a phasor parallel to the I_{in} phasor. Then V_{in} is the sum of the phasors V_{out} and R_1I_{in}. We see from Fig. 15-4*b* that now V_{out} *lags* V_{in} by the phase angle θ.

The parallel combination of R_2 and C_2 has the impedance given by the "product-over-sum" rule.

$$\frac{R_2(-jX_{C_2})}{R_2 - jX_{C_2}} = -j\frac{R_2X_{C_2}}{R_2 - jX_{C_2}}$$

Then by the voltage-divider rule we have

$$V_{out} = \frac{-j\dfrac{R_2X_{C_2}}{R_2 - jX_{C_2}}}{R_1 - j\dfrac{R_2X_{C_2}}{R_2 - jX_{C_2}}}$$

Simplifying, we have

$$V_{out} = \frac{-jR_2X_{C_2}}{R_1R_2 - jR_1X_{C_2} - jR_2X_{C_2}}V_{in} = \frac{-jR_2X_{C_2}}{R_1R_2 - j(R_1 \times R_2)X_{C_2}}V_{in}$$

Dividing both sides of the equation by V_{in} yields the high-

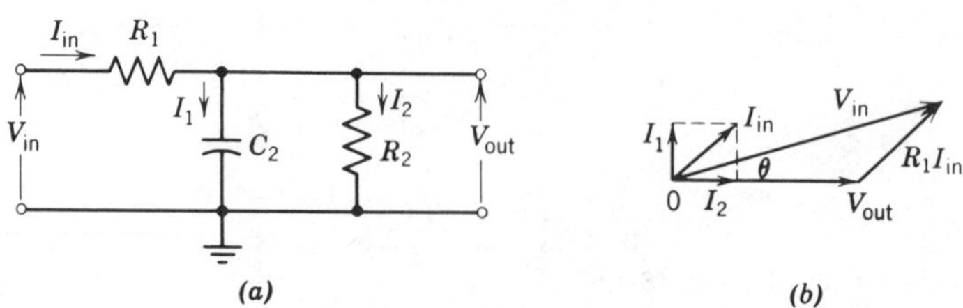

Figure 15-4 Passive circuit showing a drop in V_{out} at high frequencies. (*a*) Circuit diagram. (*b*) Phasor diagram.

frequency gain A_{HF}.

$$A_{HF} = \frac{V_{\text{out}}}{V_{\text{in}}} = \frac{-jR_2X_{C_2}}{R_1R_2 - j(R_1 + R_2)X_{C_2}}$$

Multiplying each term by j

$$A_{HF} = \frac{R_2X_{C_2}}{(R_1 + R_2)X_{C_2} + jR_1R_2}$$

At mid-band, the shunting effect of C_2 is negligible and the mid-band gain is simply the action of a resistive voltage divider.

$$A_v = \frac{R_2}{R_1 + R_2} \tag{15-2}$$

If we divide A_{HF} by A_v, we *define* K_{HF}

$$\boxed{K_{HF} \equiv \frac{A_{HF}}{A_v}} \tag{15-6}$$

where $$0 \leq K_{HF} \leq 1$$

and substituting for A_{HF} and A_v, we have

$$K_{HF} = \frac{\dfrac{R_2X_{C_2}}{(R_1 + R_2)X_{C_2} + jR_1R_2}}{\dfrac{R_2}{R_1 + R_2}} = \frac{X_{C_2}}{X_{C_2} + j\dfrac{R_1R_2}{R_1 + R_2}}$$

Now we define an equivalent resistance R_{eq} as the parallel combination of R_1 and R_2.

$$\boxed{R_{\text{eq}} = \frac{R_1R_2}{R_1 + R_2}} \tag{15-7}$$

Substituting, we have

$$K_{HF} = \frac{X_{C_2}}{X_{C_2} + jR_{\text{eq}}}$$

When each term is divided by X_{C_2}, we have

$$K_{HF} = \frac{1}{1 + j\dfrac{R_{eq}}{X_{C_2}}} = \frac{1}{\sqrt{1 + \left(\dfrac{R_{eq}}{X_{C_2}}\right)^2}} \angle -\tan^{-1}\frac{R_{eq}}{X_{C_2}} \qquad (15\text{-}8)$$

Example 15-2

Determine data for a plot of the magnitude and phase angle of K_{HF} for the circuit shown in Fig. 15-5.

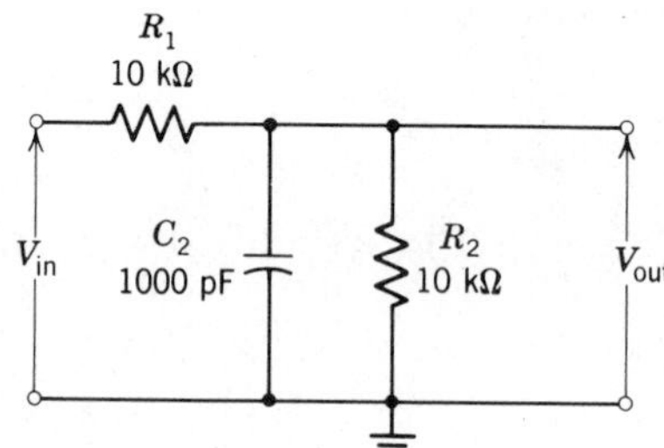

Figure 15-5 Circuit used to show high-frequency response.

Solution

By Eq. 15-7 R_{eq} is

$$R_{eq} = \frac{R_1 R_2}{R_1 + R_2} = \frac{10\ \text{k}\Omega \times 10\ \text{k}\Omega}{10\ \text{k}\Omega + 10\ \text{k}\Omega} = 5\ \text{k}\Omega \qquad (15\text{-}7)$$

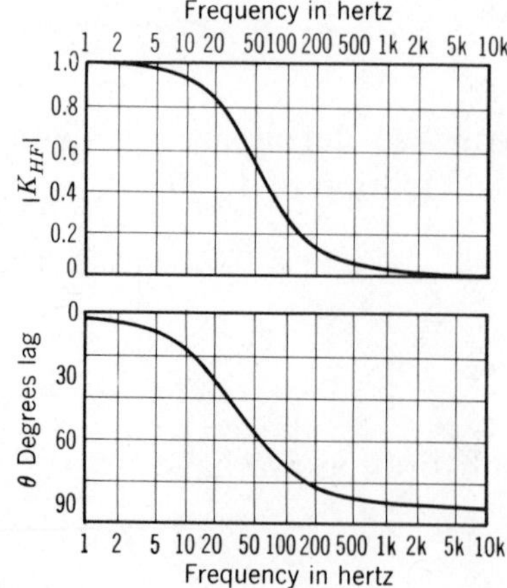

Figure 15-6 Magnitude and phase plots for K_{HF} for the circuit given in Fig. 15-5.

Table 15-2
Values for K_{HF} for the Circuit of Fig. 15-5

f (kHz)	θ (degrees)	$\|K_{HF}\|$
1	−1.8	1.000
2	−3.6	0.998
5	−8.9	0.988
10	−17.4	0.954
20	−32.1	0.847
31.8	−45.0	0.707
50	−57.5	0.537
100	−72.3	0.303
200	−81.0	0.157
500	−86.4	0.064
1000	−88.2	0.032
10000	−89.8	0.003

Using this value for R_{eq} and 1000 pF for C_2, we substitute different frequencies into Eq. 15-8 to obtain the values for the magnitude and the phase angle of K_{HF} that are listed in Table 15-2. The results are plotted in the graphs of Fig. 15-6.

Problems For each of the problems, determine sufficient data for V_{out} and for phase shift to plot the response curves.

15-2.1 C is 0.01 μF and R is 2 kΩ. Determine V_{out} and the phase shift at 25 kHz.

15-2.2 C is 1000 pF and R is 10 kΩ. Determine V_{out} and the phase shift at 60 kHz.

15-2.3 C is 20 μF and R is 100 Ω. Determine V_{out} and the phase shift at 500 Hz.

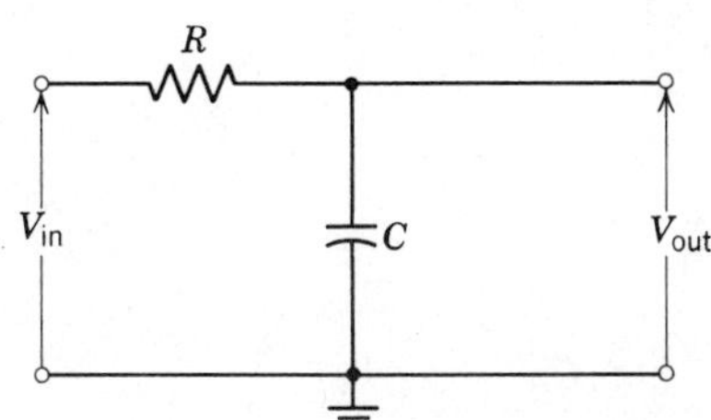

Circuit for Problems 15-2.1 to 15-2.3.

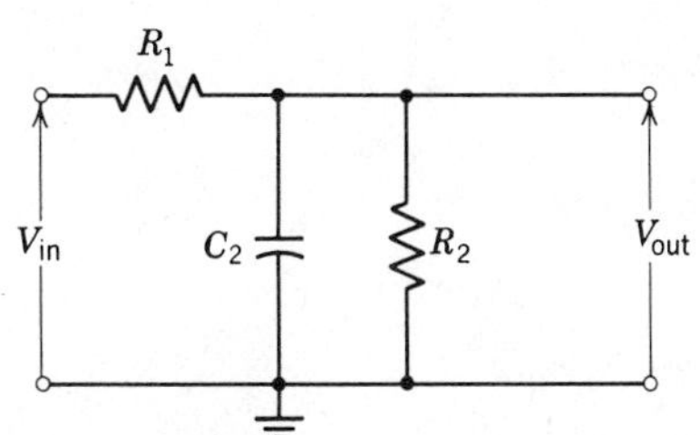

Circuit for Problems 15-2.4 to 15-2.6.

15-2.4 C_2 is 0.01 μF, R_1 is 10 kΩ, and R_2 is 500 Ω. Determine V_{out} and the phase shift at 5 kHz.
15-2.5 C_2 is 0.01 μF, R_1 is 500 Ω, and R_2 is 10 kΩ. Determine V_{out} and the phase shift at 20 kHz.
15-2.6 C_2 is 200 pF, R_1 is 33 kΩ, and R_2 is 47 kΩ. Determine V_{out} and the phase shift at 240 kHz.

Section 15-3 Bode Plots for Low-Frequency Response

The expression we obtained in Section 15-1 for K_{LF} was

$$K_{LF} = \frac{1}{1 - j\dfrac{X_{C_1}}{R_1 + R_2}} \tag{15-4}$$

Let us examine the term

$$\frac{X_{C_1}}{R_1 + R_2}$$

and replace X_{C_1} by $1/2\pi fC_1$

$$\frac{X_{C_1}}{R_1 + R_2} = \frac{1}{2\pi f(R_1 + R_2)C_1}$$

The time constant τ_1 is *defined* as

$$\boxed{\tau_1 \equiv (R_1 + R_2)C_1 \text{ seconds}} \tag{15-9}$$

The radian frequency ω_1 is *defined* as

$$\boxed{\omega_1 \equiv 2\pi f_1 \equiv \frac{1}{\tau_1} \text{ rad/s}} \tag{15-10}$$

or

$$\boxed{f_1 = \frac{1}{2\pi\tau_1} = \frac{\omega_1}{2\pi} \text{ Hz}} \tag{15-11}$$

Recalling that $2\pi f$ is ω, the term we are examining becomes

$$\frac{X_{C_1}}{R_1+R_2}=\frac{1}{2\pi f\tau_1}=\frac{\omega_1}{\omega}=\frac{f_1}{f}$$

And substituting into Eq. 15-4, we have

$$K_{LF}=\frac{1}{1-j\frac{f_1}{f}}=\frac{1}{1-j\frac{\omega_1}{\omega}}=\frac{1}{\sqrt{1+\left(\frac{f_1}{f}\right)^2}}\underline{/+\tan^{-1}\frac{f_1}{f}} \qquad (15\text{-}12)$$

In the example we used for the low-frequency response, we have

$$(R_1+R_2)=20{,}000\ \Omega$$

and

$$C_1=1\ \mu\text{F}$$

Then

$$\tau_1=(R_1+R_2)C_1=20{,}000\ \Omega\times(1\times10^{-6}\ \text{F})=0.02\ \text{s} \qquad (15\text{-}9)$$

and

$$\omega_1=\frac{1}{\tau_1}=\frac{1}{0.02}=50\ \text{rad/s} \qquad (15\text{-}10)$$

and

$$f_1=\frac{\omega_1}{2\pi}=\frac{50}{2\pi}=7.96\ \text{Hz} \qquad (15\text{-}11)$$

In Table 15-1, we see that, when K_{LF} is evaluated for 7.96 Hz, we find that the magnitude of K_{LF} is 0.707 and the phase angle is +45.0°. These are the values we should expect when (R_1+R_2) equals X_{C_1} and Eq. 15-4 and Eq. 15-12 reduce to

$$K_{LF}=\frac{1}{1-j1}$$

Let us consider Eq. 15-12 and place the restriction on it that

$$\frac{f_1}{f}\gg 1$$

Then

$$K_{LF}=\frac{1}{-j\frac{f_1}{f}}=j\frac{f}{f_1}=\frac{f}{f_1}\ \underline{/90^\circ} \qquad (15\text{-}13)$$

Now let us express the magnitude of K_{LF} from Eq. 15-13 in decibels.

$$K_{LF}, \text{dB} = 20 \log_{10} \frac{f}{f_1} \text{ dB} \tag{15-14}$$

When f equals f_1, Eq. 15-14 yields 0 dB. When f equals $\frac{1}{2}f_1$, Eq. 15-14 yields −6 dB. Note that this change in frequency of *one octave* results in a change of 6 dB in the response. When f equals $0.1f_1$, Eq. 15-14 yields −20 dB. Note that this change in frequency of *one decade* results in a change of 20 dB in the response.

The numerical values obtained for the magnitude of K_{LF} in Example 15-1 are converted into decibels by means of Eq. 15-14. These results are added to Table 15-1 to form Table 15-3.

The magnitude in decibels and the phase angle for the circuit are plotted in Fig. 15-7.

A *Bode plot* is a method of approximating the actual curves by the straight lines that are added to Fig. 15-7.

The *corner frequency* or the *break frequency* is f_1 (or ω_1) that was determined from τ_1. The gain is 0 dB for all frequencies greater than f_1 (or ω_1). Below f_1, K_{LF} *rolls off* in a straight line at a slope of *6 dB per octave* or at a slope of *20 dB per*

Table 15-3
Values for K_{LF} for the Circuit of Fig. 15-2

f (Hz)	θ (degrees)	K_{LF}	K_{LF} (dB)
1000	+0.5	1.000	0
500	+0.9	1.000	0
100	+4.6	0.997	−0.03
20	+21.7	0.929	−0.64
10	+38.5	0.782	−2.13
7.96	+45.0	0.707	−3.00
5	+57.9	0.532	−5.48
2	+75.9	0.244	−12.26
1	+82.8	0.125	−18.08
0.5	+86.4	0.063	−24.05
0.2	+88.6	0.025	−32.00
0.1	+89.3	0.013	−38.00

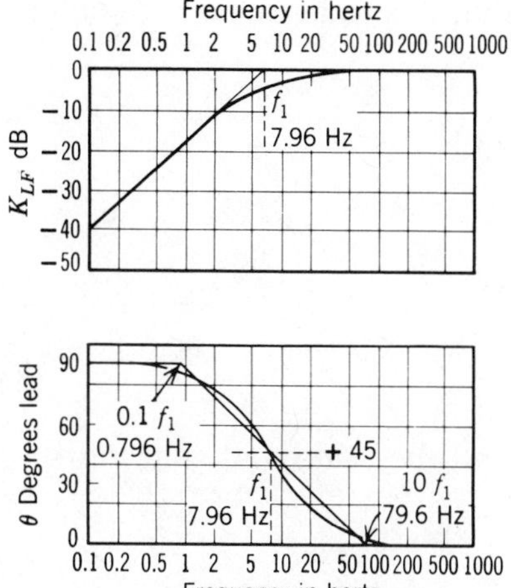

Figure 15-7 Bode plots for magnitude and phase angle for the low-frequency network given in Fig. 15-1.

decade of frequency change. Note that the actual gain at f_1 (or ω_1) is down 3 dB (−3 dB).

Reference to Table 15-3 shows that, for a frequency change from 1 Hz to 0.5 Hz (an octave change) the gain decreases from −18 dB to −24 dB, which is a change of 6 dB in a frequency change of one octave. For a frequency change from 1 Hz to 0.1 Hz (a decade change) the gain decreases from −18 dB to −38 dB or −20 dB.

The phase angle at the break frequency f_1 (or ω_1) is 45°. A straight line is drawn from 90° at $0.1f_1$ (or $0.1\omega_1$) to 0° at $10f_1$ (or $10\omega_1$). At all frequencies below $0.1f_1$ (or $0.1\omega_1$), the phase angle is 90°. At all frequencies above 10_1 (or $10\omega_1$), the phase angle is 0°.

In Example 15-1

$$A_v = \frac{10\ \text{k}\Omega}{10\ \text{k}\Omega + 10\ \text{k}\Omega} = \frac{1}{2} = -6\ \text{dB}$$

The Bode plot shows a horizontal line of −6 dB starting at f_1 (7.96 Hz), the break frequency. We start the 20 dB/decade roll-off at the coordinate (−6dB, 7.96 Hz). This is equivalent to adding −6 dB to all the decibel values in Table 15-3. It is also equivalent to moving the whole gain curve in Fig. 15-7 down 6 dB. The phase angle curve does not change.

It is much easier to draw a Bode plot than to go through all the calculations required for Table 15-1 or for Table 15-3. All that has to be done is to determine the break frequencies from

$$\tau_1 = (R_1 + R_2)C_1 \tag{15-9}$$

and

$$\omega_1 = \frac{1}{\tau_1} \tag{15-10}$$

or

$$f_1 = \frac{\omega_1}{2\pi} \tag{15-11}$$

and the mid-band gain in dB.

Example 15-3
Using the Bode plots given in Fig. 15-7, determine the values of K_{LF} and of phase angle at 1.2 Hz.

Solution
Either Eq. 15-4 or Eq. 15-5 could be used to solve this problem but we wish to show how to use the linear concepts of the Bode plots. The Bode plot for gain is linear for decibels on the Y axis and for the $\log_{10}$ of the frequency on the X axis. Also the Bode plot for phase angle is linear for phase angle on the Y axis and for the $\log_{10}$ of the frequency on the X axis.

The change of K_{LF} is 20 dB over one decade (10:1) of frequency change. The value of $\log_{10} 10/1$ is 1. In Fig. 15-7, a decade change in frequency is the change from 0.796 Hz ($0.1f_1$) to 7.96 Hz (f_1). The $\log_{10}$ of the change from 0.796 Hz to 1.2 Hz is

$$\log_{10}\frac{1.2}{0.796} = 0.178$$

Then K_{LF} at 0.796 Hz is −20 dB. The increase in dB from 0.796 Hz to 1.2 Hz is

$$20 \times 0.178 = 3.56 \text{ dB}$$

Therefore the value of K_{LF} at 1.2 Hz is

$$-20 + 3.56 = \mathbf{-16.44\ dB}$$

Figure 15-7 shows that the phase angle falls from 90° to 0° in *two* decades (from $0.1f_1$ to $10f_1$). As a $\log_{10}$ distance, two decades is 2.0.

The change from 0.796 Hz to 1.2 Hz is a $\log_{10}$ change of 0.178. The change in phase angle then is

$$\frac{0.178}{2.0} \times 90° = 8°$$

The phase angle at 1.2 Hz is

$$90° - 8° = \mathbf{82°}$$

Problems Determine the values for f_1 and ω_1, the break frequencies, and the value in dB for A_v at mid-band for each of the circuits given by the problems at the end of Section 15-1. Sketch the Bode plots for both gain and phase. Determine the K_{LF} and phase angles for the specific frequencies given for each problem by using the method explained in Example 15-3.

Section 15-4 Bode Plots for High-Frequency Response

The circuit (Fig. 15-5) that was used to determine the high-frequency response in Section 15-2 is repeated in Fig. 15-8. We determined from this circuit that

$$A_{HF} = K_{HF}A_v \tag{15-6}$$

$$A_v = \frac{R_2}{R_1 + R_2} = \frac{1}{2} \tag{15-2}$$

and

$$K_{HF} = \frac{1}{1 + j\dfrac{R_{eq}}{X_{C_2}}} \tag{15-8}$$

in which

$$R_{eq} = \frac{R_1 R_2}{R_1 + R_2} = 5\ \text{k}\Omega \tag{15-7}$$

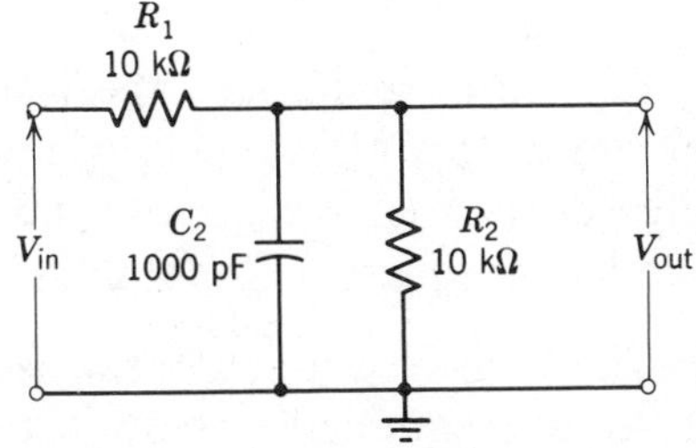

Figure 15-8 Circuit used to show high-frequency response.

As we proceeded in the last section, we consider the j term

$$\frac{R_{eq}}{X_{C_3}} = \frac{R_{eq}}{\left(\dfrac{1}{2\pi f C_2}\right)} = 2\pi f R_{eq} C_2$$

Now we *define* the time constant τ_2 as

$$\boxed{\tau_2 \equiv R_{eq} C_2 \text{ s}} \tag{15-15}$$

and we *define* ω_2 as

$$\boxed{\omega_2 \equiv \frac{1}{\tau_2} \text{ rad/s}} \tag{15-16}$$

Then

$$\boxed{f_2 = \frac{\omega_2}{2\pi} \text{ Hz}} \tag{15-17}$$

If we replace $2\pi f$ by ω, we have

$$\frac{R_{eq}}{X_{C_2}} = \frac{\omega}{\omega_2} = \frac{f}{f_2}$$

and now Eq. 15-8 becomes

$$\boxed{K_{HF} = \frac{1}{1 + j\dfrac{f}{f_2}} = \frac{1}{1 + j\dfrac{\omega}{\omega_2}} = \frac{1}{\sqrt{1 + \left(\dfrac{f}{f_2}\right)^2}} \angle -\tan^{-1}\frac{f}{f_2}} \tag{15-18}$$

Equation 15-18 has the same form as Eq. 15-12. Now, however, Eq. 15-18 shows that K_{HF} falls off as f becomes greater than f_2. We can use the same method that was used for the low-frequency Bode plots and show:

1. The high-frequency break point is f_2 (or ω_2).
2. At all frequencies below f_2 (or ω_2), the break frequency, we show the gain as a horizontal straight line.

3. At all frequencies above f_2 (or ω_2), the gain is a straight line that rolls off at the rate of 20 dB per decade or 6 dB per octave.
4. The phase shift at f_2 (or ω_2) is 45° lagging.
5. At $0.1f_2$ (or $0.1\omega_2$) and at all lower frequencies, the phase shift is zero.
6. At $10f_2$ (or $10\omega_2$) and at all higher frequencies the phase shift is 90° lag.
7. A straight line is drawn from the point locating the 0° phase shift at $0.1f_2$ (or $0.1\omega_2$) to the point locating the 90° phase shift at $10f_2$ (or $10\omega_2$).

In Example 15-2 we have

$$R_{eq} = 5\text{ k}\Omega$$

$$\tau_2 = R_{eq}C_2 = 5\text{ k}\Omega \times (1000 \times 10^{-12}\ F) = 5 \times 10^{-6}\text{ s}$$

$$\omega_2 = \frac{1}{\tau_2} = \frac{1}{5 \times 10^{-6}} = 2 \times 10^5\text{ rad/s}$$

and

$$f_2 = \frac{\omega_2}{2\pi} = \frac{2 \times 10^5}{2\pi} = 31{,}800\text{ Hz} = 31.8\text{ kHz}$$

The Bode plots showing the high-frequency response for the network are quickly sketched in Fig. 15-9. It should be

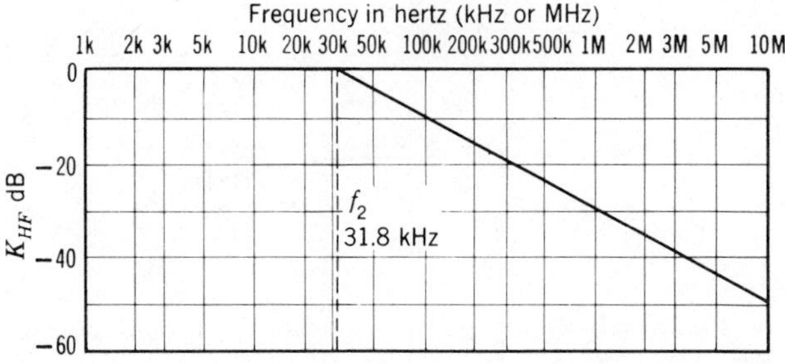

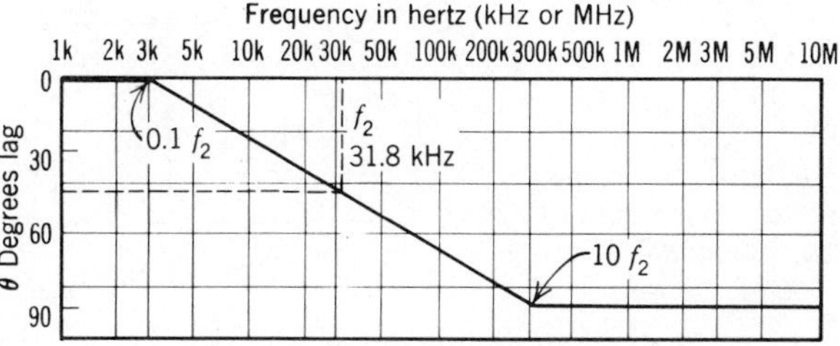

Figure 15-9 Bode plots for the high-frequency response.

remembered that the actual response at f_2 (or ω_2) is down 3 dB below the corner point.

Problems Determine the values for f_2 and ω_2, the break frequencies, and the value in decibels for A_v at mid-band for each of the circuits given by the problems at the end of Section 15-2. Sketch the Bode plots for both gain and phase. Determine the K_{HF} and phase angles for the specific frequencies given for each problem by using the method explained in Example 15-3.

Section 15-5 Accuracy of Bode Plots

The differences between the actual response and the approximations made by using Bode plots are shown in Fig. 15-10 and are listed in Table 15-4.

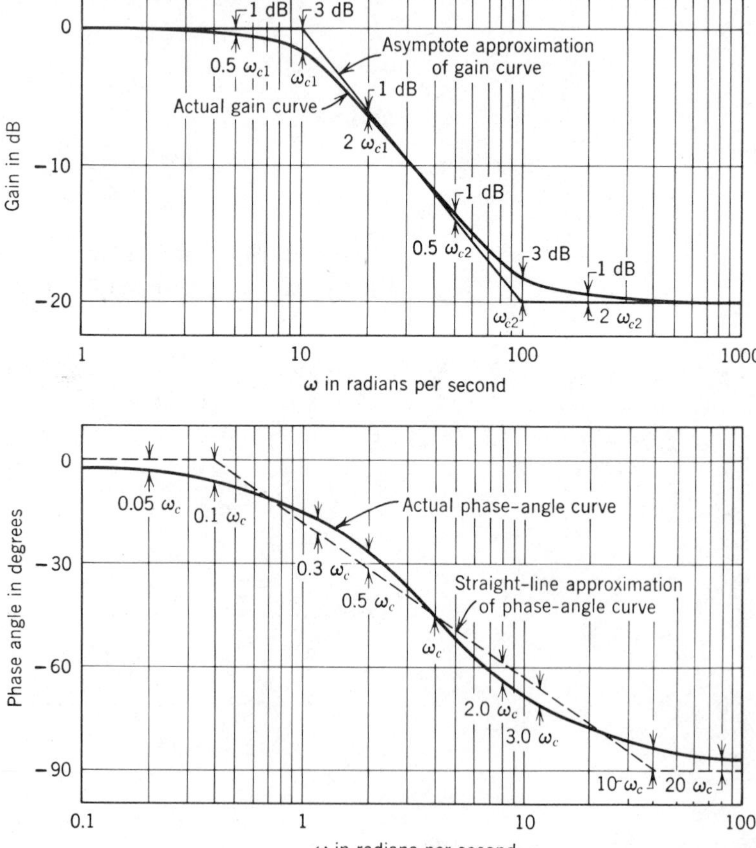

Figure 15-10 Corrections to approximate straight-line Bode plots. (*Courtesy B. I. DeRoy, Automatic Control Theory, Wiley, 1966.*)

Table 15-4
Corrections to Approximate Straight-Line, Phase-Shift, or Phase-Angle Plots

ω	Corrections
$0.05\omega_c$	$-3°$
$0.1\omega_c$	$-6°$
$0.3\omega_c$	$+5°$
$0.5\omega_c$	$+5°$
$1.0\omega_c$	$0°$
$2.0\omega_c$	$-5°$
$3.0\omega_c$	$-5°$
$10.0\omega_c$	$+6°$
$20.0\omega_c$	$+3°$

Section 15-6 Frequency Response of a Two-Stage Amplifier

The two-stage amplifier (Fig. 15-11) is formed of two stages, Stage 1 and Stage 2, that are identical. For purposes of analysis, we will separate the two stages at A. Now each stage has identical characteristics. The coupling networks (R_1, R_2, C_1, and C_2) have the same values that were used to construct the low-frequency Bode plots in Section 15-3 and the high-frequency Bode plots in Section 15-4.

The corner or break radian frequency values for each network are

$$\omega_1 = 50 \text{ rad/s}$$

and

$$\omega_2 = 2 \times 10^5 \text{ rad/s}$$

The mid-band gain of the coupling network is $\frac{1}{2}$, yielding a net

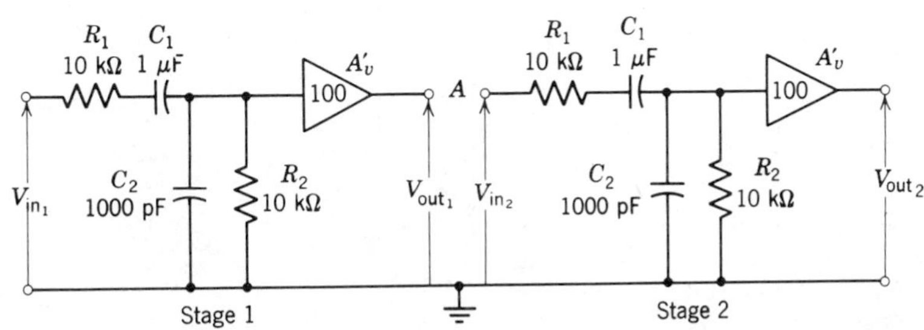

Figure 15-11 Circuit showing a two-stage amplifier.

mid-band gain for each stage of

$$A_v = \frac{1}{2}A_v' = \frac{1}{2} \times 100 = 50$$

Converting A_v into decibels, we have

$$A_v\text{dB} = 20 \log_{10} 50 = +34 \text{ dB}$$

The Bode gain plot (Fig. 15-12*a*) locates the low-frequency break or corner point at 50 rad/s (ω_1) and +34 dB. The high-frequency break or corner point is located at 2×10^5 rad/s (ω_2) and +34 dB. If we consider a frequency a decade below ω_1, the frequency is 5 rad/s. If we consider a frequency a decade above ω_2, the frequency is 2×10^6. At each of these two new frequencies, the gain is 20 dB below mid-band gain or (+34 − 20) or +14 dB. These new values locate points *A* and *B* on the gain plot. Straight lines are drawn through *A* and *B* to the corner points to complete the Bode gain plot.

In constructing the Bode phase plot (Fig. 15-12*b*) the phase shift at the corner frequencies is 45°. At ω_1 the phase shift is

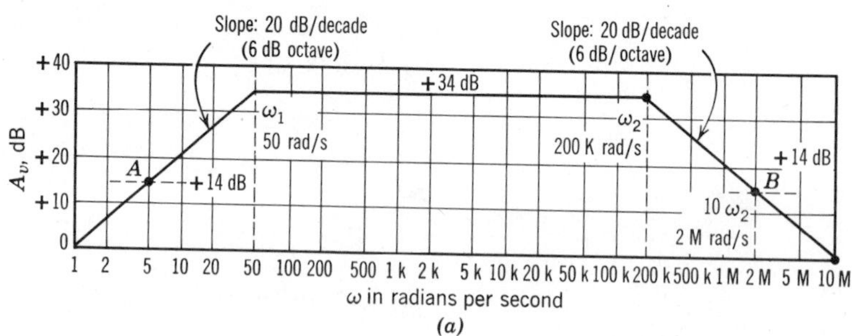

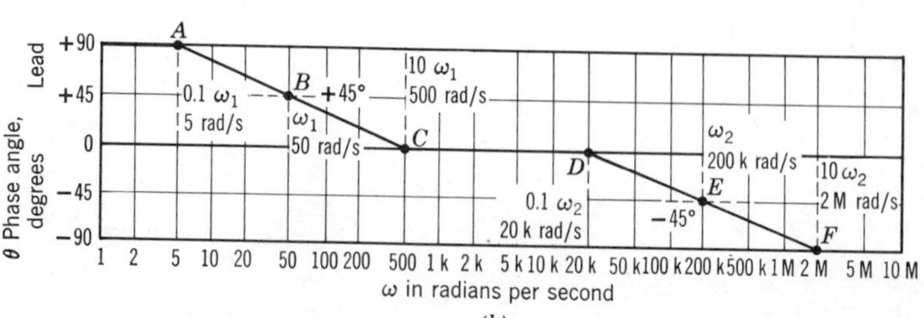

Figure 15-12 Bode plots for either Stage 1 or Stage 2 in the amplifier circuit given in Fig. 15-11. (*a*) Gain plot. (*b*) Phase plot.

leading and at ω_2, the phase shift is a lag. Points B and E are placed on the diagram. At $0.1\omega_1$ the phase shift is 90° and at $10\omega_1$ the phase shift is 0°. Points A and C are located and a straight line is drawn from A to C through B. At $0.1\omega_2$ the phase shift is 0° and at $10\omega_2$ the phase shift is 90°. Points D and F are located and a straight line is drawn from D to F through E. At frequencies less than $0.1\omega_1$ and at frequencies greater than $10\omega_2$, the phase shifts are 90°. Between C and D the phase shift is 0°.

Now the gap is closed at A in Fig. 15-11. Since we are using amplifiers with a value of zero ohms for the output resistance, making the connection at A does not change the value of V_{out_1}. Then, at mid-band, the gain is the sum of the gains of each individual stage.

$$(+34\text{ dB}) + (+34\text{ dB}) = +68\text{ dB}$$

In the single stage, the low-frequency break frequency, ω_1, occurs at 50 rad/s and the high-frequency break frequency, ω_2, occurs at 2×10^5 rad/s. When these stages are cascaded, we find that the gain is down 6 dB at ω_1 and at ω_2. However, we require new break frequencies ω_1' and ω_2' where the overall gain is down 3 dB in order to establish the corner frequencies for the Bode frequency plot. These new break frequencies ω_1' and ω_2' occur when each stage is down 1.5 dB.

Similarly, if N *identical* stages are cascaded, the new overall break frequencies f_1' and f_2' occur when the gain of each stage is down $3/N$ dB. These new frequencies can be determined by evaluating Eq. 15-12 and Eq. 15-18 for f_1/f and for f/f_2 at the values of K_{LF} and K_{HF} that correspond to the decibel values of $3/N$ dB. For example, taking the magnitude of Eq. 15-12

$$K_{LF} = \frac{1}{\sqrt{1 + \left(\frac{f_1}{f}\right)^2}} \tag{15-12}$$

and converting to dB, we have

$$-\frac{3}{N} = -20 \log_{10} \sqrt{1 + \left(\frac{f_1}{f}\right)^2}$$

Changing signs and removing the square root, we have

$$\frac{3}{N} = 10 \log_{10}\left[1 + \left(\frac{f_1}{f}\right)^2\right]$$

Dividing by 10, we find

$$\frac{0.3}{N} = \log_{10}\left[1 + \left(\frac{f_1}{f}\right)^2\right]$$

Taking antilogs, we have

$$10^{0.3/N} = (10^{0.3})^{1/N} = 1 + \left(\frac{f_1}{f}\right)^2$$

Since

$$10^{0.3} = 2$$

$$2^{1/N} = \sqrt[N]{2} = 1 + \left(\frac{f_1}{f}\right)^2$$

Solving for f_1/f, we find

$$\frac{f_1}{f} = \sqrt{\sqrt[N]{2} - 1} \tag{15-19a}$$

Similarly, we can arrive at

$$\frac{f}{f_2} = \sqrt{\sqrt[N]{2} - 1} \tag{15-19b}$$

for the high-frequency break frequency.

If we evaluate Eq. 15-19*a* and Eq. 15-19*b* for 1, 2, 3, and 4 (identical) stages, we have the results tabulated in Table 15-5.

Table 15-5
The Value of f_1' and f_2' for an N-Stage (Identical) Circuit

Stages	dB	f_1'	f_2'
1	−3	f_1	f_2
2	−1.5	$f_1/0.64$	$0.64f_2$
3	−1	$f_1/0.51$	$0.51f_2$
4	−0.75	$f_1/0.43$	$0.43f_2$

In our example, ω_1 is 50 rad/s and ω_2 is 200 k rad/s for each stage. Then the new break frequencies for the two-stage amplifier are

$$\omega_1' = \frac{\omega_1}{0.64} = \frac{50}{0.64} = 78 \text{ rad/s}$$

and

$$\omega_2' = 0.64\omega_2 = 0.64 \times 200{,}000 = 128{,}000 \text{ rad/s}$$

The Bode frequency response is shown in Fig. 15-13*a*. The corner frequencies are 78 rad/s and 128,000 rad/s. The response falls off at the rate of 40 dB/decade (12 dB/octave) for the two-stage amplifier. This 40 dB/decade (12 dB/octave) results from the 20-dB/decade (6-dB/octave) roll-off for each stage.

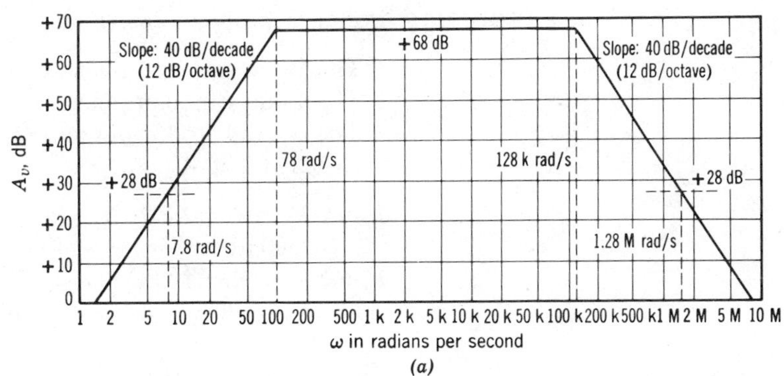

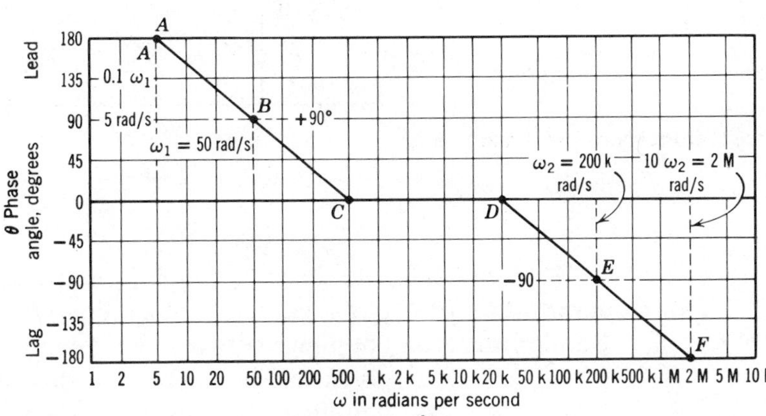

Figure 15-13 Bode plots for the two-stage amplifier circuit given in Fig. 15-11. (*a*) Gain plot. (*b*) Phase plot.

We must realize that Table 15-5 does *not* apply to the phase angle. At f_1 and at f_2 the phase angle for each stage is 45°. Cascading the two stages means that at f_1 and at f_2 the angles are now 90°. For a three-stage circuit the angles at f_1 and at f_2 would be 135°, and so on.

In the phase plot (Fig. 15-13*b*) the angles are additive.

Point *A*	$(+90°)+(+90°)=$	$+180°$	at	$0.1\,\omega_1$ (5 rad/s)
Point *B*	$(+45°)+(+45°)=$	$+90°$	at	ω_1 (50 rad/s)
Point *C*		$0°$	at	$10\,\omega_1$ (500 rad/s)
Point *D*		$0°$	at	$0.1\,\omega_2$ $(2\times 10^4$ rad/s)
Point *E*	$(-45°)+(-45°)=$	$-90°$	at	ω_2 $(2\times 10^5$ rad/s)
Point *F*	$(-90°)+(-90°)=$	$-180°$	at	$10\,\omega_2$ $(2\times 10^6$ rad/s)

These points are connected with straight lines as before to complete the phase plot.

Problems

For each problem determine:

1. The Bode gain and phase plots for Stage 1.
2. The Bode gain and phase plots for Stage 2.
3. The overall Bode gain and phase plots when Stage 1 is connected to Stage 2 at *A*.

15-6.1 $R_1 = R_3 = 2\ \text{k}\Omega$. $R_2 = R_4 = 4.7\ \text{k}\Omega$. $C_1 = C_3 = 0.5\ \mu\text{F}$. $C_2 = C_4 = 2500\ \text{pF}$. $A'_{v_1} = A'_{v_2} = 40$.

15-6.2 $R_1 = R_3 = 10\ \text{k}\Omega$. $R_2 = R_4 = 22\ \text{k}\Omega$. $C_1 = C_3 = 0.10\ \mu\text{F}$. $C_2 = C_4 = 500\ \text{pF}$. $A'_{v_1} = A'_{v_2} = 30$.

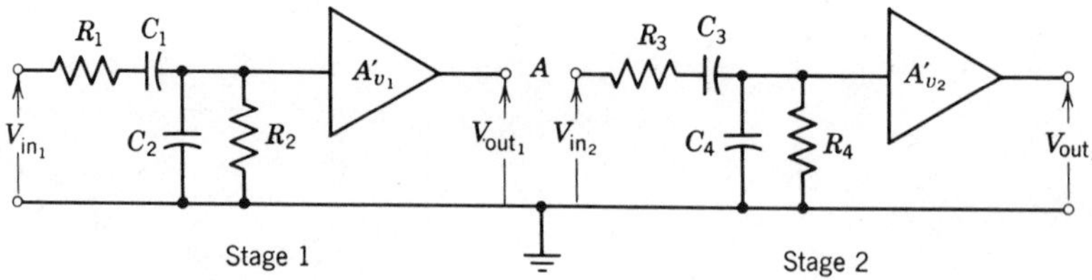

Circuit for Problems 15-6.1 and 15-6.2.

Section 15-7 The Capacitances of a Semiconductor

The passive coupling network for which we found the Bode plots earlier in this chapter is shown in Fig. 15-14*a*. The coupling network between two transistors, *Q*1 and *Q*2, is given in Fig. 15-14*b*. In this section we will consider only an amplifier stage using a common-emitter connection. We must

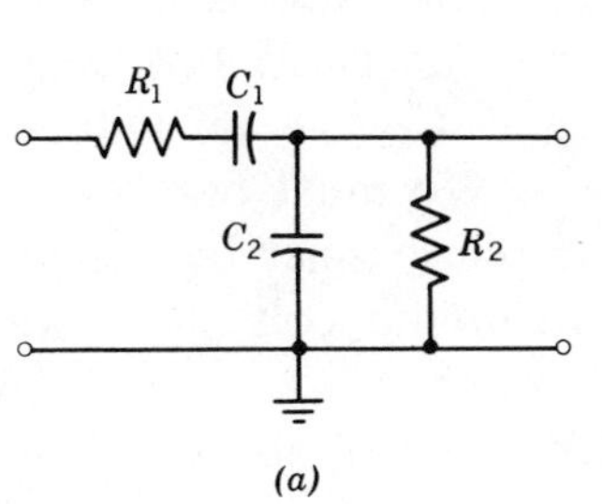

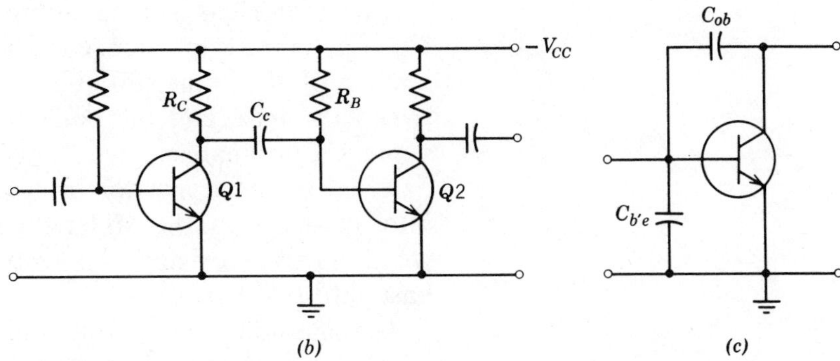

Figure 15-14 The capacitances of a transistor. (*a*) Coupling network. (*b*) Transistor amplifier circuit. (*c*) Internal capacitances of a transistor.

show how to translate the components of the transistor circuit into the circuit elements of the coupling network.

It is obvious that the coupling capacitor C_C used to connect $Q1$ and $Q2$ is identical to C_1 in the coupling network.

$$C_1 = C_C \tag{15-20a}$$

Then R_1 is the load resistor R_C in the collector of the transistor.

$$R_1 = R_C \tag{15-20b}$$

Equation 15-20*b* assumes that R_C is much smaller than the ac output resistance of $Q1$. If $Q1$ is a FET, the load in the drain circuit is R_D. Since the value of r_d is often significantly small, we must use the parallel combination of R_D and r_d for R_1.

$$R_1 = \frac{R_D r_d}{R_D + r_d} \tag{15-21}$$

R_2 is equivalent to the input resistance r'_{in}, which is the parallel combination of R_B and r_{in}. r_{in} can be obtained from Eq. 7-7, Eq. 7-13, or Eq. 7-25.

$$R_2 = r'_{\text{in}} = \frac{R_B r_{\text{in}}}{R_B + r_{\text{in}}} \tag{15-22}$$

A transistor has internal *interelectrode capacitances* $C_{b'e}$ and C_{ob}. $C_{b'e}$ is the capacitance existing between the base and the emitter of the transistor. C_{ob} is the capacitance existing between the base and the collector of the transistor. For convenience of analysis, we show $C_{b'e}$ and C_{ob} as capacitors external to the transistor in Fig. 15-14*c*. We must break C_{ob} into two parts by the Miller theorem in the manner by which we divided a resistor connected between the collector and base into two parts.

The method of applying the Miller theorem to R_B (Fig. 15-15*a*) to obtain the components in Fig. 15-15*b* applies equally to $X_{C_{ob}}$ in Fig. 15-15*c*. The input part of $X_{C_{ob}}$ by the Miller theorem shown in Fig. 15-15*d* is equal to the reactance of C'_{in} (Fig. 15-15*e*).

$$\frac{X_{C_{ob}}}{1+A_v} = \frac{1}{2\pi f C'_{\text{in}}}$$

or

$$\frac{1}{2\pi f C_{ob}(1+A_v)} = \frac{1}{2\pi f C'_{\text{in}}}$$

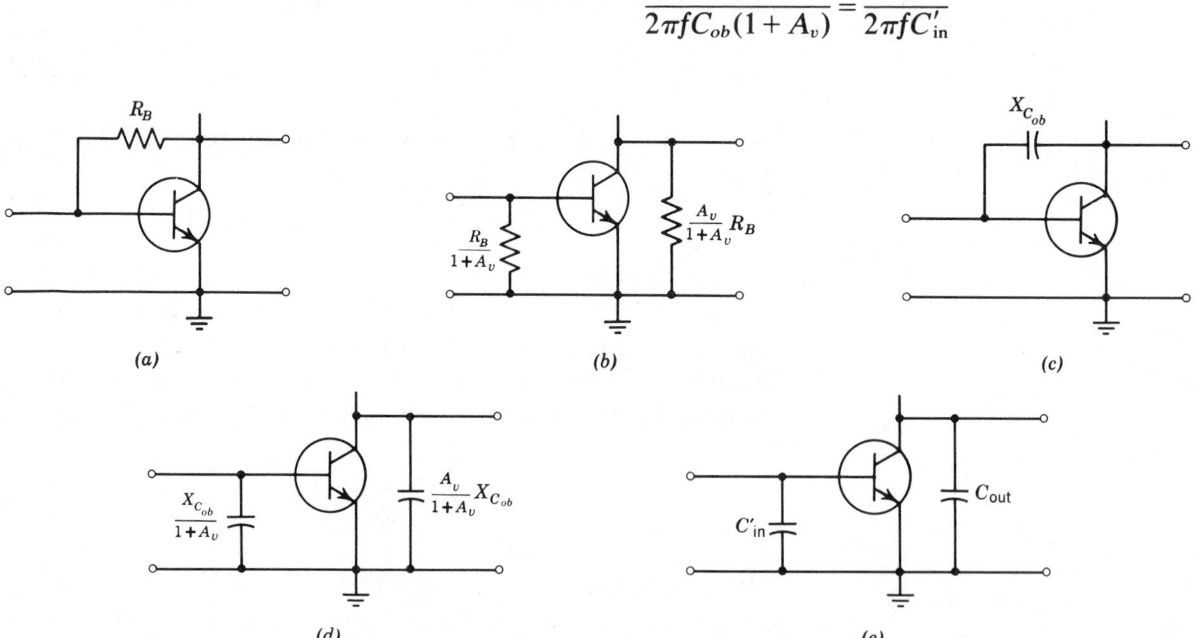

Figure 15-15 Evaluation of input and output capacitance of a transistor. (*a*) Collector-to-base feedback resistor R_B. (*b*) Miller theorem reduction of R_B. (*c*) Collector-to-base reactance $X_{C_{ob}}$ (*d*) Miller theorem reduction of $X_{C_{ob}}$ (*e*) Equivalent capacitances of Fig. 15-15*d*.

Taking reciprocals, we have

$$2\pi f C_{ob}(1 + A_v) = 2\pi f C'_{\text{in}}$$

Solving for C'_{in}, we find

$$\boxed{C'_{\text{in}} = (1 + A_v)C_{ob}} \tag{15-23}$$

Similarly, the reactances of the output capacitors are equal in Fig. 15-15*d* and Fig. 15-15*e*.

$$\frac{A_v}{1 + A_v} X_{C_{ob}} = \frac{1}{2\pi f C_{\text{out}}}$$

If A_v is large

$$\frac{A_v}{1 + A_v} \approx 1$$

and

$$X_{C_{ob}} = \frac{1}{2\pi f C_{ob}} = \frac{1}{2\pi f C_{\text{out}}}$$

Solving for C_{out}, we have

$$\boxed{C_{\text{out}} = C_{ob}} \tag{15-24}$$

In Fig. 15-14, let us call the total capacitance that shunts the coupling network between $Q1$ and $Q2$ to ground the *equivalent capacitance* C_{in}. Then

$$C_2 = C_{\text{in}} \tag{15-25}$$

Then C_{in} is the sum of the output capacitance of $Q1$ (C_{ob_1}) plus the base-to-emitter capacitance ($C_{b'e_2}$) of $Q2$ plus the reflected value (C'_{in_2}) of the collector-to-base capacitance of $Q2$ as given by the Miller theorem. Then

$$\boxed{C_{\text{in}} = C_{ob_1} + C_{b'e_2} + (1 + A_v)C_{ob_2}} \tag{15-26}$$

where A_v is the voltage gain across $Q2$. Actually, we should add the small value of wiring capacitance to C_{in}.

In a FET, we use C_{gs} for $C_{b'e}$ and C_{dg} for C_{ob}. The capacitance values of some typical transistors are:

$C_{ob} = 200$ pF	Germanium high-power audio amplifier.
$C_{ob} = 40$ pF	Germanium audio amplifier.
$C_{ob} = 7$ pF	Silicon for broadcast band use to 2 MHz.
$C_{ob} = 2.2$ pF	Silicon for FM broadcast to 100 MHz.
$C_{ob} = 0.32$ pF	Silicon high frequency to 250 MHz.
$C_{ob} = 0.55$ pF	Silicon high frequency to 500 MHz.

The interelectrode capacitances of FETs, especially MOSFETs, are very small, therefore making them especially adapted for high-frequency work. For example, in a MOSFET intended for general-purpose applications, audio, video, and high frequencies, the input capacitance between gate and source C_{gs} is 7 pF and the capacitance between gate and drain C_{dg} is 0.30 pF. This MOSFET is usable to 200 MHz.

Example 15-4

In the circuit given in Fig. 15-14*b*, we have the following parameter values

$$R_C = 8.2\ \text{k}\Omega \qquad C_C = 2\ \mu\text{F}$$

$$R_B = 470\ \text{k}\Omega \qquad r_{\text{in}} = 12\ \text{k}\Omega \qquad A_v = 45$$

and for each transistor

$$C_{ob} = 30\ \text{pF} \qquad C_{b'e} = 250\ \text{pF}$$

Determine the low-frequency and the high-frequency break (corner) frequencies.

Solution

The amplifier circuit values must be made to correspond to the values of the coupling network (Fig. 15-14*a*).

$$C_1 = C_C = 2\ \mu\text{F} \tag{15-20a}$$

$$R_1 = R_C = 8.2\ \text{k}\Omega \tag{15-20b}$$

$$R_2 = r'_{\text{in}} = \frac{R_B r_{\text{in}}}{R_B + r_{\text{in}}} = \frac{470\ \text{k}\Omega \times 12\ \text{k}\Omega}{470\ \text{k}\Omega + 12\ \text{k}\Omega} \approx 12\ \text{k}\Omega \tag{15-22}$$

$$C'_{\text{in}} = (1 + A_v)C_{ob} = (1 + 45) \times 30 = 1380 \text{ pF} \tag{15-23}$$

$$C_{\text{out}} = C_{ob} = 30 \text{ pF} \tag{15-24}$$

and

$$C_2 = C_{\text{in}} = C_{ob_1} + C_{b'e_2} + (1 + A_v)C_{ob_2}$$

$$= 30 + 250 + 1380 = 1660 \text{ pF} \tag{15-26}$$

The low-frequency break frequency is

$$\omega_1 = \frac{1}{\tau_1} = \frac{1}{(R_1 + R_2)C_1}$$

$$= \frac{1}{(8200\ \Omega + 12{,}000\ \Omega) \times (2 \times 10^{-6}\text{ F})}$$

$$= \mathbf{24.75\ rad/s} \tag{15-10}$$

or

$$f_1 = \frac{\omega_1}{2\pi} = \frac{24.75}{2\pi} = \mathbf{3.9\ Hz} \tag{15-11}$$

Before we can find the high-frequency break point, we need to evaluate R_{eq}.

$$R_{\text{eq}} = \frac{R_1 R_2}{R_1 + R_2} = \frac{8.2\ \text{k}\Omega \times 12\ \text{k}\Omega}{8.2\ \text{k}\Omega + 12\ \text{k}\Omega} = 4.87\ \text{k}\Omega \tag{15-7}$$

Then

$$\omega_2 = \frac{1}{\tau_2} = \frac{1}{R_{\text{eq}}C_2} = \frac{1}{(4870\ \Omega) \times (1660 \times 10^{-12}\text{ F})} = \mathbf{1.24 \times 10^5\ rad/s} \tag{15-16}$$

$$f_2 = \frac{\omega_2}{2\pi} = \frac{1.24 \times 10^5}{2\pi} = = \mathbf{19.7\ kHz} \tag{15-17}$$

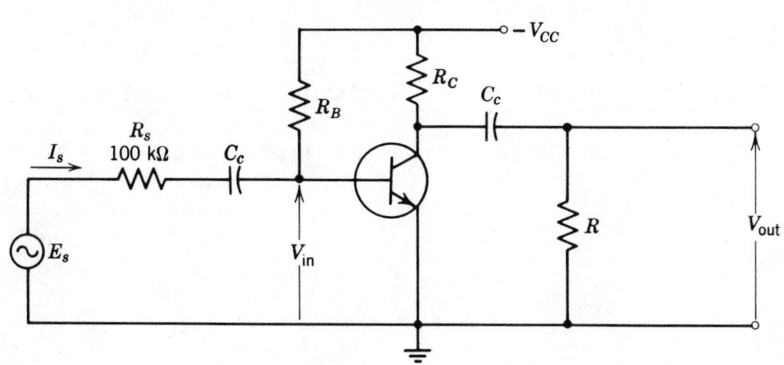

Circuit for Example 15-5.

Having the break frequencies (ω_1 and ω_2 or f_1 and f_2), we can draw the Bode plots for the circuit.

Example 15-5

In the circuit R_s is made very large so that I_s will remain constant even though the input conditions to the transistor may change. Two tests are made on the amplifier for each of two values of R. The first test is made at a low frequency at which there is negligible high-frequency roll-off. E_s, V_{in}, and V_{out} are measured. Then, without changing the value of E_s, we raise the frequency of the generator until V_{in} falls to 70.7% of its initial value (−3 dB). At this new frequency, I_s divides equally into r'_{in} and C_{in}. At this frequency r'_{in} must equal $X_{C_{in}}$. The first test yields the value of r'_{in} and the second test yields the frequency at which $X_{C_{in}}$ equals r'_{in}. Then, C_{in} can be calculated. Using the laboratory results given in Table 15-6, determine the values of $C_{b'e}$ and C_{ob} for the transistor.

Solution

Since V_{in} has the same value in Step 1 and in Step 3, r'_{in} has the same numerical value throughout the testing.

$$r'_{in} = \frac{V_{in}}{E_s - V_{in}} R_s = \frac{0.010\ \text{V}}{1.000\ \text{V} - 0.010\ \text{V}} \times 100{,}000\ \Omega$$

$$= 1000\ \Omega \qquad (7\text{-}4)$$

Then, from Step 2, r'_{in} equals the reactance of C_{in} at 2900 Hz.

$$r'_{in} = X_{C_{in}} = \frac{1}{2\pi f C_{in}}$$

Substituting values, we find

$$1000\ \Omega = \frac{1}{2\pi \times 2900\ \text{Hz} \times C_{in}}$$

Table 15-6
Test Data for the Circuit of Example 15-4

Step	Load	Frequency (Hz)	E_s (volts)	V_{in} (mV)	V_{out} (mV)	A_v
1	A	400	1.0	10	1000	100
2	A	2900	1.0	7.1	710	100
3	B	400	1.0	10	240	24
4	B	4200	1.0	7.1	170	24

Solving for C_{in}, we have

$$C_{\text{in}} = 5488 \text{ pF}$$

Then by Eq. 15-26

$$C_{\text{in}} = C_{b'e} + (1 + A_v)C_{ob} \qquad (15\text{-}26)$$

$$5488 = C_{b'e} + 101C_{ob} \qquad (1)$$

In Step 4, r'_{in} equals the reactance of C_{in} at 4200 Hz.

$$r'_{\text{in}} = X_{C_{\text{in}}} = \frac{1}{2\pi f C_{\text{in}}}$$

Substituting values, we have

$$1000\ \Omega = \frac{1}{2\pi \times 4200 \text{ Hz} \times C_{\text{in}}}$$

Solving for C_{in}, we have

$$C_{\text{in}} = 3789 \text{ pF}$$

Then by Eq. 15-26

$$C_{\text{in}} = C_{b'e} + (1 + A_v)C_{ob} \qquad (15\text{-}26)$$

$$3789 = C_{b'e} + 25C_{ob} \qquad (2)$$

If we solve (1) and (2) simultaneously we find

$$\mathbf{C_{ob} = 22.4 \text{ pF}}$$

$$\mathbf{C_{b'e} = 3230 \text{ pF}}$$

The value for C_{ob} is given in the specification data sheet for a transistor; the value for $C_{b'e}$ is not given. However, a value for f_α, the *alpha-cutoff frequency*, is given. At this frequency, the reactance of $C_{b'e}$ equals r'_e. However, r'_e is dependent on the value of the dc current I_E in the emitter. Then it is necessary to perform a dc analysis on the actual circuit to determine I_E. Then r'_e can be found from

$$\frac{25 \text{ mV}}{I_E} \leq r'_e \leq \frac{50 \text{ mV}}{I_E} \tag{7-5}$$

and using f_α, we can calculate $C_{b'e}$ for the particular bias condition for the transistor circuit.

Problems

15-7.1 R_s is 9 kΩ, R_B is 600 kΩ, R_C is 3000 Ω, and R is 2400 Ω. r'_e is 10 Ω and β is 100. Determine the mid-band gain. Find C_1 and C_2 that establish f_1 at 60 Hz for both the input and the output circuits. What is the magnitude of the gain and the phase angle at f_1?

15-7.2 R_s is 10 kΩ, R_C is 4.7 kΩ, R_E is 1500 Ω, and R is 3600 Ω, V_{CC} is +10 V, and β is 80. R_B is 510 kΩ and r'_e is 40 Ω. C_1 and C_2 are selected to give a value for f_1 at 100 Hz for each circuit. What is the mid-band gain? What is the gain and phase shift at 100 Hz and at 10 Hz?

15-7.3 Use the data for Problem 15-7.1. The value of C_{in} is 700 pF, and C_{out} is 30 pF. Find ω_2 for the input circuit to the transistor and find ω_2 for the output circuit of the transistor. What is the overall circuit gain at 200 kHz and at 2 MHz?

15-7.4 Use the data for Problem 15-7.2. The value of C_{in} is 1000 pF,and C_{out} is 30 pF. Find ω_2 for the input circuit to the transistor and find ω_2 for the output circuit of the transistor. Determine the overall gain at 200 kHz.

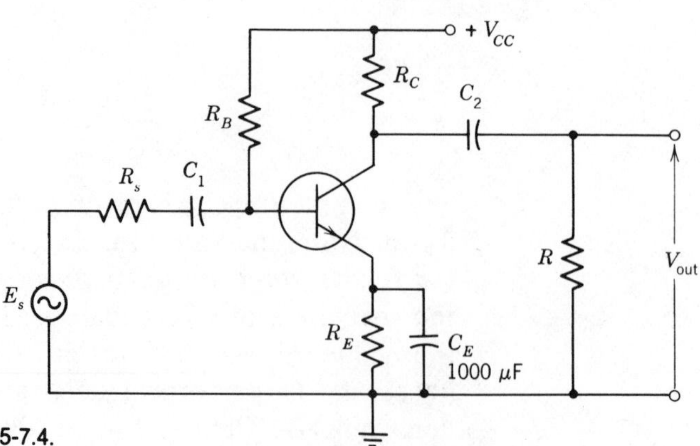

Circuit for Problems 15-7.1 through 15-7.4.

15-7.5 The data for the silicon transistor $Q1$ are

$$\beta = 100 \qquad r_e' = 50\ \Omega \qquad C_{ob} = 20\ \text{pF} \quad \text{and} \quad C_{b'e} = 200\ \text{pF}$$

a. Determine the mid-frequency gain.
b. Determine the high-frequency breakpoint, f_2.
c. Determine the circuit gain at 70 kHz and at 100 kHz.

15-7.6 The data for the germanium transistor $Q1$ are

$$\beta = 75 \qquad r_e' = 35\ \Omega \qquad C_{ob} = 30\ \text{pF} \quad \text{and} \quad C_{b'e} = 1000\ \text{pF}$$

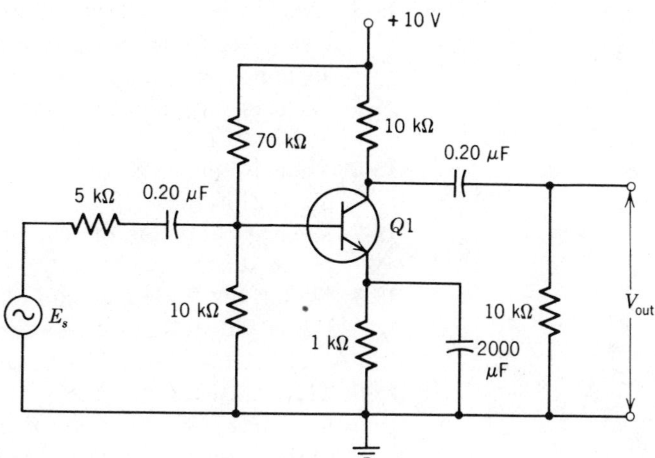

Circuit for Problems 15-7.5 and 15-7.7.

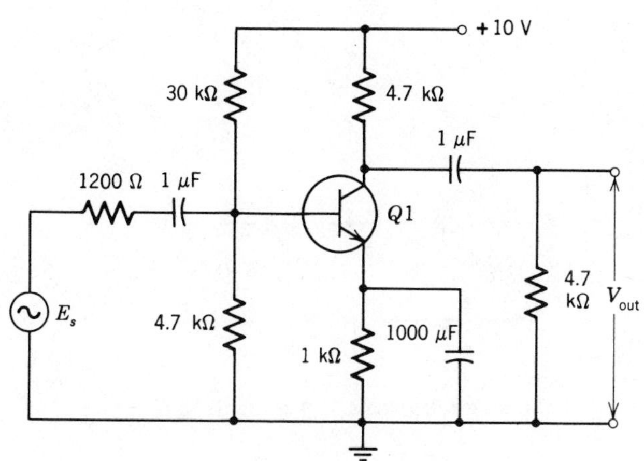

Circuit for Problems 15-7.6 and 15-7.8.

a. Determine the mid-frequency gain.
b. Determine the high-frequency breakpoint, f_2.
c. Find the circuit gain at 80 kHz and at 200 kHz.

15-7.7 Find f_α for the transistor used in Problem 15-7.5.

15-7.8 Find f_α for the transistor used in Problem 15-7.6.

Supplementary Problems

15-1 What is the value of K_{LF} at 2 Hz? Determine V_{out} and the phase shift.

15-2 What is the value of K_{LF} at 5 Hz? Determine V_{out} and the phase shift.

15-3 What is the value of K_{LF} at 8 Hz? Determine V_{out} and the phase shift.

15-4 Sketch the Bode gain and phase plots for the low frequencies. Using the Bode plots, determine the values of V_{out} and phase angle at 2 Hz, at 5 Hz, and at 8 Hz.

15-5 What is the value of K_{HF} at 1 MHz? Determine V_{out} and the phase shift.

15-6 What is the value of K_{HF} at 2 MHz? Determine V_{out} and the phase shift.

15-7 What is the value of K_{HF} at 10 MHz? Determine V_{out} and the phase shift.

15-8 Sketch the Bode gain and phase plots for the high frequencies. Using the Bode plots, determine the values of V_{out} and phase angle at 1 MHz, at 2 MHz, and at 10 MHz.

15-9 The transistor has the parameters 50 pF for C_{ob}, 300 pF for $C_{b'e}$, 50 Ω for r'_e, and 50 for β. Determine the input capacitance to the transistor. Determine f_1 and f_2. Draw the Bode gain and phase plots.

15-10 Repeat Problem 15-9 if C_E is removed from the circuit.

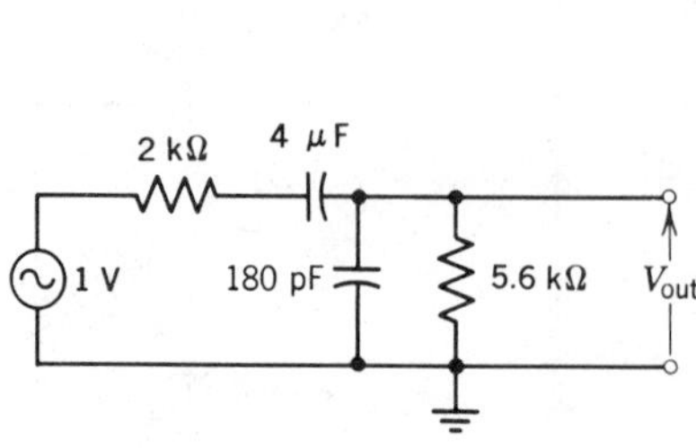

Circuit for Problems 15-1 through 15-8.

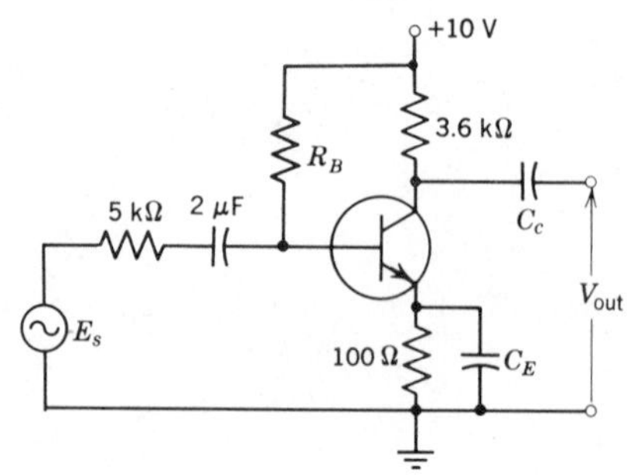

Circuit for Problems 15-9 and 15-10.

Chapter 16 Feedback

The topic of feedback is introduced by developing the fundamental feedback equation (Section 16-1). The basic properties of positive feedback (Section 16-2) and of negative feedback (Section 16-3) are examined. Negative feedback can be classified into four basic forms: voltage feedback with series input, voltage feedback with shunt input, current feedback with series input, and current feedback with shunt input (Section 16-4). The value of the feedback β_f in terms of circuit components is determined for voltage feedback by the voltage divider rule and is determined for current feedback by the ratio of two resistance values (Section 16-5). An example of circuit calculation is given for each of the four basic forms of feedback:

Voltage feedback; series input (Section 16-6).
Voltage feedback; shunt input (Section 16-7).
Current feedback; series input (Section 16-8).
Current feedback; shunt input (Section 16-9).

Section 16-1 The Fundamental Feedback Equation

For an ordinary amplifier (Fig. 16-1) the voltage gain is the output voltage divided by the input signal voltage. The signal V_{in} is amplified by the factor of A_v to the value V_{out} of output voltage. The gain A_v is often called the *open-loop gain.* If a feedback loop is added to this amplifier (Fig. 16-2), a fractional part β_f of the output voltage is fed back into the input at the *summing point.* The total input signal is the original signal plus the feedback voltage. The amplifier amplifies this total signal by the same factor A_v as in Fig. 16-1, producing the output voltage V'_{out}. The signal V_{in} is the same in each case, but the output voltages V_{out} and V'_{out} are different. The term β_f is the *feedback.* It is used as a decimal value in the equations, but, in a discussion, β_f is considered to be a percentage. For instance, 15% feedback is 0.15 when used in calculations.

The voltage fed from the output back into the input is $\beta_f V'_{out}$. The total input voltage to the amplifier is $V_{in} + \beta_f V'_{out}$. Since the input voltage times the gain is the output voltage, we

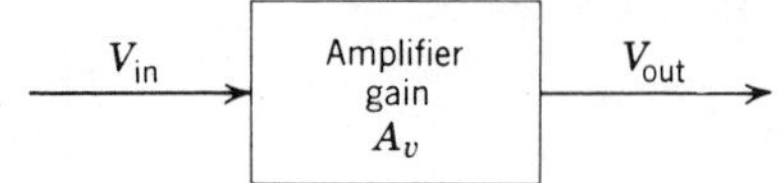

Figure 16-1 Block diagram of amplifier without feedback.

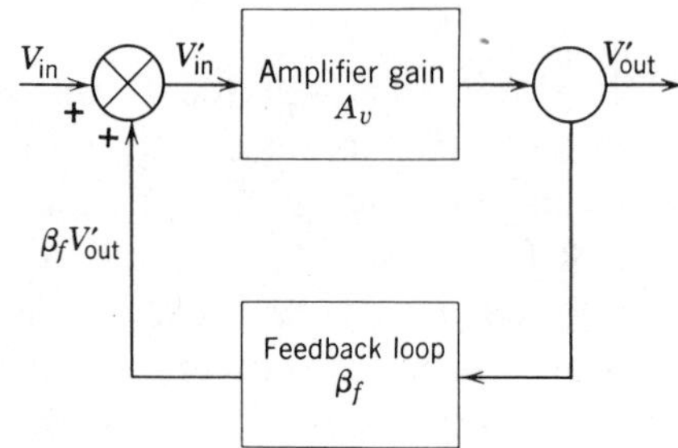

Figure 16-2 Block diagram of amplifier with feedback.

may write

$$(V_{in} + \beta_f V'_{out})A_v = V'_{out}$$

Expanding gives

$$V_{in}A_v + \beta_f A_v V'_{out} = V'_{out}$$

Rearranging, we have

$$V_{in}A_v = V'_{out} - \beta_f A_v V'_{out} = V'_{out}(1 - \beta_f A_v)$$

Then

$$\frac{A_v}{1 - \beta_f A_v} = \frac{V'_{out}}{V_{in}}$$

But V'_{out}/V_{in} is the *gain of the circuit with feedback.* Calling this gain A'_v, we have

$$\boxed{A'_v = \frac{A_v}{1 - \beta_f A_v}} \qquad (16\text{-}1)$$

where A_v is the amplifier gain without feedback, β_f is the feedback, and A'_v is the amplifier gain with feedback. The term

$\beta_f A_v$ is defined as the *feedback factor.* The *loop gain* is $(1 - \beta_f A_v)$. The gain A_v' is also referred to as the *closed-loop gain.*

Examination of Fig. 16-1 shows that

$$A_v = \frac{V_{\text{out}}}{V_{\text{in}}} \tag{16-2a}$$

and examination of Fig. 16-2 shows that

$$A_v = \frac{V'_{\text{out}}}{V'_{\text{in}}} \tag{16-2b}$$

The gain with feedback A_v' is defined from Fig. 16-2 as

$$A_v' = \frac{V'_{\text{out}}}{V_{\text{in}}} \tag{16-2c}$$

Section 16-2 Positive Feedback

In the analysis of the block diagram, we used $(V_{\text{in}} + \beta_f V'_{\text{out}})$ as the total input voltage. Purposely, no reference was made to the algebraic sign of β_f. If β_f is taken as a positive number, the feedback voltage is in phase with and adds to the incoming signal. This circuit condition is termed *positive feedback.*

An understanding of positive feedback may be obtained from a simple numerical example. Let us assume that an amplifier has a gain of 10 without feedback and let us substitute various values (Table 16-1) of positive feedback into the general equation

$$A_v' = \frac{A_v}{1 - \beta_f A_v} = \frac{10}{1 - 10\beta_f}$$

The immediate conclusion that can be drawn from the results of this table is that positive feedback increases the gain of an amplifier. For this reason, positive feedback is often called *regenerative feedback.* We will show in the next section of this chapter that positive feedback increases the distortion content of the output of an amplifier. Thus, the advantage of an increased gain must be carefully weighed against the disadvantage of an increased distortion level. As a result, we

Table 16-1
Positive Feedback

β_f (Feedback)	$\beta_f A_v$ (Feedback Factor)	$1-\beta_f A_v$ (Loop Gain)	A'_v (Closed-Loop Gain)
0	0	1	10
2%	0.20	0.80	12.5
4%	0.40	0.60	16.7
6%	0.60	0.40	25
8%	0.80	0.20	50
9%	0.90	0.10	100
9.9%	0.99	0.01	1000
9.99%	0.999	0.001	10,000
9.999%	0.9999	0.0001	100,000
10%	1.00	0	∞

do not find positive feedback used to any great extent in amplifier design.

As the feedback factor $\beta_f A_v$ approaches unity, we notice from Table 16-1 that the gain becomes infinite. Mathematically the equation shows that the gain is infinite, but electrically this does not happen. What does happen is that the circuit *oscillates*. Since the gain is infinite, the oscillator supplies its own signal for self-sustained operation. We now can state the very important and necessary conditions that must exist if a circuit is to oscillate.

1. The feedback must be positive.
2. The feedback factor must be +1.

Alternatively, these conditions may be expressed in this form.

> *In order to have an oscillator, the feedback must be positive and must be sufficient to sustain the oscillation.*

Problems Using positive feedback and the block diagram, complete the table.

Problem	V_{in}	A_v	β_f	$(1-\beta_f A_v)$	$\beta_f V'_{out}$	V'_{in}	V'_{out}
16-2.1	20 mV	20	2%				
16-2.2	20 mV	50					3 V
16-2.3		15	4%				10 V
16-2.4	20 mV	100	2%				
16-2.5	1 V	100		0.80			

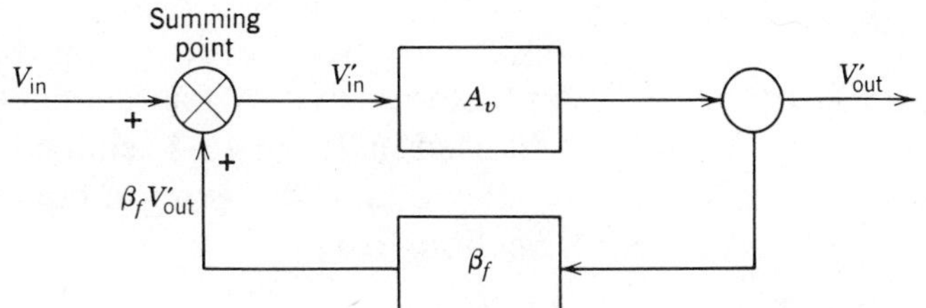

Block diagram for Problems 16-2.1 to 16-2.5.

Section 16-3 Negative Feedback

With negative feedback, the voltage $\beta_f V'_{out}$ that is fed from the output back to the input is 180° out of phase with the input. The algebraic sign of β_f for negative feedback is minus when used in the feedback equations. To illustrate negative feedback, we consider the effect of negative feedback on the amplifier used to illustrate positive feedback in Section 16-2. The amplifier without feedback has a gain of 10, and, substituting the minus sign for β_f in the feedback equation, we have

$$A'_v = \frac{A_v}{1 - \beta_f A_v} = \frac{10}{1 + 10\beta_f}$$

The results of Table 16-2 show that negative feedback reduces the overall gain of an amplifier. Since negative feedback reduces the gain, it is often called *degenerative feedback.*

Let us determine the effect of a 1% negative feedback on an amplifier that has a gain of 400 without feedback.

$$A'_v = \frac{A_v}{1 - \beta_f A_v} = \frac{400}{1 + 0.01 \times 400} = \frac{400}{1+4} = \frac{400}{5} = 80 \quad (16\text{-}1)$$

Table 16-2
Negative Feedback

β_f (Feedback)	$\beta_f A_v$ (Feedback Factor)	$1 - \beta_f A_v$ (Loop Gain)	A'_v (Closed-Loop Gain)
0	0	1	10
−1%	−0.10	1.10	9.09
−2%	−0.20	1.20	8.32
−10%	−1.00	2.00	5.00
−30%	−3.00	4.00	2.50
−40%	−4.00	5.00	2.00
−70%	−7.00	8.00	1.25
−100%	−10.00	11.00	0.909

A 1% negative feedback on this amplifier reduces the gain by a factor of five. A 1% feedback on the amplifier with a gain of 10 used in Table 16-2 reduced the gain from 10 to 9.09.

If we take the general equation for feedback developed in Section 16-1

$$A'_v = \frac{A_v}{1 - \beta_f A_v} \tag{16-1}$$

and divide each term in the numerator and denominator by A_v, we find that

$$A'_v = \frac{A_v/A_v}{1/A_v - \beta_f A_v/A_v} = \frac{1}{1/A_v - \beta_f}$$

When the value of the feedback β_f is large compared to $1/A_v$ (i.e., when a heavy negative feedback is used on a high-gain amplifier), the term $1/A_v$ may be neglected, and

$$\boxed{A'_v = -\frac{1}{\beta_f}} \tag{16-3}$$

We consider only negative feedback in this expression as we have shown that positive feedback would cause this circuit to oscillate. When a 10% negative-feedback loop is applied to an amplifier with a gain of 4000, the gain with feedback is 1/0.1 or 10. This value of gain is independent of transistor or FET variations, component changes (except in the feedback loop), and power-supply variations. The overall gain with feedback is determined by the feedback network alone provided that the feedback does not depend on the parameters of the transistor or FET. This concept is very important in a circuit use as a *decade amplifier* in servo amplifiers, in computers, and in instrument multipliers.

Distortion In Fig. 16-3, a signal is amplified by the factor A_v. At the same time, the amplifier creates a distortion D in the output. With a feedback loop (Fig. 16-4), not only is the output fed back into the input, but also the distortion is fed back. The total distortion

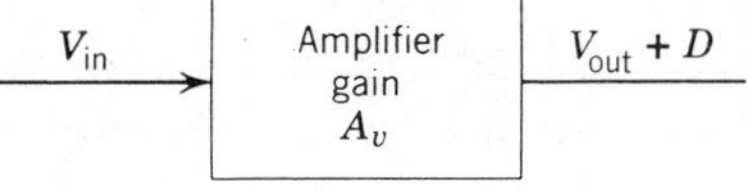

Figure 16-3 Block diagram of amplifier with distortion.

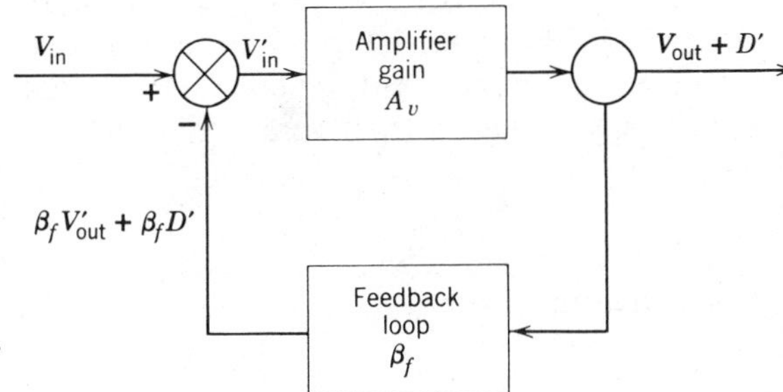

Figure 16-4 Block diagram of amplifier with feedback and distortion.

in the output D' must contain not only the amplified value of $\beta_f D'$ but also the original distortion of V_{in} that is produced by the amplifier. The input signal is so arranged that V_{out} equals V'_{out}. Then

$$D' = D + (\beta_f D')A_v$$

Rearranging,

$$D' - \beta_f A_v D' = D$$

Then

$$(1 - \beta_f A_v)D' = D$$

or

$$\boxed{D' = \frac{D}{1 - \beta_f A_v}} \qquad (16\text{-}4)$$

When the feedback is positive, the distortion with feedback becomes greater than the distortion without feedback. When the feedback is negative, D' is less than D. In other words, regenerative feedback increases distortion whereas degenerative feedback reduces distortion. Negative feedback reduces the distortion by the amount of the loop gain.

Example 16-1
An amplifier is made up of two stages. Each stage has a gain of 10 and an inherent distortion of 20%. One stage with a negative feedback of 10% is connected in cascade with a second stage that has a negative feedback of 40%. What is the overall gain and the overall distortion?

Solution
For the stage with 10% negative feedback

$$A'_{v_1} = \frac{A_v}{1 - \beta_f A_v} = \frac{10}{1 + 0.10 \times 10} = 5 \qquad (16\text{-}1)$$

and

$$D'_1 = \frac{D}{1 - \beta_f A_v} = \frac{20}{1 + 0.10 \times 10} = 10\% \qquad (16\text{-}4)$$

and for the second stage with 40% negative feedback

$$A'_{v_2} = \frac{A_v}{1 - \beta_f A_v} = \frac{10}{1 + 0.40 \times 10} = 2 \qquad (16\text{-}1)$$

and

$$D'_2 = \frac{D}{1 - \beta_f A_v} = \frac{20}{1 + 0.40 \times 10} = 4\% \qquad (16\text{-}4)$$

When the two stages are placed in cascade, the overall gain is

$$A'_v = A'_{v_1} \times A'_{v_2} = 5 \times 2 = \mathbf{10}$$

and, using decimals, the overall distortion is

$$D' = (1 + D'_1)(1 + D'_2) - 1 = 1.10 \times 1.04 - 1 = 0.144 \text{ or } \mathbf{14.4\%}$$

Example 16-2
Two stages of the amplifier of the last example are connected in cascade. An overall negative feedback of 9% is used. Determine the overall gain and the overall distortion.

Solution
Without feedback, the overall gain is

$$A_v = A_{v_1} \times A_{v_2} = 10 \times 10 = 100$$

and the overall distortion is

$$D = (1 + D_1)(1 + D_2) - 1 = 1.20 \times 1.20 - 1 = 0.44 \text{ or } 44\%$$

Now, with negative feedback

$$A_v' = \frac{A_v}{1 - \beta_f A_v} = \frac{100}{1 + 0.09 \times 100} = \mathbf{10} \qquad (16\text{-}1)$$

and

$$D' = \frac{D}{1 - \beta_f A_v} = \frac{44}{1 + 0.09 \times 100} = \mathbf{4.4\%} \qquad (16\text{-}4)$$

From the calculations developed in these two examples, we have shown that feedback over more than one stage, instead of feedback on each individual stage alone, produces the same gain but produces a much lower value of distortion. It is standard practice to use a lower percentage of feedback in a single loop over several stages rather than a larger percentage of feedback on each stage.

Frequency Response The Bode gain diagram for a single-stage amplifier without feedback is shown in Fig. 16.5*a*. At the break or corner frequencies f_1 and f_2, the actual gain is 3 dB down from the mid-band gain A_v. We *define BW, the bandwidth without feedback*, as the separation between the 3-dB frequencies f_1 and f_2.

$$\boxed{BW \equiv f_2 - f_1} \qquad (16\text{-}5)$$

When we use a negative feedback on this amplifier, the gain is reduced by the amount of the loop gain to A_v' as shown in Fig. 16-5*b*. Now, because of the negative feedback, the corner

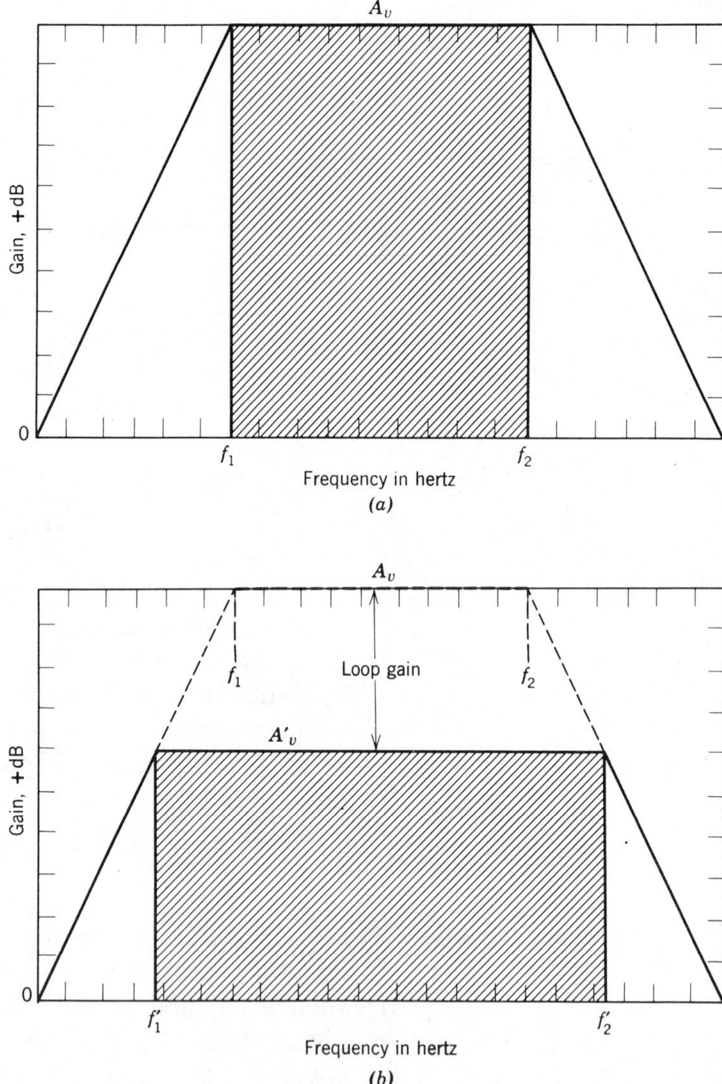

Figure 16-5 Effect of feedback on frequency response. (*a*) Without feedback. (*b*) With feedback.

or 3-dB frequencies are changed to f'_1 and f'_2. Then BW', *the bandwidth with negative feedback*, is the separation between f'_1 and f'_2.

$$BW' = f'_2 - f'_1 \tag{16-6}$$

Assume the mid-band gain of the amplifier is reduced 6 dB from A_v by negative feedback to A_v'. From the rules we developed in the last chapter, a change of 6 dB in gain along the roll-off slope represents a change in frequency of 2/1 or 1/2. Then

$$f_1' = \tfrac{1}{2}f_1 \tag{16-7}$$

$$f_2' = 2f_2 \tag{16-8}$$

A 6-dB reduction in mid-band gain means that the gain is cut in half.

$$A_v' = \tfrac{1}{2}A_v \tag{16-9}$$

Now let us form the *gain–bandwidth products,* $A_v \times BW$ and $A_v' \times BW'$ and equate them.

$$A_v \times BW = A_v' \times BW' \tag{16-10}$$

or

$$A_v \times (f_2 - f_1) = A_v' \times (f_2' - f_1') \tag{16-11}$$

in which A_v and A_v' are not in dB but are voltage ratios.

To work with numbers, assume that

$$A_v = 100 \qquad f_1 = 100\text{ Hz} \quad \text{and} \quad f_2 = 20\text{ kHz}$$

When the loop gain from negative feedback is 6 dB,

$$A_v' = 50 \qquad f_1' = 50\text{ Hz} \quad \text{and} \quad f_2' = 40\text{ kHz}$$

Thus, if we reduce the gain by 2 (6 dB), we decrease f_1 by 2 and increase f_2 by 2.

Substituting into Eq. 16-11, we have

$$A_v \times (f_2 - f_1) = A_v' \times (f_2' - f_1') \tag{16-11}$$

$$100 \times (20{,}000 - 100) = 50 \times (40{,}000 - 50)$$

$$1.99 \times 10^6 = 1.9975 \times 10^6$$

The differences between these two gain-bandwidth products is less than 0.4 of 1%.

Now let us repeat the calculation if the amount of negative feedback is increased to a loop gain of 20 dB. From the definition of the decibel ($20 \log_{10} V_2/V_1$), a reduction in gain of 20 dB is $-\log_{10} 1$ or a reduction in gain by a factor of 10. From the rules developed in considering Bode gain plots, a change of 20 dB of gain along the roll-off slope represents a decade change in frequency.

The values we used for the amplifier without feedback are

$$A_v = 100 \qquad f_1 = 100\ \text{Hz} \quad \text{and} \quad f_2 = 20\ \text{kHz}$$

Now with a 20-dB negative feedback, we have

$$A_v' = 10 \qquad f_1' = 10\ \text{Hz} \quad \text{and} \quad f_2' = 200\ \text{kHz}$$

Thus if we reduce the gain by a factor of 10, we reduce f_1 by 10 and increase f_2 by 10.

Substituting values into Eq. 16-11, we have

$$A_v \times (f_2 - f_1) = A_v' \times (f_2' - f_1') \qquad (16\text{-}11)$$

$$100 \times (20{,}000 - 100) = 10 \times (200{,}000 - 10)$$

$$1.99 \times 10^6 = 1.9999 \times 10^6$$

The difference between these two gain–bandwidth products is less than 0.05 of 1%.

If we had considered a dc amplifier or an operational amplifier (Chapters 17 and 18) that does not have a coupling capacitor, the frequency response is flat to 0 Hz. Then f_1 and f_1' are each zero and the gain–bandwidth products for all cases of the examples we used would have been

$$2 \times 10^6$$

We have used these numerical examples to show that

> *The gain–bandwidth product for any given amplifier is a constant.*

The gain–bandwidth products are indicated as being equivalent to the areas of the shaded rectangles on Fig. 16-5. A formal derivation of this result requires a complex algebraic approach that does not contribute toward the understanding of the concept of the gain–bandwidth product.

In order to obtain A_v' from A_v for negative feedback, we divide A_v by the loop gain $(1-\beta_f A_v)$. Referring back to the numerical examples, if we reduce the gain by $(1-\beta_f A_v)$, we decrease f_1 by this same amount

$$f_1' = \frac{f_1}{(1-\beta_f A_v)} \tag{16-12}$$

and we increase f_2 by this same amount.

$$f_2' = (1-\beta_f A_v) f_2 \tag{16-13}$$

Example 16-3

An amplifier has a gain of 40 and a high-frequency 3-dB break frequency of 8 kHz. Determine the gain and the break frequency when 5% negative feedback is used.

Solution

The midfrequency amplifier gain with feedback is

$$A_v' = \frac{A_v}{1-\beta_f A_v} = \frac{40}{1+0.05\times 40} = \mathbf{13.3} \qquad (16\text{-}1)$$

The new high-frequency 3-dB break frequency is

$$f_2' = f_2(1-\beta_f A_v) = 8(1+0.05\times 40) = \mathbf{24\ kHz} \qquad (16\text{-}13)$$

Stability The gain of a circuit with feedback is

$$A_v' = \frac{A_v}{1-\beta_f A_v} \tag{16-1}$$

If the percentage feedback is unchanged and if the open-loop gain changes from A_v to $A_v + \Delta A_v$, the gain of the circuit with feedback becomes

$$A''_v = \frac{A_v + \Delta A_v}{1 - \beta_f(A_v + \Delta A_v)} \qquad (16\text{-}14a)$$

The percentage change in the gain with feedback caused by the change in A_v is

$$\frac{A''_v - A'_v}{A'_v} \times 100\% \qquad (16\text{-}14b)$$

Example 16-4
The open-loop gain of an amplifier with 10% negative feedback is 30. If the open-loop gain increases by 10%, what is the percentage change in the gain with feedback?
For a 10% increase in open-loop gain, $\Delta A_v = 0.1\ A_v = 0.1 \times 30 = 3$.

Solution
We have

$$A'_v = \frac{A_v}{1 - \beta_f A_v} = \frac{30}{1 + 0.10 \times 30} = \frac{30}{4} = 7.50 \qquad (16\text{-}1)$$

and

$$A''_v = \frac{A_v + \Delta A_v}{1 - \beta_f(A_v + \Delta A_v)} = \frac{30 + 3}{1 + 0.10(30 + 3)} = \frac{33}{4.33} = 7.62 \qquad (16\text{-}14a)$$

The percentage change in gain is

$$\frac{A''_v - A'_v}{A'_v} \times 100 = \frac{7.62 - 7.50}{7.50} \times 100 = \mathbf{1.6\%} \qquad (16\text{-}14b)$$

This example shows that negative feedback materially improves the stability of an amplifier. Gain changes can result from a power-supply voltage change, from the aging of a circuit component or from a replacement transistor. In order to secure the advantages of negative feedback, it should be noted that the network that controls the negative feedback must be stable.

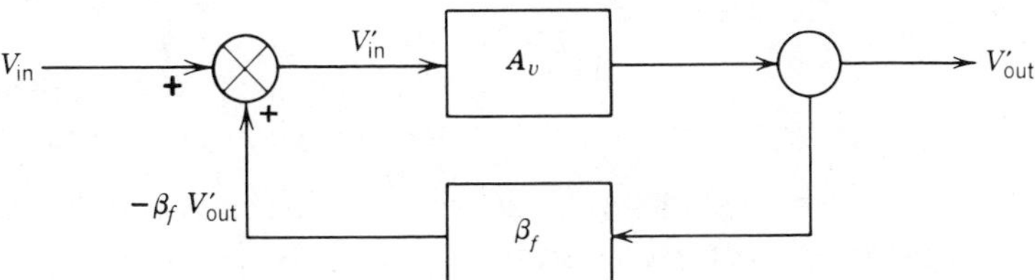

Block diagram for Problems 16-3.1 to 16-3.7.

Problems Complete the table using negative feedback.

Problem	V_{in}	A_v	β_f	$(1-\beta_f A_v)$	$\beta_f V'_{out}$	V'_{in}	V'_{out}
16-3.1	200 mV	20					1 V
16-3.2	200 mV	1000					2 V
16-3.3		50	3%				5 V
16-3.4	1 V	50	8%				
16-3.5	0.5 V		20%				2 V
16-3.6	5 V	20	100%				
16-3.7	1 V	100		4.0			

16-3.8 An audio amplifier consists of three stages.

Stage 1	$A_v = 50$	4% distortion
Stage 2	$A_v = 10$	4% distortion
Stage 3	$A_v = 20$	10% distortion

Each stage has individual feedback that reduces the gain to 10, 2, and 4, respectively. What is the overall gain and distortion with feedback? What percentage feedback is used on each stage?

16-3.9 Overall negative feedback is used to reduce the gain of the audio amplifier of Problem 16-3.8 to 80. What percentage feedback is required, and what is the overall distortion?

16-3.10 An amplifier without feedback has a mid-band gain of 200, and the 3-dB high-frequency breakpoint is 50 kHz. What is the mid-band gain and what is the high-frequency breakpoint when 10% negative feedback is used?

16-3.11 An amplifier has a mid-band gain without feedback of 200. The 3-dB frequency is 200 kHz. This amplifier is to be used as a video amplifier that requires a 5-MHz bandwidth. What gain can be obtained, and what feedback must be used? What bandwidth could be obtained if the feedback were 100%?

16-3.12 The open-loop gain of an amplifier with 10% negative feedback is 3000. If the open-loop gain increases by 10%, what is the change in the closed-loop gain?

16-3.13 The open-loop gain of an amplifier with 100% negative feedback is 30. If the open-loop gain increases to 40, what is the change in the closed-loop gain?

Section 16-4 Forms of Negative Feedback

Four arrangements for circuits using negative feedback are shown in Fig. 16-6. The output voltage provides the input to the feedback network in Fig. 16-6*a* and in Fig. 16-6*b*. The input to the feedback network is derived from the output current in Fig. 16-6*c* and in Fig. 16-6*d*.

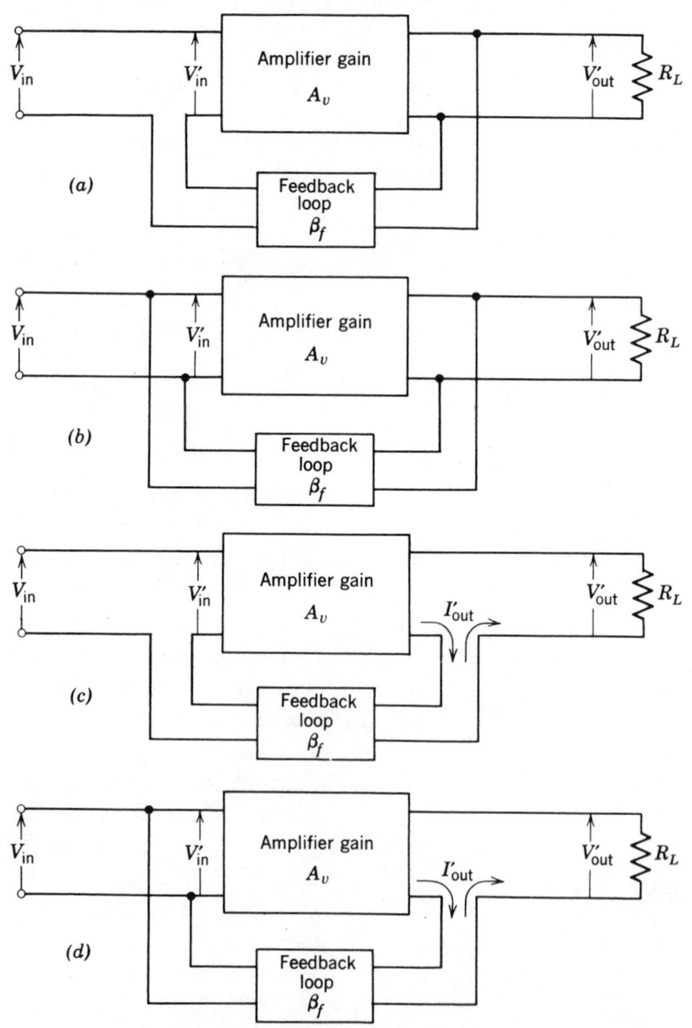

Figure 16-6 Negative feedback circuit configurations. (*a*) Voltage feedback—series input. (*b*) Voltage feedback—shunt input. (*c*) Current feedback—series input. (*d*) Current feedback—shunt input.

If we look back into the circuits of Fig. 16-6*a* and Fig. 16-6*b*, we see that the input to the feedback network is in parallel with the output of the amplifier. Therefore, as far as V'_{out} is concerned, the output resistance of the amplifier is reduced by the shunting effect of the input to the feedback network.

If we look back into the circuits of Fig. 16-6*c* and Fig. 16-6*d*, we see that the input resistance to the feedback network is in series with the output of the amplifier. Accordingly, current feedback increases the output resistance of the amplifier.

In the circuits of Fig. 16-6*a* and Fig. 16-6*c*, the output of the feedback network places both a voltage and a resistance in series with the amplifier. Therefore, V'_{in} is less than V_{in} by the amount of the voltage fed back into the input. Also, V_{in} sees two circuit elements in series, the input resistance to the amplifier and the output resistance of the feedback network. Thus, feedback causes the input resistance to the amplifier to increase.

In the circuits of Fig. 16-6*b* and Fig. 16-6*d*, V_{in} is identical to V'_{in}. However, the input to the amplifier is a parallel circuit—the amplifier input resistance in parallel with the output resistance of the feedback network. Therefore, shunt input reduces the input resistance to the circuit.

These characteristics of negative feedback amplifiers are summarized in Table 16-3.

We can show that the decrease or increase in input resistance or in output resistance with feedback as listed in Table 16-3 is by the factor of the loop gain.

Table 16-3
Negative Feedback Circuit Characteristics

Circuit Configuration	**Input Resistance**	**Output Resistance**
Voltage feedback—series input	increases	decreases
—shunt input	decreases	decreases
Current feedback—series input	increases	increases
—shunt input	decreases	increases

1. For an increase in resistance from R_1 to R_1' caused by negative feedback

$$R_1' = R_1(1 - \beta_f A_v) \tag{16-15}$$

2. For a decrease in resistance from R_2 to R_2' caused by negative feedback

$$R_2' = \frac{R_2}{1 - \beta_f A_v} \tag{16-16}$$

Eq. 16-15 and Eq. 16-16 apply as well to the general term "impedance."

Example 16-5

The open-loop gain of an amplifier (an operational amplifier) is 200,000. The amplifier has an output resistance r_{out} of 75 Ω. If the negative voltage feedback is 100%, what is the output resistance r'_{out} with feedback?

Solution

Reference to Table 16-3 shows that the output resistance decreases with feedback. The decrease is in the amount of the loop gain, Eq. 16-16.

$$r'_{\text{out}} = \frac{r_{\text{out}}}{1 - \beta_f A_v} = \frac{75}{1 + 1.00 \times 200{,}000} = \mathbf{375 \times 10^{-6}\ \Omega = 375\ \mu\Omega}$$

When a power amplifier is used to drive a loud speaker, it is very important to use negative voltage feedback to reduce the output impedance Z_{out} of the power amplifier. The impedance of the loud speaker can change radically with program music or speech. We do not want to lose power in Z_{out} nor do we want the voltage across the loud speaker to vary as the impedance of the loud speaker changes. As a result, we would never use a current feedback derived from the loud speaker (Table 16-3).

Also in a voltage regulator we use voltage feedback to reduce the output resistance. If the output resistance is zero, the load voltage is a constant.

Section 16-5 Circuit Concepts for Negative Feedback

Voltage Feedback—Series Input

Fig. 16-7 shows how a feedback voltage is derived from a voltage-divider circuit formed of R_1 and R_2. Applying the voltage divider rule to R_1 and R_2 and assuming the same current flows in R_1 and R_2, we have

$$V_f = \frac{R_1}{R_1 + R_2} V'_{out}$$

We define *voltage feedback* β_f as the fraction of the output voltage that is fed back into the input. Accordingly, for this circuit arrangement,

$$\boxed{-\beta_f = \frac{R_1}{R_1 + R_2}} \qquad (16\text{-}17)$$

Voltage Feedback—Shunt Input

In Fig. 16-8 the feedback resistor R_F is connected from the output back to the input. Now we find that we must consider

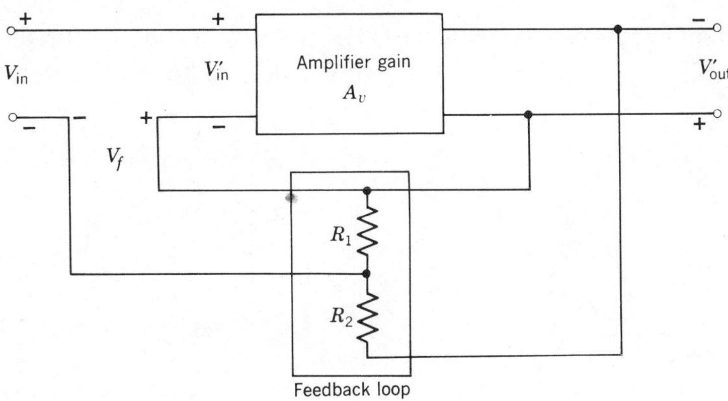

Figure 16-7 Voltage feedback—series input.

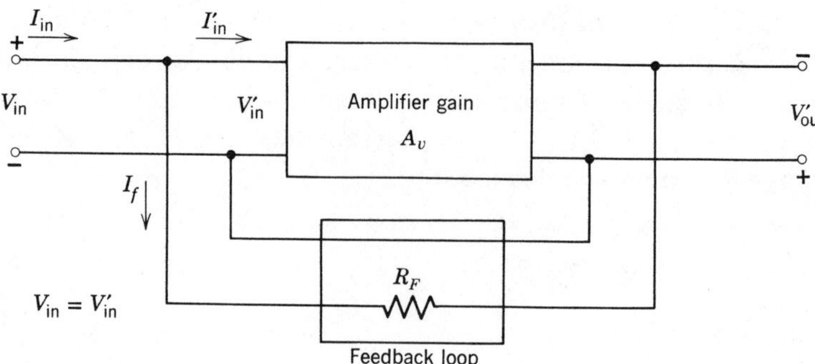

Figure 16-8 Voltage feedback—shunt input.

the Miller effect on R_F. In Section 7-7, we showed that this circuit arrangement placed an equivalent resistance

$$\frac{R_F}{1+A_v} \tag{7-28}$$

in parallel with the input to the stage and that an equivalent resistance

$$\frac{A_v}{1+A_v} R_F \tag{7-29}$$

was placed in parallel with the output. When we consider feedback in general, R_F is a resistor often connected across several stages to yield an overall feedback. In calculating gains, we must determine the equivalent resistance that the use of R_F places across the input of the amplifier.

Current Feedback—Series Input The simplest case of current feedback with series input is shown in Fig. 16-9. We have analyzed this circuit previously

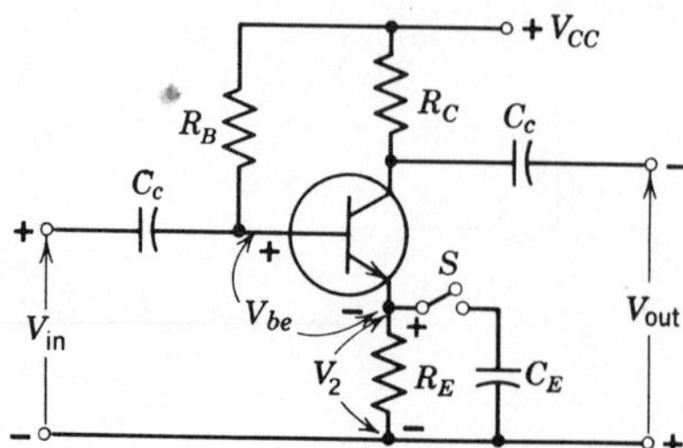

Figure 16-9 Current feedback—series input.

and have found that

1. With the switch S closed, we have no feedback.

$$r_{\text{in}} = (1+\beta)r'_e \tag{7-7}$$

and

$$A_v = \frac{R_C}{r'_e} \tag{7-8a}$$

2. With switch S open, we have emitter feedback.

$$r'_{\text{in}} = (1+\beta)(r'_e + R_E) \tag{7-25}$$

and

$$A'_v = \frac{R_C}{r'_e + R_E} \tag{7-23}$$

Now, let us use Eq. 7-8a for A_v and Eq. 7-23 for A'_v in Eq. 16-1 and solve for β_f for the negative feedback.

$$A'_v = \frac{A_v}{1 - \beta_f A_v} \tag{16-1}$$

$$\frac{R_C}{r'_e + R_E} = \frac{\left(\dfrac{R_C}{r'_e}\right)}{1 - \beta_f \dfrac{R_C}{r'_e}}$$

Simplifying, we have

$$\frac{R_C}{r'_e + R_E} = \frac{R_C}{r'_e - \beta_f R_C}$$

Inverting and cancelling R_C, we find

$$r'_e + R_E = r'_e - \beta_f R_C$$

Solving for *current feedback* β_f, we have

$$\boxed{-\beta_f = \frac{R_E}{R_C}} \tag{16-18}$$

The negative sign in Eq. 16-18 indicates that the feedback in this circuit is negative.

This last equation is NOT the voltage divider relation for β_f that we had in Eq. 16-15. A circuit of several stages must be examined very carefully to determine whether voltage feedback (Eq. 16-17) or current feedback (Eq. 16-18) should be used to calculate β_f.

To show the properties of this feedback circuit, let us determine the loop gain.

$$(1-\beta_f A_v) = 1 + \left(\frac{R_E}{R_C}\right)\left(\frac{R_C}{r_e'}\right) = 1 + \frac{R_E}{r_e'} = \frac{r_e' + R_E}{r_e'} \qquad (16\text{-}19)$$

Let us divide Eq. 7-23 by Eq. 7-8*a*.

$$\frac{A_v'}{A_v} = \frac{\left(\dfrac{R_C}{r_e' + R_E}\right)}{\left(\dfrac{R_C}{r_e'}\right)} = \frac{r_e'}{r_e' + R_E} = \frac{1}{\left(\dfrac{r_e' + R_E}{r_e'}\right)}$$

or

$$A_v' = \frac{A_v}{\left(\dfrac{r_e' + R_E}{r_e'}\right)}$$

Since A_v' is A_v divided by the loop gain, the loop gain is

$$\text{loop gain} = (1-\beta_f A_v) = \frac{r_e' + R_E}{r_e'} \qquad (16\text{-}20)$$

This is the result obtained by Eq. 16-19.

Let us divide Eq. 7-25 by Eq. 7-7.

$$\frac{r_{\text{in}}'}{r_{\text{in}}} = \frac{(1+\beta)(r_e' + R_E)}{(1+\beta)r_e'} = \left(\frac{r_e' + R_E}{r_e'}\right)$$

or

$$r_{\text{in}}' = \left(\frac{r_e' + R_E}{r_e'}\right) r_{\text{in}}$$

Thus r_{in}' is r_{in} multiplied by the loop gain.

$$r_{\text{in}}' = (1-\beta_f A_v) r_{\text{in}} \qquad (16\text{-}21)$$

Section 16-6 Voltage Feedback—Series Input

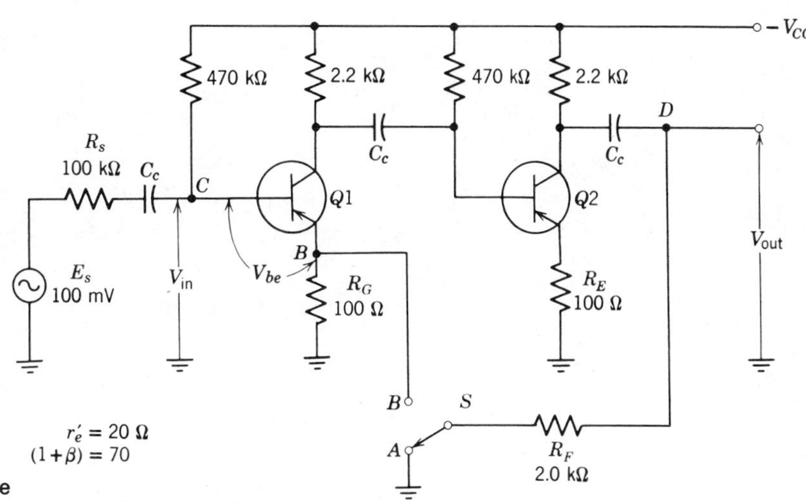

Figure 16-10 Circuit having voltage feedback—series input.

Without Feedback The two-stage amplifier circuit given in Fig. 16-10 uses voltage feedback with series input. Moving the switch S from point A to point B places the voltage feedback into the circuit.

The first procedure is to calculate the signal levels in the circuit without the negative feedback. The switch S is placed in position A. The load on $Q2$ is

$$\frac{2.2\ \text{k}\Omega \times 2.0\ \text{k}\Omega}{2.2\ \text{k}\Omega + 2.0\ \text{k}\Omega} = 1.048\ \text{k}\Omega = 1048\ \Omega$$

The voltage gain of stage $Q2$ is

$$A_{v_2} = \frac{R_L}{r'_e + R_E} = \frac{1048}{20 + 100} = 8.73 \qquad (7\text{-}23)$$

The input resistance to $Q2$ is

$$r_{\text{in}} = (1 + \beta)(r'_e + R_E) = 70(20\ \Omega + 100\ \Omega) = 8400\ \Omega \qquad (7\text{-}25)$$

The load on $Q1$ is $2.2\ \text{k}\Omega$ in parallel with $470\ \text{k}\Omega$ in parallel with $8400\ \Omega$.

$$\frac{1}{R_L} = \frac{1}{2200\ \Omega} + \frac{1}{470{,}000\ \Omega} + \frac{1}{8400\ \Omega}$$

or

$$R_L = 1737\ \Omega$$

The gain of the stage $Q1$ is

$$A_{v_1} = \frac{R_L}{r'_e + R_E} = \frac{1737\ \Omega}{20\ \Omega + 100\ \Omega} = 14.47$$

The overall gain A_v is

$$A_v = A_{v_1} \times A_{v_2} = 14.47 \times 8.73 = 126$$

The input resistance to $Q1$ is the same as the input resistance to $Q2$ or 8400 Ω. Then r'_{in} to the first stage is 8400 Ω in parallel with 470 kΩ.

$$r'_{\text{in}} = \frac{8.4\ \text{k}\Omega \times 470\ \text{k}\Omega}{8.4\ \text{k}\Omega + 470\ \text{k}\Omega} = 8.25\ \text{k}\Omega = 8250\ \Omega$$

The simplified model for the amplifier is shown in Fig. 16-11.

The input voltage to the circuit is

$$V_{\text{in}} = \frac{r'_{\text{in}}}{r'_{\text{in}} + R_s} E_s = \frac{8.25\ \text{k}\Omega}{8.25\ \text{k}\Omega + 100\ \text{k}\Omega} 100\ \text{mV} = 7.62\ \text{mV} \quad (7\text{-}2)$$

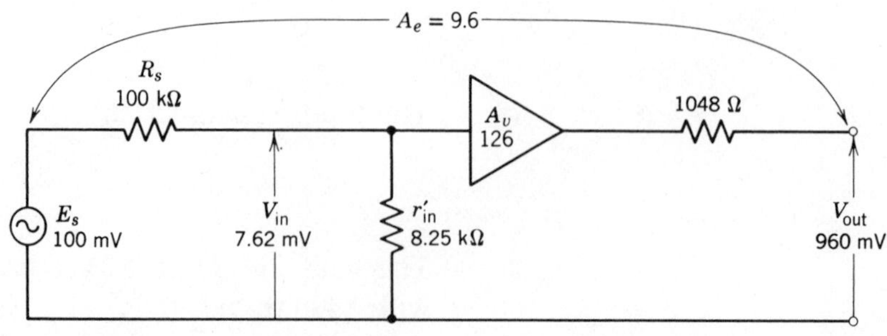

Figure 16-11 The simplified model without feedback.

The output voltage V_{out} is

$$V_{out} = V_{in}A_v = 7.62 \text{ mV} \times 126 = 960 \text{ mV} \qquad (7\text{-}1a)$$

The overall gain A_e is

$$A_e = \frac{V_{out}}{E_s} = \frac{960 \text{ mV}}{100 \text{ mV}} = 9.6 \qquad (7\text{-}1b)$$

These values are also added to the simplified model, Fig. 16-11.

With Feedback When the switch S is placed in position B (Fig. 16-10) we have a negative-feedback voltage divider network R_F and R_E placed across the output from point D to ground. The feedback β_f is

$$\beta_f = \frac{R_E}{R_E + R_F} = \frac{100 \ \Omega}{100 \ \Omega + 2000 \ \Omega} = 0.0476 \text{ or } 4.76\% \qquad (16\text{-}17)$$

The value of the gain reduction factor is the loop gain.

$$(1 - \beta_f A_v) = (1 + 0.0476 \times 126) = 7.00$$

The gain with negative feedback is

$$A'_v = \frac{A_v}{1 - \beta_f A_v} = \frac{126}{7.00} = 18.0 \qquad (16\text{-}1)$$

and the input resistance to the amplifier is increased by the loop gain from 8400 Ω to

$$r'_{in} = r_{in}(1 - \beta_f A_v) = 8400 \ \Omega \times 7.00 = 58{,}800 \ \Omega \qquad (16\text{-}15)$$

As far as E_s and R_s are concerned, this value, 58,800 Ω, is in parallel with 470 kΩ.

or
$$r''_{in} = \frac{470 \text{ k}\Omega \times 58.8 \text{ k}\Omega}{470 \text{ k}\Omega + 58.8 \text{ k}\Omega} = 52.3 \text{ k}\Omega$$

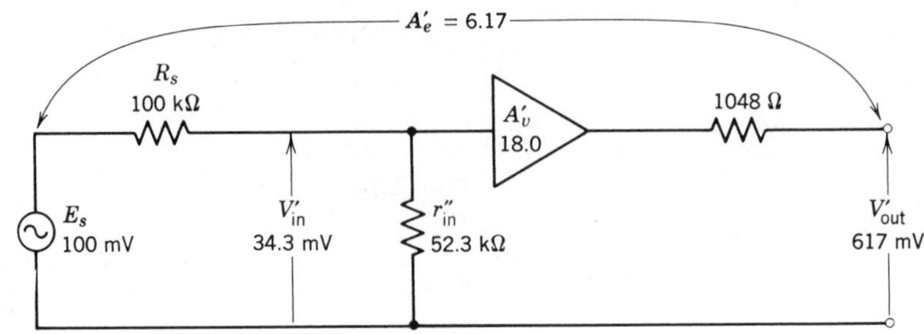

Figure 16-12 The simplified model with feedback.

Now the simplified model, Fig. 16-12, can be formed for the circuit with negative feedback.

The value of V'_{in} is determined from the input divider network.

$$V'_{in} = \frac{r''_{in}}{r''_{in} + R_s} E_s = \frac{52.3\ \text{k}\Omega}{52.3\ \text{k}\Omega + 100\ \text{k}\Omega} 100\ \text{mV} = 34.3\ \text{mV} \quad (7\text{-}2)$$

and

$$V'_{out} = V'_{in} A'_v = 34.3\ \text{mV} \times 18.0 = 617\ \text{mV} \quad (7\text{-}1a)$$

The overall gain with feedback A'_e is

$$A'_e = \frac{V'_{out}}{E_s} = \frac{617\ \text{mV}}{100\ \text{mV}} = 6.17 \quad (7\text{-}1b)$$

Problems

16-6.1 Calculate the output voltage with and without feedback for the circuit of Fig. 16-10 if R_F is 4 kΩ.

16-6.2 Repeat Problem 16-6.1 if R_F is 1500 Ω.

Section 16-7 Voltage Feedback—Shunt Input

The circuit shown in Fig. 16-13 is a three-stage amplifier. When the switch S is placed in position A, the overall feedback loop through R_F is grounded. The input resistance to each stage is

$$r_{in} = (1 + \beta)(r'_e + R_E) = (70)(20\ \Omega + 100\ \Omega) = 8400\ \Omega \quad (7\text{-}25)$$

Without Feedback

The load on E_s and R_s is r'_{in}.

$$r'_{in} = \frac{r_{in} R_B}{r_{in} + R_B} = \frac{8.4\ \text{k}\Omega \times 470\ \text{k}\Omega}{8.4\ \text{k}\Omega + 470\ \text{k}\Omega} = 8.25\ \text{k}\Omega$$

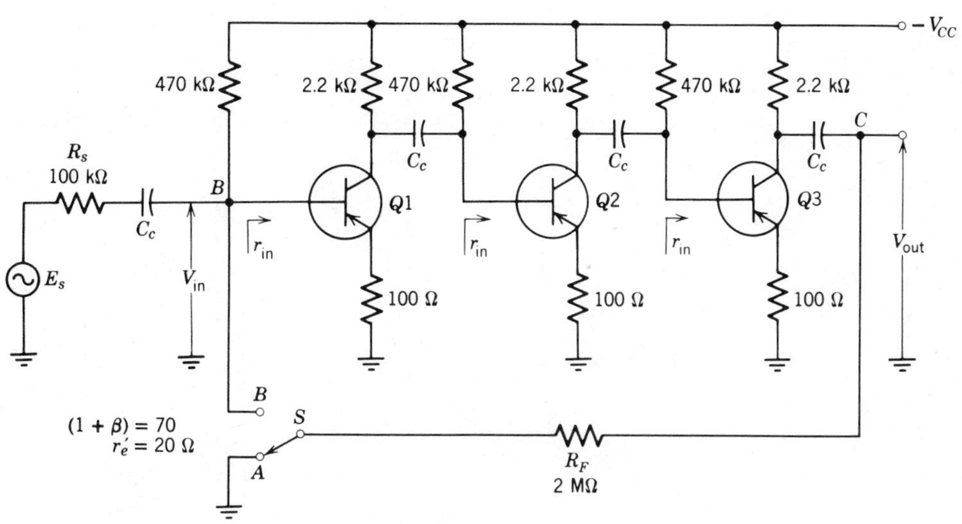

Figure 16-13 Circuit having voltage feedback—shunt input.

The load on $Q1$ and on $Q2$ is R_L.

$$\frac{1}{R_L} = \frac{1}{R_C} + \frac{1}{R_B} + \frac{1}{r_{\text{in}}} = \frac{1}{2.2\text{ k}\Omega} + \frac{1}{470\text{ k}\Omega} + \frac{1}{8.4\text{ k}\Omega}$$

or

$$R_L = 1.735\text{ k}\Omega = 1735\ \Omega$$

The voltage gains of the stages of $Q1$ and $Q2$ are

$$A_{v_1} = A_{v_2} = \frac{R_L}{r_e' + R_E} = \frac{1735\ \Omega}{20\ \Omega + 100\ \Omega} = 14.46 \qquad (7\text{-}23)$$

The load on $Q3$ is effectively 2.2 kΩ. The voltage gain of the stage of $Q3$ is

$$A_{v_3} = \frac{R_C}{r_e' + R_E} = \frac{2200\ \Omega}{20\ \Omega + 100\ \Omega} = 18.33 \qquad (7\text{-}23)$$

The overall gain from point B to point C is

$$A_v = A_{v_1} \times A_{v_2} \times A_{v_3} = 14.46 \times 14.46 \times 18.33 = 3832$$

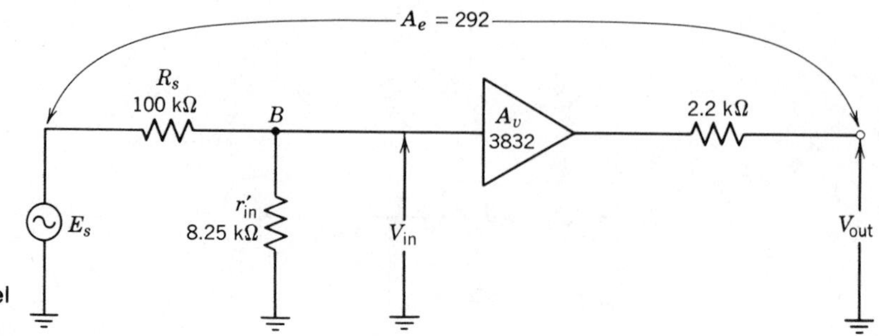

Figure 16-14 The simplified model without feedback.

The overall circuit gain A_e is

$$A_e = \frac{r'_{in}}{r'_{in} + R_s} A_v = \frac{8.25\ \text{k}\Omega}{8.25\ \text{k}\Omega + 100\ \text{k}\Omega} 3832 = 292 \qquad (7\text{-}3)$$

These values are placed on the simplified model without feedback, Fig. 16-14.

With Feedback When the switch S is placed in position B (Fig. 16-13) shunt feedback derived from the output voltage is applied to the overall circuit. The placement of R_F from point C to point B requires that the Miller effect be used to determine the shunting effect of R_F at the base of $Q1$.

$$R'_F = \frac{R_F}{1 + A_v} = \frac{2000\ \text{k}\Omega}{1 + 3832} = 0.522\ \text{k}\Omega = 522\ \Omega \qquad (7\text{-}28)$$

Now we take the simplified model without feedback (Fig. 16-14) and place R'_F in parallel with r'_{in} to obtain the simplified model with feedback given in Fig. 16-15*a*.

The reflected value of R_F, R'_F, is in parallel with r'_{in}. The equivalent value is r''_{in}.

$$r''_{in} = \frac{R'_F r'_{in}}{R'_F + r'_{in}} = \frac{522\ \Omega \times 8250\ \Omega}{522\ \Omega + 8250\ \Omega} = 491\ \Omega = 0.491\ \text{k}\Omega$$

The overall circuit gain with feedback A'_e is

$$A'_e = \frac{r''_{in}}{r''_{in} + R_s} A'_v = \frac{0.491\ \text{k}\Omega}{0.491\ \text{k}\Omega + 100\ \text{k}\Omega} 3832 = 18.7 \qquad (7\text{-}3)$$

These values are placed on the simplified model with feedback (Fig. 16-15*b*).

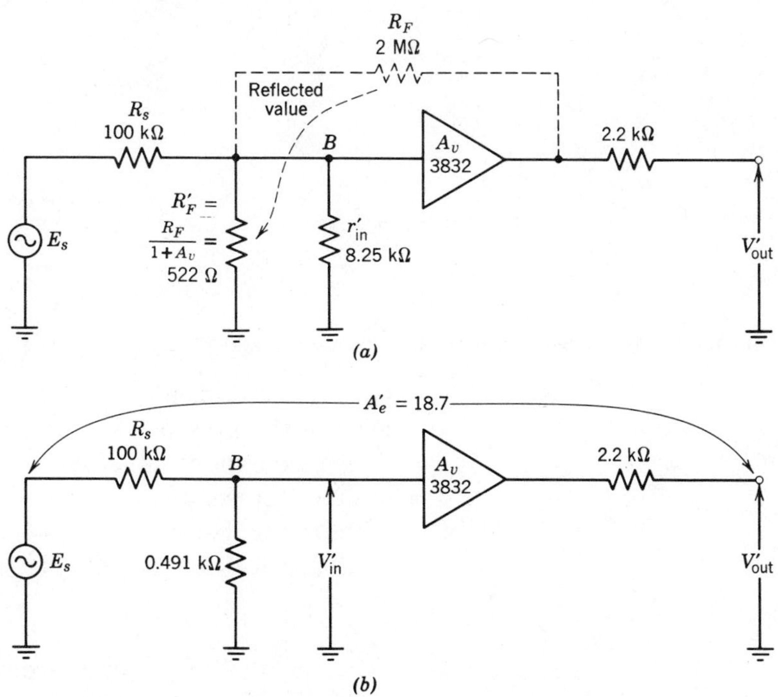

Figure 16-15 The simplified model with feedback.

Problems

16-7.1 Using the circuit of Fig. 16-13, determine A_e and A'_e if R_F is 3 MΩ.

16-7.2 Using the circuit of Fig. 16-13, determine A_e and A'_e if R_s is 47 kΩ and R_F is 1.5 MΩ.

16-7.3 Using the circuit of Fig. 16-13, determine what value of R_s yields an overall gain with feedback of 10?

Section 16-8 Current Feedback—Series Input

When the switch S is placed in position A in the circuit of Fig. 16-16, the ac value of R_E for $Q3$ is the parallel combination of 120 Ω and 600 Ω.

Without Feedback

$$R_E = \frac{120 \times 600}{120 + 600} = 100\ \Omega$$

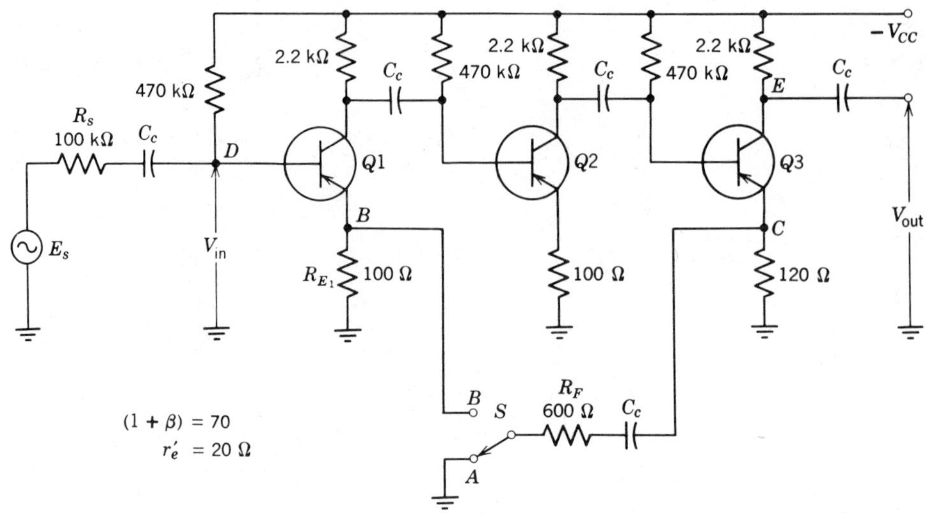

Figure 16-16 Circuit having current feedback—series input.

Now the circuit has the same open-loop values as the circuit used in the previous section, Fig. 16-13. Therefore, the open-loop calculations obtained in Section 16-7 are valid for this circuit. Consequently, we can use the simplified model without feedback (Fig. 16-14) as the simplified model without feedback for this circuit, Fig. 16-17.

With Feedback Now switch S is changed from position A to position B (Fig. 16-16). In doing this we have placed (600 Ω + 100 Ω) in parallel with 120 Ω in the emitter of $Q3$. The ac value of R_{E_3} is

$$R_{E_3} = \frac{700\ \Omega \times 120\ \Omega}{700\ \Omega + 120\ \Omega} = 102\ \Omega \approx 100\ \Omega$$

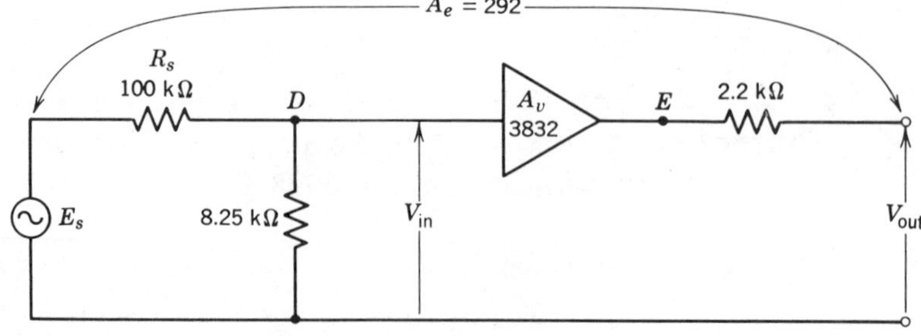

Figure 16-17 The simplified model without feedback.

The current feedback β_f' in $Q3$ is

$$\beta_f' = \frac{R_E}{R_C} = \frac{100\ \Omega}{2200\ \Omega} \tag{16-18}$$

However, this feedback is not applied directly to point B of $Q1$; there is a voltage divider formed by R_F and R_{E_1} to make the effective feedback β_f equal to

$$\beta_f = \frac{R_{E_1}}{R_{E_1} + R_F}\,\beta_f' = \frac{100\ \Omega}{100\ \Omega + 600\ \Omega} \times \frac{100\ \Omega}{2200\ \Omega}$$
$$= 0.00649 = 0.649 \text{ of } 1\%$$

The value of the loop gain is

$$(1 - \beta_f A_v) = (1 + 0.00649 \times 3832) = 25.88$$

The value of A_v' is

$$A_v' = \frac{A_v}{1 - \beta_f A_v} = \frac{3832}{25.88} = 148 \tag{16-1}$$

The input resistance to $Q1$ is increased by the loop gain $(1 - \beta_f A_v)$ to

$$r_{\text{in}}' = (1 - \beta_f A_v) r_{\text{in}} = 25.88 \times 8.4 = 217.3\ \text{k}\Omega \tag{16-15}$$

Placing r_{in}' in parallel with R_B (470 kΩ), we have

$$r_{\text{in}}'' = \frac{r_{\text{in}}' R_B}{r_{\text{in}}' + R_B} = \frac{217.3\ \text{k}\Omega \times 470\ \text{k}\Omega}{217.3\ \text{k}\Omega + 470\ \text{k}\Omega} = 148\ \text{k}\Omega$$

The overall gain A_e' is

$$A_e' = \frac{r_{\text{in}}''}{r_{\text{in}}'' + R_s}\,A_v' = \frac{148\ \text{k}\Omega}{148\ \text{k}\Omega + 100\ \text{k}\Omega}\,148 = 88.3 \tag{7-3}$$

Now we have numerical values to form the simplified model for the circuit with feedback (Fig. 16-18).

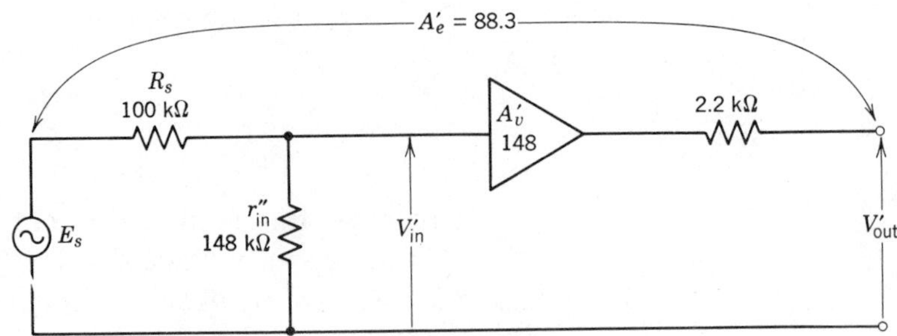

Figure 16-18 The simplified model with feedback.

Problems

16-8.1 If E_s in Fig. 16-16 is 10 mV, find V_{in} and V_{out} without and with feedback. Find A_e and A_e'.

16-8.2 If R_F is 100 Ω and E_s is 10 mV in Fig. 16-16, find V_{in} and V_{out} without and with feedback. Find A_e and A_e'.

16-8.3 If R_F is 500 Ω and E_s is 20 mV in Fig. 16-16, find V_{in} and V_{out} without and with feedback. Find A_e and A_e'.

Section 16-9 Current Feedback—Shunt Input

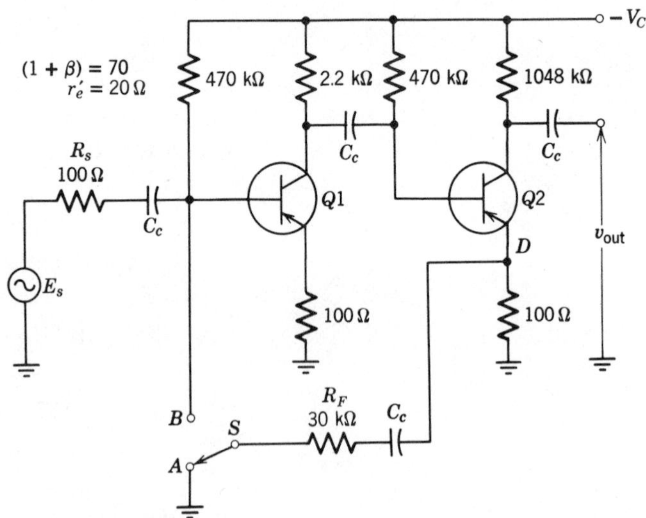

Figure 16-19 Circuit having current feedback—shunt input.

Without Feedback The circuit (Fig. 16-10) used in Section 16-6 for voltage feedback—series input is used in this section to illustrate current feedback with shunt input. In Fig. 16-10, the load on $Q2$ without feedback is the parallel combination of R_C (2.2 kΩ) and R_F (2.0 kΩ). The value of the parallel com-

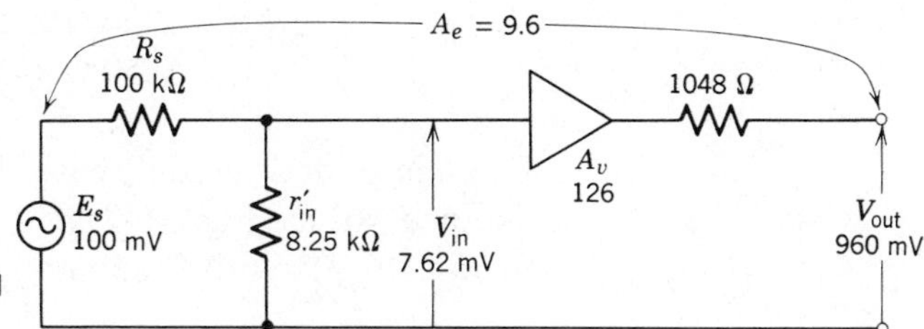

Figure 16-20 The simplified model without feedback.

bination is 1048 Ω. Therefore, in Fig. 16-19, the value for R_C for $Q2$ is 1048 Ω. Accordingly, the calculations obtained for the circuit without feedback in Section 16-6 are also valid for this section. The simplified model for the circuit without feedback, Fig. 16-20, is identical to Fig. 16-11.

With Feedback When switch S is in position B, there is a feedback placed in parallel with the base of $Q1$. The Miller effect must be used to evaluate the magnitude of the shunting resistance caused by R_F.

In this circuit R_F is *not* placed across the circuit where the voltage gain A_v is 126. The gain to be considered for the Miller effect is a value reduced from A_v because R_F is taken, not from the collector of $Q2$, but from the emitter of $Q2$. Then the current feedback ratio in $Q2$ must be taken into account. In $Q2$

$$\beta_f = \frac{R_E}{R_C} = \frac{100\ \Omega}{1048\ \Omega} = 0.0954 = 9.54\% \qquad (16\text{-}18)$$

Then the gain to be used for the Miller effect in this case, A''_v, is

$$A''_v = \beta_f A_v = 0.0954 \times 126 = 12.0$$

Then, by the Miller effect, the input resistance part of R_F is R'_F.

$$R'_F = \frac{R_F}{1 + A'_v} = \frac{30{,}000\ \Omega}{1 + 12.0} = 2300\ \Omega \qquad (7\text{-}28)$$

The input resistance r''_{in} to be used on the simplified model with feedback is the parallel combination of R'_F and r'_{in}.

$$r''_{\text{in}} = \frac{r'_{\text{in}}R'_F}{r'_{\text{in}} + R'_F} = \frac{8250\ \Omega \times 2300\ \Omega}{8250\ \Omega + 2300\ \Omega} = 1800\ \Omega = 1.8\ \text{k}\Omega$$

The simplified model with feedback is drawn using this resistance value (Fig. 16-21).

In Fig. 16-21, with feedback, if E_s is 100 mV

$$V'_{\text{in}} = \frac{r''_{\text{in}}}{r''_{\text{in}} + R_s} E_s = \frac{1.8\ \text{k}\Omega}{1.8\ \text{k}\Omega + 100\ \text{k}\Omega} 100\ \text{mV} = 1.77\ \text{mV} \qquad (7\text{-}3)$$

and

$$V'_{\text{out}} = V'_{\text{in}}A_v = 1.77\ \text{mV} \times 126 = 223\ \text{mV} \qquad (7\text{-}1a)$$

and

$$A'_e = \frac{V'_{\text{out}}}{E_s} = \frac{223\ \text{mV}}{100\ \text{mV}} = 2.23 \qquad (7\text{-}1b)$$

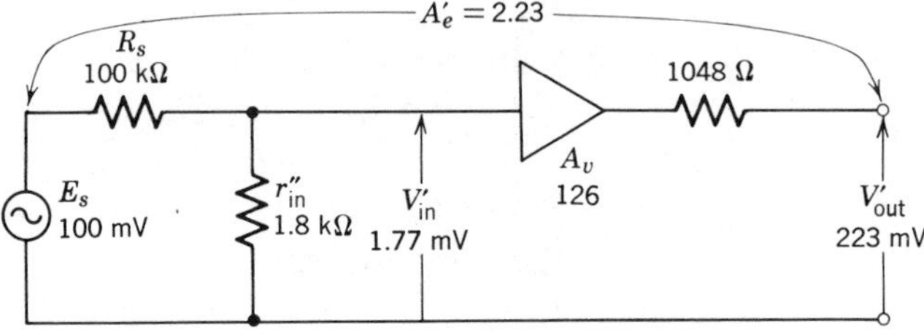

Figure 16-21 The simplified model with feedback.

Problems **16-9.1** If R_F is 20 kΩ in Fig. 16-19, determine V_{in} and V_{out} both without and with feedback when E_s is 20 mV.

16-9.2 Repeat Problem 16-9.1 if R_F is 40 kΩ.

Supplementary Problems **16-1** The voltage at point C is 20 V and A_v is 50. If the circuit is an ideal oscillator, what are the values of voltage at points A, B, and D and what is the value of β_f in percent?

16-2 The voltage at point A is 50 mV, A_v is 40, and the positive feedback is 2%. What are the voltages at points B, C, and D?

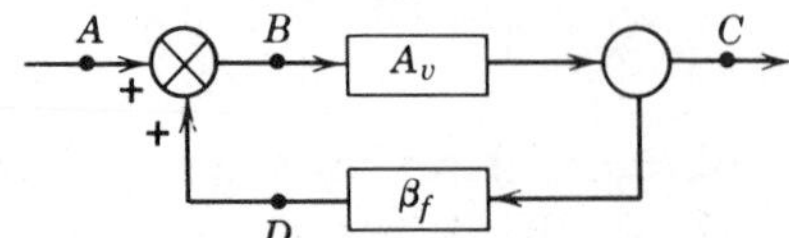

Circuit for Problems 16-1 through 16-4.

16-3 A_v is 80 and the feedback is 1% negative. If the voltage at point C is 15 V, what are the voltages at points A, B, and D?

16-4 The voltage at point C is 20 V and A_v is 50. If the feedback is 20% negative, what are the values of voltage at points A, B, and D?

16-5 Two stages are connected in cascade. The first stage has a gain of 40 and a distortion of 5%. The second stage has a gain of 50 and a distortion of 10%. What percentage of overall negative feedback is required to reduce the overall gain to 200? Now what is the overall distortion?

16-6 The feedback is negative. At full power, without feedback, the distortion is 8%. After feedback, a distortion of 0.3 of 1% is required. Determine values for R, E_s, and E_s'. What is β_f in percent?

16-7 The open-loop voltage gain of an amplifier has a nominal value of 2000. Its gain can vary from 1200 to 2800. If enough negative feedback is used to reduce the gain to 50 (nominal value), what is the expected variation in closed-loop gain when the open-loop gain ranges from 1200 to 2800?

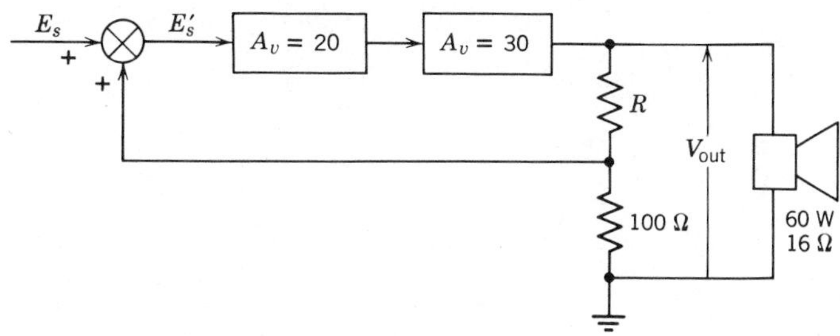

Circuit for Problem 16-6.

16-8 An amplifier has an open-loop gain of 60 dB. The value of f_1 is 30 Hz, and the value of f_2 is 10 kHz. The distortion is 10%. An overall feedback loop is added to provide 2% negative feedback. What are the new values for the gain in dB. What are the new values for f_1', f_2', and D'?

16-9 An amplifier uses 10% negative voltage feedback with shunt input. The open-loop data for the amplifier are: A_v is 30, the input resistance is 4000 Ω, the output resistance is 200 Ω, and the distortion is 15%. What are the values of these four parameters with the feedback?

16-10 Each amplifier has an input resistance of 10 kΩ, a gain of 50, and an output resistance of 10 kΩ. A negative feedback loop of 8% is added to each stage.

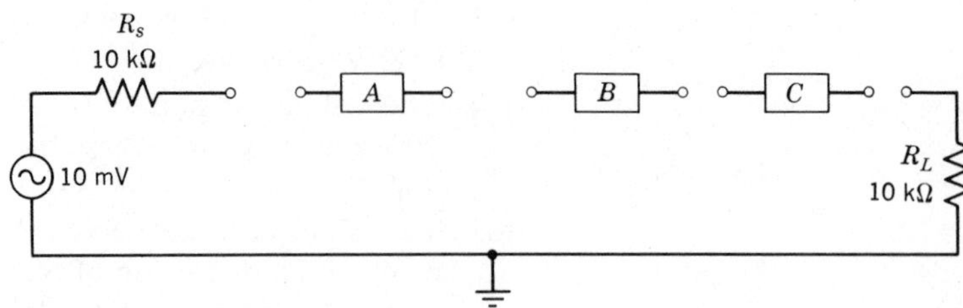

Circuit for Problem 16-10.

Stage A is voltage feedback with shunt input.
Stage B is current feedback with series input.
Stage C is voltage feedback with series input.

When the three stages are placed in cascade with the 10-mV source, what is the voltage across R_L?

Chapter 17 The Operational Amplifier

The characteristics and properties of an operational amplifier are detailed (Section 17-1). One of the two basic operational amplifier circuits is the inverting amplifier (Section 17-2) and the other is the noninverting amplifier (Section 17-3). The basic derived circuits are presented: the voltage follower, the inverting adder, the noninverting adder, and the subtractor.

Section 17-1 The Ideal Operational Amplifier

In Section 12-7, we considered the basic circuit of the operational amplifier (Fig. 12-11) as an application of the differential amplifier. The practical operational amplifier (*op amp*) uses this basic circuit but also includes several additional transistors to provide stabilizing feedback circuits. In this chapter, we consider the complete op amp circuit as a "black box" (Fig. 17-1*a*).

The op amp is a device available in many different packagings. The most common arrangements are the three types shown in Fig. 17-1: the DIP, the TO-5 case, and the flat pack. Many op amps are available in any one of several different packaging arrangements.

The "black box" shown in Fig. 17-1*a* has five terminals. The usual op amp requires both a positive power-supply connection ($+V_{CC}$) and a negative power supply connection ($-V_{CC}$). The common return (ground) required for the two power supplies is obtained from the external circuitry. There is a signal output terminal (V_{out}). As we pointed out in Section 12-7, there are two input terminals: the inverting input terminal (−) and the noninverting input terminal (+). The relative polarity of the output terminal is positive (+) although it is not labeled.

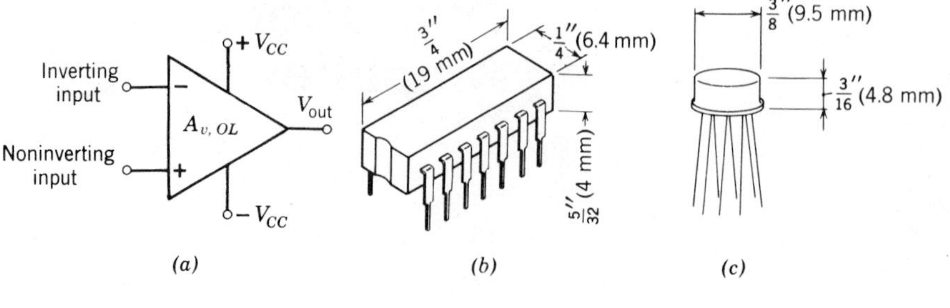

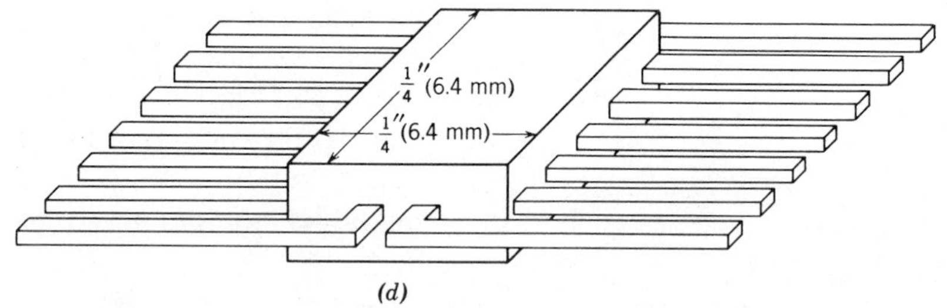

Figure 17-1 The op amp. (*a*) Circuit symbol and connections. (*b*) Dual-in-line package (DIP). (*c*) Eight-lead TO-5 case. (*d*) Flat package. (*Dimensions are approximate only*).

The ideal op amp has the following characteristics:

1. The open-loop voltage gain $A_{v,OL}$ of the op amp is extremely high and ideally approaches infinity.
2. The *intrinsic input resistance* r_i, measured between the inverting input terminal and the noninverting input terminal, is extremely high and ideally approaches an open circuit (infinite ohms).
3. The *intrinsic output resistance* r_o, seen looking back into the output terminal, is very low and ideally approaches zero ohms.

If an input signal V_i is placed across the input terminals to the op amp (Fig. 17-2*a*), a center point is required to provide a ground return. When V_i is reduced to zero, both input terminals (+ and −) are at ground potential. The ideal op amp is perfectly balanced and the output voltage V_{out} is zero. This condition locates point C at the origin on the transfer characteristic (Fig. 17-2*b*). When V_i is increased from zero with the polarity shown on the circuit, the output voltage rises linearly

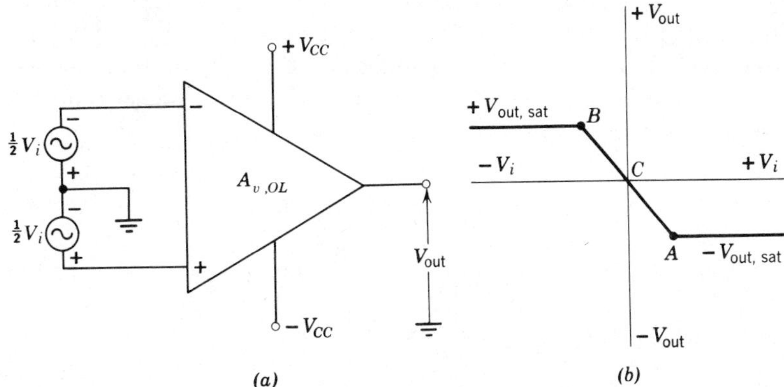

Figure 17-2 Saturation of the op amp. (*a*) Circuit. (*b*) Transfer characteristic.

in a positive direction. The maximum possible positive output voltage obtainable from the op amp is $+V_{out,sat}$, point B. Likewise, when the polarity of V_i is reversed, the op amp reaches a negative saturation, $-V_{out,sat}$, at point A.

The maximum possible values of $+V_{out}$ and $-V_{out}$ are usually about 2 V less than the supply voltages, $+V_{CC}$ and $-V_{CC}$. The slope of the transfer characteristic from point A to point B is the open-loop gain of the op amp, $A_{v,OL}$.

Example 17-1

The manufacturer's specification for the μA741C general-purpose op amp states that the saturation values of V_{out} are ± 13 V when the supply voltages are ± 15 V. The large signal differential voltage gain A_{VD} is typically 200 V/mV with a minimum value of 20 V/mV. Determine the typical and the minimum values of $A_{v,OL}$ and determine the corresponding values of V_i required to saturate the op amp.

Solution

A_{VD} is specified with the units volts/millivolts whereas the definition of voltage gain from Eq. 7-1*a* requires that the output voltage and the input voltage must be in the same units. If we change A_{VD} to volts/volts or millivolts/millivolts, we have the required values for $A_{v,OL}$.

$$200 \frac{\text{V}}{\text{mV}} = \frac{200 \times 1000 \text{ mV}}{1 \quad \text{mV}} = \frac{200 \text{ V}}{0.001 \text{ V}} = \mathbf{200{,}000}\ A_{v,OL} \text{ typical}$$

and

$$20 \frac{\text{V}}{\text{mV}} = \frac{20 \times 1000 \text{ mV}}{1 \quad \text{mV}} = \frac{20 \text{ V}}{0.001 \text{ V}} = \mathbf{20{,}000}\ A_{v,OL} \text{ minimum}$$

The output saturation specification of the op amp is ±13 V. Therefore, from Fig. 17-2*b*, the maximum peak-to-peak value of undistorted output V_{out} is 26 V. Beyond this output saturation occurs. Then, for the typical value, the corresponding input signal required to saturate the op amp is

$$V_{i,p\text{-}p} = \frac{V_{\text{out,sat},p\text{-}p}}{A_{v,OL}} = \frac{26}{200{,}000} = 0.00013\ V_{\text{p-p}}$$

$$= \mathbf{130\ \mu V_{\text{p-p}}} \qquad (7\text{-}1a)$$

and for the minimum value

$$V_{i,p\text{-}p} = \frac{V_{\text{out,sat},p\text{-}p}}{A_{v,OL}} = \frac{26}{20{,}000} = 0.0013\ V_{\text{p-p}}$$

$$= \mathbf{1.3\ mV_{\text{p-p}}} \qquad (7\text{-}1a)$$

The cost of the μA741C op amp is less than $1. Very high gain chopper-stabilized op amps are available at about $50 each or more that provide a minimum open-loop gain of 140 dB ($A_{v,OL} = 10^7$).

Section 17-2 The Inverting Amplifier

The *inverting amplifier*, Fig. 17-3, has the noninverting input terminal (+) connected to the ground return. A resistor R_1 connects the input signal to the inverting input. A feedback resistor R_f is connected from the output back to the inverting input. At first there may seem to be an inconsistency in the polarity markings. We must remember that the (−) and the (+) markings on the op amp merely designate which terminal is the inverting input (*I*) and which terminal is the noninverting input (*NI*).

The polarity of V_i is determined by the polarity of the circuit input voltage V_{in}. The polarity of V_{out} is reversed from

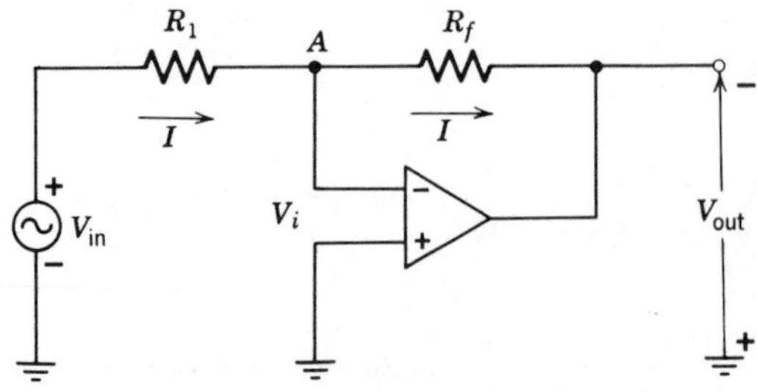

Figure 17-3 The inverting amplifier.

the polarity of V_{in}. Consequently, this circuit gives a 180° phase inversion to the signal. We have placed the polarity notations on V_{in} and V_{out} for convenience to show relative current directions.

If the op amp is ideal, the magnitude of V_i is zero. Also, in the ideal op amp, the intrinsic input resistance r_i is extremely high (an open circuit). Then the input current to the op amp is zero. Thus, the summing point, point A, is ideally at ground potential. The voltage across $(R_1 + R_f)$ is $(V_{in} + V_{out})$ and a current I flows from the input terminal to the output terminal and into the op amp. Since V_i is assumed to be zero,

$$V_{out} = -IR_f$$

and

$$V_{in} = IR_1$$

If we divide these two equations, we obtain the equation for the gain of the inverting amplifier.

$$A_v = \frac{V_{out}}{V_{in}} = \frac{-IR_f}{IR_1}$$

$$\boxed{A_v = -\frac{R_f}{R_1}} \tag{17-1}$$

Now let us examine this circuit from the viewpoint of a negative feedback amplifier. Since V_{out} is (−) and V_{in} is (+), we have negative feedback. The fundamental feedback equation developed in Chapter 16 is

$$A'_v = \frac{A_v}{1 - \beta_f A_v} \tag{16-1}$$

In this chapter A_v, the open-loop gain, will be labeled $A_{v,OL}$. Also, since we are concerned only with negative feedback, we will change the (−) sign to a (+) sign. We will indicate the phase inversion of the amplifier by placing a (−) sign before the gain equation. To keep this discussion in accordance with conventional practice, we will use A_v in place of A'_v. Now Eq.

16-1 becomes

$$A_v = -\frac{A_{v,OL}}{1+\beta_f A_{v,OL}} \tag{17-2}$$

When each term in Eq. 17-2 is divided by $A_{v,OL}$, we have

$$A_v = -\frac{1}{\dfrac{1}{A_{v,OL}}+\beta_f} \tag{17-3}$$

Since

$$\frac{1}{A_{v,OL}} \ll \beta_f$$

Eq. 17-3 becomes

$$A_v = -\frac{1}{\beta_f} \tag{17-4}$$

If we examine Fig. 17-3, we say that V_{in} produces a current I in R_1. This current flows through R_f and produces a voltage V_{out} equal to IR_f. If we reverse this logic, we can say that V_{out} produces a current I in R_f that produces a voltage drop in R_1 that must equal V_{in}. Therefore the output voltage produces a *current feedback*. If we had used the technique developed in Section 16-5, we could have looked at the circuit of Fig. 17-3 and have stated that it was obviously current feedback and not voltage feedback. Then we could have written immediately from the definition of current feedback that

$$\beta_f = \frac{R_1}{R_f} \tag{17-5}$$

and then from Eq. 17-4 we could have written the closed-loop gain as

$$A_v = -\frac{1}{\beta_f} = -\frac{R_f}{R_1} \tag{17-1}$$

Example 17-2

A μA741C op amp is used in the circuit given in Fig. 17-4. Determine the output voltage:

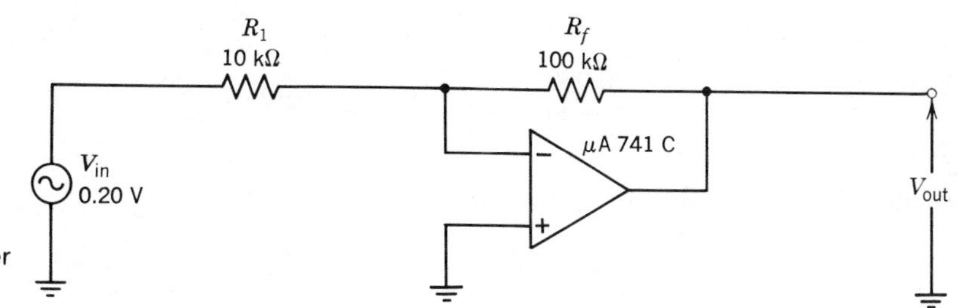

Figure 17-4 The inverting amplifier with numerical values.

Case I. For the ideal op amp.
Case II. For $A_{v,OL} = 200{,}000$ (the typical value).
Case III. For $A_{v,OL} = 20{,}000$ (the minimum value).

What errors are made by assuming that the op amp is ideal?

Solution

Case I. If the op amp is ideal, the gain is

$$A_v = \frac{V_{out}}{V_{in}} = -\frac{R_f}{R_1} = -\frac{100{,}000\ \Omega}{10{,}000\ \Omega} = -10 \qquad (17\text{-}1)$$

and

$$V_{out} = A_v V_{in} = -10 \times 0.20 = \mathbf{-2.00\ V}$$

Case II. If the op amp has a gain of 200,000,

$$A_v = -\frac{1}{\dfrac{1}{A_{v,OL}} + \beta_f} = -\frac{1}{\dfrac{1}{A_{v,OL}} + \dfrac{R_1}{R_f}}$$

$$= -\frac{1}{\dfrac{1}{200{,}000} + \dfrac{10{,}000\ \Omega}{100{,}000\ \Omega}} = -9.9995 \qquad (17\text{-}3)$$

and

$$V_{out} = -A_v V_{in} = -9.9995 \times 0.20 = \mathbf{-1.9999\ V}$$

Case III. If the op amp has a gain of 20,000

$$A_v = -\frac{1}{\dfrac{1}{A_{v,OL}} + \dfrac{R_1}{R_f}} = -\frac{1}{\dfrac{1}{20{,}000} + \dfrac{10{,}000}{100{,}000}} = -9.995 \quad (17\text{-}3)$$

and

$$V_{out} = A_v V_{in} = -9.995 \times 0.20 = \mathbf{-1.999\ V}$$

If the op amp actually has a gain of 200,000, the error is

$$2.00 - 1.9999 = 0.0001\ \text{V} = \mathbf{100\ \mu V}$$

or

$$\frac{0.0001}{2.00} \times 100 = 0.005 \text{ of } 1\%$$

If the op amp actually has a gain of 20,000, the error is

$$2.00 - 1.999 = 0.001 \text{ V} = \mathbf{1\ mV}$$

or

$$\frac{0.001}{2.00} \times 100 = 0.05 \text{ of } 1\%$$

This example shows that the error made by assuming that the op amp is ideal is very much less than the precision of the resistors used to form the feedback network of R_f and R_1. Consequently, in all future cases we will assume that the op amp is ideal and we will derive the equations for A_v by the simplest possible approach.

Problems

17-2.1 If R_f is 1 MΩ and R_1 is 10 kΩ, calculate the gain and the error made by assuming that the op amp is ideal. Use the op amp data given in Example 17-2.

17-2.2 Use the data given in Example 17-2. Assume the op amp is ideal. If R_f and R_1 are resistors with a ±10% tolerance, what are the possible variations in the output from the ideal?

17-2.3 Repeat Problem 17-2.2 if the resistors have a tolerance of ±5%.

17-2.4 Repeat Problem 17-2.2 if the resistors have a tolerance of ±1%.

17-2.5 Repeat Problem 17-2.2 if the resistors have a tolerance of ±0.1 of 1%.

Section 17-3 Other Basic Operational-Amplifier Circuits

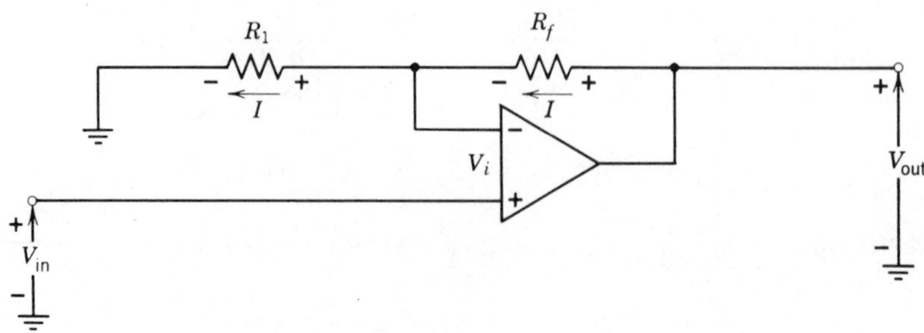

Figure 17-5 The noninverting amplifier.

The frequency f_2 is located at 0 dB at point A on the response curve of the op amp (Fig. 19-5b). Through point A we draw a straight line that has a roll-off of 20 dB/decade (6 dB/decade). Figure 19-5b shows that the integrator is unconditionally stable; it can never oscillate. On the other hand, as we increase the frequency of the incoming signal, the output of the integrator decreases. This explains why the peak-to-peak output of the integrator decreases as the frequency of an incoming square waveform increases.

Problems

19-1.1 The given waveform is the input voltage to an integrator in which R is 100 kΩ and C is 1 μF. Find the output voltage waveform.

19-1.2 Repeat Problem 19-1.1 for the second input voltage waveform.

19-1.3 Repeat Problem 19-1.1 for the third input voltage waveform.

19-1.4 R is 20 kΩ and C is 200 pF in an integrator. If the offset voltage is 15 mV, how long does it take for the op amp to saturate at ±13 V?

19-1.5 Repeat Problem 19-1.4 if R is 2 MΩ and if C is 10 μF.

Section 19-2 The Differentiator

From Fig. 19-6, we see that the output voltage of the circuit is

$$v_{out} = -Ri \tag{19-6a}$$

The current in the capacitor C produced by v_{in} is

$$i = C\frac{dv_{in}}{dt} \tag{19-6b}$$

Substituting Eq. 19-6b into Eq. 19-6a, we have

$$\boxed{v_{out} = -RC\frac{dv_{in}}{dt}} \tag{19-7}$$

Equation 19-7 shows that the output of the circuit is the derivative of the input signal. Consequently, the circuit is called the *differentiator*.

The Noninverting Amplifier The circuit for the *noninverting amplifier* is given in Fig. 17-5. Inspection of the circuit shows that the polarity of V_{out} is the same as the polarity of V_{in}; that is, V_{out} is in phase with V_{in}. The direction of the current I through R_f and R_1 is shown on the diagram. Now we place polarity markings on R_1 and on R_f. The input voltage V_i to the op amp is zero (the ideal case). If we express V_i as the difference between the two input voltages V_{in} and IR_1, we have

$$V_i = V_{in} - IR_1 = 0$$

or

$$V_{in} = IR_1$$

The output voltage is

$$V_{out} = I(R_1 + R_f)$$

Dividing the two equations yields the gain A_v.

$$A_v = \frac{V_{out}}{V_{in}} = \frac{I(R_1 + R_f)}{IR_1}$$

Then

$$A_v = \frac{V_{out}}{V_{in}} = \frac{R_1 + R_f}{R_1}$$

$$\boxed{A_v = 1 + \frac{R_f}{R_1}} \tag{17-6}$$

An examination of the circuit of Fig. 17-5 shows that this is a negative voltage feedback. From the technique we developed in Section 16-5, the feedback is

$$\beta_f = -\frac{R_1}{R_1 + R_f} \tag{16-7}$$

In developing Eq. 17-4, we showed that

$$A_v = -\frac{1}{\beta_f} \tag{17-4}$$

Then we have directly

$$A_v = \frac{R_1 + R_f}{R_1} = 1 + \frac{R_f}{R_1} \tag{17-7}$$

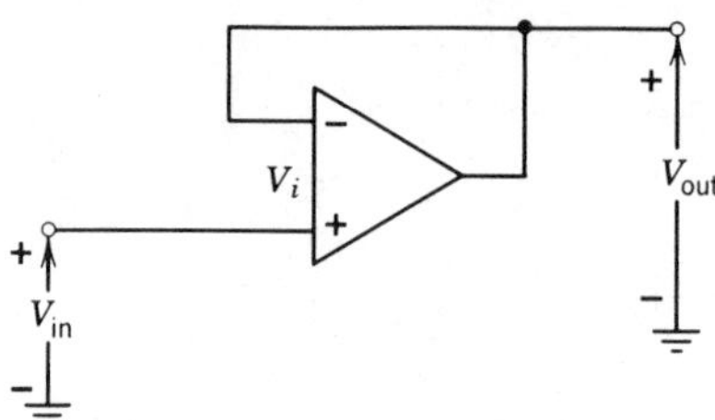

Figure 17-6 The voltage follower.

The Voltage Follower The circuit for the *voltage follower* is given in Fig. 17.6. Inspection of this circuit shows that V_{out} is in phase with V_{in}. Then, if V_i is to be zero V_{in} must be identical to V_{out}.

$$V_i = V_{in} - V_{out} = 0$$

or

$$\boxed{V_{out} = V_{in}} \tag{17-8}$$

The input resistance to the voltage follower is ideally infinite (an open circuit) and the output resistance is ideally zero. This circuit with unity gain is often used to make a source impedance independant of the load resistance. The circuit is called a *buffer amplifier* when it is used for this objective.

The Inverting Adder The circuit for the *inverting adder* or the *inverting summer* is shown in Fig. 17-7. The connections to the amplifier show that there is phase inversion in the amplifier.

We will derive the equation for the output voltage using the superposition theorem.

In the ideal amplifier, V_i is zero (0 V). Since the noninverting input (+) is connected directly to ground and since V_i is zero (0 V), the voltage from point S, the summing point, to ground must be 0 V. Therefore, point S is effectively at ground potential. We call point S a *virtual ground* to describe

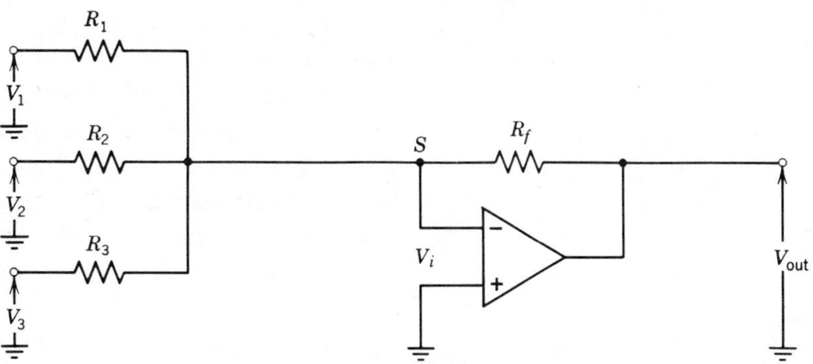

Figure 17-7 The inverting adder.

this condition. The current in R_1 is produced by V_1 alone. The current in R_1 is not affected by V_2, by R_2, by V_3, or by R_3. The output V_{out} is merely the sum of the output voltages produced by each of V_1, V_2, and V_3 independently. We can write immediately

$$V_{out_1} = -\frac{R_f}{R_1}V_1$$

$$V_{out_2} = -\frac{R_f}{R_2}V_2$$

and

$$V_{out_3} = -\frac{R_f}{R_3}V_3$$

By the superposition theorem, the output voltage is the sum of the output voltages produced by the individual input voltages.

$$V_{out} = V_{out_1} + V_{out_2} + V_{out_3}$$

Then

$$\frac{V_{out}}{R_f} = -\frac{V_1}{R_1} - \frac{V_2}{R_2} - \frac{V_3}{R_3}$$

Solving for V_{out}, we have

$$V_{out} = -R_f\left[\frac{V_1}{R_1} + \frac{V_2}{R_2} + \frac{V_3}{R_3}\right] \qquad (17\text{-}9a)$$

or

$$V_{out} = -\left[\frac{R_f}{R_1}V_1 + \frac{R_f}{R_2}V_2 + \frac{R_f}{R_3}V_3\right] \qquad (17\text{-}9b)$$

The nature of Eq. 17-9*b* shows why this circuit is often called a *scaling adder.*

One application of this circuit is its use as an *audio mixer.* Three microphones are used as the inputs to V_1, V_2, and V_3. The combined output is $-V_{out}$. Another application is the use for a control circuit. Three continuously varying control signals are fed into V_1, V_2, and V_3. Each control input is multiplied by a different factor and the combined weighted signal is the output.

If all three resistors at the input are equal

$$R_1 = R_2 = R_3 = R$$

the output becomes

$$V_{out} = -\frac{R_f}{R}(V_1 + V_2 + V_3) \tag{17-10}$$

and if R_f is also equal to R,

$$V_{out} = -(V_1 + V_2 + V_3) \tag{17-11}$$

If this circuit has n inputs and all input resistors have the value R,

$$V_{out} = -\frac{R_f}{R}(V_1 + V_2 + \cdots + V_n) \tag{17-12}$$

If each input resistor has the value R and, if with n inputs

$$R_f = \frac{R}{n}$$

then Eq. 17-12 becomes

$$V_{out} = -\frac{\left(\frac{R}{n}\right)}{R}(V_1 + V_2 + \cdots + V_n)$$

or

$$V_{out} = -\frac{V_1 + V_2 + \cdots + V_n}{n} \tag{17-13}$$

The circuit represented by Eq. 17-13 is called an *averager* since the output voltage is the average value of the input voltages.

The Noninverting Adder The circuit for the *noninverting adder* is shown in Fig. 17-8. The summing point S in Fig. 17-8 is not a virtual ground. It would be a virtual ground only if the noninverting input (+) were at ground potential. Therefore when we consider V_1 for the superposition theorem and when we short out V_2, the voltage applied to the noninverting input (+) of the op amp is determined by the voltage divider rule.

$$\frac{R_B}{R_A + R_B} V_1$$

and the resulting output voltage from V_1 is

$$V_{\text{out}_1} = \left(\frac{R_B}{R_A + R_B} V_1\right)\left(1 + \frac{R_f}{R_1}\right)$$

Similarly, the voltage at the noninverting input (+) of the op amp produced by V_2 when V_1 is shorted is

$$\frac{R_A}{R_A + R_B} V_2$$

and the resulting output voltage from V_2 is

$$V_{\text{out}_2} = \left(\frac{R_A}{R_A + R_B} V_2\right)\left(1 + \frac{R_f}{R_1}\right)$$

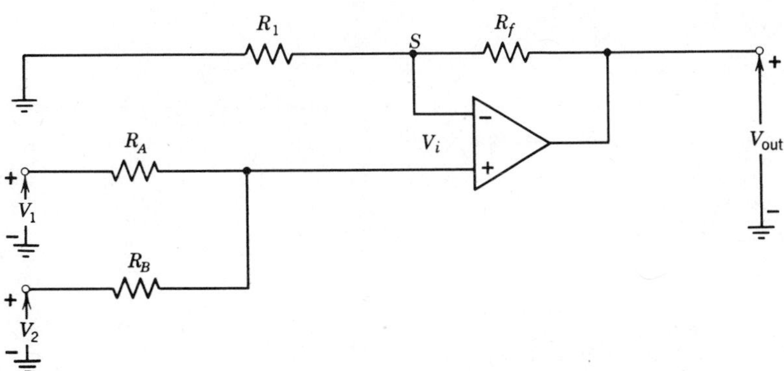

Figure 17-8 The noninverting adder.

Then, combining the results according to the superposition theorem, we have

$$V_{\text{out}} = V_{\text{out}_1} + V_{\text{out}_2}$$
$$= \left(\frac{R_B}{R_A + R_B} V_1\right)\left(1 + \frac{R_f}{R_1}\right) + \left(\frac{R_A}{R_A + R_B} V_2\right)\left(1 + \frac{R_f}{R_1}\right)$$

Collecting the terms, we have

$$V_{\text{out}} = \left(1 + \frac{R_f}{R_1}\right)\left(\frac{R_B V_1 + R_A V_2}{R_A + R_B}\right) \tag{17-14}$$

If $R_A = R_B$

$$V_{\text{out}} = \left(1 + \frac{R_f}{R}\right)\frac{V_1 + V_2}{2} \tag{17-15}$$

The Subtractor The circuit for the *subtractor* is given in Fig. 17-9. If we use the superposition theorem, we find that

$$V_{\text{out}} = V'_{\text{out}} + V''_{\text{out}}$$

where V'_{out} is the output produced by V_1 and where V''_{out} is the output produced by V_2. By Eq. 17-1

$$V'_{\text{out}} = A_v V_{\text{in}} = -\frac{R_f}{R_1} V_1$$

and by Eq. 17-6

$$V''_{\text{out}} = A_v V_{\text{in}} = \left(1 + \frac{R_f}{R_1}\right) V_2$$

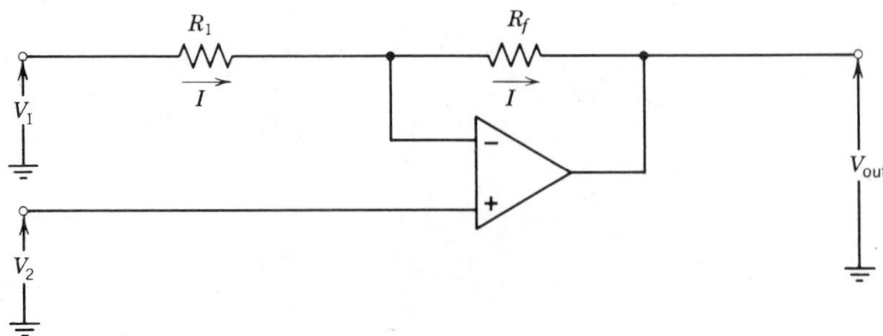

Figure 17-9 The subtractor.

Then

$$V_{out} = V'_{out} + V''_{out} = -\frac{R_f}{R_1}V_1 + \left(1 + \frac{R_f}{R_1}\right)V_2$$

or

$$V_{out} = \left(1 + \frac{R_f}{R_1}\right)V_2 - \frac{R_f}{R_1}V_1 \qquad (17\text{-}16)$$

Example 17-3

Find the value of V_{out} obtained from the given circuit.

Solution

If we consider the approach by the superposition theorem, the gain of the op amp as an inverting amplifier is

$$A_v = -\frac{R_f}{R_1} = -\frac{10{,}000\ \Omega}{5000\ \Omega} = -2 \qquad (17\text{-}1)$$

and the output voltage produced from V_1 is

$$V_{out_1} = A_v V_1 = (-2)(+5) = -10\ \text{V}$$

If we consider the other input to the circuit by the superposition theorem, the gain of the circuit as a noninverting amplifier is

$$A_v = 1 + \frac{R_f}{R_1} = 1 + \frac{10{,}000\ \Omega}{5{,}000\ \Omega} = 3 \qquad (17\text{-}6)$$

and the output voltage produced from V_2 is

$$V_{out_2} = A_v V_1 = (3)(4) = 12\ \text{V}$$

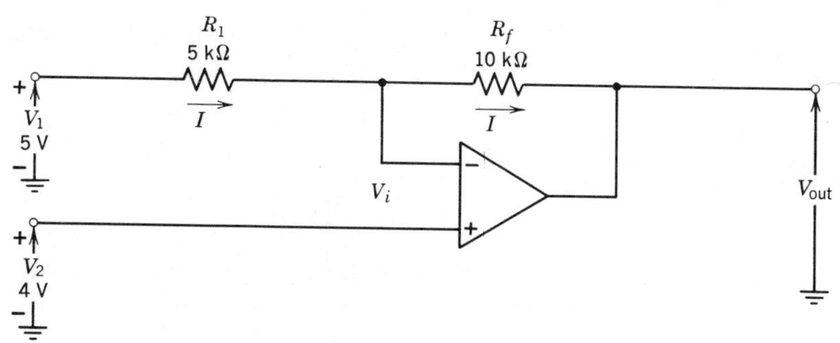

Circuit for Example 17-3.

The actual output voltage by adding the partial values is

$$V_{out} = V_{out_1} + V_{out_2} = -10 + 12 = \mathbf{+2\ V}$$

If we substitute directly into Eq. 1-16, we have

$$V_{out} = \left(1 + \frac{R_f}{R_1}\right)V_2 - \frac{R_f}{R_1}V_1 = \left(1 + \frac{10{,}000\ \Omega}{5{,}000\ \Omega}\right)4 - \frac{10{,}000\ \Omega}{5{,}000\ \Omega}5 = \mathbf{+2\ V}$$

If we take the numerical values for the circuit components used in Example 17-3 and place them into Eq. 17-16, we have

$$V_{out} = \left(1 + \frac{10{,}000\ \Omega}{5{,}000\ \Omega}\right)V_2 - \frac{10{,}000\ \Omega}{5{,}000\ \Omega}V_1$$

or

$$V_{out} = 3V_2 - 2V_1$$

Now this is the *equation* for the circuit of Example 17-3. V_1 and V_2 were voltage values in Example 17-3. They could also be voltage waveforms. In that case, we would multiply the amplitude of the V_1 waveform by 2 and subtract it point by point from the waveform obtained by multiplying the amplitude of V_2 by 3.

Let us take this basic circuit and require that it must perform the function

$$V_{out} = V_2 - V_1$$

We must divide V_2 by 3 and divide V_1 by 2. Thus, we place input voltage dividers into the input circuits that make these reductions. A method of accomplishing this is shown in Fig. 17-10. Numerical values are placed at the different points on the circuit to show how the subtraction is accomplished. Note in Fig. 17-10 that R_1 (5000 Ω) should be reduced by 50 Ω, which is the value of 100 Ω in parallel with 100 Ω if high accuracy is required.

Other subtractor-type circuits are shown in Fig. 17-11 and in Fig. 17-12.

For Fig. 17-11

$$V_{out} = \frac{R_f}{R_1}\left[\frac{1 + \dfrac{R_1}{R_f}}{1 + \dfrac{R_A}{R_B}}V_2 - V_1\right] \tag{17-17}$$

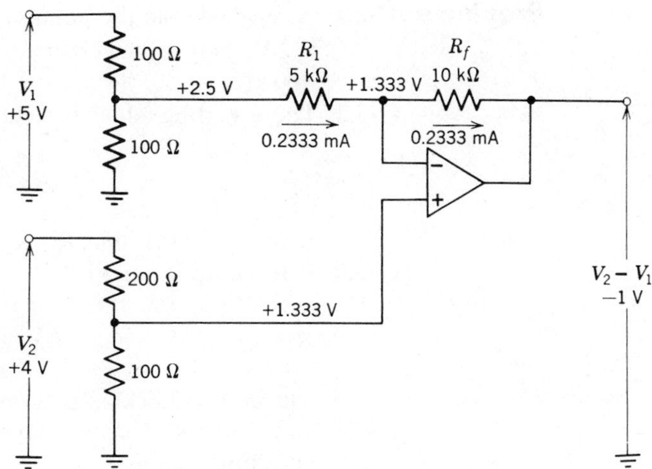

Figure 17-10 Circuit that performs $V_{out} = V_2 - V_1$.

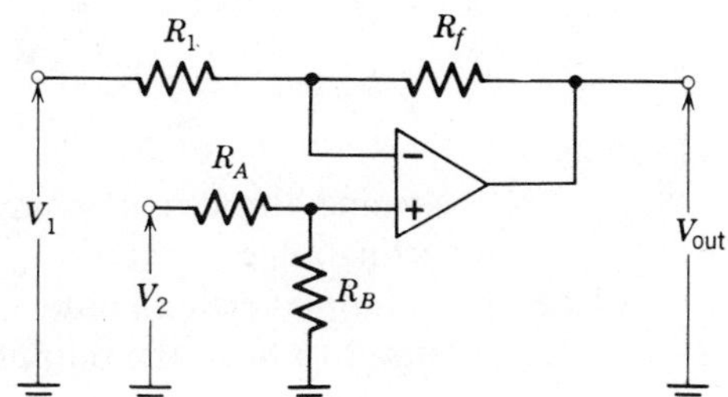

Figure 17-11 Subtractor circuit.

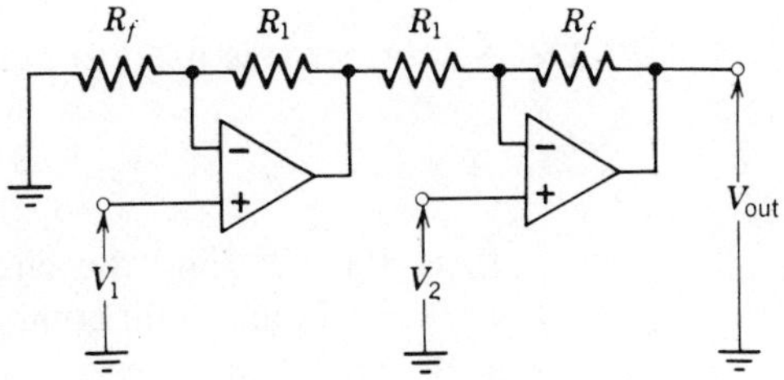

Figure 17-12 Subtractor circuit.

When $R_1 = R_A$ and $R_f = R_B$, Eq. 17-17 reduces to

$$V_{out} = \frac{R_f}{R_1}[V_2 - V_1] \tag{17-18}$$

For Figure 17-12 $V_{out} = \left[1 + \frac{R_f}{R_1}\right](V_2 - V_1)$ (17-19)

Problems

17-3.1 R_f is 100 kΩ in the noninverting amplifier (Fig. 17-5). If V_{in} is 0.2 V, plot a curve for V_{out} for a range of R_1 from 5 kΩ to 100 kΩ.

17-3.2 The inverting adder is required to perform the function

$$-V_{out} = 6V_1 + 4V_2 + 3V_3$$

If R_f is 150 kΩ, find R_1, R_2, and R_3.

17-3.3 If the output must be

$$V_{out} = 6V_1 + 4V_2 + 3V_3$$

and R_f is 330 kΩ, determine the required circuit and its components. Use an inverting adder in cascade with an inverting amplifier.

17-3.4 If the output must be

$$-V_{out} = \frac{V_1 + V_2 + V_3}{3}$$

determine the circuit components if R_f is 120 kΩ. Use the inverting adder circuit.

17-3.5 In a noninverting adder, R_f is 100 kΩ. Find the circuit required to have the output

$$V_{out} = V_1 + V_2$$

17-3.6 A subtractor is used to perform

$$V_{out} = 3V_2 - 2V_1$$

R_f is 100 kΩ. Find the circuit components required for Fig. 17-9. Find the circuit components required for Fig. 17-11 if R_B is 10 kΩ.

17-3.7 Repeat Problem 17-3.6 if the required equation is

$$V_{out} = 2V_1 - 3V_2$$

Cascading may be required.

17-3.8 Repeat Problem 17-3.6 if the required equation is

$$V_{out} = 3V_1 - 5V_2$$

17-3.9 Derive Eq. 17-17.

17-3.10 Derive Eq. 17-18.

17-3.11 Derive Eq. 17-19.

17-3.12 Derive the equation for V_{out}. What is V_o i $R_1 = R_2 = R_3$?

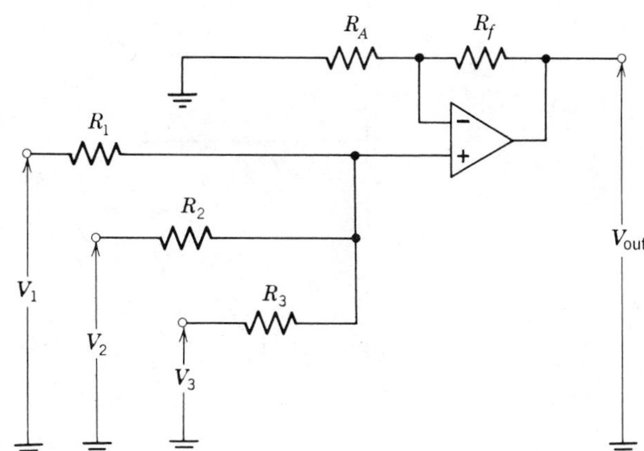

Circuit for Problem 17-3.12.

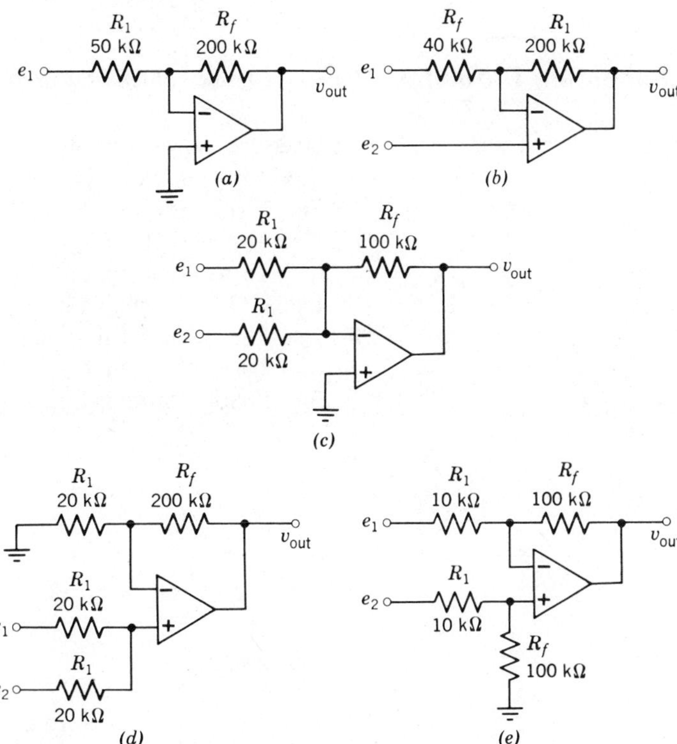

Figure 17-13 Circuits (*a*), (*b*), (*c*), (*d*), and (*e*) for Supplementary Problems for chapter 17.

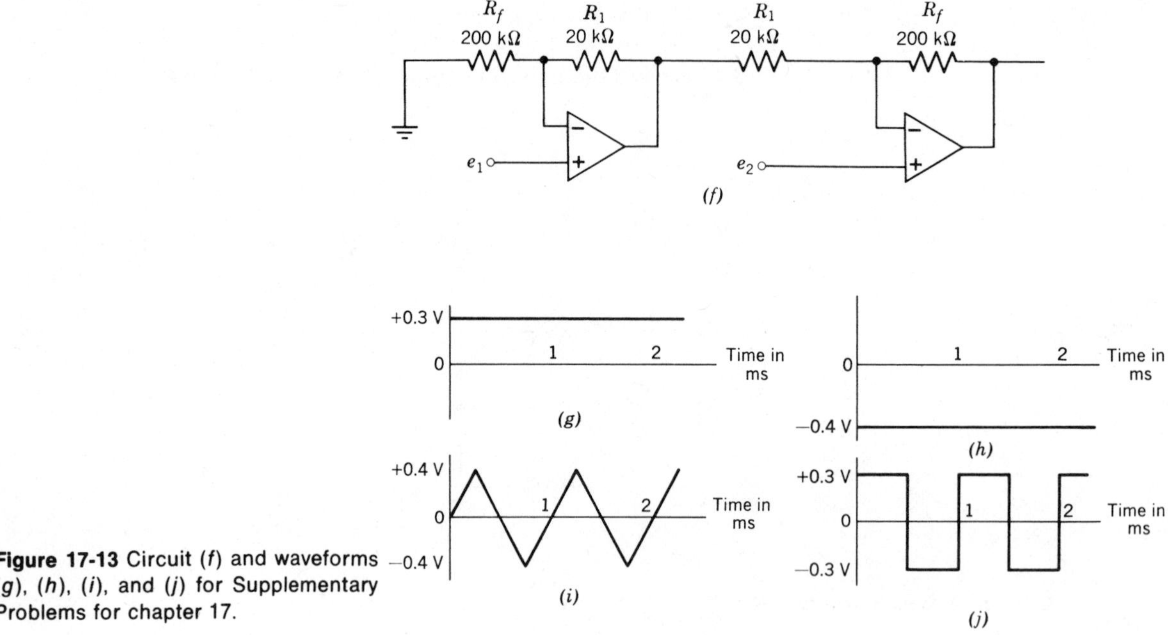

Figure 17-13 Circuit (*f*) and waveforms (*g*), (*h*), (*i*), and (*j*) for Supplementary Problems for chapter 17.

Supplementary Problems

Hint: Use peak values and then add waveforms point by point.

17-1 Use the waveform given in Fig. 17-13*g* for e_1 and use the waveform given in Fig. 17-13*h* for e_2 in each of the circuits in Fig. 17-13*a* through Fig. 17-13*f* and determine v_{out}.

17-2 Use the waveform given in Fig. 17-13*h* for e_1 and use the waveform given in Fig. 17-13*i* for e_2 in each of the circuits in Fig. 17-13*a* through Fig. 17-13*f* and determine v_{out}.

17-3 Use the waveform given in Fig. 17-13*i* for e_1 and use the waveform given in Fig. 17-13*j* for e_2 in each of the circuits in Fig. 17-13*a* through Fig. 17-13*f* and determine v_{out}.

Chapter 18 The Practical Operational Amplifier

As a result of internal unbalances in the circuitry of the operational amplifier, we find there are an offset voltage, an input bias current and an input offset current that often require compensation circuits (Section 18-1). A precision operational amplifier can oscillate under certain gain and signal frequency conditions unless proper frequency compensation is made (Section 18-2). The slew-rate specification can limit both the frequency and the available output signal level from an operational amplifier (Section 18-3).

Section 18-1 Characteristics of the Nonideal Operational Amplifier

The connections to the terminals of typical op amps are shown in Fig. 18-1. The μA741C op amp is a commercial-grade general-purpose unit. The μA741M is the version that meets military specifications. The μA777C is a commercial-

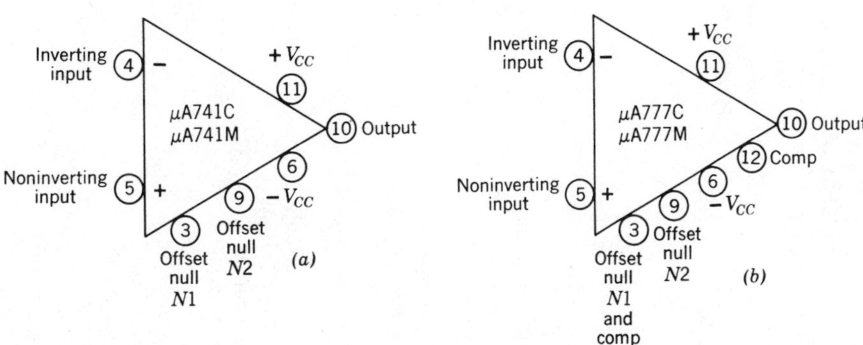

Figure 18-1 14-Pin dual-in-line-package (DIP) connections for typical operational amplifiers. (*a*) General-purpose grade. (*b*) High-performance grade. (*Courtesy Texas Instruments Inc.*)

Table 18-1
Performance Characteristics of Typical Operational Amplifiers

Letter Symbol	Name	μA741C (General-Purpose Commercial Grade)			μA741M (General-Purpose Military Grade)			μA777C (High-Performance Commercial Grade)			μA777M (High-Performance Military Grade)			Units
		Min	Typical	Max	Min	Typical	Max	Min	Typical	Max	Min	Typical	Max	
$\pm V_{CC}$	Supply Voltage			±18			±22			±22			±22	V
	Voltage from $N1$ or $N2$ to $-V_{CC}$			±0.5			±0.5			−0.5 to 2.0			−0.5 to 2.0	V
	Duration of output short circuit		Unlimited			Unlimited			Unlimited			Unlimited		
P_D	Dissipation in free air 25°C			500			500			500			500	mW
T_A	Operating free air temp.		0 to 70			−55 to 125			0 to 70			−55 to 125		°C
V_{io}	Input offset voltage		1	6		1	5		0.7	5		0.5	2	mV
αV_{io}	Temp. coeff. of V_{io}								4	30		2.5	15	μV/°C
ΔV_{io}	Offset voltage adjust range		±15			±15								mV
I_{ib}	Input bias current		80	500		80	500		25	100		8	25	nA
I_{io}	Input offset current		20	200		20	200		0.7	20		0.25	3	nA
$\alpha_{I_{io}}$	Temp. coeff. of I_{io}								20	600		6.5	150	pA/°C
$A_{v,OL}$	Open-loop gain	20	200		50	200		25	250		50	250		V/mV
r_i	Input resistance	0.3	2		0.3	2		1	2		2	10		MΩ
r_o	Output resistance		75			75			100			100		Ω
CMRR	Common-mode rejection ratio	70	90		70	90		70	95		80	95		dB
I_{os}	Short-circuit output current		±25	±40		±25	±40		±25			±25		mA
I_{CC}	Supply current		1.7	2.8		1.7	2.8		1.9	3.3		1.9	2.8	mA
t_r	Risetime		0.3			0.3			0.3			0.3		μs
C_i	Input capacitance		1.4			1.4			3			3		pF
SR	Slew rate		0.5*			0.5*			0.5**			0.5**		V/μs
SR	Slew rate								5.5***			5.5***		V/μs
	Overshoot factor		5*			5*			5**			5**		%

* $V_{in} = 10$ V; $A_v = 1$; $R_L = 2$ kΩ; $C_L = 100$ pF.
** $V_{in} = 20$ mV; $A_v = 1$; $R_L = 2$ kΩ; $C_L = 100$ pF; $C_C = 30$ pF (connected $N1$ to Comp).
*** $V_{in} = 10$ mV; $A_v = 10$; $R_L = 2$ kΩ; $C_L = 100$ pF; $C_C = 3.5$ pF (connected $N1$ to Comp).

Test waveform: V_{in}, 0, Time

***Source*:** Courtesy Texas Instruments Inc.

grade high-performance op amp. The μA777M is the version that meets military specifications. Table 18-1 compares the ratings of these four op amps in detail.

Offset Voltage In Section 12-7, Fig. 12-11, the differential amplifier was given as the basic circuit for the op amp. When the input signals to the differential amplifier are zero, the output of the differential amplifier should be zero. Unless there is an exact balance between all circuit components including the transistors, there will be a small *output offset voltage.* When a potentiometer is connected between the emitters of the differential amplifier and when the negative supply is fed into the variable arm, the unbalance can be reduced to zero. The two terminals used for the correction potentiometer on an op amp are marked *offset null* $N1$ and *offset null* $N2$. A 10-kΩ potentiometer is usually used to obtain an output null in such op amps as the μA741 and the μA777 (Fig. 18-2).

The circuit is connected as a voltage follower with noninverting input tied to ground, Fig. 18-2. A dc voltmeter or oscilloscope is connected from the output terminal to ground and the potentiometer is adjusted for zero voltage. The potentiometer should be able to adjust the null output from a positive value through zero to a negative value.

As a general practice, we assume that the op amp has been nulled before signals are applied to the circuit.

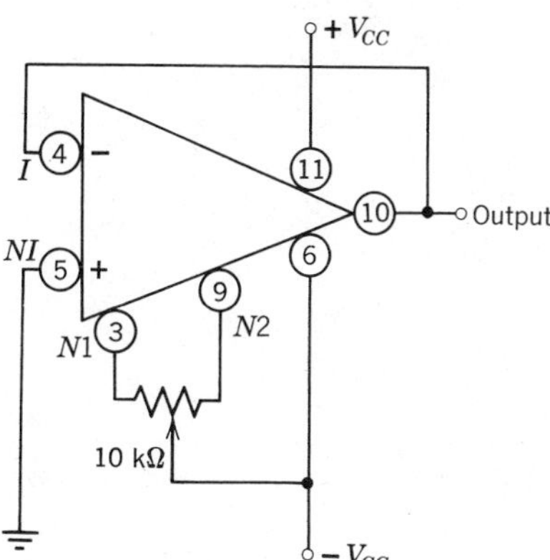

Figure 18-2 Connection used to null the output.

Input Bias Current and Input Offset Current

In the ideal op amp, the resistance between the inverting input (−) and the noninverting input (+) is an open circuit (∞Ω) and the voltage between these two terminals is zero. In an actual op amp, the internal circuit between the inverting input and the noninverting input can create small currents that are called *input bias currents*—I_{ib-} in the inverting input and I_{ib+} in the noninverting input. The combined effect of the input bias currents is called the *input offset current* I_{io}.

Let us consider the testing circuit shown in Fig. 18-3. This circuit is a voltage follower with unity gain. We assume that the op amp has been nulled previously for a minimum. The capacitors are placed in the circuit to short out noise pulses and other transient effects. R_1 equals R_2 and each resistor must be at least 10 MΩ.

Four measurements are performed:

1. Switch 1 and switch 2 are closed. The circuit is now a voltage follower. If there is an *input offset voltage* V_{io} caused by internal unbalances in the input, V_{io} will appear as an output voltage V_{out_1}, measured from the output terminal to ground.

$$V_{out_1} = V_{io}$$

2. Switch 2 is open and switch 1 is closed. Now the output voltage V_{out_2} is V_{io} plus the voltage drop in R_1 produced by the current I_{ib+}.

$$V_{out_2} = R_1 I_{ib+} + V_{io}$$

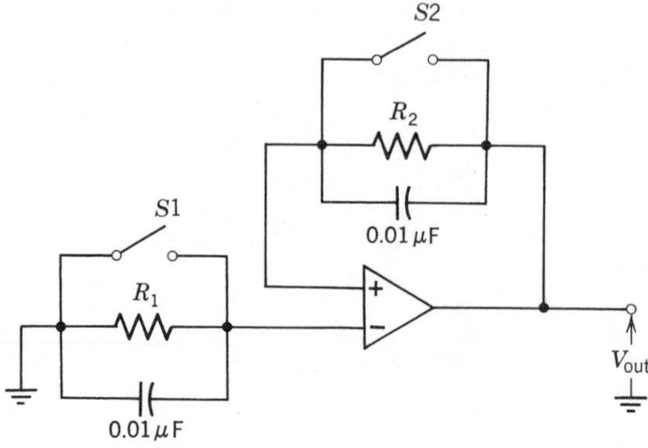

Figure 18-3 Testing circuit.

Solving for I_{ib+}, we have

$$I_{ib+} = \frac{V_{\text{out}_2} - V_{io}}{R_1}$$

3. Switch 1 is open and switch 2 is closed. Now the output voltage V_{out_3} is V_{io} plus the voltage drop in R_2 produced by I_{ib-}.

$$V_{\text{out}_3} = R_2 I_{ib-} + V_{io}$$

Solving for I_{ib-}, we have

$$I_{ib-} = \frac{V_{\text{out}_3} - V_{io}}{R_2}$$

4. When switch 1 is open and switch 2 is open, we find that the effects of I_{ib+} and I_{ib-} do not balance out. We do find a net current in the input circuit that is called the *input offset current* I_{io} that produces the output voltage V_{out_4}.

$$V_{\text{out}_4} = R_1 I_{io} + V_{io}$$

or

$$V_{\text{out}_4} = R_2 I_{io} + V_{io}$$

Solving for I_{io}, we have

$$I_{io} = \frac{V_{\text{out}_4} - V_{io}}{R_1} = \frac{V_{\text{out}_4} - V_{io}}{R_2}$$

These test procedures define the values given in Table 18-1 for V_{io}, I_{ib}, and I_{io}. It should be noted that Table 18-1 gives temperature coefficient values for V_{io} and I_{io}, $\alpha_{V_{io}}$, and $\alpha_{I_{io}}$, for high-performance op amps. Then temperature compensation can be made when precision is required for instrumentation.

The input bias current I_{ib-} divides at junction A, Fig. 18-4*a*. By using the current divider rule, we find

$$I_1 = \frac{R_f}{R_1 + R_f} I_{ib-}$$

and

$$I_2 = \frac{R_1}{R_1 + R_f} I_{ib-}$$

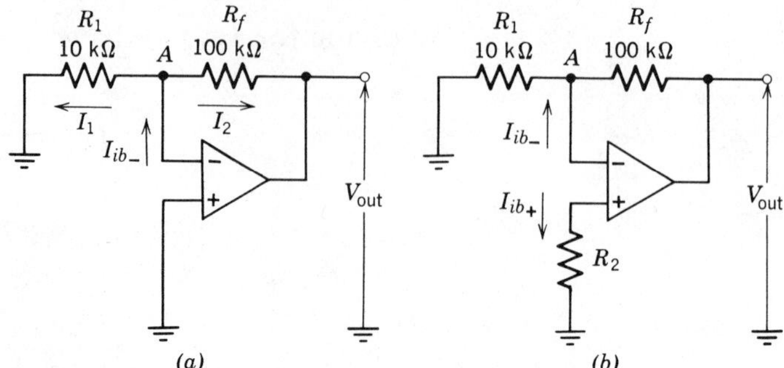

Figure 18-4 Effects of I_{ib} and I_{io}. (*a*) Without compensation. (*b*) With compensation.

The voltage drop across R_1 is I_1R_1. The circuit amplifies this voltage drop by the gain A_v.

$$V'_{out} = -A_v V_{in} = -\frac{R_f}{R_1}\left[\left(\frac{R_f}{R_1 + R_f} I_{ib-}\right)R_1\right] = -R_f \frac{R_f}{R_1 + R_f} I_{ib-}$$

The voltage drop across R_f is I_2R_f and this voltage drop is the output voltage V''_{out}.

$$V''_{out} = -I_2R_f = -\left(\frac{R_1}{R_1 + R_f} I_{ib-}\right)R_f = -R_1 \frac{R_f}{R_1 + R_f} I_{ib-}$$

The total output voltage produced by the input bias current I_{ib-} is

$$V_{out} = V'_{out} - V''_{out} = -R_f \frac{R_f}{R_1 + R_f} I_{ib-} - R_1 \frac{R_f}{R_1 + R_f} I_{ib-} = -R_f I_{ib}$$

or

$$|V_{out}| = R_f I_{ib-} \tag{18-1}$$

The voltage at junction A, V_A, caused by the input bias current I_{ib-} is

$$V_A = \frac{R_1 R_f}{R_1 + R_f} I_{ib-} \tag{18-2}$$

If we call R_2 the equivalent of R_1 and R_f in parallel,

$$V_A = R_2 I_{ib-} \tag{18-3}$$

Example 18-1
Determine the maximum output voltage caused by using the μA741C op amp and the μA777M op amp in the amplifier circuit of Fig. 18-4*a*.

Solution
For the μA741C, we have from Table 18-1

$$I_{ib,\max} = 500 \text{ nA} = 500 \times 10^{-9} \text{ A}$$

Then by Eq. 18-1

$$|V_{out}| = R_f I_{ib} = (100{,}000\ \Omega) \times (500 \times 10^{-9} \text{ A}) = 0.050 \text{ V} = \mathbf{50\ mV}$$

For the μA741M, we have from Table 18-1

$$I_{ib,\max} = 25 \text{ nA} = 25 \times 10^{-9} \text{ A}$$

Then by Eq. 18-1

$$|V_{out}| = R_f I_{ib} = (100{,}000\ \Omega) \times (25 \times 10^{-9} \text{ A}) = 0.0025 \text{ V} = \mathbf{2.5\ mV}$$

In order to attempt to compensate for I_{ib}, a resistor R_2 equal to the parallel combination of R_1 and R_f is placed between the noninverting input and the ground return (Fig. 18-4*b*).

$$\boxed{R_2 = \frac{R_1 R_f}{R_1 + R_f}} \tag{18-4}$$

Unfortunately, the values of I_{ib} in each lead are not equal. The difference is I_{io}, the *input offset current*. Now we must use I_{io} in Eq. 18-1 to determine the effect of I_{io} on the output voltage when we compensate the circuit with R_2 between the noninverting input and ground.

Example 18-2
Determine the compensation resistor required for the circuit of Fig. 18-4*b*. Determine the maximum output voltage for the μA741C and for the μA777M op amps in this circuit.

Solution

The value of the compensation resistor is determined from Eq. 18-4.

$$R_2 = \frac{R_1 R_f}{R_1 + R_f} = \frac{10{,}000\ \Omega \times 100{,}000\ \Omega}{10{,}000\ \Omega + 100{,}000\ \Omega} = \mathbf{9091\ \Omega} \qquad (18\text{-}4)$$

For the μA741C, we have from Table 18-1

$$I_{io,\max} = 200\ \text{nA} = 200 \times 10^{-9}\ \text{A}$$

Then by Eq. 18-1

$$|V_{\text{out}}| = R_f I_{io} = (100{,}000\ \Omega) \times (200 \times 10^{-9}\ \text{A}) = 0.020\ \text{V} = \mathbf{20\ mV}$$

For the μA777M, we have from Table 18-1

$$I_{io,\max} = 3\ \text{nA} = 3 \times 10^{-9}\ \text{A}$$

Then, by Eq. 18-1

$$|V_{\text{out}}| = R_f I_{io} = (100{,}000\ \Omega) \times (3 \times 10^{-9}\ \text{A}) = 0.0003\ \text{V} = \mathbf{300\ \mu V}$$

The results of Example 18-1 and Example 18-2 show why compensation for I_{io} is normally required for the op amp circuits that were developed in Sections 17-2 and 17-3. Also, the results of these two examples show the differences between the two op amps—why one is called *general-purpose* and the other is called *high-performance.*

Typical compensation circuits are shown in Fig. 18-5. The circuit for the μA741C given in Fig. 18-5*a* shows both the nulling potentiometer and the compensation for I_{ib}. The compensation resistor for the inverting amplifier (Fig. 18-5*b*) requires that all the input resistors are considered. The compensation circuits of Fig. 18-5*c* and Fig. 18-5*d* show that two compensations are part of the circuit. The compensation resistor for I_{ib} as determined by Eq. 18-1 is used. Also, there is a variable dc voltage applied to the circuit to develop an exact compensation for I_{io}. Each circuit must be individually adjusted to set V_{out} to zero without input signals. A temperature compensation can be incorporated in the R_4–R_5 circuit to yield an exact compensation over a range of temperatures.

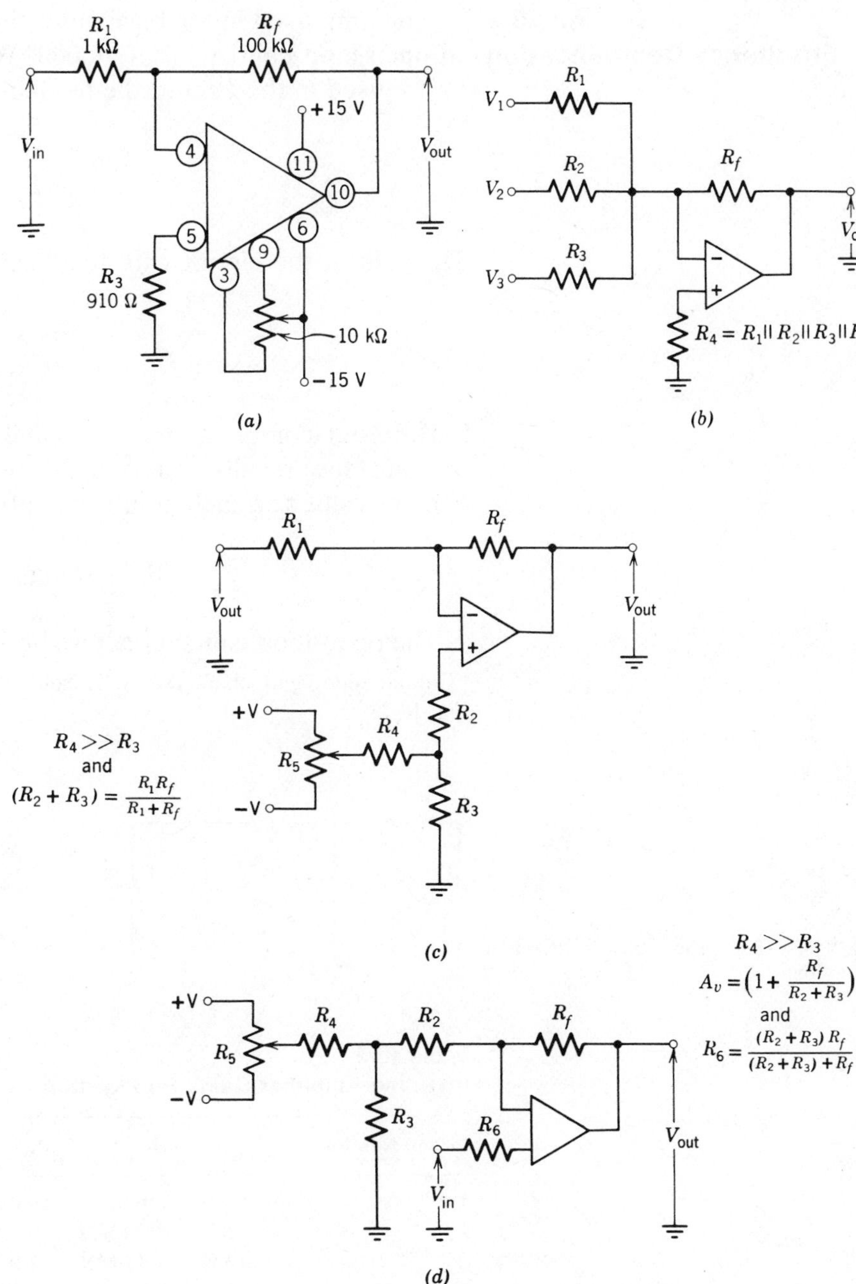

Figure 18-5 Compensation circuits. (*a*) Compensation circuit for the μA741C op amp. (*b*) Compensation for the inverting adder. (*c*) Compensation for the inverting amplifier. (*d*) Compensation for the noninverting amplifier.

Section 18-2 Frequency Compensation

The op amp used in the basic inverting amplifier, Fig. 18-6, has an open-loop gain $A_{v,OL}$ of 10,000. When the resistors R_1 and R_f are used in the circuit, the feedback is negative and is given by

$$\beta_f = -\frac{R_1}{R_f} \tag{17-1}$$

The gain of the circuit with feedback is given by

$$A_v = \frac{A_{v,OL}}{1 - \beta_f A_{v,OL}} \tag{17-2}$$

Different combinations of R_1 and R_f are used to calculate A_v and the results listed in Table 18-2. Additionally, the decibel value for each gain is calculated and listed by using

$$\text{dB} = 20 \log_{10} A_v \tag{18-5}$$

The open-loop gain characteristic* is shown in the form of a

*The open-loop gain characteristic is called the *uncompensated response* on Fig. 18-7.

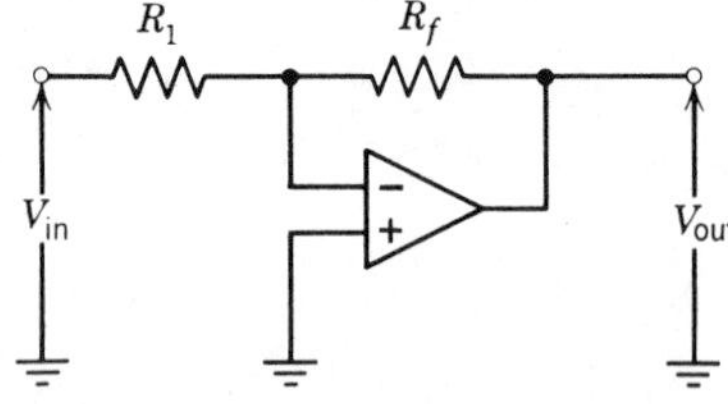

Figure 18-6 The basic inverting amplifier.

Table 18-2
Inverting Amplifier Gains for Fig. 18-6

Circuit Number	R_1	R_f	A_v	dB
1	1 kΩ	Open	10 000	+80
2	1 kΩ	4.26 MΩ	3 162	+70
3	1 kΩ	1.11 MΩ	1 000	+60
4	1 kΩ	326 kΩ	316	+50
5	10 kΩ	316 kΩ	31.6	+30
6	10 kΩ	100 kΩ	10	+20
7	10 kΩ	31.6 kΩ	3.16	+10

Bode gain plot in Fig. 18-7. The manufacturer usually furnishes an open-loop response curve as a part of the specifications for the particular op amp. The gain is flat at +80 dB from dc to 1 kHz, point *A*. The frequency at point *A*, f_A, is the first 3-dB corner frequency. The response rolls off at the rate of 20 dB/decade (6 dB/octave) to point *B*. Point *B*, at 3 kHz, is the second 3-dB corner frequency f_B. Now the response rolls off at the rate of 40 dB/decade (12 dB/octave) to point *C*. Point *C*, at 10 kHz, is the third 3-dB corner frequency f_C. From point *C*, the response rolls off at the rate of 60 dB/decade (18 dB/octave). The roll-off continues at this rate toward unity gain (0 dB) at 68 kHz, f_D. The gain values tabulated in Table 18-2 are shown as horizontal dashed lines on Fig. 18-7.

The fact that this op amp has three break or corner frequencies, f_A, f_B, and f_C, indicates that the internal circuit of the op amp has three stages, each of which has a different high-frequency break point.

At all frequencies below $0.1 f_A$, the phase relation between V_{in} and V_{out} is 180° (Fig. 18-8*a*). From the results of Section 15-2, as the frequency increases from $0.1 f_A$ toward f_A, the first break frequency, a lagging phase angle θ, is introduced.

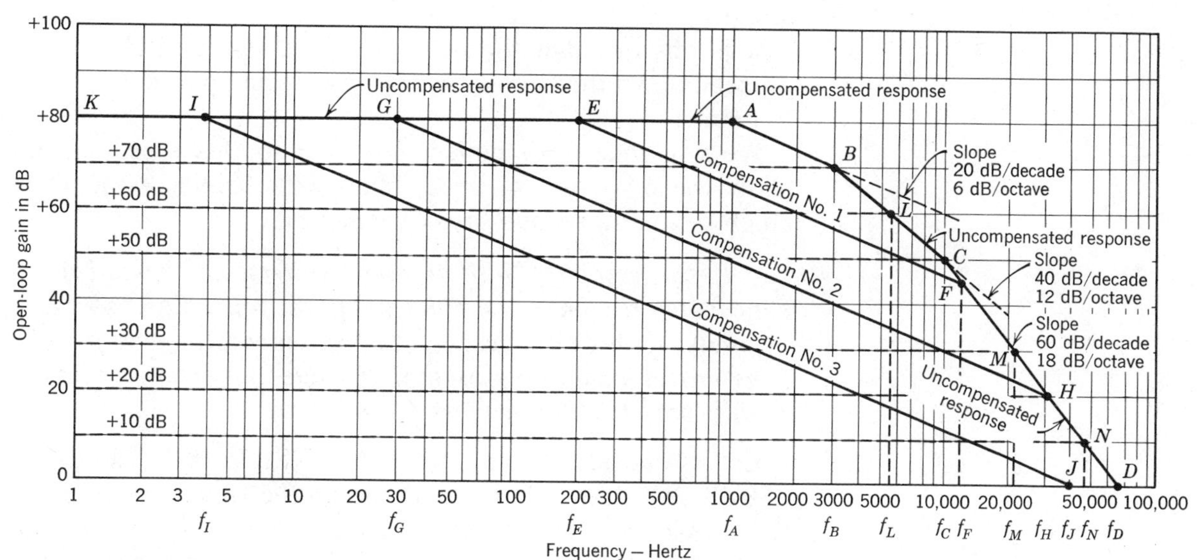

Figure 18-7 Frequency response.

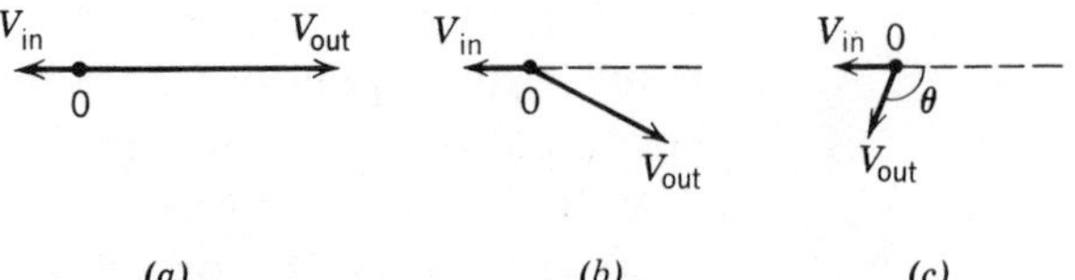

Figure 18-8 Phasor diagrams. (*a*) At very low frequencies. (*b*) At f_A. (*c*) At f_B.

At f_A, the angle becomes 45° (Fig. 18-8*b*). At frequencies higher than f_A, the phase angle increases toward 90°. However, since we have a second break frequency at f_B, we introduce a further phase shift caused by f_B. At f_B, the total lagging phase angle is the angle generated by f_A plus the 45° caused by f_B. Now the total lagging phase angle must be greater than 90° (Fig. 18-8*c*).

If we take V_{out} in Fig. 18-8*c* and break it into components in-phase with and out-of-phase with V_{out}, we see that the in-phase component is in the same direction as V_{in}. In Fig. 18-8*a* and in Fig. 18-8*b* the output has no component that can be in phase with V_{in}. In these cases the circuit is completely stable; there can be no condition of a positive feedback that could cause an oscillation. When there is an in-phase component (Fig. 18-8*c*), we have a positive feedback condition and an oscillation is very likely to occur. We must avoid this situation to prevent the circuit from becoming unstable (oscillation).

It is obvious that the op amp becomes less stable as we proceed down the open-loop response from point *B* to point *C* to point *D*.

We will take a simplified approach to this topic and not allow the operation of an op amp with feedback to extend much beyond a second break frequency.

> *If the response with feedback intersects the uncompensated response at a point where the roll-off has a slope greater than* 40 *dB/decade* (12 *dB/octave*), *the circuit will be unstable and may oscillate because the internal phase shift may be* 180° *or more.*

Let us consider Circuit No. 3 from Table 18-2. The gain is +60 dB. We draw a horizontal line on the frequency response (Fig. 18-7) at +60 dB. The response is flat to point *L* at 5500 Hz, f_L. However the slope of the uncompensated response is 40 dB/decade (12 dB/octave) at point *L*. Therefore this circuit is unstable and may oscillate. Figure 18-7 shows that any circuit that has a gain less than +70 dB may become unstable and oscillate.

In order to use the op amp, we must *compensate* the op amp against this unstable condition. Let us assume we can compensate the op amp by any one of three different external circuits. The compensation circuit introduces a new first corner frequency at a lower frequency than point *A* on the open-loop response (Fig. 18-7). Also the roll-off introduced by the compensating circuit must be 20 dB/decade (6 dB/octave). The responses of the three compensation circuits are drawn on Fig. 18-7.

If we reconsider Circuit No. 3 from Table 18-2, which yields a gain of +60 dB, we find:

A If the circuit is uncompensated (Fig. 18-7), it is unstable.
B If we use Compensation Network No. 3 (Fig. 18-7), the gain is +60 dB at all frequencies up to 38 Hz.
C If we use Compensation Network No. 2 (Fig. 18-7), the gain is +60 dB at all frequencies up to 310 Hz.
D If we use Compensation Network No. 1 (Fig. 18-7), the gain is +60 dB at all frequencies up to 2000 Hz.

Now let us consider Circuit No. 7 (Table 18-2) which has a gain of +10 dB. We find from examining Fig. 18-7 that:

A If the circuit is uncompensated, the circuit is unstable, since the +10-dB gain extends to point *N*, 47 kHz.
B If we use Compensation Circuit No. 1, the +10-dB gain still extends to point *N* and the circuit is unstable.
C If we use Compensation Circuit No. 2, the +10-dB gain still extends to point *N* and the circuit is unstable.
D If we use Compensation Circuit No. 3, the +10-dB gain extends to 13 kHz and the circuit is stable.

In Chapter 15 we studied the method of displaying the gain response by a Bode plot. The Bode plot for the R_x-C_x network given in Fig. 18-9*a* has the Bode gain response shown in Fig. 18-9*b*. The corner frequency is ω_x or f_x where

$$\omega_x = \frac{1}{R_x C_x} \text{ rad/s} \quad \text{or} \quad f_x = \frac{1}{2\pi R_x C_x} \text{ Hz} \qquad (18\text{-}6)$$

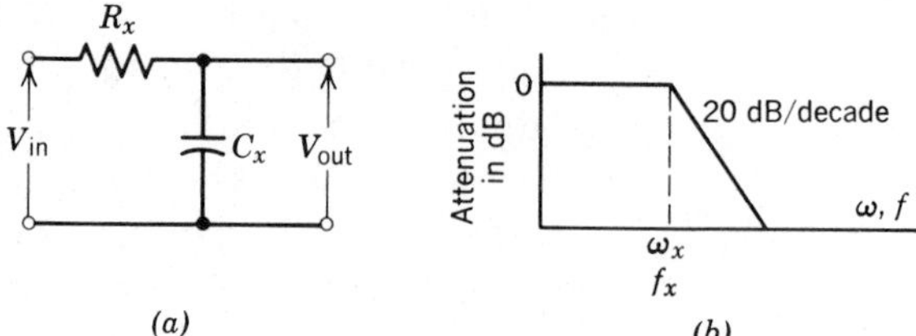

Figure 18-9 R-C network. (*a*) Circuit. (*b*) Bode response.

When this circuit is applied to the op amp, the circuit is called a *lag-phase compensation.*

The need for compensation is seen from the open-loop response characteristic for the op amp given in Fig. 18-7. If we introduce a lag-phase compensation at point *I*, *G*, or *E* on this characteristic, we deliberately introduce a 20-dB/decade (6-dB/octave) roll-off at each of these points (*I*, *G*, or *E*). The roll-off reduces the frequency response but we find we can avoid an unstable condition where the amplifier may oscillate. The circuits shown in Fig. 18-10 accomplish this compensation. It is evident from Eq. 18-6 that the time constant of Compensation No. 3 on Fig. 18-7 is greater than the time constant of Compensation No. 2. Also the time constant of Compensation No. 2 is greater than the time constant of Compensation No. 1.

The recommended compensation method is part of the manufacturer's specification sheet for a particular op amp. The μA741 op amp is internally compensated and does not require an external circuit. The typical condition data for the

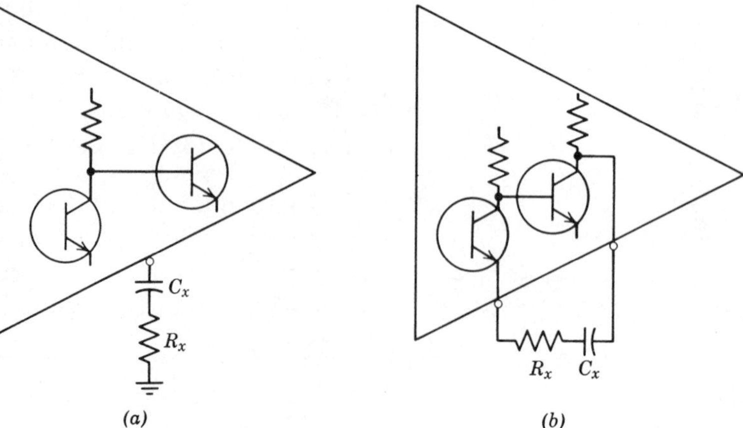

Figure 18-10 Lag-phase compensation.

μA741 op amp show that the open-loop gain is 200,000 to 7 Hz. The gain rolls off at a rate of 20 dB/decade (6 dB/octave) to unity gain (0 dB) at 1 MHz. The μA777 op amp requires the external compensation methods suggested in the data sheets.

Problems Refer to Fig. 18-7 for all problems.

18-2.1 If the gain with feedback at 2 Hz is +70 dB, what compensation circuits insure stability?
18-2.2 Repeat Problem 18-2.1 for a +50-dB gain.
18-2.3 Repeat Problem 18-2.1 for a +30-dB gain.
18-2.4 Repeat Problem 18-2.1 for a +20-dB gain.

Section 18-3 Slew Rate

The *slew rate, SR,* defines the maximum rate of change of output voltage that an op amp can accept because of a charge–discharge capacitive effect within the op amp. If the rate of change of the input signal is greater than the slew rate, the op amp circuit cannot produce an output voltage that "keeps up" with the input signal. The units of the slew rate are given in the specification sheets of an op amp as volts per microsecond, V/μs.

A square wave signal v_{in} (Fig. 18-11) is applied to an op amp circuit that has a gain of unity. The output voltage v_{out} rises from point A to point B in a finite time determined by the slew rate. The output voltage may overshoot the final value at point B by a small amount.

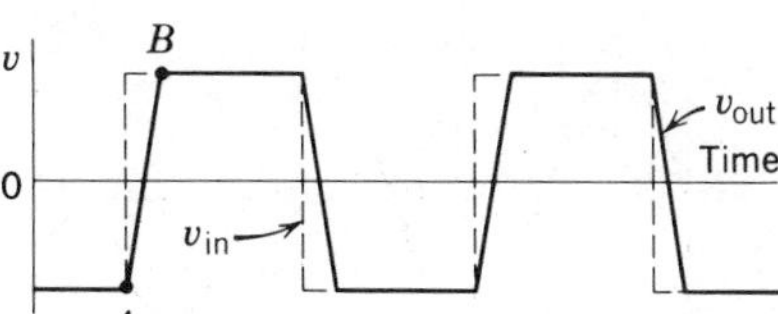

Figure 18-11 Effect of slew rate.

Example 18-3

The input signal v_{in} to an op amp circuit is a square wave, Fig. 18-11. The peak values of the input square wave are ±6 V. The gain of the op amp circuit is unity and the value of SR is 0.5 V/μs. Determine the time for the output to rise from point A to point B.

Solution

The total voltage change from point A to point B is 12 V. Then the time required for the change in the output is

$$\frac{V_{\text{out, peak-peak}}}{SR} = \frac{12\text{ V}}{0.5\text{ V}/\mu\text{s}} = \mathbf{24\ \mu s}$$

A sinusoidal input signal v_{in} is shown in Fig. 18-12. The slope of the sinusoid, dv/dt, is zero at points B, D, F, and H. The slope has a maximum positive value at points A, E, and I. The slope has a maximum negative value at points C and G. If we plot these values and values for intermediate points, we have the waveform shown in Fig. 18-12b.

If we take a mathematical approach, the equation for the input voltage waveforms is

$$v_{\text{out}} = V_{\text{out,max}} \sin \omega t$$

When the derivative is taken by methods of calculus, we have

$$\frac{dv_{\text{out}}}{dt} = \omega V_{\text{out,max}} \cos \omega t = 2\pi f V_{\text{out,max}} \cos \omega t \qquad (18\text{-}7)$$

When the slope is a maximum positive value, $\cos \omega t$ must be numerically $+1$. When the slope is a maximum negative value, $\cos \omega t$ must be numerically -1. In either case the

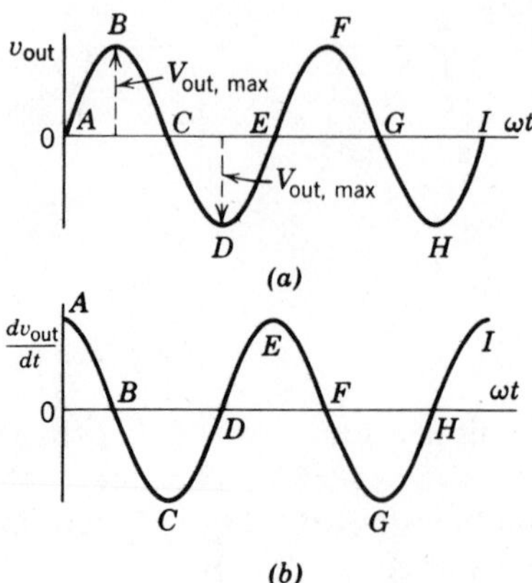

Figure 18-12 Sinusoidal voltage waveforms. (a) Input voltage waveform. (b) The derivative of the input voltage waveform.

magnitude of the maximum value of the slope is

$$\omega V_{\text{out,max}} = 2\pi f V_{\text{out,max}}$$

In the op amp, the maximum allowable slope is determined by the slew rate. Therefore

$$\boxed{SR = \omega V_{\text{out,max}} = 2\pi f V_{\text{out,max}}} \tag{18-8}$$

Solving this equation for frequency, we find that the maximum sinusoidal frequency that can be used for a particular value of $V_{\text{out,max}}$ is

$$f_{\text{max}} = \frac{SR}{2\pi V_{\text{out,max}}} \tag{18-9a}$$

or

$$\omega_{\text{max}} = \frac{SR}{V_{\text{out,max}}} \tag{18-9b}$$

Example 18-4

Consider an op amp circuit with feedback having a gain of +10 dB. The frequency response is given in Fig. 18-7. Compensation No. 3 is used. If the *SR* is 0.5 V/μs and if the output signal level has a peak value of 10 V, what is the highest frequency that can be reproduced by the circuit without distortion?

Solution

Using Eq. 18-9*a*

$$f_{\text{max}} = \frac{SR}{2\pi V_{\text{out,max}}} = \frac{0.5\ \text{V}/\mu\text{s}}{2\pi \times 10\ \text{V}} = \frac{0.00796}{10^{-6}} = \mathbf{7.96\ kHz}$$

Example 18-5

Use the data given in Example 18-4. What is the maximum possible value of $V_{\text{out,max}}$ if the full flat frequency response at +10-dB gain is to be realized?

Solution

Figure 18-7 shows that the +10-dB horizontal gain line intersects Compensation No. 3 at 12,500 Hz. This is the value of f_{max} to be used in Eq. 18-9*a*.

$$f_{max} = \frac{SR}{2\pi V_{out,max}}$$

$$12{,}500 = \frac{0.5\ \text{V}/\mu\text{s}}{2\pi V_{out,max}} = \frac{0.5}{2\pi V_{out,max} \times 10^{-6}}$$

$$V_{out,max} = \frac{0.5}{2\pi \times 12{,}500 \times 10^{-6}} = \mathbf{6.37\ V}$$

Example 18-4 and Example 18-5 show that the response curves given in Fig. 18-7 cannot be used without consideration of the slew rate. The intersection of the compensation curve with the horizontal gain line gives the highest frequency that can be obtained at that gain level. However, there is a definite limit to the output signal level at this frequency that is less than the maximum peak-to-peak output of the op amp. If the maximum allowable peak-to-peak output voltage of the op amp must be realized, the maximum permissible frequency can be calculated from the slew rate specification.

Problems Data for Problems 18-3.1 through 18-3.3: The slew rate is 0.5 V/μs. The output saturation voltage for the op amp is ±13 V. The op amp has the frequency response shown in Fig. 18-7.

18-3.1 What is the maximum frequency at which the full output voltage for a sinusoidal signal, 26 V peak to peak, can be obtained without distortion caused by the slew rate specification?

18-3.2 Plot a curve of maximum possible sinusoidal signal output voltage without distortion caused by slew rate against frequency for the op amp in a circuit with 0-dB gain. Compensation No. 3 is used.

18-3.3 Can the op amp using compensation No. 3 and having a gain of +10 dB be used to obtain an undistorted sinusoidal output signal of 20 V peak to peak at 10 kHz?

18-3.4 An op amp is in a circuit with a gain of +40 dB. Slew rate limits the output to 6 V at 20 kHz. The external resistors R_1 and R_f are changed to yield a gain of +20 dB. Now, at what frequency is the output voltage limited to 6 V?

Chapter 19 Applications of the Operational Amplifier

The integrator (Section 19-1) and the differentiator (Section 19-2) are applications suited to the operational amplifier. Representative nonlinear applications for the operational amplifier that are covered in Section 19-3 are: the ideal rectifier, the ideal full-wave rectifier (the absolute value circuit), the comparator, and the Schmitt trigger. The chapter concludes with an examination of a typical operational amplifier designed for use as a complete audio amplifier (Section 19-4).

Section 19-1 The Integrator

In the first course in dc and ac circuit analysis, we defined capacitance as

$$C \equiv \frac{Q}{V} \text{ farad} \qquad (19\text{-}1a)$$

where Q is the charge on the capacitor and V is the voltage across the capacitor. If Eq. 19-1a is solved for V, we have

$$V = \frac{1}{C}Q \qquad (19\text{-}1b)$$

The charge Q on the capacitor is the total accumulation of current times time in the capacitor. This concept is represented mathematically by using the integration symbol, $\int$, as

$$Q = \int i\,dt \qquad (19\text{-}1c)$$

If we substitute Eq. 19-1c into Eq. 19-1b, we have, using an

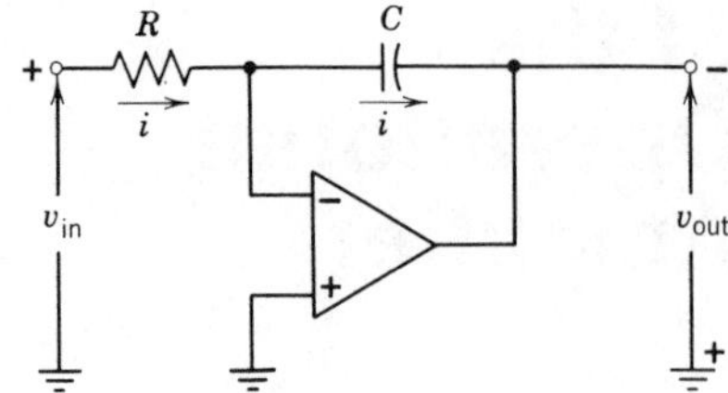

Figure 19-1 The integrator.

instantaneous voltage v for the voltage

$$v = \frac{1}{C} \int i \, dt \tag{19-2}$$

The circuit for the *integrator* is given in Fig. 19-1. The current i in the resistor R is

$$i = \frac{v_{in}}{R}$$

The voltage across the capacitor is v_{out}. Then, substituting into Eq. 19-2, we have

$$-v_{out} = \frac{1}{C} \int \frac{v_{in}}{R} \, dt$$

or

$$\boxed{v_{out} = -\frac{1}{RC} \int v_{in} \, dt} \tag{19-3}$$

We introduce the minus sign because the op amp is an inverting amplifier.

Example 19-1

The integrator, Fig. 19-1, has 1 μF for C and 100 kΩ for R. The input to the integrator is the ±10-V, 250-Hz square wave shown in Fig. 19-2*a*. Determine the output voltage waveform.

Solution

The time constant of the circuit, RC, is

$$RC = (10^{-5}\ \Omega) \times (1 \times 10^{-6}\ \text{F}) = 0.1\ \text{s}$$

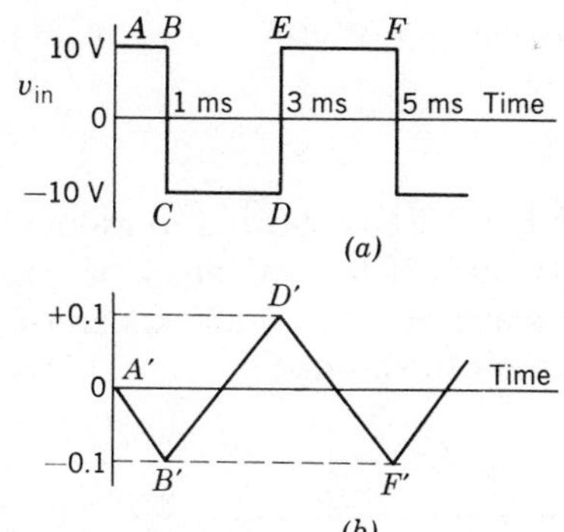

Figure 19-2 Integrator waveforms. (*a*) Input voltage waveform. (*b*) Output voltage waveform.

Substituting the time constant value into Eq. 19-3, we have

$$v_{\text{out}} = -\frac{1}{RC}\int v_{\text{in}}\,dt = -10\int v_{\text{in}}\,dt$$

The area under the curve is $\int v_{\text{in}}\,dt$. If we start at A and go toward B on the square wave, the area increases linearly. The final area at B is

$$\int v_{\text{in}}\,dt = (10\text{ V}) \times (0.001\text{ s}) = 0.01$$

Then

$$v_{\text{out}} = -10\int v_{\text{in}}\,dt = (-10) \times (0.01) = -0.1\text{ V}$$

Thus the voltage changes linearly from zero at A' to -0.1 V at B'.

The area from C to D is negative and the magnitude of the area increases linearly from C to D. The total area from C to D is

$$-(10\text{ V}) \times (0.002\text{ s}) = -0.02$$

Then

$$v_{\text{out}} = -10\int v_{\text{in}}\,dt = (-10) \times (-0.02) = +0.2\text{ V}$$

At B', v_{out} is -0.1 V. The linear change of v_{out} from B' to D' is $+0.2$ V. Thus the voltage at D' is

$$-0.1 + 0.2 = +0.1\text{ V}$$

The area under the input voltage waveform increases linearly from E to F. This increasing area causes the voltage to fall linearly from D' to F'.

Thus, if a square wave is fed into the input of an integrator, the output voltage waveform is the triangular waveform shown in Fig. 19-2*b*.

Example 19-2

The input to the integrator circuit of Fig. 19-1 is shorted to ground. The op amp has an offset voltage of 15 mV. If the op amp saturates at ±13 V, what is the output voltage waveform? R is 100 kΩ and C is 1 μF. How long does it take the circuit to saturate?

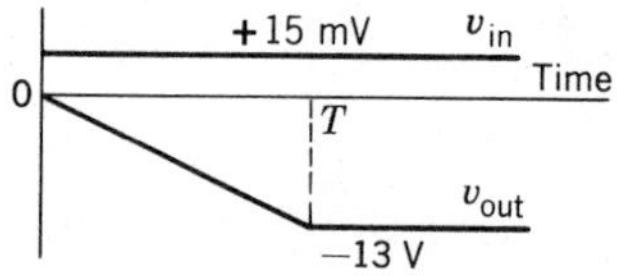

Figure 19-3 Effect of an offset voltage on the operation of an integrator.

Solution

The offset voltage waveform is the horizontal line shown in Fig. 19-3. As time increases from zero, the area under the input voltage waveform increases linearly with time. The output voltage is given by Eq. 19-3.

$$v_{out} = -\frac{1}{RC}\int v_{in}\,dt = -10\int v_{in}\,dt$$

Thus the magnitude of the output voltage increases linearly with time until saturation occurs at time T.

$$-13 = (-10) \times (0.015\,T)$$

Solving for T, we find

$$T = \mathbf{86.7\ s}$$

Example 19-2 shows that any offset voltage present in an integrator eventually saturates the op amp. As a result, in a practical integrator we must provide some means for the discharge of the capacitor. The short-circuiting switch in Fig. 19-4*a* removes the charge produced by the offset voltage. The compensating resistor R_S must be used to minimize the effects of the offset currents. A resistor R_C can be used (Fig. 19-4*b*), but its use adversely affects the operation of the circuit at low frequencies.

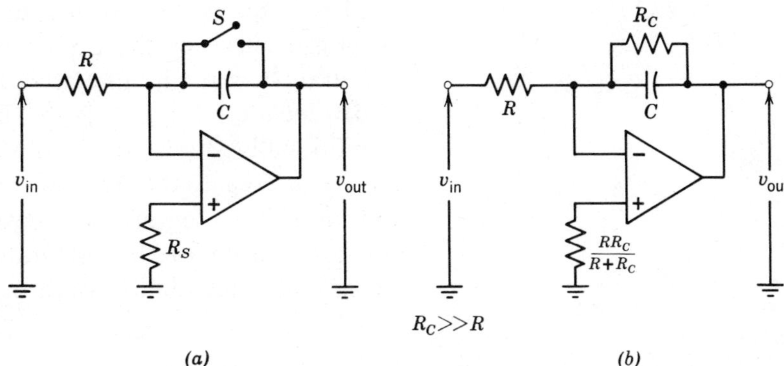

Figure 19-4 Methods of discharging the capacitor in an integrator. (*a*) Discharge switch. (*b*) Discharge resistor.

If we apply Eq. 17-1 ($A_v = -R_f/R_1$) to the circuit of the integrator (Fig. 19-5*a*), we have

$$A_v = -\frac{\left(-j\frac{1}{2\pi fC}\right)}{R} = j\frac{1}{2\pi fRC} \tag{19-4}$$

Equation 19-4 shows that the gain of the circuit decreases as the frequency increases. If we express the gain in decibels, the gain in decibels falls off at the rate of the 20 dB/decade (6 dB/octave).

The magnitude of the gain A_v from Eq. 19-4 is unity at the frequency f_2 when

$$2\pi f_2 RC = 1$$

Then

$$f_2 = \frac{1}{2\pi RC} \tag{19-5a}$$

or

$$\omega_2 = \frac{1}{RC} \tag{19-5b}$$

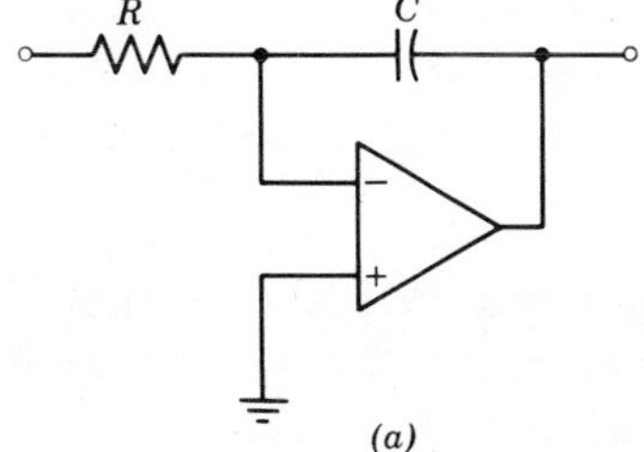

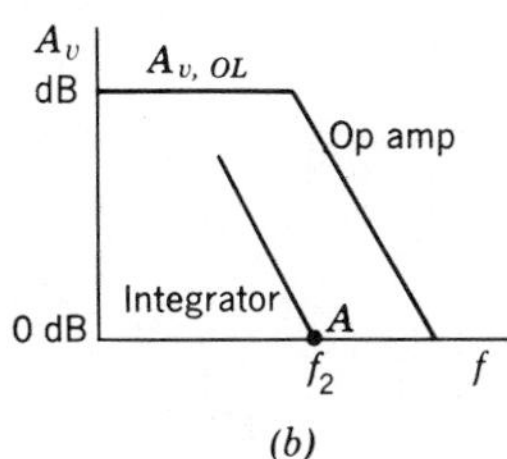

Figure 19-5 The integrator. (*a*) Circuit. (*b*) Frequency response.

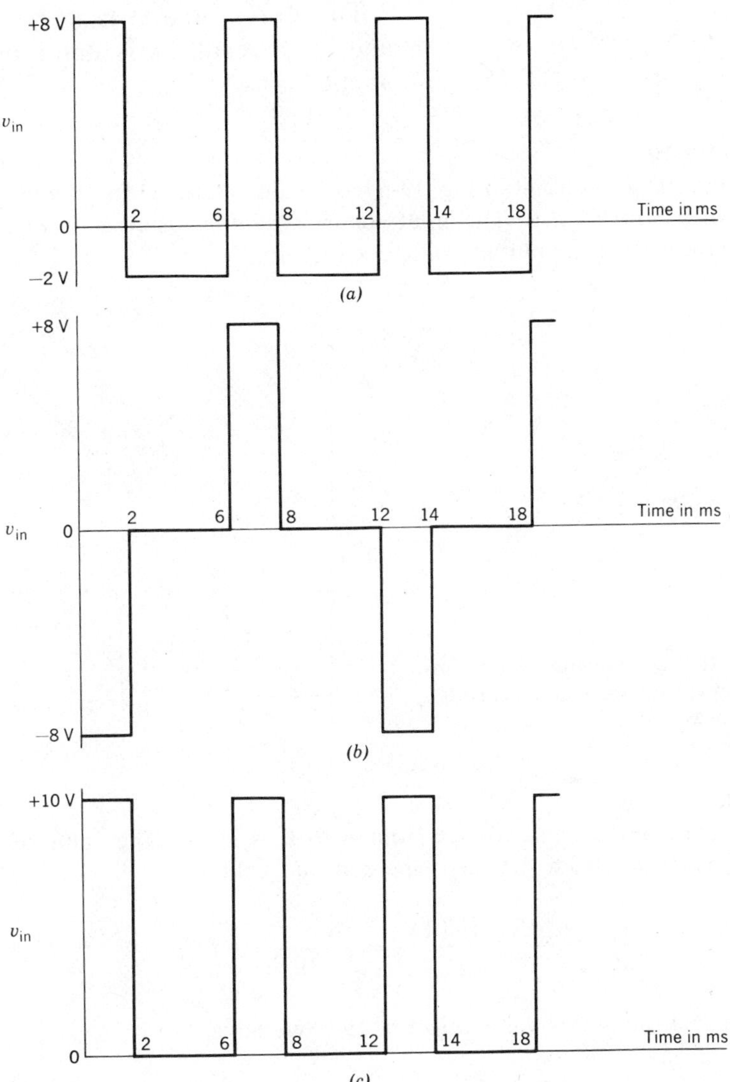

(*a*) Waveform for Problem 19-1.1. (*b*) Waveform for Problem 19-1.2. (*c*) Waveform for Problem 19-1.3.

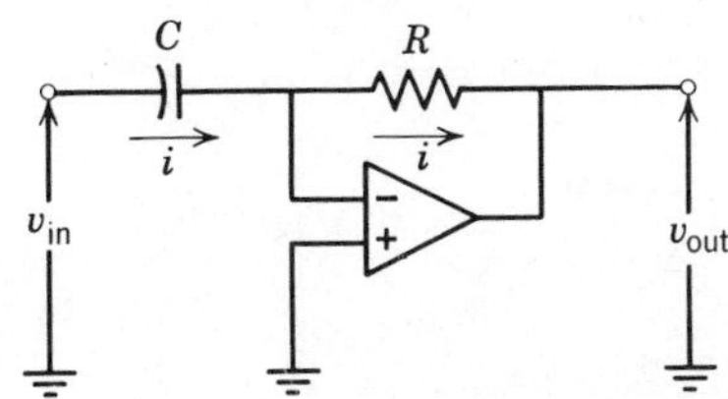

Figure 19-6 The differentiator.

The differentiator is very sensitive to short noise pulses and, as a result, it is not a preferred circuit to use in most applications.

Example 19-3

The differentiator circuit in Fig. 19-6 has a value of 10 kΩ for R and a value of 0.001 μF for C. The input voltage waveform is given in Fig. 19-7*a*. Determine the output voltage waveform.

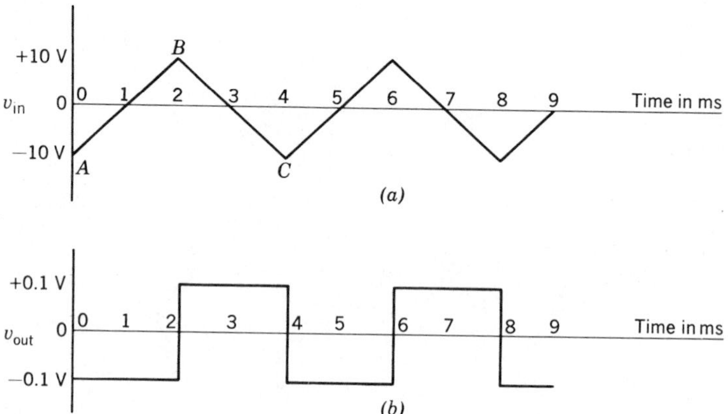

Figure 19-7 Differentiator waveforms. (*a*) Input voltage waveform. (*b*) Output voltage waveform.

Solution

The change in the input voltage from A to B is linear. The value of the slope dv_{in}/dt from A to B is the constant value

$$\frac{dv_{in}}{dt} = \frac{+10\text{ V} - (-10\text{ V})}{2\text{ ms}} = \frac{20}{2 \times 10^{-3}} = 10^4\text{ V/s}$$

and, substituting this value into Eq. 19-7, we have

$$v_{out} = -RC\frac{dv_{in}}{dt}$$

$$= -(10^4\ \Omega)(0.001 \times 10^{-6}\text{ F})(10^4\text{ V/s}) = -0.1\text{ V} \qquad (19\text{-}7)$$

The slope of the input voltage waveform from B to C is

$$\frac{dv_{in}}{dt} = \frac{-10\text{ V} - (+10\text{ V})}{2\text{ ms}} = \frac{-20\text{ V}}{2 \times 10^{-3}\text{ s}} = -10^4\text{ V/s}$$

and, substituting this value into Eq. 19-7, we have

$$v_{out} = -RC\frac{dv_{in}}{dt} = -(10^4\ \Omega)(0.001 \times 10^{-6}\ \text{F})(-10^4\ \text{V/s}) = +0.1\ \text{V} \tag{19-7}$$

The output voltage is the square wave shown in Fig. 19-7*b*.

Example 19-4

Using the circuit values of Example 19-3, find the output voltage waveform if the input voltage waveform is the square wave given in Fig. 19-8*a*.

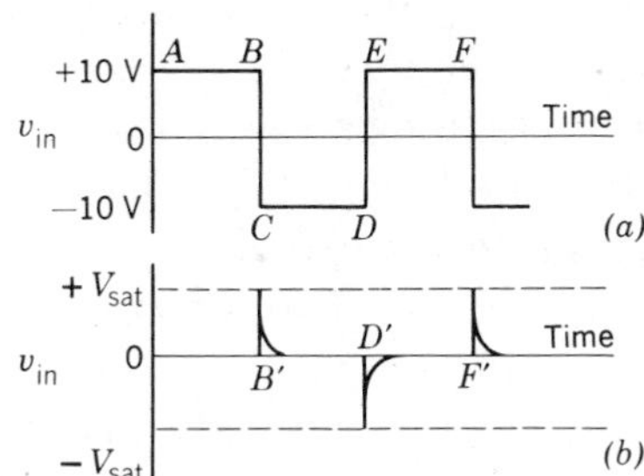

Figure 19-8 Differentiator waveforms. (*a*) Input voltage waveform. (*b*) Output voltage waveform.

Solution

The change in voltage from A to B, from C to D, and from E to F is zero. Therefore the output voltage is zero through these intervals. The change in voltage from B to C is 20 V in a negative direction in zero time. Mathematically the slope is $-\infty$ but the op amp is limited to a saturation condition. Thus, at B' and at F', the output voltage is a very short spike rising to $+V_{sat}$. From D to E, the slope mathematically is $+\infty$. Therefore the output voltage at D' is a sharp negative spike rising to $-V_{sat}$. We do find a finite width to the output spikes caused by rise and decay times in the circuit. The output voltage waveform is shown in Fig. 19-8*b*.

If we apply Eq. 17-1 to the circuit of the differentiator (Fig. 19-6) we have

$$A_v = -\frac{R}{j\dfrac{1}{\omega C}} = j\omega RC = j2\pi fRC \tag{19-8}$$

Equation 19-8 shows that the gain of the circuit increases as frequency increases. The increase is linear with frequency and consequently, on a decibel scale, we have an increase in gain at the rate of 20 dB/decade (6 dB/octave).

The gain is unity or 0 dB when the frequency is f_1.

$$1 = 2\pi f_1 RC$$

or

$$f_1 = \frac{1}{2\pi RC} \quad \text{and} \quad \omega_1 = \frac{1}{RC} \tag{19-9}$$

Figure 19-9*a* shows the frequency response for an op amp. To superimpose the frequency-response characteristic of the differentiator on this curve, we locate the 0-dB frequency f_1 as determined from Eq. 19-9 at point *A*. Then from point *A* we extend the response as an increasing gain at the rate of 20 dB/decade. This response for the differentiator intersects the open-loop response curve of the op amp at point *B*, at frequency f_2. The overall response of the circuit then increases from *A* to *B* and falls toward *E*. At point *B* the slope changes from +20 dB/decade to −20 dB/decade. This indicates a total phase change at point *B* of a possible 180°. Therefore the circuit of Fig. 19-6 is unstable and may oscillate.

The principle of compensating a differentiator against oscillation is shown in Fig. 19-10. A capacitor C_1 that is small compared to C is placed in parallel with R. At high frequencies the reactance of C_1 becomes small as compared to the resistance of R. Now we can neglect R and say that the gain

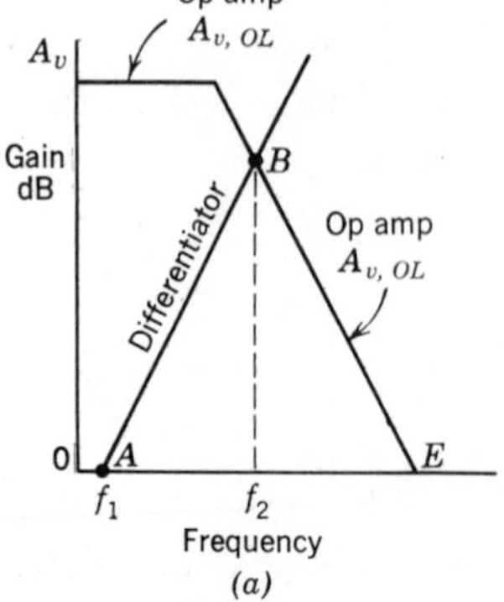

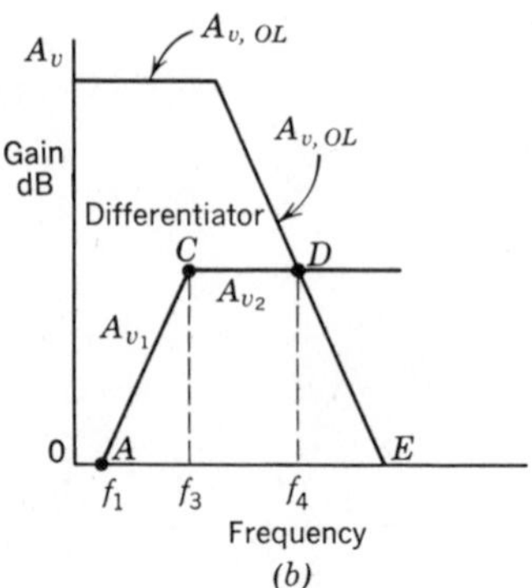

Figure 19-9 Frequency response of the differentiator. (*a*) Without compensation. (*b*) With compensation.

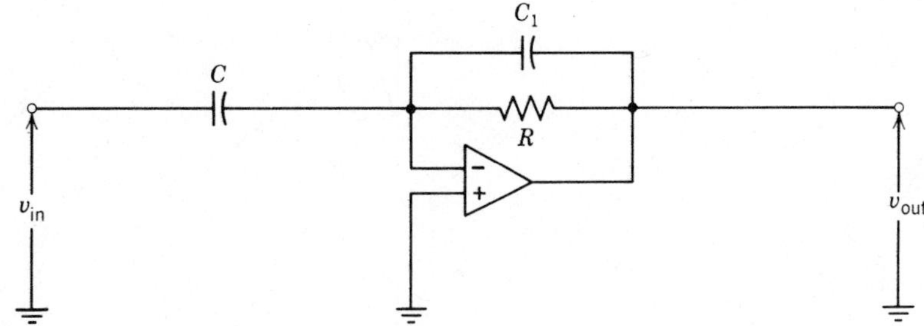

Figure 19-10 The compensated differentiator.

of the circuit is

$$A_{v2} = \frac{-jX_{C_1}}{-jX_C} = \frac{X_{C_1}}{X_C} = \frac{\dfrac{1}{2\pi fC_1}}{\dfrac{1}{2\pi fC}} = \frac{C}{C_1} \tag{19-10}$$

Equation 19-10 shows that the gain does not vary with frequency and therefore plots as a horizontal line on the frequency response of the differentiator (Fig. 19-9*b*). The point of intersection of this straight line with the rising straight line of the differentiator at low frequencies is f_3. The frequency f_3 is the break frequency determined when the reactance of C_1 equals R.

$$\frac{1}{2\pi f_3 C_1} = R$$

or

$$f_3 = \frac{1}{2\pi RC_1} \quad \text{or} \quad \omega_3 = \frac{1}{RC_1} \tag{19-11}$$

Thus the frequency response of the differentiator rises from point A to point C, remains flat from point C to the intersection with the open-loop response of the op amp at f_4, point D, and then follows the op amp response toward E. Now we do not have the abrupt 180° phase change and the circuit will be stable.

Problems

19-2.1 The differentiator (Fig. 19-6) uses 6.8 kΩ for R and 0.002 μF for C. What is the output voltage waveform if the input voltage has the given waveform?

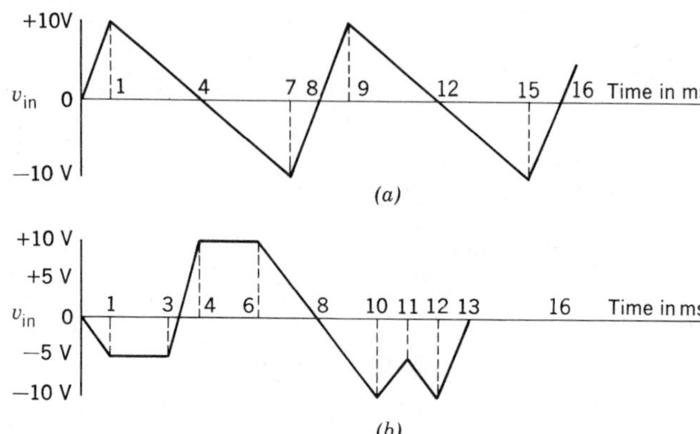

(*a*) Input voltage waveform for Problem 19-2.1. (*b*) Input voltage waveform for Problem 19-2.2.

19-2.2 Repeat Problem 19-2.1 for the second input voltage waveform.

19-2.3 Repeat Problem 19-2.1 if the capacitor is changed to 750 pF.

19-2.4 Repeat Problem 19-2.2 if the capacitor is changed to 100 pF and the resistor is changed to 1000 Ω.

Section 19-3 Nonlinear Applications

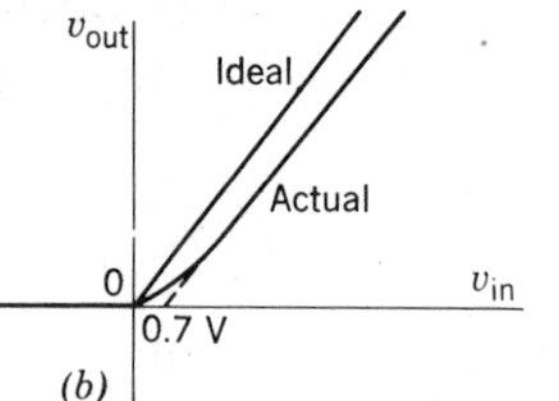

Figure 19-11 The simple half wave rectifier. (*a*) Circuit. (*b*) Response.

The Ideal Rectifier

In Section 2-3 and in Section 2-4 we considered a number of diode circuits that were used to show the applications of diodes to wave-shaping circuits. At that point, we assumed that the diodes were ideal in the forward direction by stating that the input voltage was sufficiently high so that we could neglect the forward diode voltage drop V_F. In many applications the incoming signal levels are smaller than the value of V_F. A simple half-wave rectifier using a silicon diode is shown in Fig. 19-11*a*. The characteristic of this circuit is given in Fig. 19-11*b* and shows how V_F (0.7 V) causes a deviation from the ideal characteristic.

Two identical diodes, *D*1 and *D*2, are used with an op amp in the circuit of Fig. 19-12*a*. The output of the circuit is not

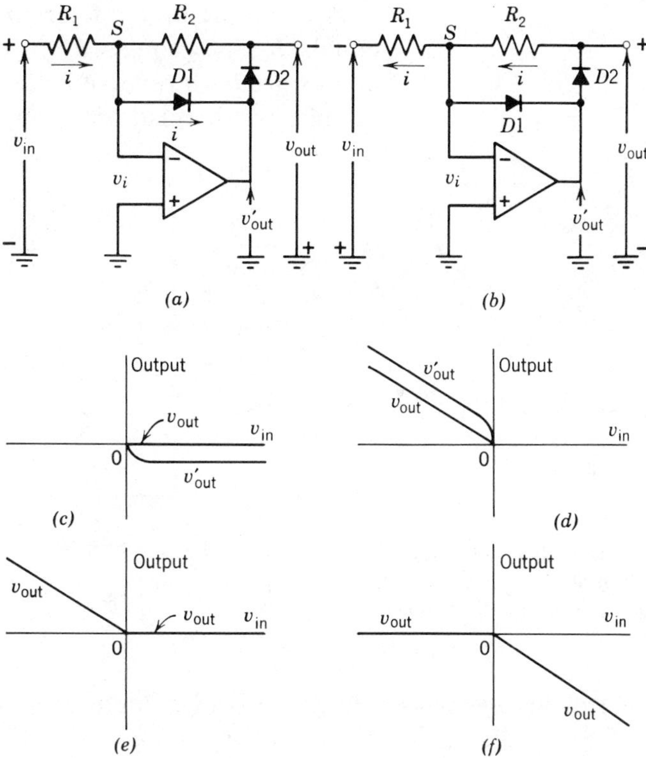

Figure 19-12 The ideal rectifier. (*a*), (*b*) Circuit. (*c*) Positive input voltage response. (*d*) Negative input voltage response. (*e*) Combined response for v_{out}. (*f*) Response with diodes reversed.

v'_{out}, but is v_{out} taken from the junction of R_2 and D_2. The polarity assigned to v_{in} causes current to flow into R_1. At the summing point S three possible paths exist for i. Current cannot go into the inverting terminal of the op amp because of its very high input resistance. Current cannot go into R_2 because $D2$ presents a blocking reverse connection to i. Therefore the current i must flow through $D1$ into the output terminal of the op amp. The input voltage to the op amp v_i is zero. Then, v_{out} must also be zero since the IR drop in R_2 is zero. Then, v'_{out} must be below zero by the amount of the forward voltage drop across $D1$. This characteristic is shown in Fig. 19-12*c*.

When the polarity of v_{in} is reversed (Fig. 19-12*b*), the current i reverses direction through R_1. Now $D1$ blocks the current flow and i must flow through $D2$ and R_2. Since v_{out} is taken from the junction of $D2$ and R_2, V_{out} is exactly iR_2. This characteristic is shown in Fig. 19-12*d*.

The complete characteristic is shown in Fig. 19-12*e*. If R_1 equals R_2, v_{out} equals v_{in} in the forward direction. When the diodes are both reversed, we obtain the forward characteristic in the fourth quadrant (Fig. 19-12*f*).

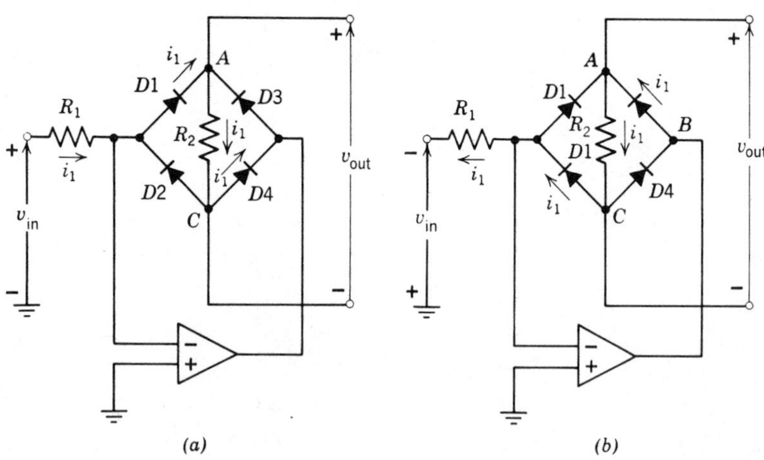

Figure 19-13 The ideal full-wave rectifier. (*a*) Current directions when v_{in} is positive. (*b*) Current directions when v_{in} is negative.

The Ideal Full-Wave Rectifier The ideal bridge rectifier is shown in Fig. 19-13*a*. The input current to the circuit is

$$i_1 = \frac{v_{in}}{R_1} \tag{19-12}$$

This current is blocked by $D2$ and passes through $D1$. The current flows through R_2 and through diode $D4$ to the output of the op amp. When the polarity of v_{in} reverses (Fig. 19-13*b*), the direction of i_1 reverses. Now i_1 flows through $D3$, R_2, and $D2$. The current is blocked by $D4$.

The value of the output voltage v_{out} is

$$v_{out} = R_2 i_1 \tag{19-13}$$

Substituting Eq. 19-12 into Eq. 19-13, we have

$$v_{out} = \frac{R_2}{R_1} \left| v_{in} \right| \tag{19-14}$$

We must modify Eq. 19-14 to show the effect of rectification.

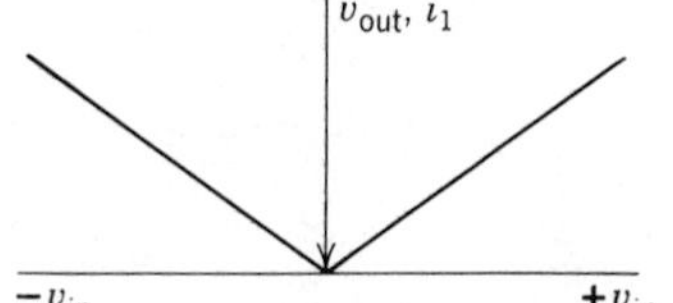

Figure 19-14 Transfer characteristic of the ideal full-wave rectifier.

The polarity of v_{out} is always positive. Therefore

$$v_{out} = \frac{R_2}{R_1}|v_{in}| \tag{19-15}$$

The transfer characteristic of this ideal full-wave rectifier is shown in Fig. 19-14. The use of the op amp causes this characteristic to be ideally linear. Also the current through R_2 is exactly proportional to v_{in}. The circuit can be used to serve as a linear ac voltmeter. Alternatively, the circuit can be used as an *absolute value circuit.*

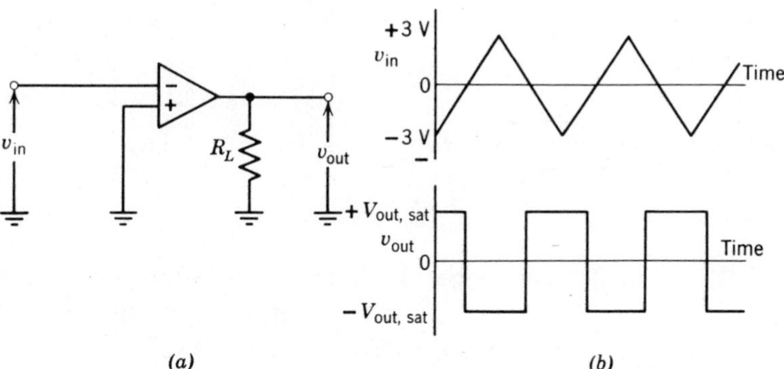

Figure 19-15 The comparator. (*a*) Circuit. (*b*) Waveforms.

The Comparator The circuit given in Fig. 19-15*a* is used as a *voltage comparator* or *comparator.* The op amp is used in the open-loop mode. A very small input signal drives the output into saturation. Therefore, the output exists in either of two modes: $+V_{out,SAT}$ or $-V_{out,SAT}$. Typical input and output voltage waveforms are shown in Fig. 19-15*b*. If a dc voltage is introduced into the noninverting input, as in Fig. 19-16, the switching points of the output waveform are changed as shown in Fig. 19-16.

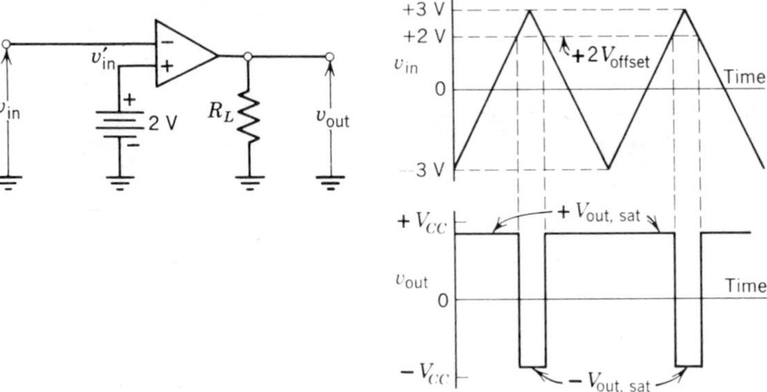

Figure 19-16 Circuit and waveforms for comparator using an offset voltage.

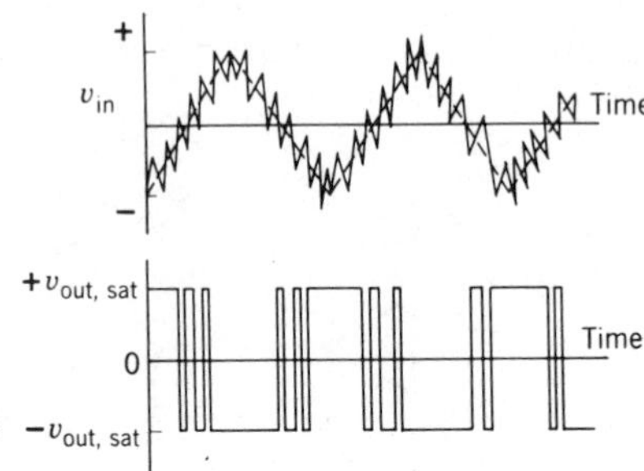

Figure 19-17 Comparator waveforms with noise.

The Schmitt Trigger If there is a noise or other disturbing signal on the input signal to the comparator, false switching can occur at the output as shown in Fig. 19-17.

The *Schmitt trigger*, Fig. 19-18, is a circuit designed to eliminate this problem of false switching. In Fig. 19-18*a*

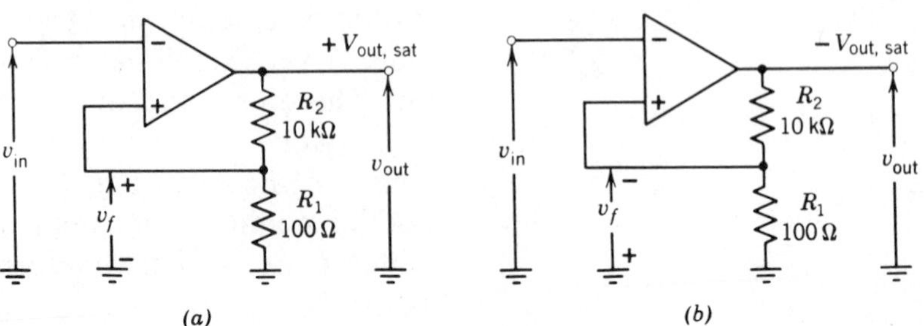

Figure 19-18 The Schmitt trigger. (*a*) Positive saturation. (*b*) Negative saturation.

assume that v_{out} is saturated at +10 V. Then the positive feedback voltage v_f fed to the noninverting terminal is

$$v_f = \frac{R_1}{R_1 + R_2} V_{out,SAT} = \frac{100\ \Omega}{100\ \Omega + 10{,}000\ \Omega} 10\ \text{V} = 0.099\ \text{V}$$
$$= +99\ \text{mV}$$

Therefore the output can be at positive saturation $+V_{out,SAT}$ only if

$$v_{in} < +99\ \text{mV}$$

Correspondingly, in Fig. 19-18*b*, when the output is saturated at −10 V, the positive feedback voltage v_f at the noninverting input terminal is

$$v_f = \frac{R_1}{R_1 + R_2}(-V_{out,SAT}) = \frac{100\ \Omega}{100\ \Omega + 10{,}000\ \Omega}(-10\ \text{V})$$
$$= -0.099\ \text{V} = -99\ \text{mV}$$

Therefore, the output voltage can be at negative saturation $-V_{out,SAT}$ only if

$$v_{in} > -99\ \text{mV}$$

The transfer characteristic for the Schmitt trigger is shown in Fig. 19-19. If the signal v_{in} is initially a high positive value, the output is $-V_{out,SAT}$. The input voltage must be reduced to −99 mV before the output switches to $+V_{out,SAT}$. If the signal v_{in} is initially a large negative value, the output is $+V_{out,SAT}$. Now v_{in} must be increased to +99 mV before the output switches to $-V_{out,SAT}$. The overlap from +99 mV to −99 mV is

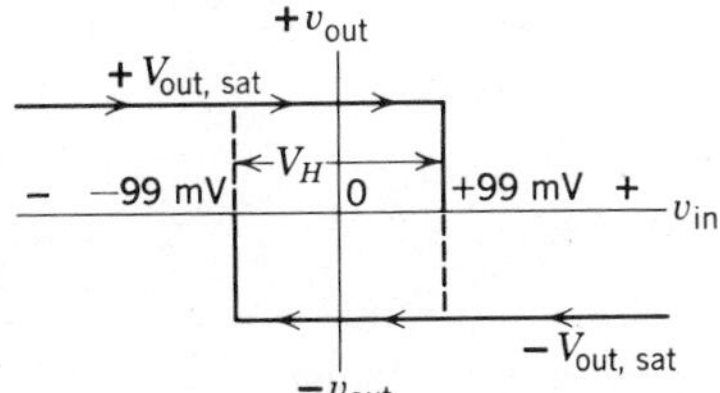

Figure 19-19 Schmitt trigger transfer characteristic.

198 mV. This overlap is called the *hysteresis voltage*, V_H. If the noise level is less than 198 mV, the circuit will not erratically switch back and forth.

Problems **19-3.1** Determine the waveform for v_1 and the waveform for v_2 for the given circuit.

19-3.2 Find the transfer characteristic for the given circuit.

19-3.3 Find the transfer characteristic for the given circuit.

19-3.4 Find the transfer characteristic for the given circuit.

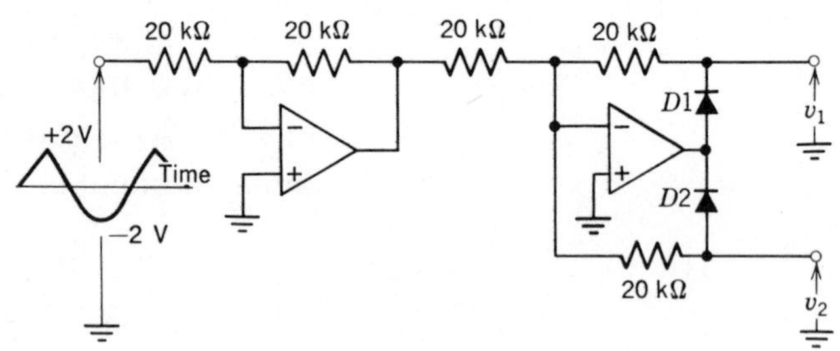

Circuit for Problem 19-3.1.

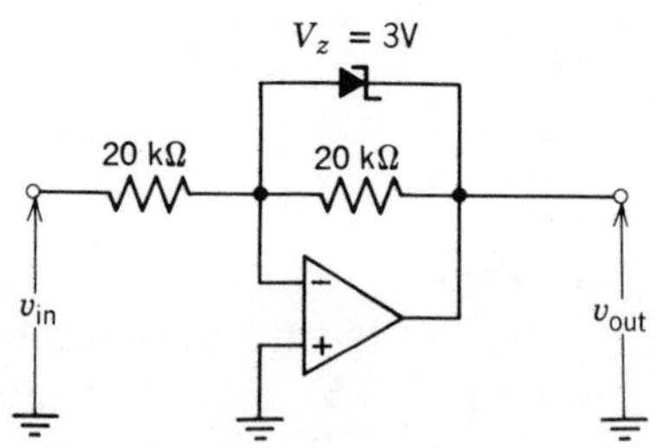

Circuit for Problem 19-3.2.

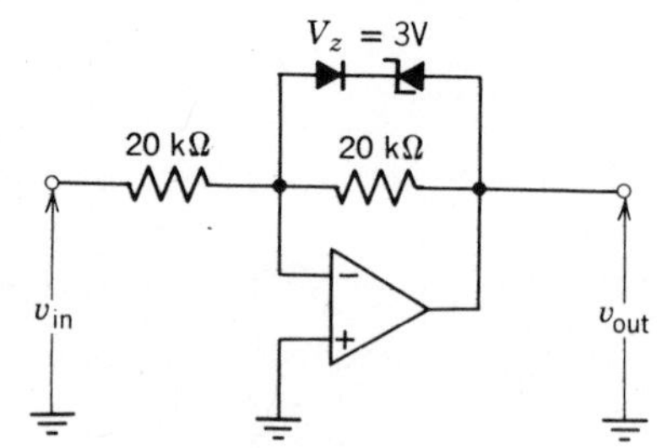

Circuit for Problem 19-3.3.

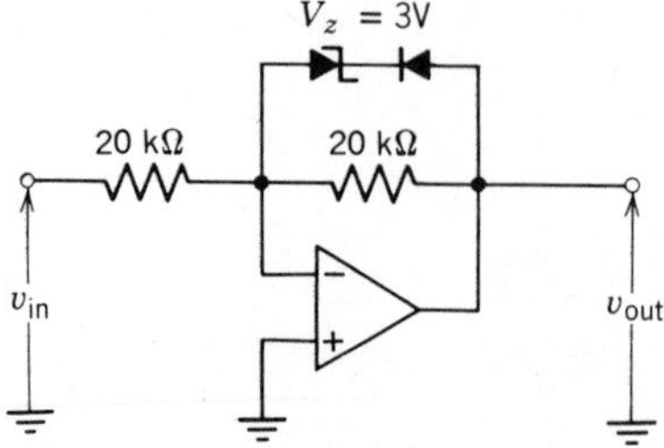

Circuit for Problem 19-3.4.

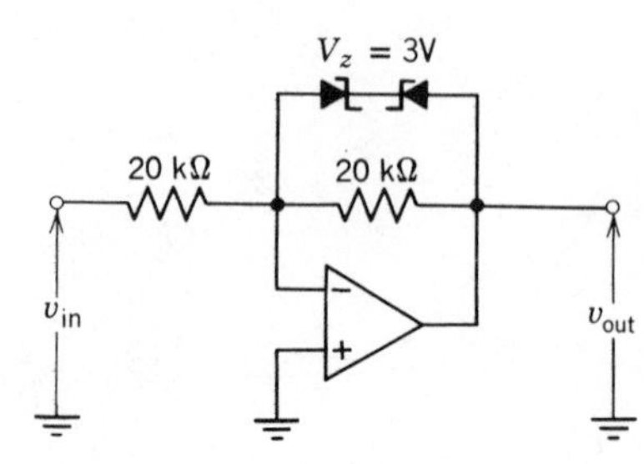

Circuit for Problem 19-3.5.

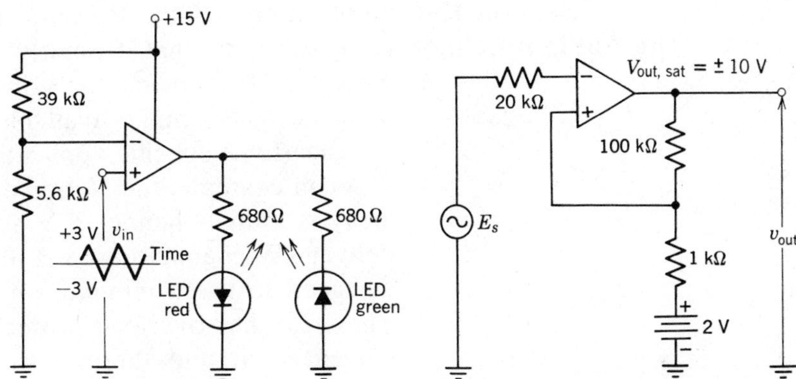

Circuit for Problem 19-3.6.

Circuit for Problems 19-3.7 and 19-3.8.

19-3.5 Find the transfer characteristic for the given circuit.

19-3.6 The frequency of the given input voltage waveform is 10 Hz. How long is the green light on and how long is the red light on?

19-3.7 Determine the transfer characteristic and the value of V_H for the given circuit.

19-3.8 Determine the transfer characteristic and the value of V_H for the given circuit if the 1-kΩ resistor is changed to 2 kΩ and if the 2-V offset battery is changed to 1 V.

19-3.9 Show how the circuit yields the absolute value of the input signal; that is, show that

$$v_{out} = |v_{in}|$$

Determine the waveforms for v_1 and v_{out} for the given input waveform. What are the functions of diodes, $D1$ and $D2$?

19-3.10 What is the effect on the circuit operation when the diodes, $D1$ and $D2$, are reversed?

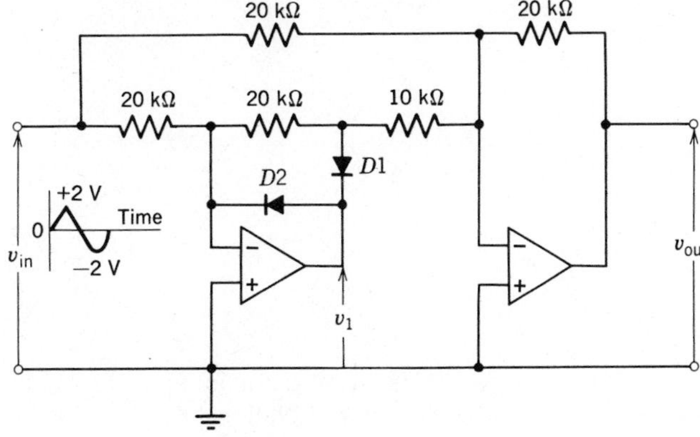

Circuit for Problems 19-3.9 and 19-3.10.

Section 19-4 The Audio Amplifier

Applications of the op amp have been extended to include many circuits that formerly were made of discrete components. Now, using the construction techniques of the op amp, designers make available a complex circuit as a single package for a specific application.

As an example, the National Semiconductor* LM 377 linear array is a dual-channel 2-W audio amplifier. The LM 377 will deliver 2W/channel into 8-Ω or 16-Ω loads. The amplifiers are designed to use a minimum number of external components. The array has overload protection consisting of both internal current limit and thermal shutdown. Figure 19-20 lists the absolute maximum ratings and shows the terminal connections to the dual-in-line (DIP) package. Table 19-1 lists the detailed electrical specifications. Figure 19-21 shows typical performance characteristics and Fig. 19-22 gives some representative applications for the amplifier.

Supply voltage	26 V
Input voltage	0 V–V_{supply}
Operating temperature	0°C to +70°C
Storage temperature	−65°C to +150°C
Junction temperature	150°C
Lead temperature (soldering, 10 seconds)	300°C

Pin	Function	Pin	Function
1	Bias	14	V+
2	Output 1	13	Output 2
3	GND	12	GND
4	GND	11	GND
5	GND	10	GND
6	Input 1	9	Input 2
7	Feedback 1	8	Feedback 2

Figure 19-20 Absolute maximum ratings and connection diagram. (*Courtesy National Semiconductor Inc.*)

*The data given in Figs. 19-20, 19-21, and 19-22 and in Table 19-1 are courtesy of National Semiconductor Inc.

Table 19-1
Electrical Characteristics

$V_S = 20$ V, $T_{TAB} = 25°C$, $R_L = 8\ \Omega$, $A_V = 50$ (34 dB), unless otherwise specified.

Parameter	Conditions	Min	Typ	Max	Units
Total supply current	$P_{out} = 0$ W		15	50	mA
	$P_{out} = 1.5$ W/Channel		430	500	mA
DC output level			10		V
Supply voltage		10		26	V
Output power	T.H.D. = <5%	2	2.5		W
T.H.D.	$P_{out} = 0.05$ W/Channel, $f = 1$ kHz		0.25		%
	$P_{out} = 1$ W/Channel, $f = 1$ kHz		0.07	1	%
	$P_{out} = 2$ W/Channel, $f = 1$ kHz		0.10		%
Offset voltage			15		mV
Input bias current			100		nA
Input impedance		3			MΩ
Open-loop gain	$R_S = 0\ \Omega$	66	90		dB
Channel separation	$C_F = 250\ \mu$F, $f = 1$ kHz	50	70		dB
Ripple rejection	$f = 120$ Hz, $C_F = 250\ \mu$F	60	70		dB
Current limit			1.5		A
Slew rate			1.4		V/μs
Equivalent input noise voltage	$R_S = 600\ \Omega$, 100 Hz – 10 kHz		3		μVrms

Source: Courtesy National Semiconductor Inc.

Note 1: For operation at ambient temperatures greater than 25°C the LM377 must be derated based on a maximum 150°C junction temperature using a thermal resistance which depends upon device mounting techniques.

Note 2: Dissipation characteristics are shown for four mounting configurations:

a. Infinite sink—13.4°C/W.

b. P.C. board $+V_7$ sink—21°C/W. P.C. board is $2\frac{1}{2}$ square inches. Staver V_7 sink is 0.02 inch thick copper and has a radiating surface area of 10 square inches.

c. P.C. board only—29°C/W. Device soldered to $2\frac{1}{2}$ square inch P.C. board.

d. Free air—58°C/W.

Note 3: T.H.D. is total-harmonic distortion.

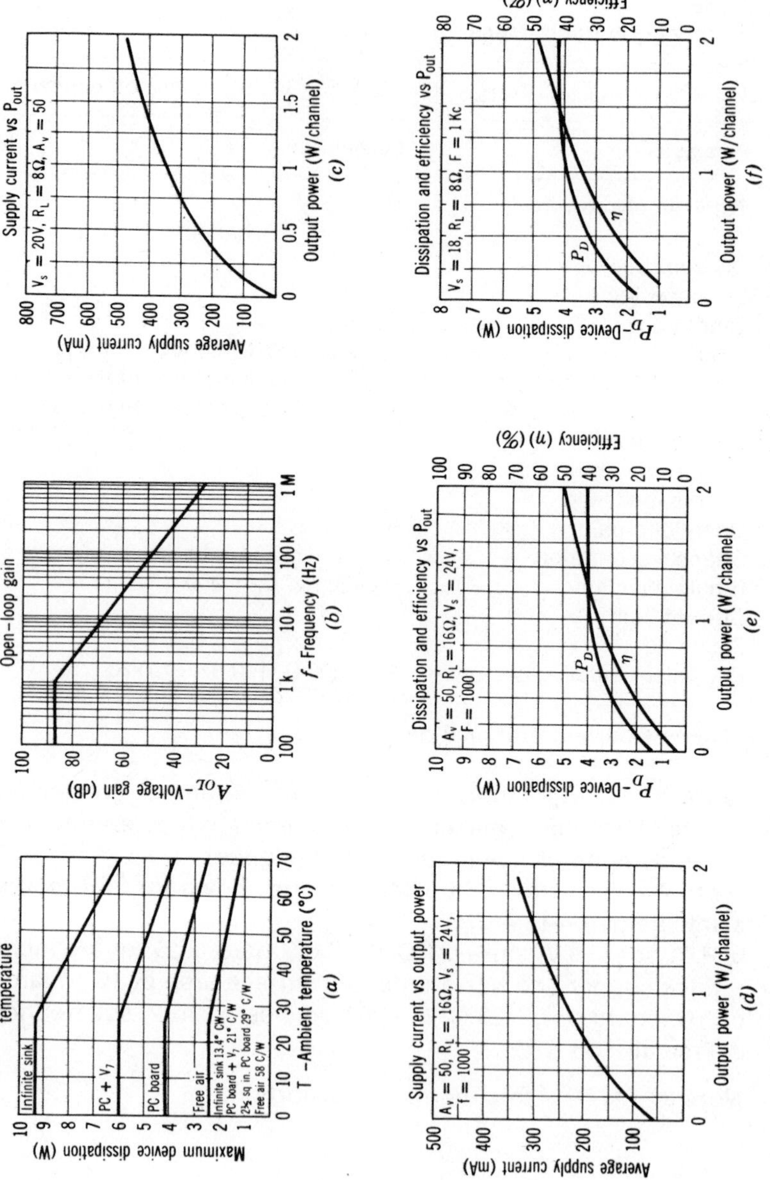

Figure 19-21 Typical performance characteristics (*Courtesy National Semiconductor Inc.*)

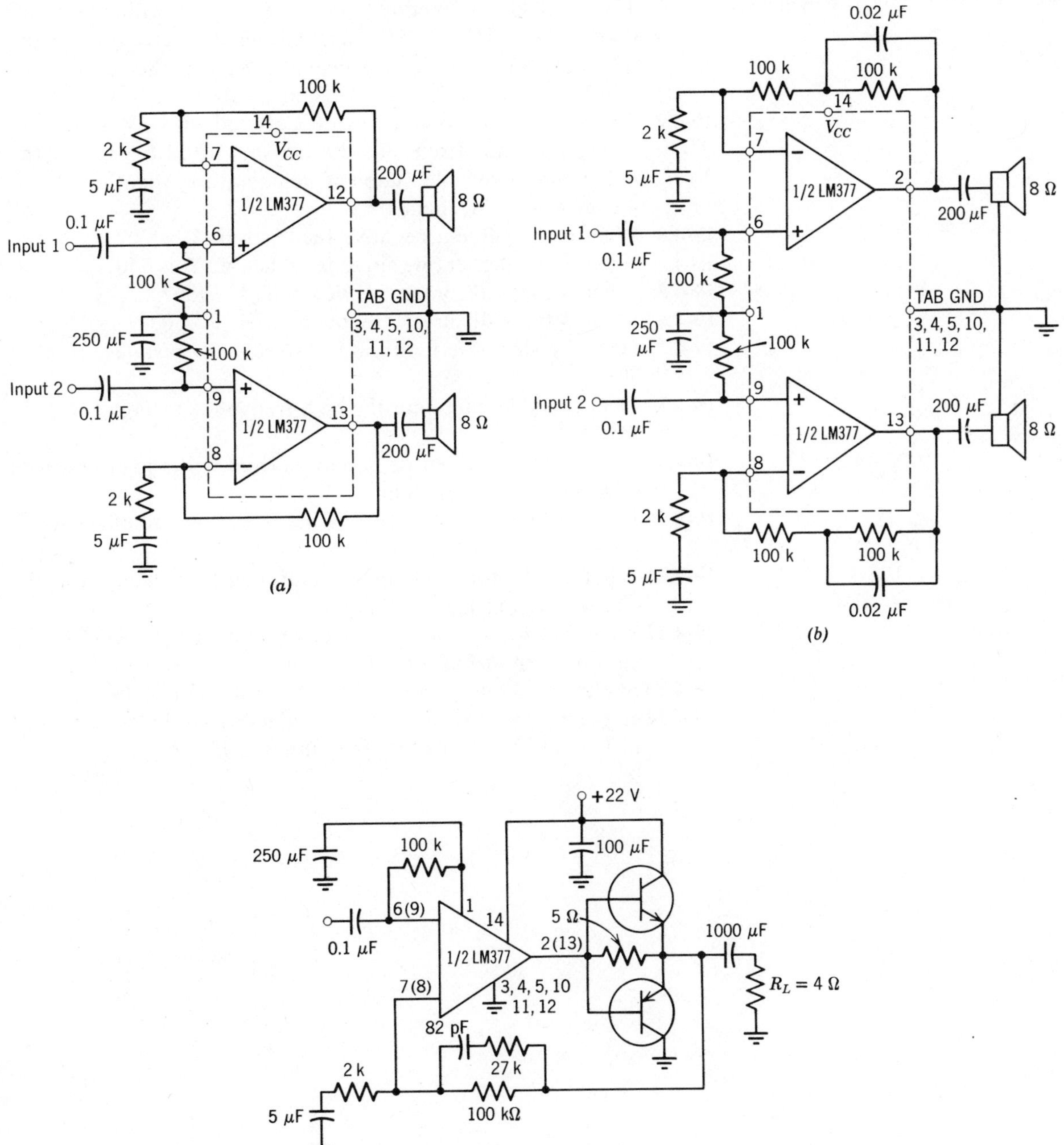

(a)

(b)

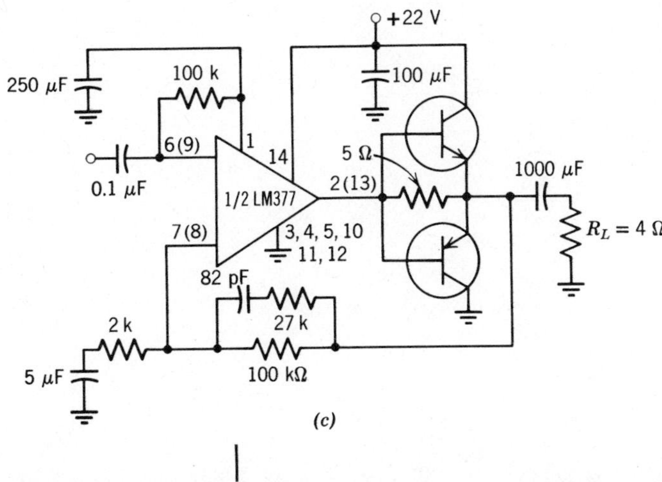

(c)

Figure 19-22 Typical applications. (*a*) Simple stereo amplifier. (*b*) Simple stereo amplifier with bass boost. (*c*) 10 W per channel amplifier. (*Courtesy National Semiconductor Inc.*)

Problems

19-4.1 The amplifier is operated on a P.C. board without an additional heat sink in a 55°C ambient temperature. What is the maximum permissible P_{out}/channel? What is the supply current? Use Fig. 19-21*a*, 19-21*d*, and 19-21*e*.

19-4.2 The amplifier is mounted on a P.C. board with a V_7 heat sink. Using the data from Fig. 19-21*a* and 19-21*d*, what is the maximum allowable ambient temperature for the output power to be 1 W/channel?

19-4.3 What is the percent negative feedback if A_v' is 50?

19-4.4 What is the percent negative feedback if A_v' is 250?

19-4.5 What is the 3-dB break frequency if A_v is 50?

19-4.6 What is the 3-dB break frequency if A_v is 250?

19-4.7 Does the slew rate limit the high-frequency response if A_v' is 50?

19-4.8 Does the slew rate limit the high-frequency response if A_v' is 250?

19-4.9 What is the required dc supply current if the load power is 0.75 W/channel? Use Fig. 19-21*f*.

19-4.10 Use the data of Table 19-1. What is the overall efficiency if both channels are operated at 1.5 W?

19-4.11 What is A_v for each channel at mid-frequencies for the circuit given in Fig. 19-22*a*?

19-4.12 What is A_v for each channel at mid-frequencies for the circuit given in Fig. 19-22*b*?

19-4.13 Sketch the Bode gain plot for the circuit of Fig. 19-22*b*.

19-4.14 Explain how 10 W/channel is obtained from the circuit given in Fig. 19-22*c*. Prove that the output is 10 W/channel.

Chapter 20 Voltage Regulators

The elementary form of the voltage regulator is the shunt regulator (Section 20-1). The series regulator uses a pass transistor to obtain large load currents (Section 20-2). The use of an operational amplifier provides a closely regulated reference voltage that may be either larger or smaller than the Zener voltage of the diode from which the reference voltage is derived (Section 20-3). A voltage regulator can be designed to offer several different features such as current limiting, short-circuit shutdown, or fold-back (Section 20-4). The flexibility of a typical commercial precision voltage regulator is examined (Section 20-5). A simpler package is available that can be used without external components to provide a specific regulated output voltage (Section 20-6).

Section 20-1 Shunt Regulators

A *shunt regulator* is placed in parallel with the load R_L as shown in Fig. 20-1. The input voltage V_{in} for this and for other regulator circuits usually is the output of a full-wave rectifier using a capacitor filter.

In Section 2-2, we showed how the Zener diode is used to maintain the load voltage at a nearly constant value for a changing load current or for a changing input voltage. The

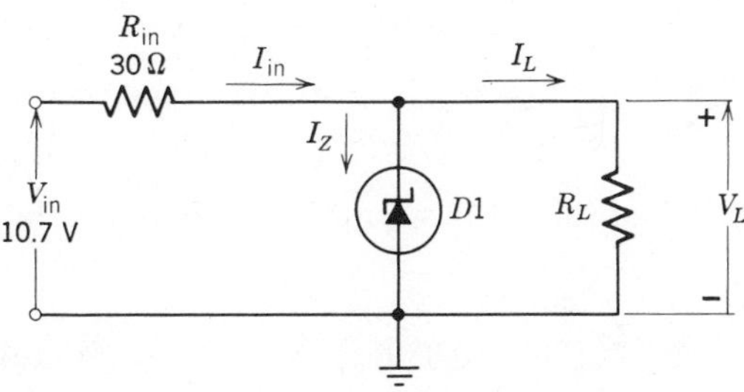

Figure 20-1 The shunt regulator.

equations that can be used for the Zener circuit of Fig. 20-1 are

$$I_L = I_{in} - I_z \qquad (20\text{-}1)$$

and

$$V_L = V_{in} - I_{in}R_{in} \qquad (20\text{-}2)$$

Example 20-1
The Zener used in Fig. 20-1 has the following ratings.

$$V_Z = 6.2\ \text{V} \quad \text{at} \quad I_Z = 50\ \text{mA}$$

$$Z_Z = 2\ \Omega \quad \text{at} \quad I_Z = 50\ \text{mA}$$

$$I_{Z,\text{min}} = 5\ \text{mA} \quad \text{and} \quad I_{Z,\text{max}} = 150\ \text{mA}$$

Determine the exact values of Zener current and load voltage for the following values of I_L:

150 mA, 145 mA, 105 mA, 101 mA, 100 mA, 99 mA, 95 mA, and 0 mA

Solution
Let us call the reference values of the Zener diode V_Z (6.2 V) and I_Z (50 mA). Also let us call the Zener current and voltage at any other condition V'_Z and I'_Z. Since I_{in} is 150 mA at all times,

$$I_L = I_{in} - I'_Z = 150 - I'_Z\ \text{mA} \qquad (20\text{-}1)$$

or

$$I'_Z = 150 - I_L\ \text{mA} \qquad (20\text{-}3a)$$

The load voltage is

$$V_L = V'_Z = V_Z + Z_Z(I'_Z - I_Z) \qquad (20\text{-}3b)$$

Substituting values, we have

$$V_L = V'_Z = 6.2 + 2(I'_Z - 0.050)\ \text{V} \qquad (20\text{-}3c)$$

when $I_L = 145$ mA

$$I'_Z = 150 - I_L = 150 - 145 = \mathbf{5\ mA} \qquad (20\text{-}3a)$$

and

$$V_L = V'_Z = 6.2 + 2(0.005 - 0.050) \qquad (20\text{-}3c)$$

$$= 6.2 + 2(-0.045) = 6.2 - 0.090 = \mathbf{6.110\ V}$$

Table 20-1
Zener Voltage and Current Values

I_L (mA)	I'_Z (mA)	$V'_Z = V_L$ (volts)	Change in V_L from 6.2 V (millivolts)
145	5	6.110	−90
105	45	6.190	−10
101	49	6.198	−2
100	50	6.200	0
99	51	6.202	+2
95	55	6.210	+10
0	150	6.400	+200

The other required values for I_L are substituted into Eq. 20-3*a* and into Eq. 20-3*c* and the results are listed in Table 20-1. Additionally, in Table 20-1, we show the change in the load (or Zener) voltage from the reference value of V_Z at 50 mA.

When the load current changes over its maximum range from 145 mA to 0 mA, the load voltage changes a total of (200 + 90) or 290 mV. Using 6.2 V as the reference value, we find that the percentage change in V_L is

$$\frac{290 \text{ mV}}{6.2 \text{ V}} \times 100 = \frac{0.290 \text{ V}}{6.2 \text{ V}} \times 100 = 4.7\%$$

For many applications, a change in load voltage of 4.7% is completely acceptable. For other applications, this high value of *voltage regulation* is intolerable. In Section 20-3, we discuss circuits in which the voltage regulation can be reduced to a value of less than 1%.

The ac equivalent circuit for the Zener shunt regulator used in Fig. 20-1 is shown in Fig. 20-2. At a load current of 100 mA,

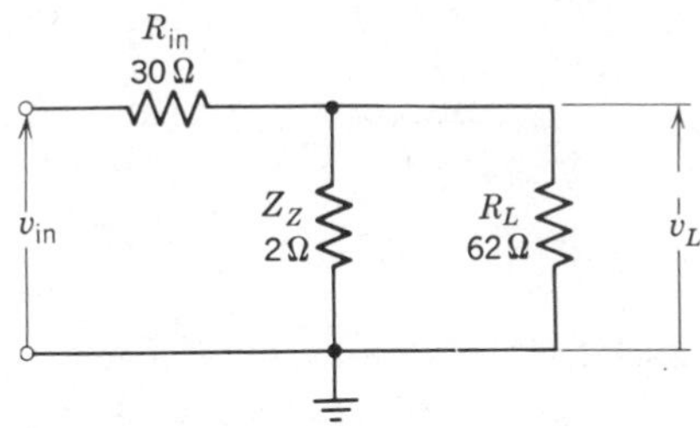

Figure 20-2 ac equivalent circuit for the Zener shunt regulator.

the load resistance R_L is 6.2 V/100 mA or 62 Ω. The equivalent resistance of the parallel combination of Z_Z and R_L is approximately equal to Z_Z. Using the voltage divider rule, we have

$$v_L = \frac{Z_Z}{Z_Z + R_{\text{in}}} v_{\text{in}} \tag{20-4a}$$

Solving for v_L/v_{in} and using numerical values, we have

$$\frac{v_L}{v_{\text{in}}} = \frac{Z_Z}{Z_Z + R_{\text{in}}} = \frac{2}{2+30} = \frac{1}{16} \tag{20-4b}$$

Thus this circuit not only maintains a nearly constant load voltage, but also reduces the ripple voltage, in this example, by a factor of 16.

The regulator shown in Fig. 20-3 uses a transistor $Q1$ in parallel with the Zener diode. The load voltage across the load is the Zener voltage V_Z plus the base-to-emitter voltage drop of $Q1$, V_{BE}

$$V_L = V_Z + V_{BE} \tag{20-5}$$

As long as the Zener is maintained in a reverse conduction condition, the circuit regulates. The change in I_L now can be much greater than the allowable change in Zener current. The change in I_L is limited only by the allowable emitter (collector) current and by the allowable power dissipation for $Q1$.

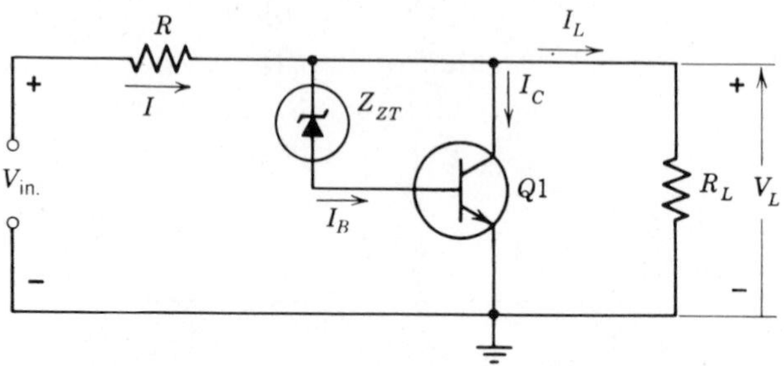

Figure 20-3 Shunt voltage regulator.

Problems **20-1.1** If the shunt regulator used in Example 20-1 must be limited to a 1% voltage regulation, what is the maximum variation in I_L?

20-1.2 Repeat Problem 20-1.1 if the voltage regulation must be limited to $\frac{1}{2}$ of 1%.

20-1.3 The shunt regulator shown in Fig. 20-1 has a fixed load resistance, 62 Ω. How far can V_{in} be varied if V_L must be constant within ±30 mV?

20-1.4 Repeat Problem 20-1.3 with the restriction that V_L must be maintained constant with ±1%.

Section 20-2 Series Regulators

A series voltage regulator is shown in Fig. 20-4. The load voltage V_L is the Zener voltage V_Z less the voltage drop across the transistor from base to emitter, V_{BE}

$$V_L = V_Z - V_{BE} \tag{20-6}$$

The load current I_L is the emitter current I_E in the *pass transistor* $Q1$. Then

$$I_L = I_E = (1 + \beta)I_B \tag{20-7}$$

The current through R_1 is

$$I_Z + I_B = \frac{V_{in} - V_Z}{R_1} \tag{20-8}$$

The value of I_Z is established at a value within the normal current range of the Zener. When R_L is reduced, since the

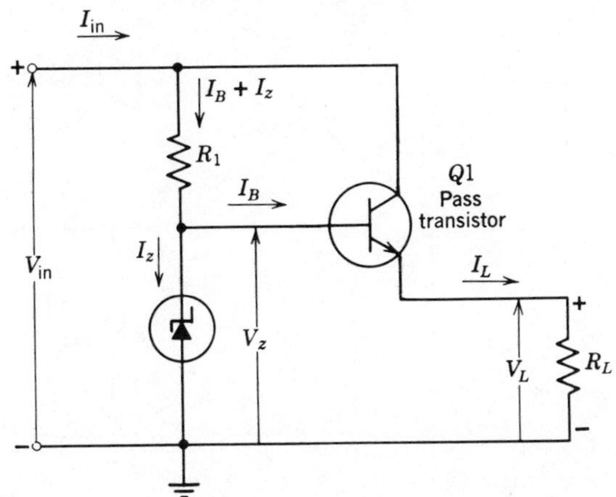

Figure 20-4 Series voltage regulator.

load voltage is fixed by Eq. 20-6, the load current must increase. The Zener current decreases by a corresponding amount. The range of load current that can be regulated is dependent on the range of I_z, on the current limitation of $Q1$, and on the power dissipation capability of $Q1$. The power dissipation in $Q1$ is

$$P_C = (V_{in} - V_L)I_C = (V_{in} - V_L)(I_{in} - I_Z)\text{ W} \tag{20-9}$$

The regulated power supply shown in Fig. 20-5 has the additional feature of an output voltage control. The voltage control is the potentiometer $(R_1 + R_2)$. The various currents are labeled on the diagram. For simplicity of analysis, assume that

$$I \gg I_{B_2}$$

Then the current in R_1 is I. By the voltage-divider rule

$$V_Z + V_{BE_2} = \frac{R_2}{R_1 + R_2} V_L$$

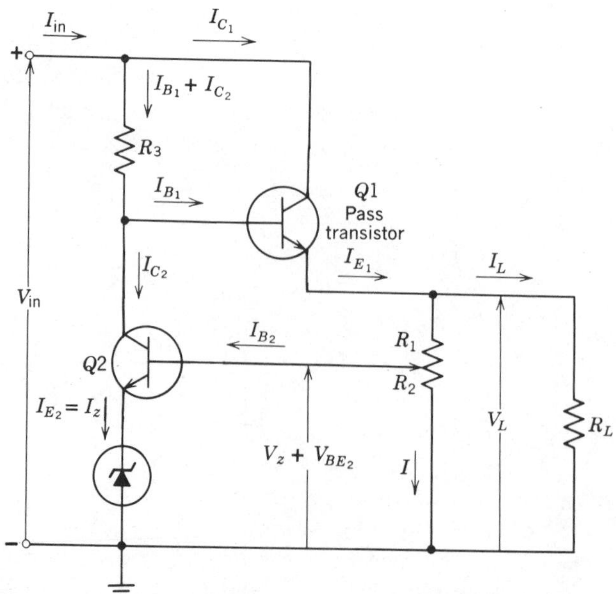

Figure 20-5 Regulated power supply.

Solving for V_L, we have

$$V_L = \frac{R_1 + R_2}{R_2}(V_Z + V_{BE_2}) \tag{20-10}$$

In Eq. 20-10, $(V_Z + V_{BE_2})$ and $(R_1 + R_2)$ are both constant values. Therefore, V_L is dependent upon R_2, which is the setting of the potentiometer. If the potentiometer is turned so R_2 increases, V_L decreases. If the potentiometer is turned so R_2 is decreased, V_L increases.

Consider the case in which R_L decreases. Then I_L increases and V_L decreases. This falling value of V_L reduces the forward bias on $Q2$. The value of V_{CE_2} rises. This rise of V_{CE_2} increases the forward bias on $Q1$. The rise of forward bias on $Q1$ causes V_{CE_1} to decrease. Now V_L is returned to its original value.

An alternative approach is to state that a change in load voltage ΔV_L that is due to either loading or ripple is amplified by $Q2$ with a phase reversal and applied to the control transistor $Q1$. This 180° phase inversion is a negative feedback and the circuit gain is adjusted so the action of $Q1$ cancels out the change in the load. In this circuit the load voltage is monitored. Circuits can be made that monitor the load current or both load current and load voltage.

Problems

20-2.1 Use the data given in Table 20-1 for the Zener diode. In Fig. 20-4 V_{in} is 15 V and R_1 is selected to set I_Z to 10 mA when I_B is zero. The value of β for the transistor is 60 and the value of V_{BE} is 0.6 V. What is the possible range of load current and load voltage?

20-2.2 If I_Z in Problem 20-2.1 is raised to 20 mA, what is the possible range of load current and load voltage?

Section 20-3 Operational-Amplifier Regulators

In Section 20-1, we showed that the Zener diode used for the data in Table 20-1 had a voltage regulation of 4.7% when the Zener current varied from 5 to 150 mA. If the change in Zener current is limited to 50 ± 1 mA, the change in V_Z is ± 2 mV.

The voltage regulation is

$$\frac{4\text{ mV}}{6.2\text{ V}} \times 100 = \frac{0.004\text{ V}}{6.2\text{ V}} \times 100 = 0.06 \text{ of } 1\%$$

In the circuit used in Fig. 20-6, an op amp used as a voltage follower is connected to the Zener diode. Since the output impedance of the op amp is effectively zero and since the gain of the voltage follower is unity, the output voltage of the op amp is identical to V_Z. If the current in the Zener can be held to 50 ± 1 mA, changes in V_L are held to ± 2 mV over the range of I_L within the current capability of the op amp.

By this means, we have a very stable value of V_L that is held or *clamped* to V_Z. When we use this circuit to secure a very stable value of V_L, we call the load voltage the *reference voltage*, V_{ref}.

Many variations of circuits using a Zener diode and an op amp are possible. Three variations are shown in Fig. 20-7. We can use the circuit given in Fig. 20-7*b* to illustrate a concept that applies to all voltage regulator circuits. In the normal operation of the circuit V_i is zero. If the output falls by an amount $-\Delta V_{out}$, the fall in V_1 is the input signal V_i to the op amp.

$$\Delta V_1 = V_i = \frac{R_1}{R_1 + R_f}(-\Delta V_{out})$$

The op amp amplifies V_i by the factor $A_{v,OL}$ and inverts the polarity at the output. This output signal causes V_{out} to return

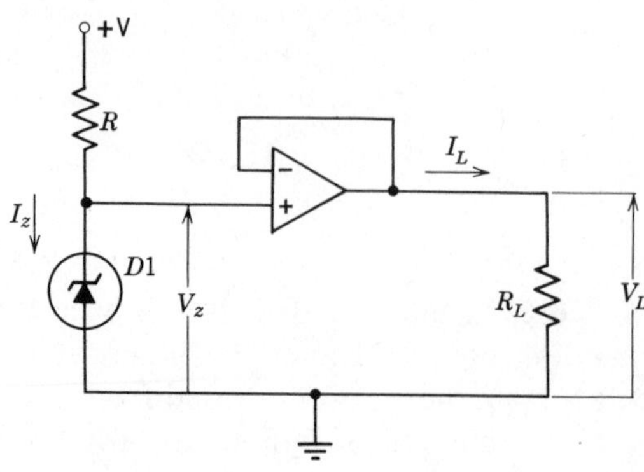

Figure 20-6 A Zener reference voltage.

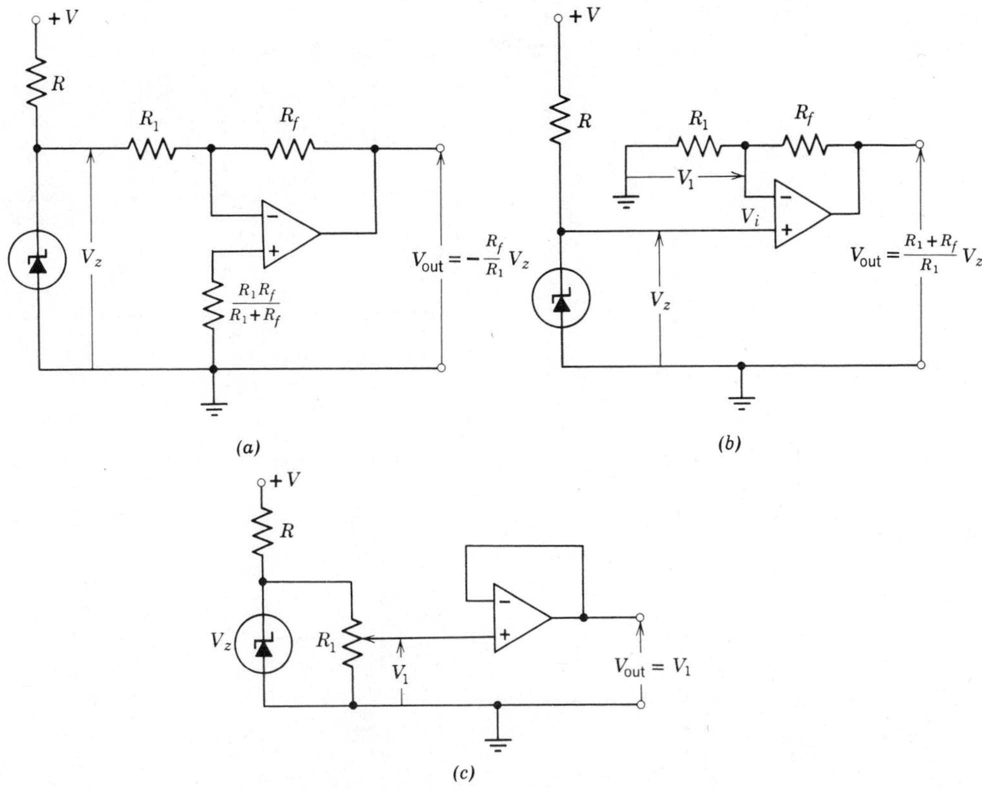

Figure 20-7 Zener reference sources. (*a*) Inverting, usually $|-V_{out}| > |V_Z|$. (*b*) Noninverting, usually $V_{out} > V_Z$. (*c*) Variable reference, $V_{out} < V_Z$.

to its original value. As a result of this action, the op amp is often called an *error amplifier.*

If the circuit given in Fig. 20-8*a* is operating properly,

$$V_1 = \frac{R_1}{R_1 + R_f} V_L \tag{20-11a}$$

Since V_i is zero in this ideal op amp, V_1 must equal V_Z.

$$V_1 = V_Z \tag{20-11b}$$

Substituting Eq. 20-11*a* into Eq. 20-11*b*, we have

$$V_z = \frac{R_1}{R_1 + R_f} V_L$$

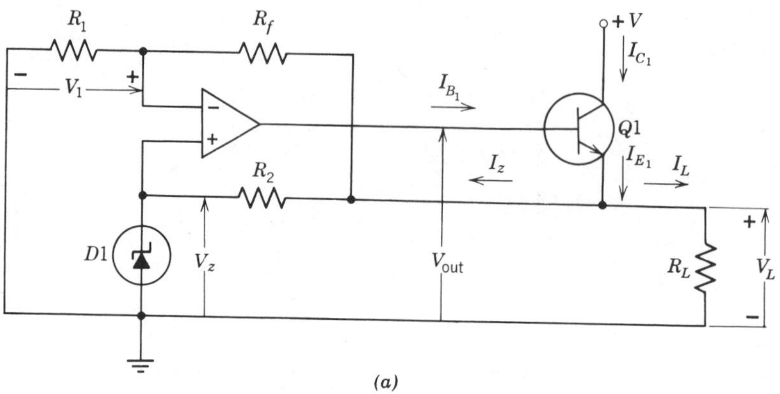

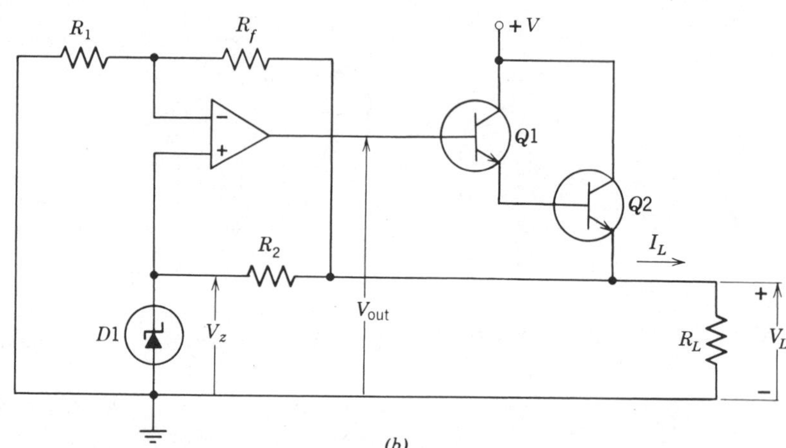

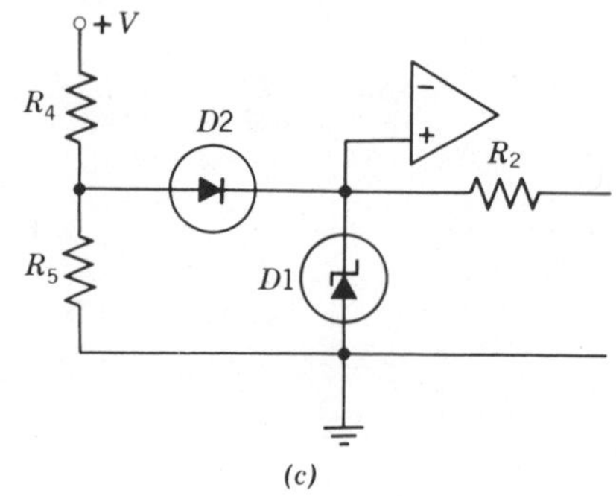

Figure 20-8 Basic op amp regulator. (*a*) With single pass transistor. (*b*) With two pass transistors. (*c*) Zener start-up circuit to be added to circuits of Figs. 20-8*a* and *b*.

Solving for V_L, we find

$$V_L = \frac{R_1 + R_f}{R_1} V_Z \tag{20-12}$$

Eq. 20-12 shows that V_L is a constant value that is independent of I_L. The output of the op amp is

$$V_{\text{out}} = V_{BE_1} + V_L \tag{20-13}$$

The currents in $Q1$ are indicated on the circuit diagram. Inspection of the circuit shows that

$$I_{E_1} = (1 + \beta)I_{B_1} \tag{20-14}$$

and

$$I_{E_1} = I_Z + I_L \tag{20-15}$$

The op amp produces an output current that supplies base current to $Q1$. If I_L is zero, I_{E_1} is I_Z, the current required to operate the Zener diode. When a load R_L is placed on the circuit, the emitter current of $Q1$ must increase. The op amp must produce a sufficient base current I_{B_1} in order to keep V_1 equal to V_Z. If the maximum output current of the op amp is 5 mA and if β for $Q1$ is 80, the maximum available current through $Q1$ is

$$I_{E_1} = (1 + \beta)I_{B_1} = (1 + 80)5 = 405 \text{ mA}$$

If I_Z is 5 mA, then the maximum load current I_L is 400 mA. The pass transistor $Q1$ must have an ability to dissipate

$$P_C = (V - V_L)I_C = (V - V_L)\beta I_B \text{ W} \tag{20-16}$$

When larger currents are required from the regulator, two pass transistors are connected in a Darlington circuit as shown in Fig. 20-8*b*. The previous equations are valid except Eq. 20-13, which must be changed to

$$V_{\text{out}} = V_{BE_1} + V_{BE_2} + V_L \tag{20-17}$$

Very large currents can be obtained from a regulator circuit by placing additional transistors in parallel with $Q2$.

When the circuit of Fig. 20-8*a* is turned off, both V_L and V_{out} are zero. The pass transistor $Q1$ is zero biased and cut off. If $+V$ is applied to the circuit, V_{out} and V_L remain at zero. In order to have this circuit operate properly when the circuit is turned on, a supplementary circuit (Fig. 20-8*c*) must be used to insure that the Zener diode reaches its operating value. The diode $D2$ supplies current to the Zener diode to raise V_Z from zero to its operating value. This arrangement is called a *start-up circuit.*

Problems

20-3.1 In Fig. 20-7*a*, V_Z is 6.2 V and I_Z is 10 mA. $+V$ is 15 V and V_{out} must be -9.6 V. R_1 is 10 kΩ. Determine the circuit component values.

20-3.2 In Fig. 20-7*b*, V_Z is 6.2 V and I_Z is 10 mA. $+V$ is 15 V and V_{out} must be 8.7 V. If R_1 is 10 kΩ, find R and R_f.

20-3.3 In Fig. 20-7*c*, V_Z is 6.2 V and I_Z is 10 mA. $+V$ is 15 V and R_1 is 20 kΩ. Find R and the range of V_{out}.

20-3.4 In Fig. 20-8*a*, V_Z is 5.0 V and I_Z is 8 mA. $+V$ is 15 V and V_L is 10 V. The op amp output is limited to ±20 mA. If β is 40 and if V_{BE} is 0.6 V for the transistor, what is the maximum range of regulated I_L? What are the values of R_f and R_2 if R_1 is 20 kΩ?

20-3.5 Repeat Problem 20-3.4 for the circuit given in Fig. 20-8*b*. Assume both transistors are identical.

Section 20-4 Regulator Features

Current Limiting

The current-limiting circuit shown in Fig. 20-9 is the basic op amp regulator circuit given in Fig. 20-8*a* with the addition of resistor R and transistor $Q2$. If the voltage drop across R, $I_{E_1}R$, is less than 0.6 V or 0.7 V, $Q2$ is cut off. If the voltage drop across R is greater than 0.6 V or 0.7 V, transistor $Q2$ is turned on. When transistor $Q2$ is turned on, the output current of the op amp is diverted from the base of the pass transistor $Q1$ into $Q2$. Therefore, the load current is prevented from increasing beyond a predetermined level.

Different values of R can be used to provide an adjustable current limit. In any case, even if a short circuit is placed on the load, the short-circuit current cannot exceed the predetermined value established by R. A heat sink must be

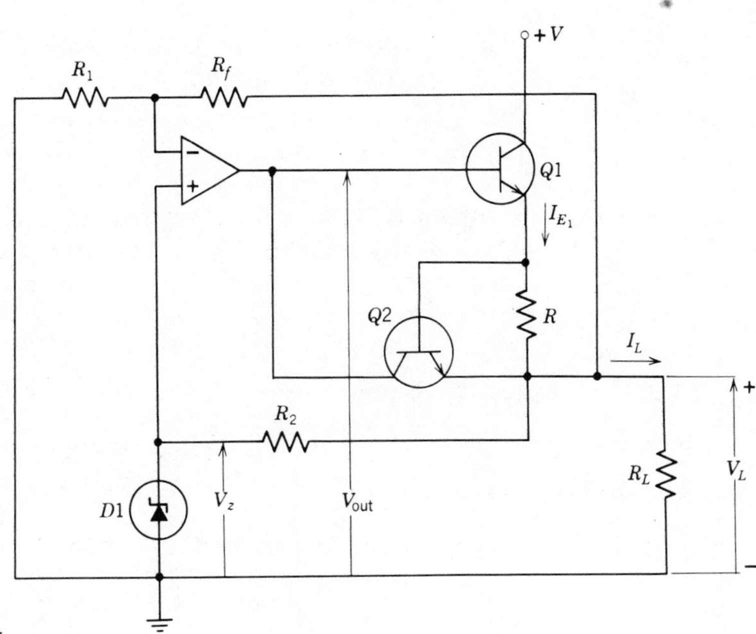

Figure 20-9 The current limiter.

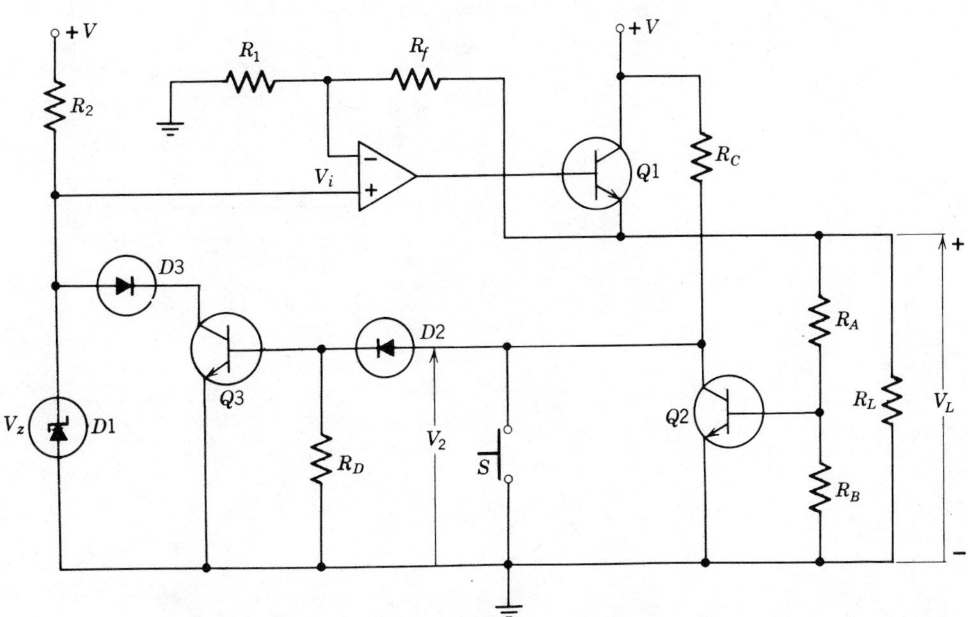

Figure 20-10 Circuit to provide short-circuit shutdown.

provided for $Q1$ to dissipate the full power of the limited short-circuit current.

$$P_C = VI_{sc} \text{ W} \tag{20-18}$$

Short-Circuit Shutdown The basic op amp regulator circuit of Fig. 20-8*a* is modified by returning R_2 to the supply voltage as shown in Fig. 20-10.

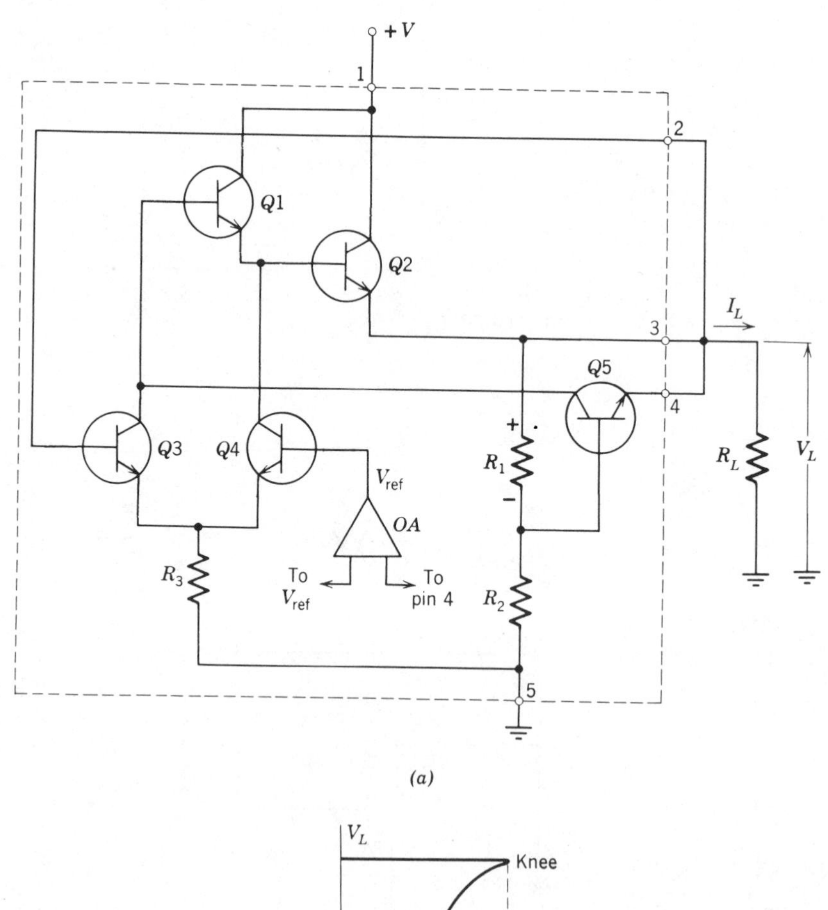

Figure 20-11 Fold-back. (*a*) Regulator without fold-back. (*b*) Regulator with fold-back. (*c*) Fold-back characteristic.

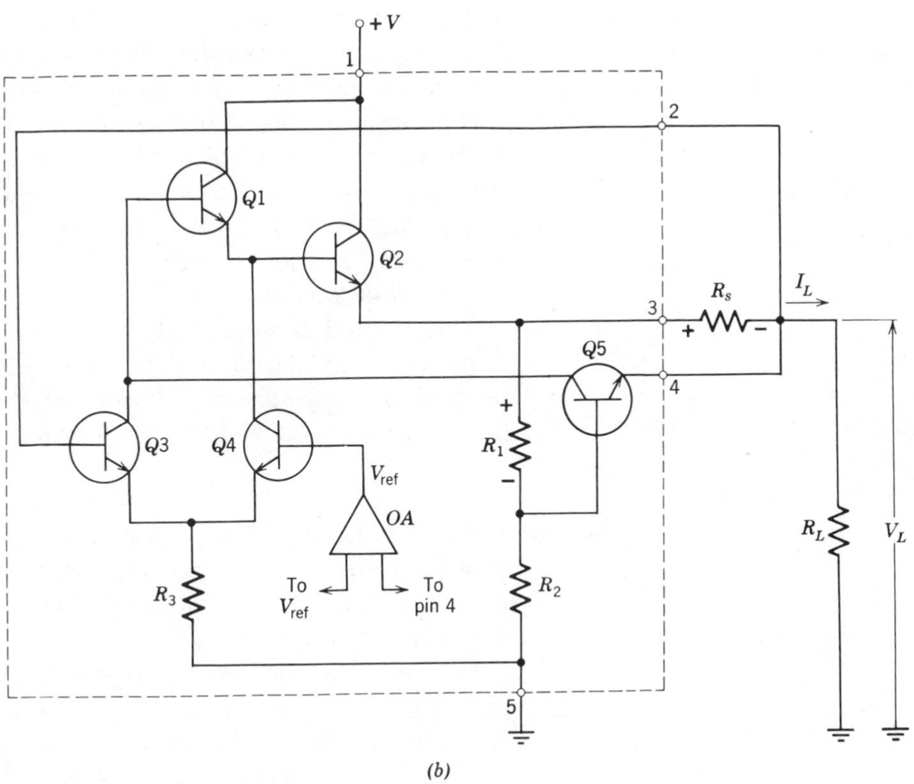

(b)

Components $Q2$, $Q3$, $D2$, $D3$, R_A, R_B, R_C, and R_D are added to the circuit.

When the regulator is operating normally, there is a base current in $Q2$ produced by the network of R_A and R_B. This base current saturates $Q2$. The collector voltage is sufficiently low so that the current in $D2$ and in the base of $Q3$ is zero. Then $Q3$ is cut off and its high impedance does not provide a shunting path across the Zener diode $D1$. Therefore the Zener diode $D1$ operates normally.

If a short circuit occurs across the load, the base current in $Q2$ falls to zero. Then $Q2$ is cut off and its collector voltage rises toward $+V$. Now current flows through $D2$ into the base of $Q3$. Current flows through $D2$ and $Q3$ saturates. The Zener diode becomes inoperative because the shunting voltages of $D3$ and $Q3$ are very low. This low voltage brings the noninverting input terminal of the op amp toward zero. Then, the inverting input terminal of the op amp is forced toward zero because V_i must remain very small. Then the output of the op amp is also forced toward zero. The base current in $Q1$ goes to zero and I_L goes to zero.

This *shutdown* is a stable condition and, as a result, the output V_L and I_L remain at zero. If the push-button switch S is momentarily depressed, $Q2$ is shorted out and the circuit returns to a normal operation if the short circuit on the load has been removed. Switch S is labeled *reset* on the panel.

Fold-Back A modified diagram for a regulator is shown in Fig. 20-11*a*. $Q1$ and $Q2$ are the series pass transistors that deliver the current I_L to the load. The circuit symbol labeled OA is the circuitry that provides the reference voltage V_{ref} to the base of $Q4$. The values of R_2 and R_1 are set so that the voltage drop across R_1 places a cutoff bias across the base and emitter of $Q5$. The collector current in $Q5$ is zero.

The circuit is now connected to a load as shown in Fig. 20-11*b*. The voltage drop I_LR_s across R_s has the polarity shown on the diagram. This polarity conteracts the polarity of the voltage drop across R_1. At a predetermined value of load current I'_L, transistor $Q5$ becomes forward biased. Current now flows in the collector of $Q5$.

The collector current in $Q5$ reduces the base current of $Q1$. This reduction in the base current of $Q1$ reduces both I_L and V_L. Since R_s, R_1, and R_2 are all in series across V_L, $Q5$ stays in conduction as I_L and V_L fall. The load current continues to fall until it reaches a final value at I''_L. The load voltage at this final point may be zero or V''_L. The path of operation of the voltage and current *fold-back* is shown in Fig. 20-11*c*.

At this point, we must summarize what we have done in this section to present a clear picture of the objectives of these circuits.

1. A current limiting circuit (Fig. 20-9) is used to establish a maximum current value that can be delivered to a load. This maximum current value can flow continuously without damage to the regulator.
2. The short-circuit shutdown (Fig. 20-10) is used to turn off the regulator completely at a predetermined value of load current. The power dissipation of the regulator is effectively zero after shutdown. A reset button must be pressed to turn the regulator back on.
3. A current or voltage fold-back circuit (Fig. 20-11*b*) reduces the output current to a low value as soon as a predetermined load current is reached. The power dissipation of the regulator after fold-back must be a safe value for continuous operation. After the cause of the overload is removed, the circuit returns to its normal output voltage level.

Section 20-5 The Precision Voltage Regulator

The precision voltage regulator* is a monolithic linear integrated circuit available in different physical packages (Fig. 20-12). The circuit (Fig. 20-13) incorporates a temperature-compensated voltage reference source, an op amp circuit used

*The data for the μA723 used in this section are available by courtesy of Texas Instruments Inc.

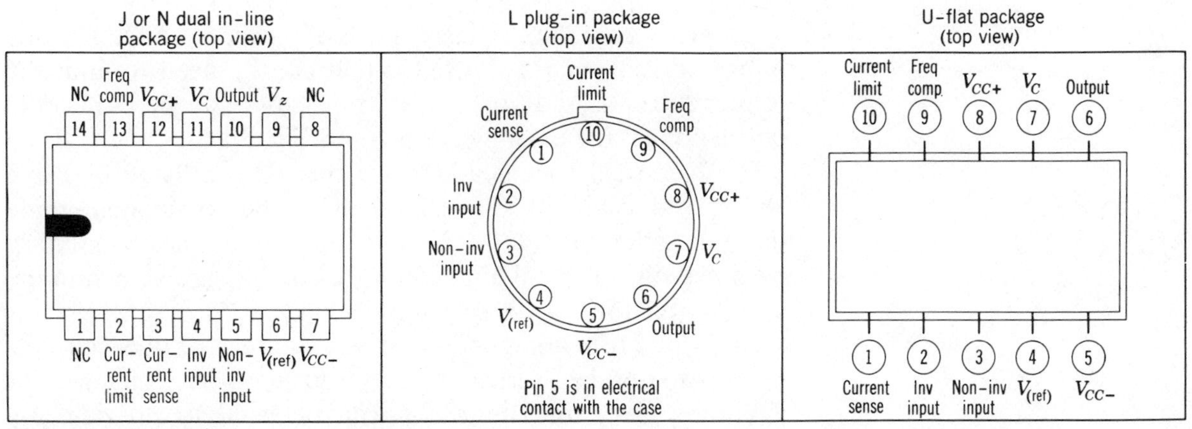

Figure 20-12 Available packaging for the μA723 precision voltage regulator. (*Courtesy Texas Instruments Inc.*)

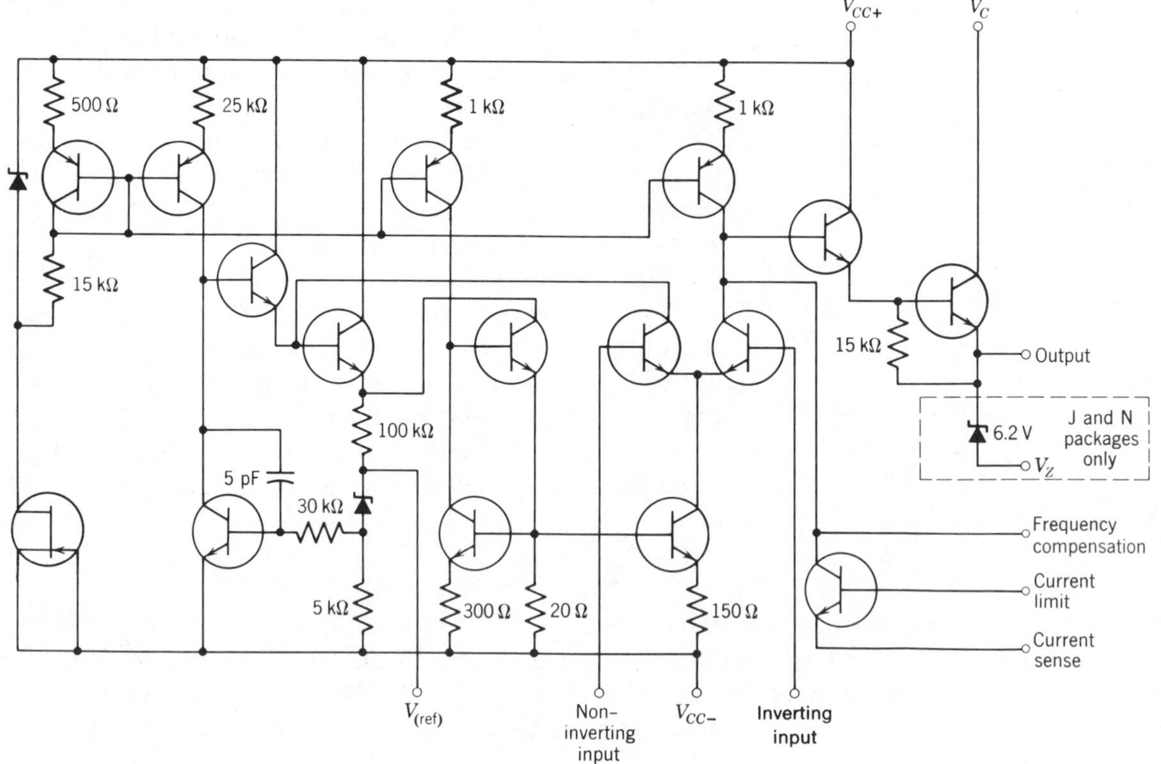

Figure 20-13 Schematic for the μA 723 precision voltage regulator. (*Courtesy Texas Instruments Inc.*)

as an error amplifier, a pass transistor capable of a 150-mA output current, and a transistor that can be used to limit the output current. Additional pass transistors can be connected externally as, for example, shown on Fig. 20-17.

The electrical specifications for this IC are listed in Table 20-2 and in Table 20-3. The μA723C is the commercial-grade unit and the μA723M is the unit that meets military specifications. Typical applications of this IC are shown in the circuits given in Fig. 20-14 through Fig. 20-25.

Table 20-4 lists component values required for these circuits in order to obtain a series of standardized output voltages. The equations given in Table 20-5 can be used to provide any output voltage. Instructions are also given to provide a variable output voltage.

Table 20-2
Electrical Specifications

Absolute Maximum Ratings over Operating Free-Air Temperature Range (Unless Otherwise Noted)

Peak voltage from V_{CC+} to V_{CC-} ($t_w \leq 50\ ms$)	50 V
Continuous voltage from V_{CC+} to V_{CC-}	40 V
Input-to-output voltage differential	40 V
Differential input voltage to error amplifier	±5 V
Voltage between noninverting input and V_{CC-}	8 V
Current from V_Z	25 mA
Current from $V_{(ref)}$	15 mA
Continuous total dissipation at (or below) 25°C free-air temperature	
J or N package	1000 mW
L package (see Note 1)	800 mW
U package	675 mW
Operating free-air temperature range: μA723M Circuits	−55°C to 125°C
μA723C Circuits	0°C to 150°C
Storage temperature range	−65°C to 150°C
Lead temperature $\frac{1}{16}$ in. from case for 60 s, J, L, or U package	300°C
Lead temperature $\frac{1}{16}$ in. from case for 10 s, N package	260°C

***Note*:** 1. This rating for the L package requires a heat sink that provides a thermal resistance from case to free-air, θ_{CA}, of not more than 105°C/W.

Recommended Operating Conditions

	Min	Max	Unit
Input voltage, V_I	9.5	40	V
Output voltage, V_O	2	37	V
Input-to-output voltage differential, $V_C - V_O$	3	38	V
Output current, I_O		150	mA

***Source*:** Texas Instruments Inc.

Problems

Assume the maximum permissible output current of the IC is 150 mA. Also P_C is limited to 800 mW for all problems. In all circuits use 10 kΩ for R_1.

20-5.1 Use Fig. 20-14. The load voltage is 4.2 V and $+V_1$ is +10 V. Determine the external circuit components required and determine the maximum permissible value of I_L.

20-5.2 Use Fig. 20-15. The load voltage is 11.5 V and $+V_1$ is 20 V. Determine the external circuit components required and determine the maximum permissible value of I_L.

Table 20-3

Electrical Characteristics at Specified Free-Air Temperature

Parameter	Test Conditions*		μA723M Min	μA723M Typ	μA723M Max	μA723C Min	μA723C Typ	μA723C Max	Unit
Input regulation	$V_I = 12$ V to $V_I = 15$ V	25°C		0.01%	0.1%		0.01%	0.1%	
	$V_I = 12$ V to $V_I = 40$ V	25°C		0.02%	0.2%		0.1%	0.5%	
	$V_I = 12$ V to $V_I = 15$ V	Full range			0.3%			0.3%	
Ripple rejection	$f = 50$ Hz to 10 kHz, $C_{(ref)} = 0$	25°C		74			74		dB
	$f = 50$ Hz to 10 kHz, $C_{(ref)} = 5\ \mu$F	25°C		86			86		
Output regulation	$I_O = 1$ mA to $I_O = 50$ mA	25°C		−0.03%	−0.15%		−0.03%	−0.2%	
		Full range			−0.6%			−0.6%	
Reference voltage, $V_{(ref)}$		25°C	6.95	7.15	7.35	6.8	7.15	7.5	V
Standby current	$V_I = 30$ V, $I_O = 0$	25°C		2.3	3.5		2.3	4	mA
Temperature coefficient of output voltage		Full range		0.002	0.015		0.003	0.015	%/°C
Short-circuit output current	$R_{sc} = 10\ \Omega$, $V_O = 0$	25°C		65			65		mA
Output noise voltage	$BW = 100$ Hz to 10 kHz, $C_{(ref)} = 0$	25°C		20			20		μV
	$BW = 100$ Hz to 10 kHz, $C_{(ref)} = 5\ \mu$F	25°C		2.5			2.5		

Source**:** Texas Instruments Inc.

*Full range for μA723M is −55°C to 125°C and for μA723C is 0°C to 70°C.

Note**:** For all values in this table the device is connected as shown in Figure 20-14 with the divider resistance as seen by the error amplifier ⩽10 kΩ. Unless otherwise specified, $V_I = V_{CC+} = V_C = 12$ V, $V_{CC-} = 0$, $V_O = 5$ V, $I_O = 1$ mA, $R_{sc} = 0$, and $C_{(ref)} = 0$.

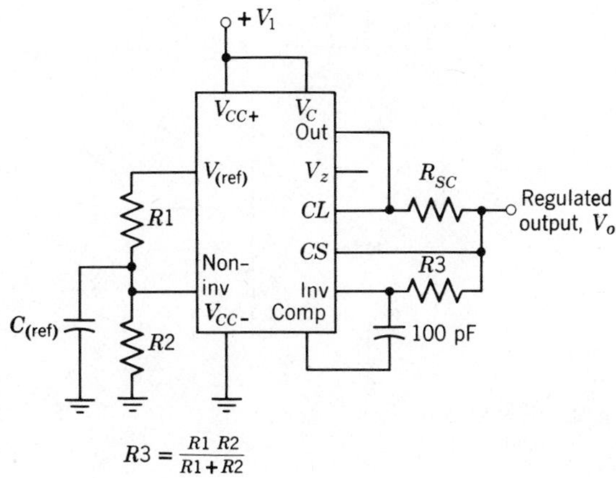

Figure 20-14 Basic low-voltage regulator ($V_0 = 2$ to 7 volts). (*Courtesy Texas Instruments Inc.*)

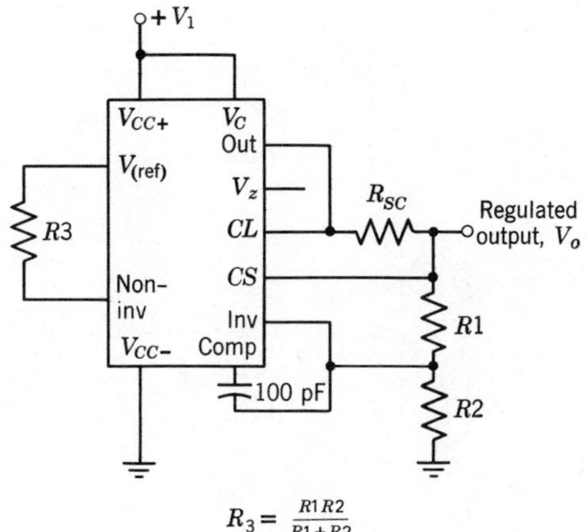

Figure 20-15 Basic high-voltage regulator ($V_o = 7$ to 37 volts). (*Courtesy Texas Instruments Inc.*)

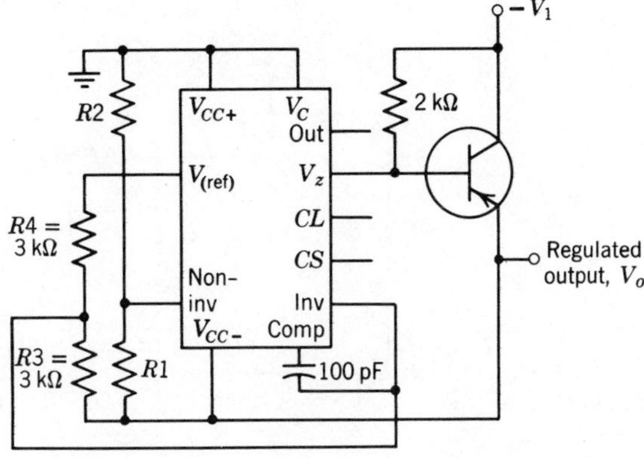

Figure 20-16 Negative voltage regulator. If the magnitude of $-V_i$ is less than 9 V, connect V_{CC+} and V_C to a positive supply such that V_{CC+} to V_{CC-} is greater than 9 V. (*Courtesy Texas Instruments Inc.*)

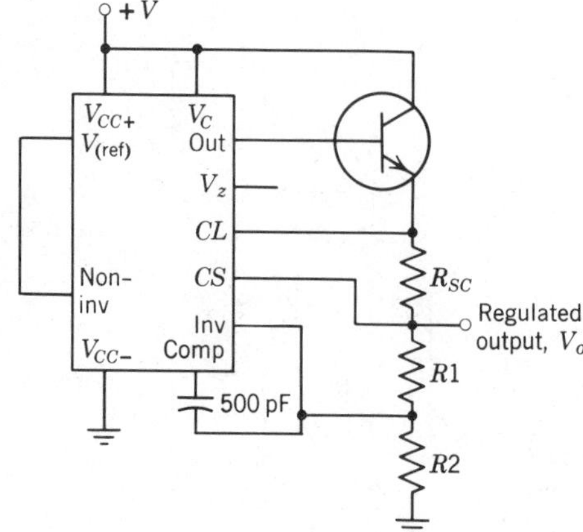

Figure 20-17 Positive voltage regulator (external *N-P-N* pass transistor). (*Courtesy Texas Instruments Inc.*)

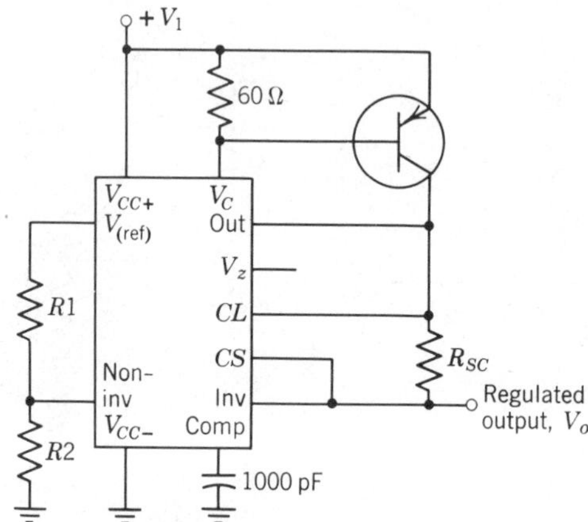

Figure 20-18 Positive voltage regulator (external *P-N-P* pass transistor). (*Courtesy Texas Instruments Inc.*)

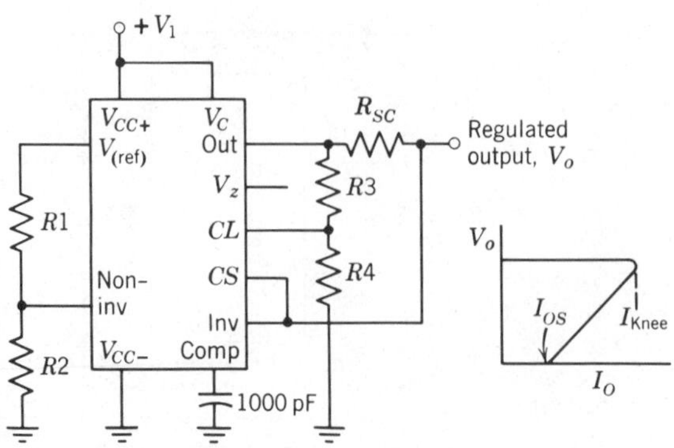

Figure 20-19 Fold-back current limiting. (*Courtesy Texas Instruments Inc.*)

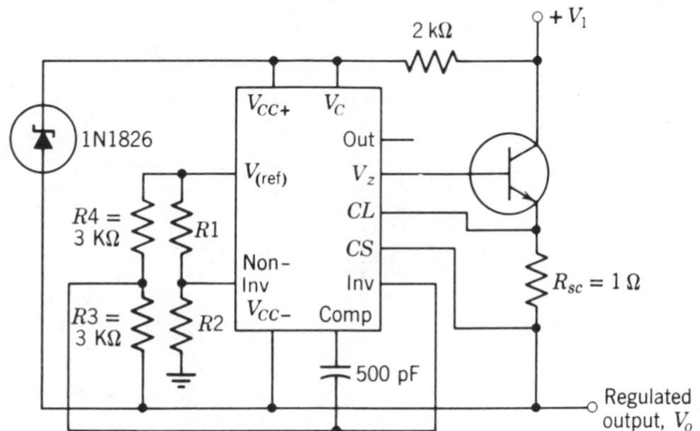

Figure 20-20 Positive floating regulator. (*Courtesy Texas Instruments Inc.*)

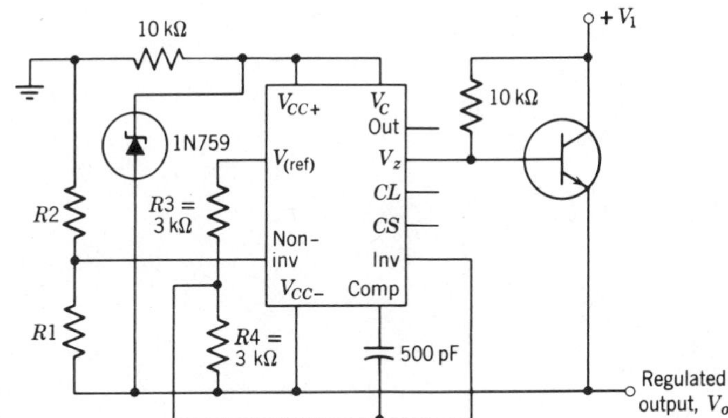

Figure 20-21 Negative floating regulator. (*Courtesy Texas Instruments Inc.*)

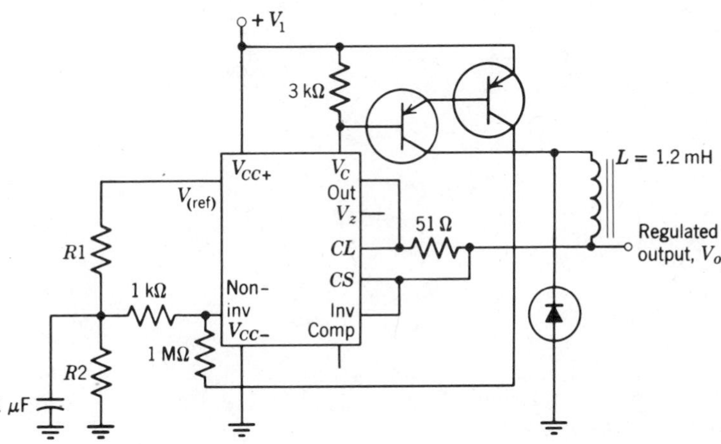

Figure 20-22 Positive switching regulator. (*Courtesy Texas Instruments Inc.*)

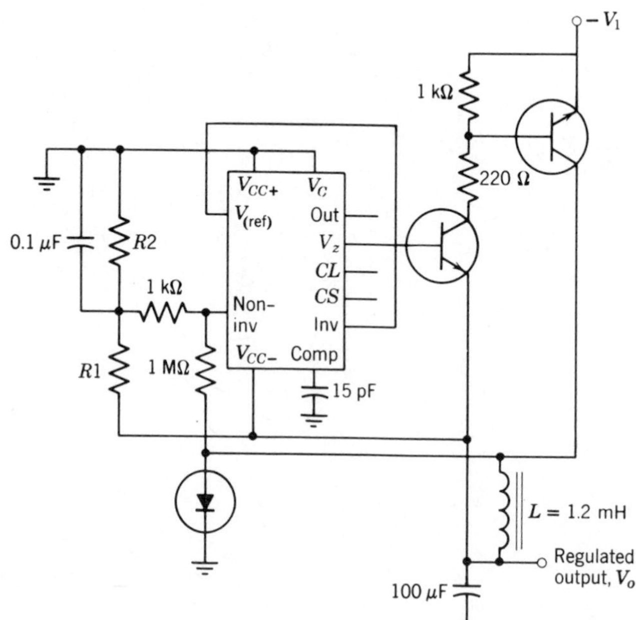

Figure 20-23 Negative switching regulator. (*Courtesy Texas Instruments Inc.*)

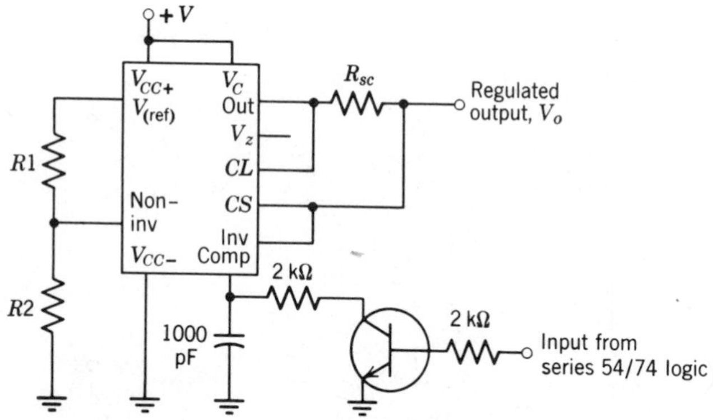

Figure 20-24 Remote shutdown regulator with current limiting. (*Courtesy Texas Instruments Inc.*)

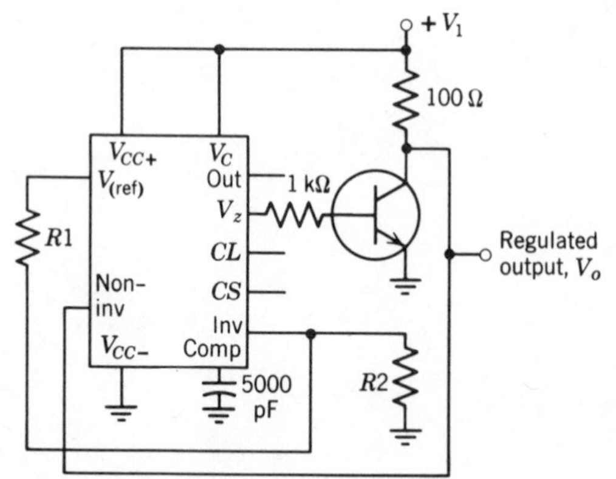

Figure 20-25 Shunt regulator. (*Courtesy Texas Instruments Inc.*)

Table 20-4
Resistor Values (kΩ) for standard output voltages

Output Voltage	Applicable Figures	Fixed Output (±5%)		Output Adjustable (±10%)		
		R1 (kΩ)	R2 (kΩ)	R1 (kΩ)	P1 (kΩ)	R2 (kΩ)
+3.0	20-14, 20-18, 20-19, 20-24, 20-25	4.12	3.01	1.8	0.5	1.2
+3.6	20-14, 20-18, 20-19, 20-24, 20-25	3.57	3.65	1.5	0.5	1.5
+5.0	20-14, 20-18, 20-19, 20-24, 20-25	2.15	4.99	0.75	0.5	2.2
+6.0	20-14, 20-18, 20-19, 20-24, 20-25	1.15	6.04	0.5	0.5	2.7
+9.0	20-15, 20-17, 20-18, 20-19, 20-22, 20-25	1.87	7.15	0.75	1.0	2.7
+12	20-15, 20-17, 20-18, 20-19, 20-22, 20-25	4.87	7.15	2.0	1.0	3.0
+15	20-15, 20-17, 20-18, 20-19, 20-22, 20-25	7.87	7.15	3.3	1.0	3.0
+28	20-15, 20-17, 20-18, 20-19, 20-22, 20-25	21.0	7.15	5.6	1.0	2.0
+45	20-20	3.57	48.7	2.2	10	39
+75	20-20	3.57	78.7	2.2	10	68
+100	20-20	3.57	105	2.2	10	91
+250	20-20	3.57	255	2.2	10	240
−6	20-16, 20-23	3.57	2.43	1.2	0.5	0.75
−9	20-16, 20-23	3.48	5.36	1.2	0.5	2.0
−12	20-16, 20-23	3.57	8.45	1.2	0.5	3.3
−15	20-16, 20-23	3.57	11.5	1.2	0.5	4.3
−28	20-16, 20-23	3.57	24.3	1.2	0.5	10
−45	20-21	3.57	41.2	2.2	10	33
−100	20-21	3.57	95.3	2.2	10	91
−250	20-21	3.57	249	2.2	10	240

20-5.3 Use Fig. 20-16. The load voltage is -10 V and $-V_1$ is -15 V. The β for the transistor is 50. What is the maximum value of I_L? Determine the external circuit components and the value of P_C for the transistor.

20-5.4 Use Fig. 20-17. Repeat Problem 20-5.3 for this circuit. Assume all voltages are now positive.

20-5.5 Use Fig. 20-18. Repeat Problem 20-5.4 for this circuit.

20-5.6 Use Fig. 20-19. V_L is $+18$ V and $+V_1$ is $+25$ V. What is the maximum allowable value of I_L. Then I_{OS} must be 30% of the knee current. What are the external circuit component values that are required?

Table 20-5
Formulas for Output Voltage

Outputs from +2 to +7 volts (Figs. 20-14, 20-18, 20-19, 20-22, 20-24, 20-25, 20-17*) $V_O = V_{(ref)} \times \frac{R2}{R1 + R2}$	Outputs from +4 to +250 volts (Fig. 20-20) $V_O = \frac{V_{(ref)}}{2} \times \frac{R2 - R1}{R1}$; $R3 = R4$	Current limiting $I_{(limit)} \approx \frac{0.65\text{ V}}{R_{SC}}$
Outputs from +7 to +37 volts (Figs. 20-15, 20-17, 20-18*, 20-19*, 20-22*, 20-24*, 20-25*) $V_O = V_{(ref)} \times \frac{R1 + R2}{R2}$	Outputs from −6 to −250 volts (Figs. 20-16, 20-21, 20-23) $V_O = -\frac{V_{(ref)}}{2} \times \frac{R1 + R2}{R1}$; $R3 = R4$	Fold-back current limiting (Fig. 20-19) $I_{(knee)} \approx \frac{V_O R3 + (R3 + R4)0.65\text{ V}}{R_{SC} R4}$; $I_{OS} \approx \frac{0.65\text{ V}}{R_{SC}} \times \frac{R3 + R4}{R4}$

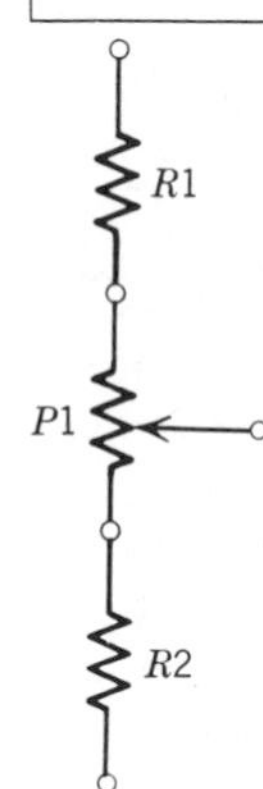

Notes:

1. Figures 1 through 12 show the $R1/R2$ divider across either V_O or $V_{(ref)}$. Figure numbers with * may be used if the $R1/R2$ divider is placed across the other voltage ($V_{(ref)}$ or V_O) that it was not placed across in the figures without *.

2. To make the voltage adjustable, the $R1/R2$ divider shown in the figures must be replaced by the divider shown at the right.

3. For negative output voltages less than 9 V, V_{CC+} and V_C must be connected to a positive supply such that the voltage between V_{CC+} and V_{CC-} is greater than 9 V.

4. When 10-lead uA723 devices are used in applications requiring V_Z, an external 6.2-V regulator diode must be connected in series with the V_O terminal.

Section 20-6 The Complete Voltage Regulator

Fixed-voltage monolithic IC voltage regulators are available to provide various output currents and voltages. They have three terminals:

1. Common. **2.** Input. **3.** Output.

These units can be connected into a circuit without a need for additional external components beyond the usual input and output capacitors that are used to suppress intermittent noise pulses. The units have internal circuits similar to that of the precision voltage regulator discussed in Section 20-5. The units also have internal current limiting and thermal shut-

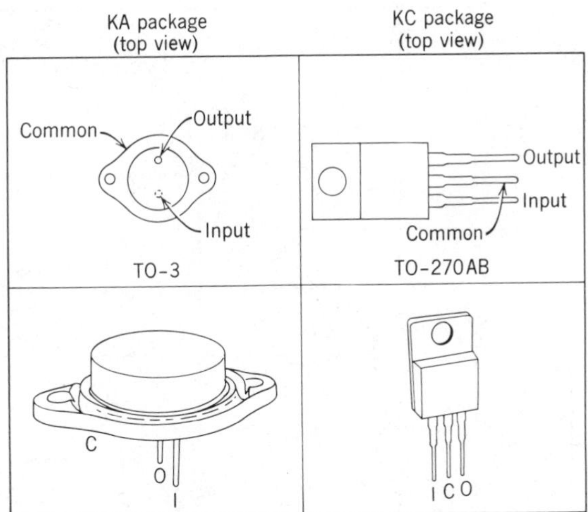

Figure 20-26 Outline drawings for the μA7808 positive-voltage regulator.

down, which make the regulator immune to an overload or to a short circuit.

The μA7808C is the commercial version* and the μA7808M is the military version of a positive-voltage regulator that delivers up to 1.5 A at +8 V to a load. The outline drawings for the available packaging are given in Fig. 20-26. The internal circuit is shown in Fig. 20-27. The electrical specifications are listed in Table 20-6 and in Table 20-7. The derating curves are given in Fig. 20-28.

*The data for the μA7808 used in this section are available by courtesy of Texas Instruments Inc.

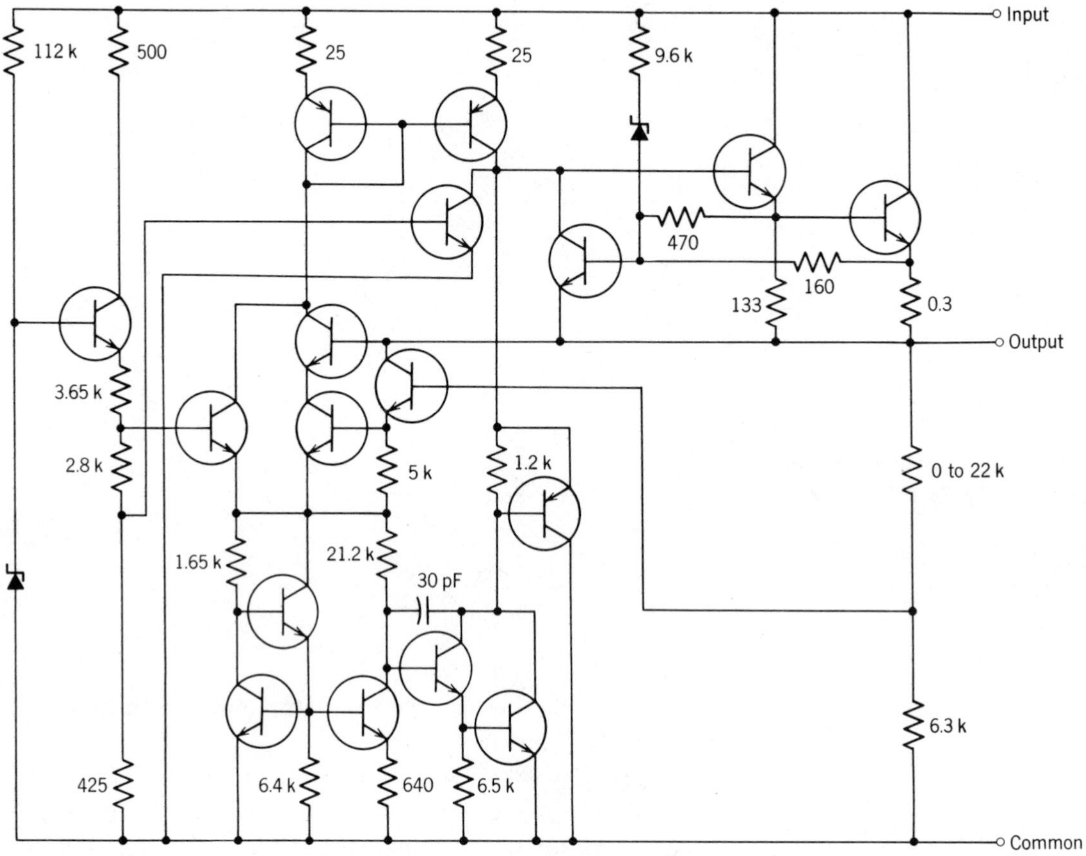

Figure 20-27 Schematic diagram for the μA7808 positive-voltage regulator.

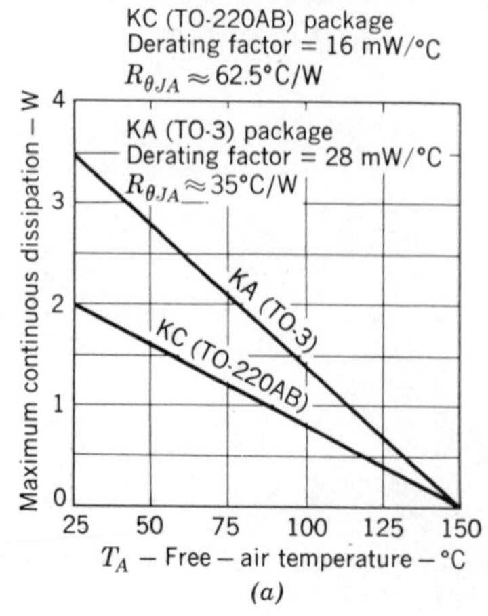

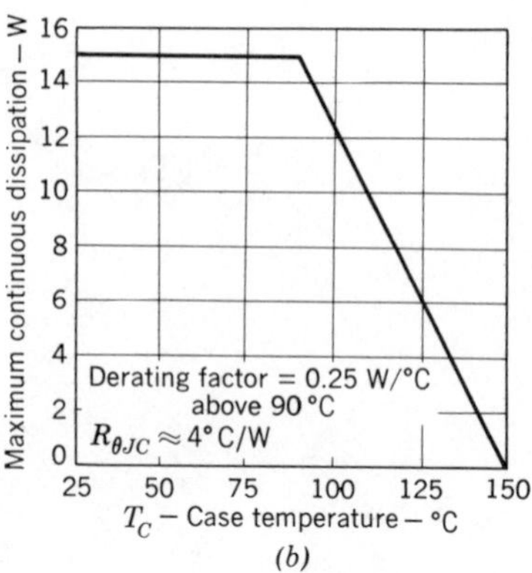

Figure 20-28 Dissipation derating curves. (*a*) Free air. (*b*) Case temperature for both TO-3 and TO-220AB packages. (*Courtesy Texas Instruments Inc.*)

Table 20-6
Absolute Maximum Ratings over Operating Temperature Range

		μA7808M	μA7808C	Unit
Input voltage		35	35	V
Continuous total dissipation at 25°C free-air temperature	KA (TO-3) package	3.5	3.5	W
	KC (TO-220AB) package		2	W
Continuous total dissipation at (or below) 25°C case temperature		15	15	W
Operating free-air, case, or virtual junction temperature range		−55 to 150	0 to 150	°C
Storage temperature range		−65 to 150	−65 to 150	°C
Lead temperature 1/16 in. from case for 60 s	KA (TO-3) package	300	300	°C
Lead temperature 1/16 in. from case for 10 s	KC (TO-220AB) package		260	°C

(*Courtesy Texas Instruments Inc.*).

Table 20-7
Electrical Characteristics at Specified Virtual Junction Temperature, $V_I = 14$ V, $I_O = 500$ mA

Parameter	Test Conditions*			μA7808M Min	μA7808M Typ	μA7808M Max	μA7808C Min	μA7808C Typ	μA7808C Max	Unit
Output voltage			25°C	7.7	8	8.3	7.7	8	8.3	V
	$I_O = 5$ mA to 1 A,	$V_I = 11.5$ V to 23 V	−55°C to 150°C	7.6		8.4				
	$P \leqslant 15$ W	$V_I = 10.5$ V to 23 V	0°C to 125°C				7.6		8.4	
Input regulation	$V_I = 10.5$ V to 25 V		25°C		6	80		6	160	mV
	$V_I = 11$ V to 17 V				2	40		2	80	
Ripple rejection	$V_I = 11.5$ V to 21.5 V, f = 120 Hz		−55°C to 150°C	62	72					dB
			0°C to 125°C				56	72		
Output regulation	$I_O = 5$ mA to 1.5 A		25°C		12	80		12	160	mV
	$I_O = 250$ mA to 750 mA				4	40		4	80	
Output resistance	$f = 1$ kHz		−55°C to 150°C		0.016					Ω
			0°C to 125°C					0.016		
Temperature coefficient of output voltage	$I_O = 5$ mA		0°C to 150°C		−0.8					mV/°C
			0°C to 125°C					−0.8		
Output noise voltage	$f = 10$ Hz to 100 kHz		25°C		52			52		μV
Dropout voltage	$I_O = 1$ A		25°C		2.0			2.0		V
Bias current			25°C		4.3	6		4.3	8	mA
Bias current change	$V_I = 11.5$ V to 25 V		−55°C to 150°C			0.8				mA
	$V_I = 10.5$ to 25 V		0°C to 125°C						1	
	$I_O = 5$ mA to 1 A		−55°C to 150°C			0.5				
			0°C to 125°C						0.5	
Short-circuit output current			25°C		450			450		mA
Peak output current			25°C		2.2			2.2		A

*All characteristics are measured with a capacitor across the input of 0.33 μF and a capacitor across the output of 0.1 μF and all characteristics except noise voltage and ripple rejection ratio are measured using pulse techniques ($t_w \leq 10$ ms, duty cycles $\leq 5\%$). Output voltage changes due to changes in internal temperature must be taken into account separately. (*Courtesy Texas Instruments Inc.*)

Supplementary Problems

The μA723C precision voltage regulator is to be used to construct a +5-V power supply that can deliver up to 1 A, maximum. The rectifier supplies 10 V. In each problem, determine the power dissipation of the μA723C regulator (maximum, 800 mW), the power dissipation of the pass transistor, and the values of all resistors to be used in the external circuit. R_1 is 10 kΩ. The pass transistor has a β of 50.

20-1 The circuit shown in Fig. 20-14 is to be used.

20-2 The circuit shown in Fig. 20-15 is to be used.

20-3 The circuit shown in Fig. 20-17 is to be used.

20-4 The circuit shown in Fig. 20-18 is to be used.

20-5 The circuit shown in Fig. 20-19 is to be used. I_{OS} is 50 mA. The foldback current-limiting equations must be solved for R_{SC}.

20-6 The specifications are the same as before except that a −5-V supply is required using a rectifier that supplies −10 V. The circuit shown in Fig. 20-16 is to be used.

For Problems 20-7 through 20-10, the specifications are the same as before except that a +15-V, 0.5-A regulator is required from a 20-V rectifier.

20-7 The circuit shown in Fig. 20-15 is to be used.

20-8 The circuit shown in Fig. 20-17 is to be used.

20-9 The circuit shown in Fig. 20-18 is to be used.

20-10 The circuit shown in Fig. 20-19 is to be used. I_{OS} is 50 mA. The foldback current-limiting equations must be solved for R_{SC}.

20-11 The specifications are the same as before except that a −15-V, 0.5-A regulator is required from a 20-V rectifier. The circuit shown in Fig. 20-16 is to be used.

20-12 Give examples where the circuit shown in Fig. 20-20 could be used.

20-13 Give examples where the circuit shown in Fig. 20-21 could be used.

Chapter 21 Breakdown Devices

The unijunction transistor (Section 21-1) is used as the active element in the relaxation oscillator (Section 21-2). The four-layer diode, the Shockley diode (Section 21-3), can be modified with a gate electrode to form the silicon controlled rectifier (Section 21-4). The four-layer arrangement of the silicon controlled rectifier is modified into the ac controlled device, the triac (Section 21-5).

Section 21-1 The Unijunction Transistor

One type of construction of the *double-based diode* or *unijunction transistor* (*UJT*) has a small rod of P material extending into a block of N material to form a P–N junction (Fig. 21-1*a*). Two metallic contacts, called *ohmic contacts*, Base 1 ($B1$) and Base 2 ($B2$), are welded to the N material *without* creating P-N junctions at the points of the welds. The electrode $B1$ is the common return for the circuit. The circuit symbols for the UJT are given in Figs. 21-1*b* and 21-1*c*. The letter symbols for the currents and voltages are shown in Fig. 21-1*d*.

When a positive voltage V_{BB} is applied from $B1$ to $B2$, there is a uniform voltage drop through the N material that has a linear *interbase resistance* R_{BB} ranging from 4.7 kΩ to 9.1 kΩ, measured when the emitter is open. The physical location of the emitter is at some point between $B1$ and $B2$. The voltage at that location can be determined by the voltage divider rule and is given by

$$\eta V_{BB}$$

The value of η, the *intrinsic standoff ratio*, is a decimal value that ranges between 0.50 and 0.85 in commercial units.

Let us assume that 25 V is applied between $B1$ and $B2$. When the emitter is grounded (connected to $B1$), a small

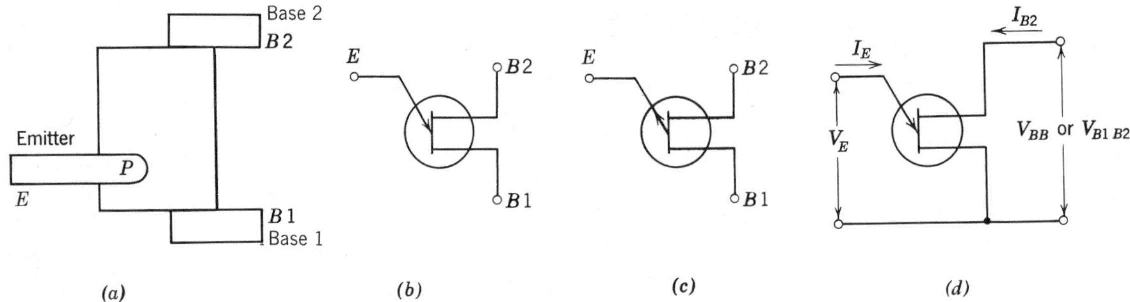

Figure 21-1 The unijunction transistor. (*a*) Construction. (*b*) Symbol for the UJT with an *N*-type. (*c*) Symbol for the UJT with a *P*-type base. (*d*) Nomenclature.

leakage current I_{EO} of the order of 1 μA flows in the emitter (Fig. 21-2*a*). The emitter junction now has a reverse voltage equal to ηV_{BB} volts. The junction is reverse biased as long as the voltage from emitter to ground ($B1$) is less than V_P.

$$V_P = \eta V_{BB} + V_D \tag{21-1}$$

where V_D is the barrier (junction) voltage for a silicon junction (approximately 0.6 V).

When the voltage on the emitter equals V_P or $(\eta V_{BB} + V_D)$, the emitter junction becomes forward biased. At this value of V_P (the *peak voltage*), the forward current is I_P (the *peak current*) and has a value of the order of 0.4 to 5 μA. This small current injects enough current carriers into the N material to reduce R_{BB} to a very low value. Now the emitter current is limited primarily to the value determined by the resistance in the external circuit. The emitter voltage falls from the peak point V_P toward zero as I_E becomes large. However, the circuit, after it *fires*, acts like a forward-biased diode. We find that the junction voltage plus the *IR* drop in the bulk resistance produces a voltage that increases as I_E increases. This increasing diode voltage drop is labeled $V_{E,\text{sat}}$. Consequently, we find that the sum of the two effects results in a minimum voltage called the *valley point*, V_V. The current corresponding to V_V is the *valley current* I_V, as shown in Fig. 21-2*a*.

Between $V_P(I_P)$ and $V_V(I_V)$, the voltage decreases with an

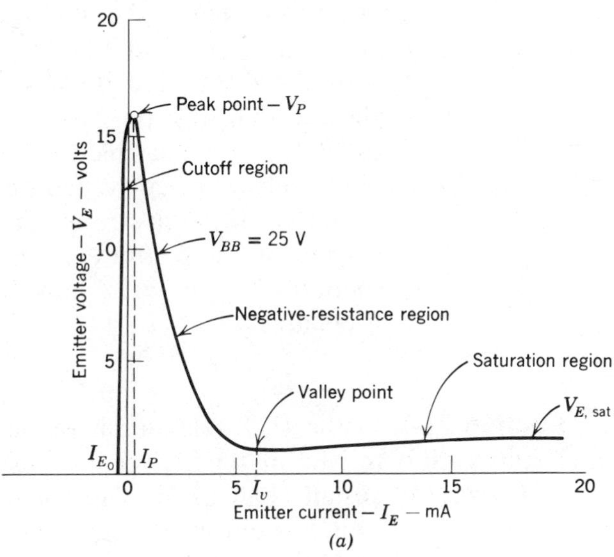

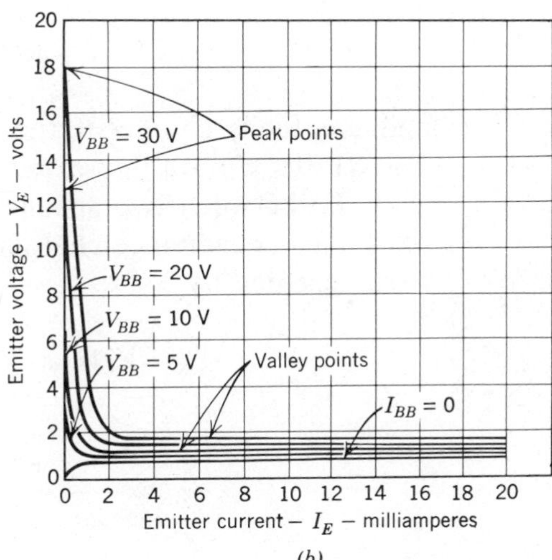

Figure 21-2 Unijunction transistor characteristics. (*a*) Expanded characteristic. (*b*) Characteristic for a typical UJT.

increase in current. We define the ac emitter resistance r_e as

$$r_e \equiv \frac{\Delta V_E}{\Delta I_E} \tag{21-2}$$

where ΔV_E is a change in emitter voltage and ΔI_E is a change in emitter current. As the emitter current increases ($+\Delta I_E$), the emitter voltage decreases ($-\Delta V_E$). Accordingly r_e is a *negative resistance.* Thus, the UJT has a negative-resistance region between $V_P(I_P)$ and $V_V(I_V)$.

In the discussion on positive feedback, Section 16-2, we showed that, if a positive feedback is sufficient, the amplifier oscillates. It is possible to show that a circuit that oscillates is effectively a negative resistance. Then, conversely, we can state that, if an electronic circuit has the property of negative resistance, the circuit can be used as an oscillator. In the next section we will show how the UJT is used as a relaxation oscillator.

Section 21-2 The UJT Relaxation Oscillator

In the UJT relaxation oscillator we make use of the characteristic of the charge and discharge of a capacitor in an R-C circuit (Fig. 21-3). The capacitor begins to charge when the switch is closed at zero time. The voltage across the capacitor builds up with time as a function of the time constant

$$\tau = RC \text{ seconds}$$

and eventually charges to the asymptotic value V as shown in Fig. 21-3*b*. We are interested primarily in the time difference $(T_2 - T_1)$ between two specific voltage levels V_1 and V_2. The resistance-capacitance charging curve for the exponential rise is represented by

$$v_c = V[1 - \epsilon^{-t/RC}] = V - V\epsilon^{-t/RC}$$

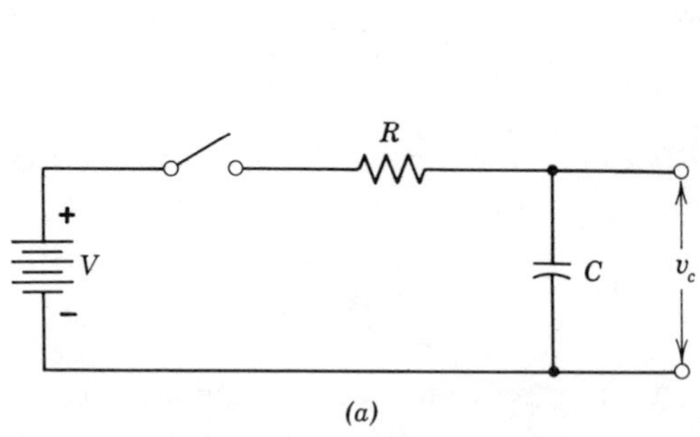

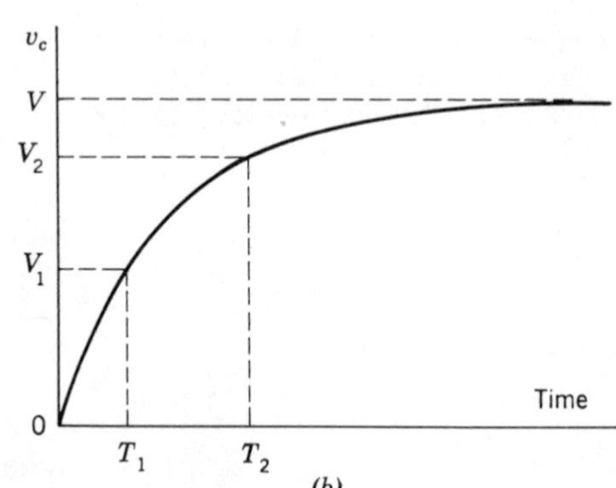

Figure 21-3 R-C charging circuit. (*a*) Circuit. (*b*) Response.

The potential V_1 at T_1 is

$$V_1 = V - V\epsilon^{-T_1/RC}$$

and the potential V_2 at T_2 is

$$V_2 = V - V\epsilon^{T_2/RC}$$

Solving each for T_1 and T_2, we have

$$\epsilon^{-T_2/RC} = \frac{V - V_2}{V} \quad \text{and} \quad \epsilon^{-T_1/RC} = \frac{V - V_1}{V}$$

Then

$$-\frac{T_2}{RC} = \ln\frac{V - V_2}{V} = \ln(V - V_2) - \ln V$$

and

$$-\frac{T_1}{RC} = \ln\frac{V - V_1}{V} = \ln(V - V_1) - \ln V$$

Subtracting the first from the second gives

$$\frac{T_2}{RC} - \frac{T_1}{RC} = \ln(V - V_1) - \ln(V - V_2)$$

Then

$$T_2 - T_1 = RC\ln\frac{V - V_1}{V - V_2} \tag{21-3a}$$

When each term in the fraction is divided by the source voltage V, we have

$$T_2 - T_1 = RC\ln\frac{1 - V_1/V}{1 - V_2/V} \tag{21-3b}$$

Example 21-1
In Fig. 21-3

$$R = 2\text{ k}\Omega, \quad C = 3.0\ \mu\text{F}, \quad V = 20\text{ V}, \quad V_2 = 10\text{ V}, \text{ and } V_1 = 0.7\text{ V}$$

Determine the rise time of the circuit from T_1 to T_2.

Solution

Substituting numerical values into Eq. 21-3*b*, we have

$$T_2 - T_1 = RC \ln \frac{1 - V_1/V}{1 - V_2/V} = (2000\ \Omega)(3.0 \times 10^{-6}\ \text{F}) \ln \frac{1 - \frac{0.7}{20}}{1 - \frac{10}{20}}$$

$$= \mathbf{3.9 \times 10^{-3}\ s = 3.9\ ms} \qquad (21\text{-}3b)$$

Now let us apply these concepts to the UJT circuit shown in Fig. 21-4*a*. The charging circuit is V_{EE} applied to the circuit of C_E and R_E. The emitter voltage v_e increases exponentially to V_P (Fig. 21-4*b*). When the emitter voltage reaches V_P, the emitter-to-base junction becomes forward biased. As soon as the emitter current reaches I_P, this diode breaks down and the emitter-to-base resistance drops to a very low value. The

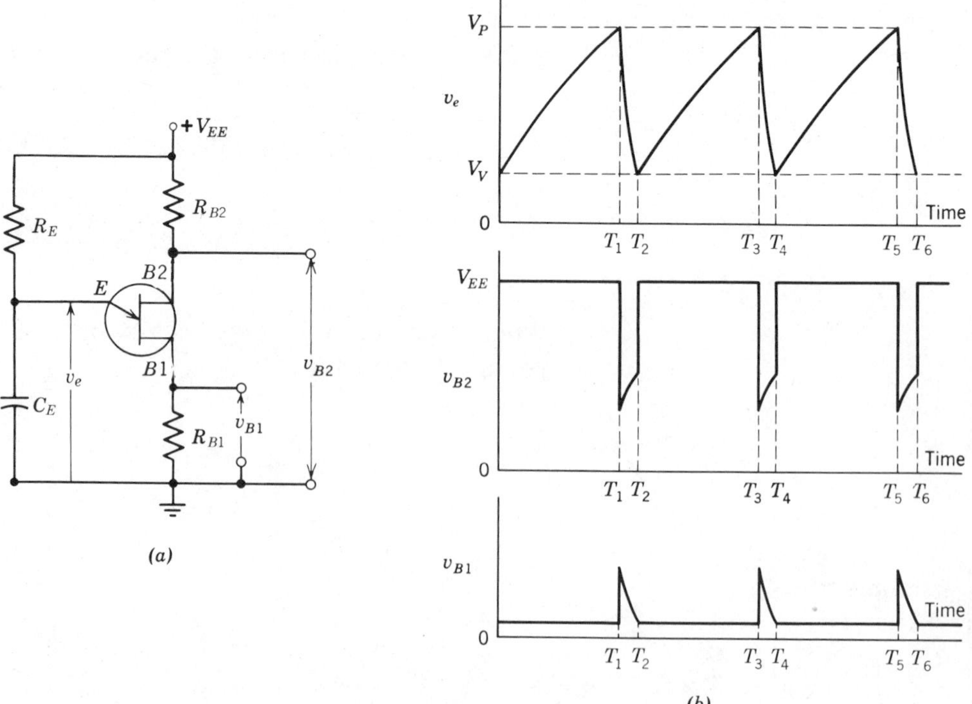

Figure 21-4 Unijunction relaxation oscillator. (*a*) Circuit. (*b*) Waveforms.

capacitor is now quickly discharged from V_P to V_V since the sum of the diode resistance and R_{B1} is much less than R_E.

In order for the circuit to break down or "fire," the current in the emitter determined by

$$(V_{EE} - V_P)/R_E$$

must be greater than I_P. Practically, then, the maximum value of R_E is limited to about 3 MΩ. In order for the circuit to "turn off," the operation of the circuit must be confined to the negative-resistance region. Therefore, when the capacitor becomes discharged after firing, the current in the emitter must be less than the valley current I_V or

$$\left(I_E = \frac{V_{EE}}{R_E}\right) < I_V$$

Practically, the minimum value of R_E is approximately 3 kΩ. Then the allowable range of R_E is approximately

$$3\ \text{k}\Omega < R_E < 3\ \text{M}\Omega$$

The range of values used for R_{B2} is from 4.7 kΩ to about 9.1 kΩ. R_{B1} is less than 100 Ω and is typically 47 Ω.

The waveforms obtained from the circuit are shown in Fig. 21-4*b*. Most applications use the relaxation oscillator circuit to provide the waveform of voltage v_{B1} taken across R_{B1}. This output voltage provides sharp pulses of voltage that are used to fire or "trigger" another circuit. The interval between pulses is approximately the time given by Eq. 21-3*b*. When we substitute into Eq. 21-3*b* voltage values using the notation for the UJT, Eq. 21-3*b* becomes

$$T = T_3 - T_2 = T_5 - T_4 = R_E C_E \ln \frac{1 - V_V/V_{EE}}{1 - V_P/V_{EE}} \qquad (21\text{-}4a)$$

where T is the period of the ac waveform in seconds. We defined V_P/V_{EE} as the intrinsic standoff ratio, η. Also, if we assume that V_V is much smaller than V_{EE}, the term V_V/V_{EE}

can be neglected. Then, Eq. 21-4*a* becomes

$$T = \frac{1}{f} = R_E C_E \ln \frac{1}{1-\eta} \text{ s} \tag{21-4b}$$

where f is the approximate value of the frequency of oscillation in hertz.

When the intrinsic standoff ratio η is 0.632, the value of $\ln 1/(1-\eta)$ is unity. Then Eq. 21-4*b* becomes

$$T = R_E C_E \text{ s} \tag{21-4c}$$

and

$$\boxed{f = \frac{1}{R_E C_E} \text{ Hz}} \tag{21-4d}$$

Since the values of the intrinsic standoff ratio are of the order of 0.632, Eq. 21-4*c* and Eq. 21-4*d* give a quick evaluation of the approximate operating values of a circuit.

Example 21-2

The windshield wiper motor of an automobile is controlled by a UJT circuit. The capacitor for C_E is 50 μF. The resistor for R_E is the series combination of a 51-kΩ resistor and a 510-kΩ potentiometer. The value of η is 0.632. What is the minimum-to-maximum range of the number of blade strokes per minute?

Solution

The least value of the time constant (or period) is

$$T = R_E C_E = (51{,}000\ \Omega)(50 \times 10^{-6}\text{ F}) = 2.6\text{ s} \tag{21-4c}$$

and the maximum value of the time constant (or period) is

$$T = R_E C_E = (51{,}000\ \Omega + 510{,}000\ \Omega)(50 \times 10^{-6}\text{ F}) = 28.1\text{ s} \tag{21-4c}$$

Thus the potentiometer can adjust the windshield wiper to give a range of number of strokes per minute of

$$\frac{60\text{ s}}{28.13\text{ s}} = \mathbf{2.1} \quad \text{to} \quad \frac{60\text{ s}}{2.6\text{ s}} = \mathbf{23}$$

Problems 21-2.1 A UJT relaxation oscillator has the following values.

$$V_{EE} = 12\ \text{V} \qquad R_{B2} = 4.7\ \text{k}\Omega \qquad R_{B1} = 47\ \Omega$$
$$R_E = 47\ \text{k}\Omega \qquad \eta = 0.70$$

What value of C_E is required to obtain a frequency of 440 Hz?

21-2.2 Use the data of Problem 21-2.1. The UJT is replaced with a UJT having a value for η of 0.60. What is the new frequency of operation?

21-2.3 Using the data of Problem 21-2.1, determine what the size of the capacitor is to obtain a frequency of 4 kHz?

21-2.4 It is necessary to vary the frequency of the circuit of Problem 21-2.1 between 60 Hz and 10 kHz using a fixed value of 0.01 μF for C_E. What is the range of R_E that is required?

21-2.5 through **21-2.8** For each problem 21-2.1 through 21-2.4 assume that η is 0.632 so that Eq. 21-4*d* can be used. Using this simplified equation, what is the frequency and what is the error in percent when compared to the exact equation, Eq. 21-4*b*.

Section 21-3 Thyristor Concepts

The *thyristor* is a member of the family of semiconductor devices that have two stable states of operation: one stable state has very low and often negligible current, and the other stable state has a very high current that is limited only by the resistance of the external circuit.

The fundamental thyristor is the *NPNP* semiconductor (Fig. 21-5*a*). There are three junctions, each of which forms an equivalent diode (Fig. 21-5*b*). When the anode is positive, the *NPNP* semiconductor is termed as being *biased in the forward direction.* Now diodes 1 and 3 are forward biased and diode 2 is reverse biased. When the *NPNP* is reverse biased (the cathode positive), diodes 1 and 3 block.

This four-layer diode, often referred to as a *Shockley diode,* is formally described as a *reverse blocking diode thyristor.* The characteristic (Fig. 21-5*c*) shows that an avalanche breakdown occurs in the forward direction at V_{BO}, the *break-over voltage.* After breakdown the diode voltage drop falls to a very small value and can permit a large current flow. By having two blocking diodes in series in the reverse direction, we find that the peak inverse voltage rating is usually much larger in magnitude than V_{BO}. When the reverse voltage exceeds the

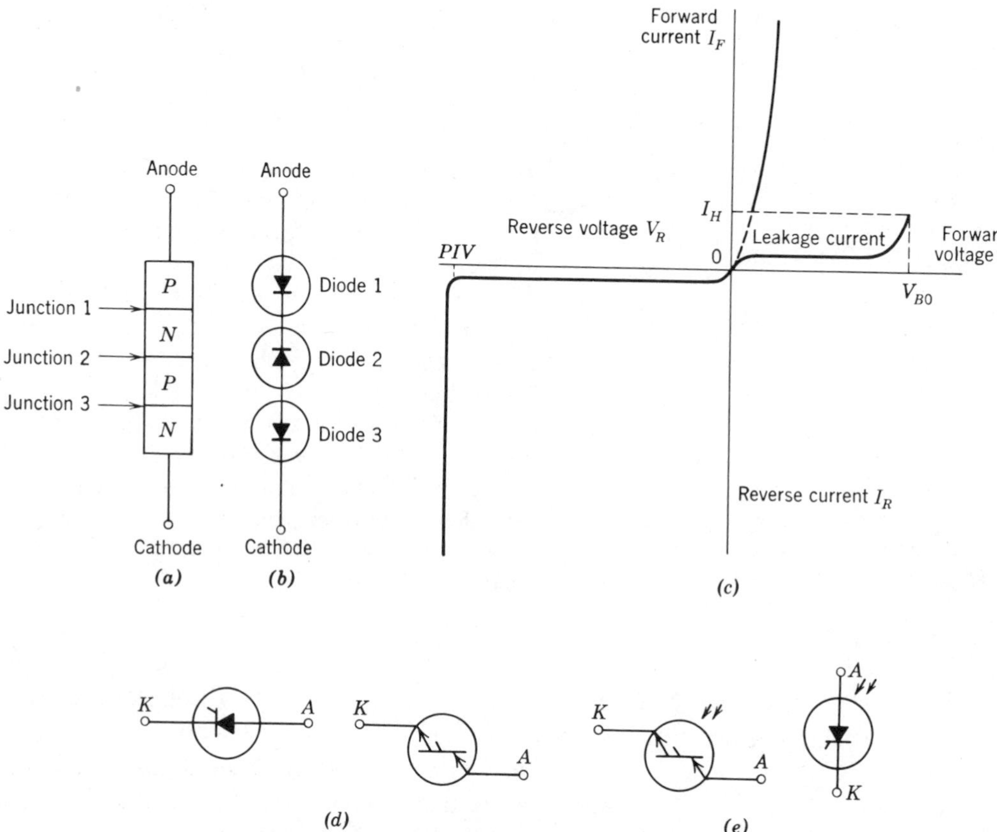

Figure 21-5 The *NPNP* transistor. (*a*) Construction. (*b*) Diode model. (*c*) Breakdown characteristics. (*d*) Graphic symbols. (*e*) Graphic symbols for the light-activated switch.

peak inverse voltage, a Zener breakdown occurs. Commercial Shockley diodes are available to PIV ratings of 1200 V and forward current ratings of 300 A peak. The circuit symbols for the Shockley diodes are shown in Fig. 21-5*d*.

A modification of the Shockley diode is the *light-activated switch* (*LAS*). Light entering an optical window breaks down covalent bonds within the diode allowing break-over voltage to become a function of the intensity of the incident illumination. The symbol for the LAS is shown in Fig. 21-5*e*. The LAS is available in ratings to 200 V and 0.5 A.

If the *NPNP* diode has the two inner layers "divided" as shown in Fig. 21-6*a*, we can redraw this divided diode into

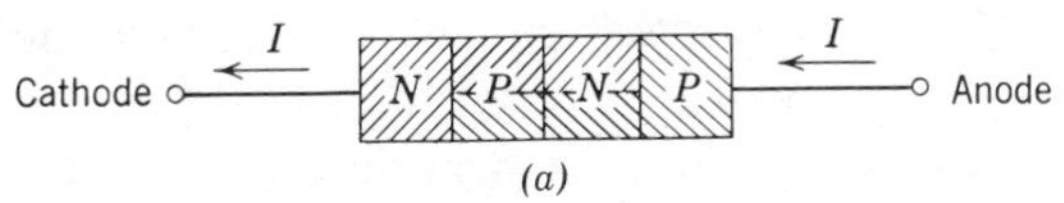

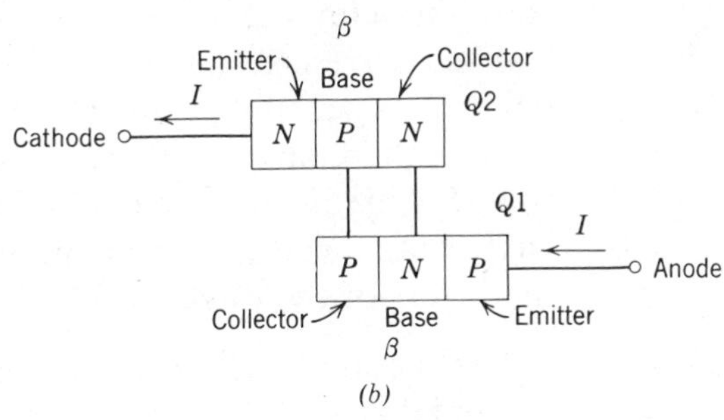

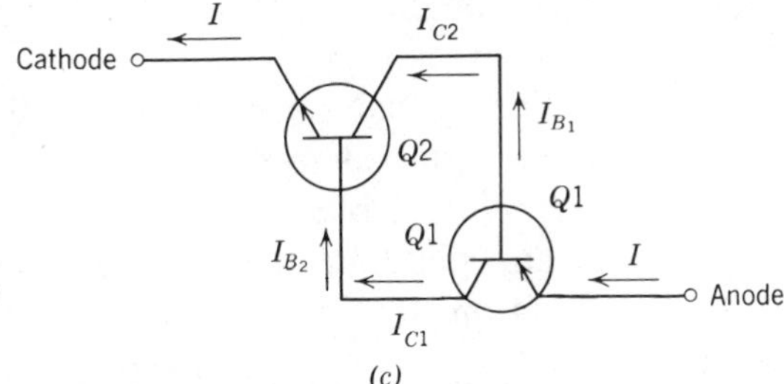

Figure 21-6 The *NPNP* semiconductor as the equivalent of an *NPN* transistor and a *PNP* transistor. (*a*) *NPNP* diode. (*b*) Division into two transistors. (*c*) Circuit.

two parts (Fig. 21-6*b*). Each of these two parts is now the equivalent of a transistor. Labels for transistor electrodes are given on Fig. 21-6*b*. Now the circuit is redrawn in Fig. 21-6*c* using standard transistor circuit symbols.

Any collector current in $Q2$ is the base current in $Q1$. Any collector current in $Q1$ is the base current in $Q2$.

If we have a collector current I_{C_1} in $Q1$, this current enters $Q2$ as the base current I_{B_2}. Transistor $Q2$ multiplies this base current by β to give I_{C_2}. But I_{C_2} is the base current in $Q1$ and transistor action immediately multiplies I_{C_2} by the β of $Q1$. Now the new collector current in $Q1$ is

$$I'_{C_1} = \beta(\beta I_C) = \beta^2 I_C$$

This action occurs simultaneously in each transistor, $Q1$ and $Q2$, and the line current I goes to infinity. Actually, in a design application, I must be limited by the resistance of the external circuit to keep I below its maximum permissible value.

From a practical point of view, we find that the normal leakage current in the silicon *NPNP* diode is insufficient to start this cumulative buildup. Any method of increasing the leakage current to the critical level by breaking more covalent bonds sets off this cumulative buildup.

Thus there are two stable states in the *NPNP* diode, ON or OFF. In this manner the *NPNP* diode serves as a *solid-state switch.* Any method of creating an increase in current within the layers of the diode will initiate this cumulative breakdown.

1. The application of a voltage sufficiently high to cause breakdown.
2. A sufficient increase in temperature that breaks covalent bonds.
3. The release of electrons by the action of incident light.
4. An induced transistor action by the creation of a forward transistor bias.
5. The generation of current within the diode by capacitive action.

Let us examine this last method of "turn on." When a forward voltage (the anode is positive and the cathode is negative) is placed across the Shockley diode, a reverse voltage exists across Junction 2 (Fig. 21-5*a*). As we developed in Section 2-7, a reverse voltage across a junction forms a capacitor by the action of depletion. When the voltage is changed by the amount ΔV volts in ΔT second, there is a current flow in the capacitor as given by the fundamental equation

$$i = C\frac{\Delta V}{\Delta T} \tag{21-5}$$

If this current i is sufficient to initiate the cumulative buildup, then the large current I flows in the circuit. Thus a sufficient voltage change across the diode can *trigger* a large current flow. This can be a useful method to obtain a breakdown, or it may be necessary that a particular circuit be protected against unwanted voltage transients.

After a thyristor breaks down, the device can be restored to

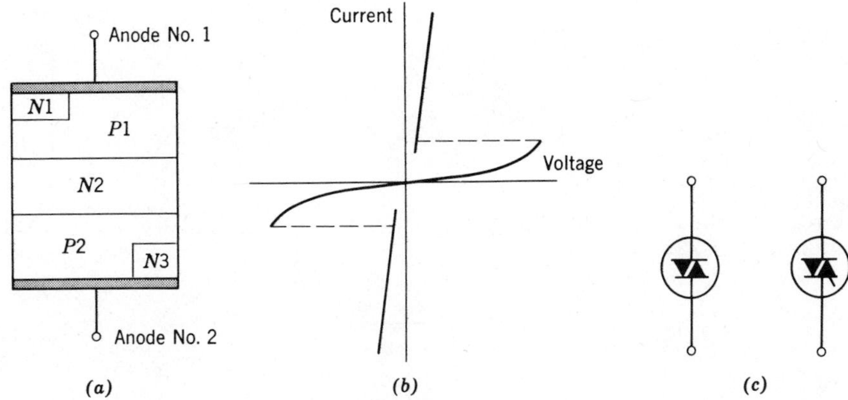

Figure 21-7 Bidirectional thyristors. (*a*) Layer structure. (*b*) Characteristic. (*c*) Diac symbols.

the OFF state by any one or by the combination of several methods.

1. Remove the external source voltage.
2. Reduce the external source voltage to the point where the current falls below the *holding current* value, I_H (Fig. 21-5*c*).
3. Reverse the polarity of the externally applied voltage as in an ac supply.

There is a finite time required for charges to redistribute within the four-layer diode in order to switch from ON to OFF. In most cases this time is in the order of microseconds.

The four-layer diode can be made bidirectional by changing its structure (Fig. 21-7*a*). When Anode No. 1 is positive, the path is $P1$-$N2$-$P2$-$N3$. When Anode No. 2 is positive, the path is $P2$-$N2$-$P1$-$N1$. The characteristic of the diode, then, shows a breakdown characteristic both in the first quadrant and in the third quadrant, yielding symmetrical properties in both forward and reverse directions (Fig. 21-7*b*). This device is termed a *bidirectional dipole thyristor* and is given the acronym *diac*. The symbol is given in Fig. 21-7*c*.

Problems

21-3.1 Assume for the purpose of calculation that the characteristic is formed of straight-line segments. The supply voltage is 117 V rms. Determine R_L to limit the current to the maximum

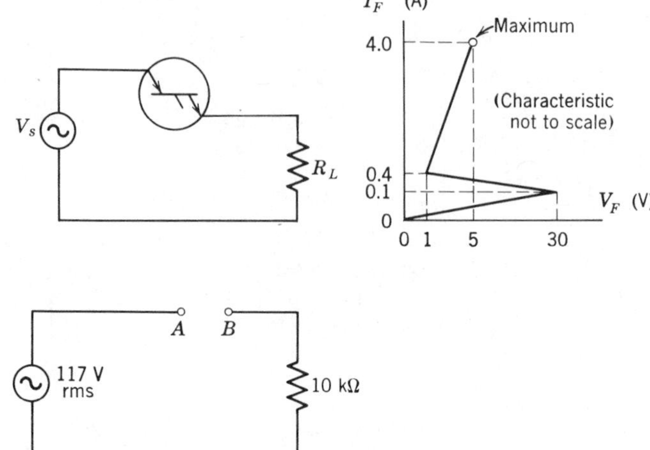

Circuit and characteristic for Problems 21-3.1 and 21-3.2.

Circuit for Problems 21-3.3 to 21-3.5.

value. Determine the value to which e_s must be reduced to turn the diode off. What is the ON-state forward resistance of the diode?

21-3.2 If R_L is 25 Ω, what is the maximum allowable value of e_s?

21-3.3 A Shockley diode is connected between terminals *A* and *B*. The diode has a value for V_{BO} of 30 V and for BV_R of 200 V. Sketch the voltage waveform across the load.

21-3.4 A light-activated switch is connected between terminals *A* and *B*. Sketch the voltage waveform across the load for different conditions of incident light.

21-3.5 A diac is connected between terminals *A* and *B*. The diac has a value for V_{BO} of 30 V and for BV_R of 200 V. Sketch the voltage waveform across the load.

Section 21-4 The Silicon Controlled Rectifier

The *silicon controlled rectifier* (*SCR*) is a modification of the Shockley diode. A *gate* connection is formed to the lower *P* layer of the *NPNP* structure (Fig. 21-8*a*). When the gate is forward biased with respect to the cathode, the *PN* junction is biased in the forward direction, and there is a current flow that creates a breakdown between anode and cathode, as explained for the Shockley diode. The circuit symbol is given in Fig. 21-8*b*.

The operating characteristic of the SCR is shown in Fig. 21-9. When the gate current is zero (I_{G1}), the characteristic is identical to the characteristic of the Shockley diode. After

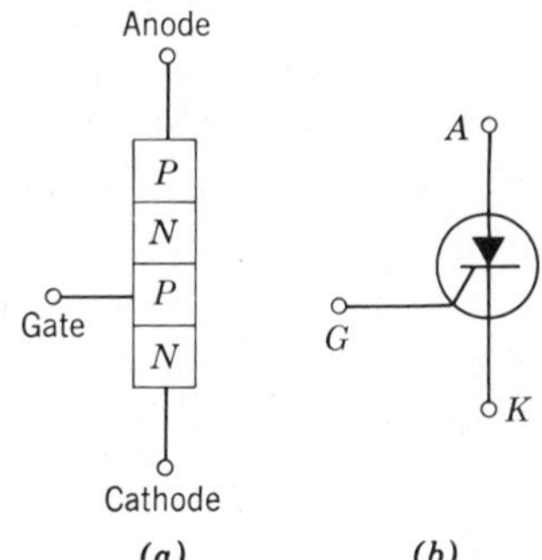

Figure 21-8 The silicon controlled rectifier. (*a*) Construction. (*b*) Symbol.

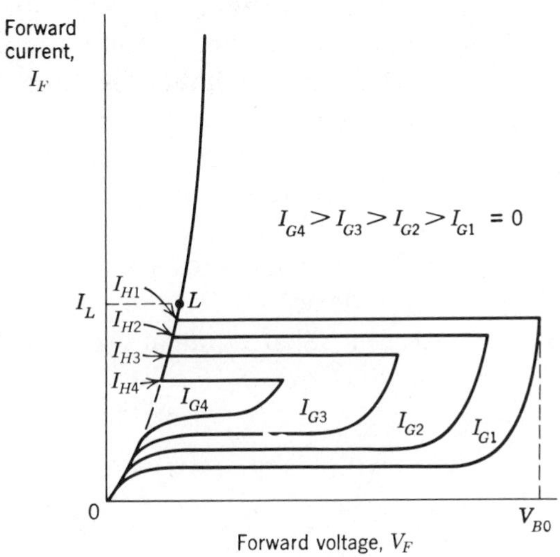

Figure 21-9 SCR characteristics.

"firing," the anode current I_F is limited only by the dc resistance in the anode circuit. The least anode current that sustains the fired state is the *holding current* I_{H1}. When the gate current is raised from zero, the SCR will break down or fire at voltage values lower than V_{BO} (Fig. 21-9). In applications of the SCR, anode voltages applied to the SCR are much lower than V_{BO} to insure that the firing is controlled only by injection of gate current. As long as the anode current is at least the *latching current* I_L (Fig. 21-9), the SCR will remain fired after the gate current triggering pulse falls to zero. The output pulse obtained from a UJT relaxation oscillator is ideally suited to trigger the SCR.

The SCR is manufactured in ratings up to several thousand volts (V_{BO}) and hundreds of amperes average load current (I_F). A number of modifications of the SCR are commercially available. One form has an optical window in addition to the conventional gate contact. Thus, it can be triggered by a combined action of gate current and incident illumination. This device is called a *light-activated SCR* (*LASCR*) (Fig. 21-10*a*). When a second gate is added to an SCR, the device is called a *silicon controlled switch* (*SCS*) (Fig. 21-10*b*). The required triggering pulse on the lower gate is positive. The upper gate is connected to *N* material and, therefore, negative triggering pulses are needed for firing.

A circuit giving an application of an SCR that is controlling a load in an automobile circuit is shown in Fig. 21-11. The 680-Ω resistor limits the gate current to 18 mA. A momentary closing of the start contact turns the SCR on. The stop contact bypasses the load current around the anode circuit of the SCR and permits the SCR to regain its blocked or OFF state. When the stop contact is released, the load current is zero.

The circuit shown in Fig. 21-12 uses an SCR to protect a tape deck or a CB radio from theft. The switch *S* is located at a concealed point in the car. The switch is kept closed. The

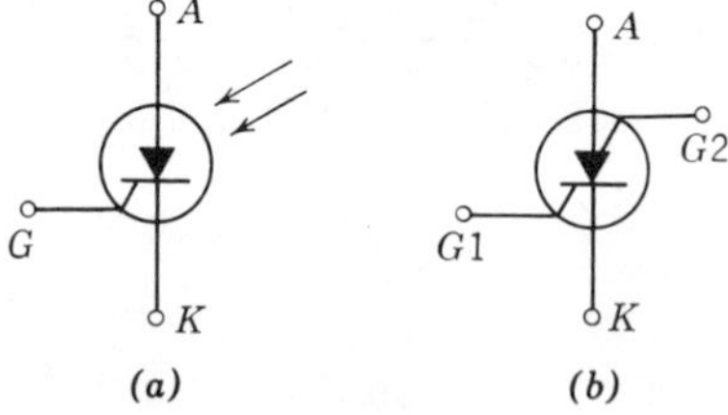

Figure 21-10 Thyristor devices. (*a*) Light-activated SCR (LASCR). (*b*) Silicon controlled switch (SCS).

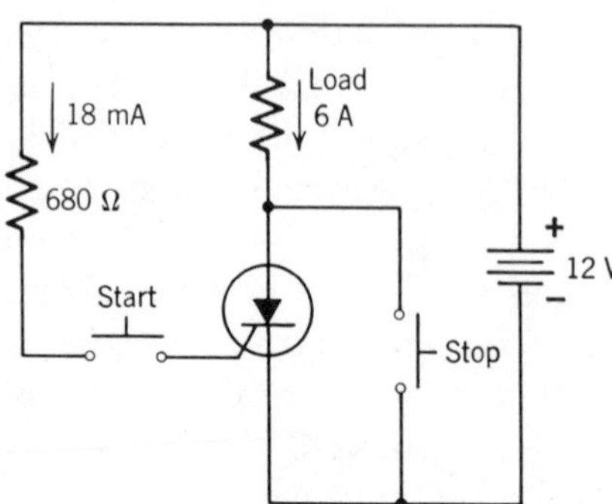

Figure 21-11 SCR application.

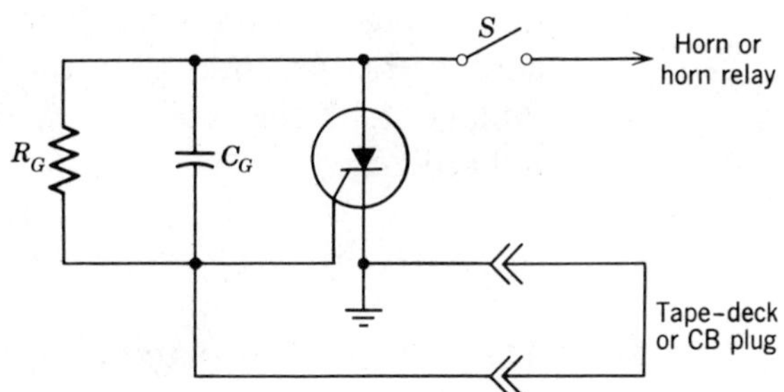

Figure 21-12 Protective circuit for an automobile.

gate is grounded through the tape-deck or CB plug. Therefore, the SCR is off. If the tape deck or CB radio is removed, the gate is no longer grounded. Now the gate is connected to the car battery through R_G. The resulting gate current fires the SCR. The horn blows and continues to blow until switch S is opened.

An SCR can be used as a relaxation oscillator (Fig. 21-13*a*). The waveform of the cathode voltage is shown in Fig. 21-13*b*. When the supply voltage V_{AA} is first impressed on the circuit,

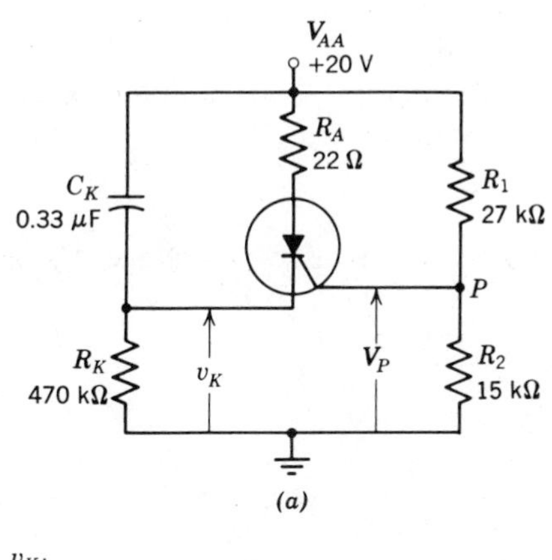

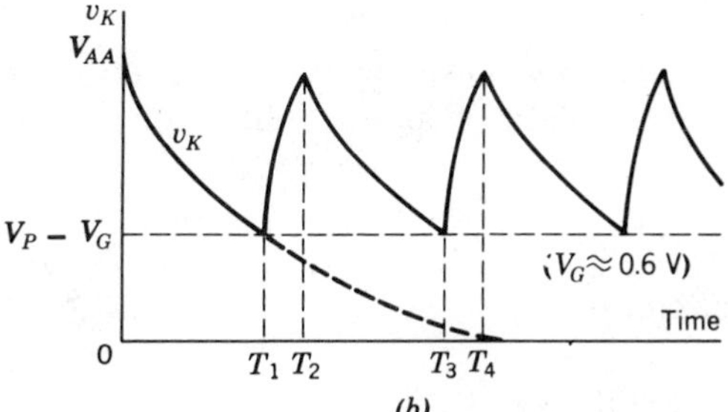

Figure 21-13 Relaxation oscillator using an SCR. (*a*) Circuit. (*b*) Voltage waveform at the cathode.

the voltage across C_K is zero. The whole supply voltage is across R_K. As time goes on, C_K charges toward the supply voltage and the voltage across R_K falls off exponentially toward zero.

$$v_K = V_{AA}\epsilon^{-t/R_K C_K} \tag{21-6}$$

The voltage V_P is determined by the voltage divider ratio.

$$V_P = \frac{R_2}{R_1 + R_2} V_{AA} \tag{21-7}$$

When the voltage across R_K falls below V_P by the amount of the voltage V_G required to trigger the SCR, the SCR fires at T_1.

When the SCR fires, R_A and the SCR effectively short circuit and discharge C_K. The voltage across R_K returns to its initial value, V_{AA}. After C_K discharges, the current through the SCR is V_{AA}/R_K. As long as this current is less than I_H, the SCR turns off and the cycle starts again.

Example 21-3
Determine the approximate frequency of oscillation using the numerical values given for the circuit of Fig. 21-13*a*.

Solution
The voltage divider of R_1 and R_2 places a fixed voltage V_P on the gate.

$$V_P = \frac{R_2}{R_1 + R_2} V_{AA} \tag{21-7}$$

$$= \frac{15{,}000\ \Omega}{27{,}000\ \Omega + 15{,}000\ \Omega} 20\ \text{V} = 7.1\ \text{V}$$

Assuming that the gate voltage (V_G) must be 0.6 V positive with respect to the cathode, the SCR will fire when v_K falls to

$$V_K = V_P - V_G = 7.1 - 0.6 = 6.5\ \text{V}$$

The time constant of the circuit is

$$R_K C_K = (470{,}000\ \Omega)(0.33 \times 10^{-6}\ \text{F}) = 0.155\ \text{s}$$

Substituting these values into the exponential equation, we have

$$v_K = V_{AA}\epsilon^{-t/R_K C_K} \qquad (21\text{-}6)$$

$$6.5 = 20\epsilon^{-T_1/0.155}$$

or

$$\epsilon^{-T_1/0.155} = 0.325$$

Taking the natural logarithm (ln) of both sides

$$-T_1/0.155 = -1.12$$

$$T_1 = 0.174 \text{ s}$$

The frequency is approximately

$$f = \frac{1}{T_1} = \mathbf{5.7\ Hz}$$

Section 21-5 The Triac

The *triac* was developed to extend the concept of an SCR to a device that can be triggered into conduction regardless of the polarity of the voltage on the anode and regardless of the polarity of the triggering pulses. Since the unit responds to both positive and negative voltages on the anode, the concept of the cathode (K) used for the SCR is dropped. Both electrodes are called anodes, one $A1$ and the other $A2$.

The cross-sectional representation of the triac is shown in Fig. 21-14*a*. When Anode No. 2 is positive, the path of current flow is $P1$-$N1$-$P2$-$N2$. Junctions $P1$-$N1$ and $P2$-$N2$ are forward biased, and junction $N1$-$P2$ is blocked. A positive gate (with respect to Anode No. 1) biases the junction $P2$-$N2$ in the forward direction, and breakdown occurs as a normal SCR operation. A negative gate (with respect to Anode No. 1) biases the junction $P2$-$N3$ in the forward direction, and the injected current carriers into P-2 turn on the four-layer diode. When Anode No. 1 is positive, the path of current flow is $P2$-$N1$-$P1$-$N4$. Junctions $P2$-$N1$ and $P1$-$N4$ are forward biased and junction $N1$-$P1$ is blocked. A positive gate (with respect to Anode No. 1) injects carriers by forward biasing $P2$-$N2$, and a negative gate injects current carriers by forward biasing $P2$-$N3$. The circuit symbol is shown in Fig. 21-14*b* and typical characteristics are given in Fig. 21-14*c*.

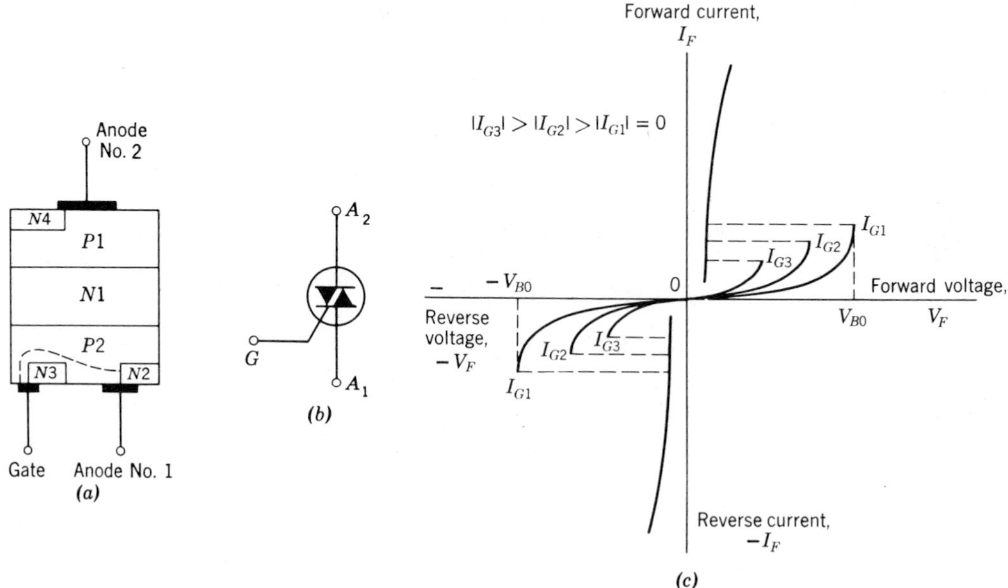

Figure 21-14 The triac. (*a*) Cross-sectional model. (*b*) Symbol. (*c*) Characteristic.

Since the triac can be turned on by any one of the four conditions, it can be used directly on an ac line to control the amount of current in a load without rectifying. The disadvantage of this device is that a relatively long time is required for it to recover to the OFF state. Accordingly, it is limited to use on 50, 60, or 400 Hz applications.

Questions

21-1 Define or explain each of the following terms: (*a*) double-based, (*b*) peak point, (*c*) valley point, (*d*) intrinsic standoff ratio, (*e*) negative resistance, (*f*) UJT, (*g*) break-over voltage, (*h*) holding current, (*i*) diac, (*j*) (SCR, (*k*) gate, (*l*) trigger, and (*m*) triac.

21-2 Explain why a UJT cannot break down if the emitter is zero biased.

21-3 How does a UJT return to the OFF state after breakdown in a relaxation oscillator circuit?

21-4 Compare a Shockley diode with an ordinary diode.

21-5 How is breakdown accomplished in a thyristor?

21-6 What is the principle behind the action of breakdown in a Shockley diode?

21-7 Compare a Shockley diode with a Zener diode.

21-8 Compare a Shockley diode with a diac.

21-9 Explain the means used to break down an SCR.

21-10 How can light be used to cause breakdown in an SCR?

21-11 How can an SCR be returned to an OFF state after it has been triggered?

21-12 A relaxation oscillator uses a UJT. Does it function on a charging exponential curve or on a discharging exponential curve?

21-13 A relaxation oscillator uses an SCR. Does it function on a charging exponential curve or on a discharging exponential curve?

21-14 Is leakage current used to trigger an SCR?

21-15 Which is larger, latching or holding current? Explain.

21-16 Compare the SCR with the triac.

21-17 Which anode of the triac could be called the *cathode*?

Chapter 22 Controlled Rectifiers

The load current and the load voltage in a controlled rectifier are established by controlling the point in the ac input cycle at which the circuit is turned on (Section 22-1). The phase-shift circuit (Section 22-2) is commonly used to establish the turn-on point in the ac cycle. A number of circuits including a unijunction transistor relaxation oscillator are used to illustrate silicon controlled rectifier circuits (Section 22-3). The triac is used to vary the power in an ac load (Section 22-4).

Section 22-1 Analysis of Load Voltage and Load Current

In an ordinary diode rectifier circuit, current flows in the diode whenever the instantaneous ac supply voltage is greater than the voltage across the load at that instant. When the load on a simple diode circuit is a resistive load, load current flows at all times during the half of the ac cycle that the anode is positive. In a controlled rectifier with a resistive load (Fig. 22.1*a*) the load current is zero at all times unless a control signal is applied to the device to initiate anode current flow. The application of a control signal turns on the rectifier at a specific point, *A*, in the cycle (Fig. 22-1*b*). Point *A* corresponds to an angle, θ_1, which is a point later than the start of the positive half of the ac cycle.

Once the rectifier is turned on, the rectifier remains in conduction until point *B* near the end of the positive half cycle. At this time located at θ_2 degrees, the current falls to zero in the rectifier. In a silicon controlled rectifier, when the anode current falls below the holding current I_H, the conduction current ceases. The value of θ_2 is a function of the rectifier characteristic and is *not* determined by the control signal.

Thus, point *A*, the firing point, is determined by the angle of delay in the application of the firing signal in the control

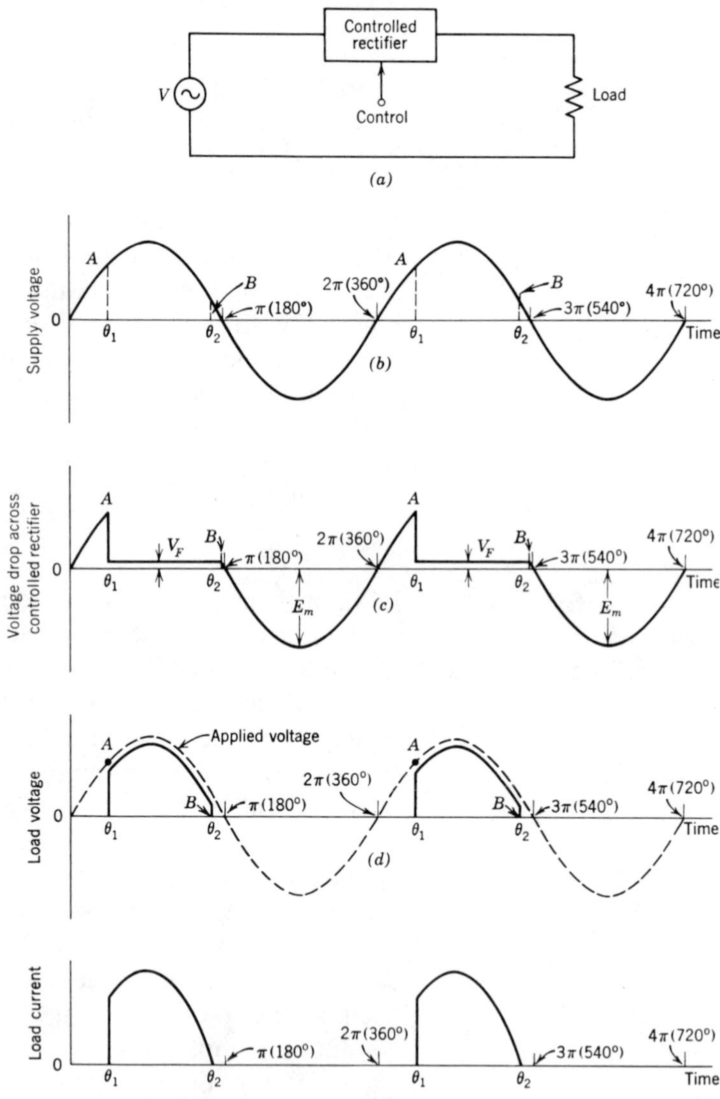

Figure 22-1 The controlled rectifier. (*a*) Block diagram. (*b*) Supply voltage. (*c*) Voltage drop across controlled rectifier. (*d*) Load voltage. (*e*) Load current.

circuit. As the delay angle θ_1 increases, point A occurs later and later in the cycle and the current in the load decreases.

When the rectifier is turned on at A, the forward voltage drop across the rectifier falls to the level indicated by V_F on Fig. 22-1*c* and remains at that value until the rectifying action ceases at θ_2. During the negative half of the cycle the inverse voltage having a peak value of E_m volts appears across the rectifier in the same manner as in an ordinary diode rectifier circuit. The voltage drop across the load (Fig. 22-1*d*) is the ac supply voltage between θ_1 and θ_2 less the forward voltage drop V_F of the diode. Since the load current (Fig. 22-1*e*) follows Ohm's law, its waveform is proportional to the load voltage waveform. The dc voltage across the load and the dc current in the load are the *average* of the values in the waveform over the *full* cycle from 0 to 2π.

The average value of the load voltage is given by

$$V_L \equiv \frac{1}{2\pi}\int_0^{2\pi} v_L \, d\theta \tag{22-1}$$

When we use methods of calculus to evaluate Eq. 22-1 for the load voltage waveform given in Fig. 22-1*d*, we find the load voltage is

$$V_L = \frac{V_m}{2\pi}(\cos\theta_1 - \cos\theta_2) - \frac{\theta_2 - \theta_1}{360} V_F \tag{22-2a}$$

and the load current is

$$I_L = \frac{V_m}{2\pi R_L}(\cos\theta_1 - \cos\theta_2) - \frac{\theta_2 - \theta_1}{360}\frac{V_F}{R_L} \tag{22-2b}$$

In most applications of controlled rectifiers, the peak value of the line voltage V_m is very much greater than the forward voltage drop V_F. For example, the peak voltage of a 117-V circuit is 166 V whereas a typical value of V_F for a silicon controlled rectifier is 1 V. Under this condition, the second term may be neglected to simplify the equations. When V_m is large, θ_2 approaches 180°. When 180° is substituted for θ_2, the error is small. Since the numerical value of cos 180° is −1, the equations become

$$\boxed{V_L = \frac{V_m}{2\pi}(1 + \cos\theta_1)} \tag{22-3}$$

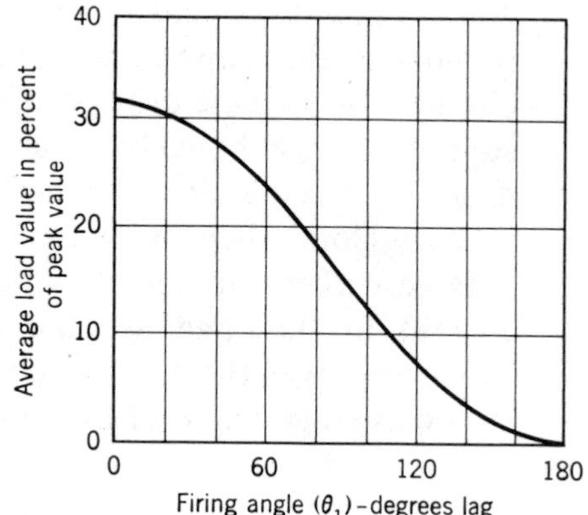

Figure 22-2 Average value of load voltage and load current in a phase-controlled circuit.

and

$$I_L = \frac{V_m}{2\pi R_L}(1+\cos\theta_1) \tag{22-4}$$

If θ_1 is allowed to go to zero, the equations become

$$V_L = V_m/\pi \qquad \text{and} \qquad I_L = V_m/\pi R_L$$

These equations are the same as the equations for a half-wave rectifier with a resistive load in which the conduction was assumed for the full positive half of the ac cycle.

The curve given in Fig. 22-2 is a normalized plot of either Eq. 22-3 or Eq. 22-4. From this curve, it is seen that there is a smooth control of the output in the load from a maximum value to zero. Typical waveforms that may be observed on an oscilloscope for different values of control angle θ_1 are shown in Fig. 22-3.

Problems

22-1.1 The supply voltage is 200 V peak. The forward voltage drop of the rectifier is 2 V. When the load resistance is 1000 Ω, what is the range of dc load current with firing at 0°, at 45°, at 90°, and at 135°?

22-1.2 The supply voltage is 30 V rms, and the forward voltage drop of the rectifier is 2 V. When the load resistance is 10 Ω, what

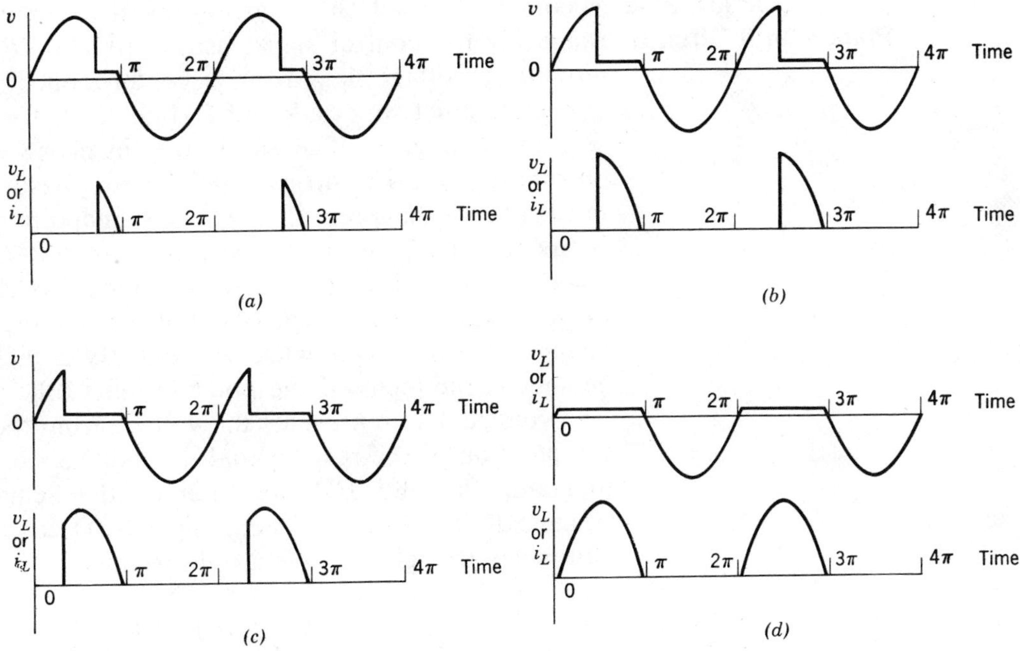

Figure 22-3 Controlled rectifier waveforms for anode voltage (v), for load voltage (v_L), and for load current (i_L) at different angles of turn-on. (a) 135°. (b) 90°. (c) 45°. (d) 0°.

is the range of dc load current with firing at 0°, at 45°, at 90°, and at 135°?

22-1.3 A 10-Ω load is connected to a 117-V rms supply through a controlled rectifier. The load power must be varied between 90 W and 20 W. What is the angular firing control that is required? Assume that V_F is 2 V.

22-1.4 A load is connected to a 117-V rms supply through a controlled rectifier. The maximum load power is 50 W and the load power must be controlled down to 20 W. What is the value of the load resistance and over what angular range must firing control be maintained? Assume that V_F is 2 V.

22-1.5 A load is connected to a 117-V rms source through a controlled rectifier. The peak current in the load is 4.0 A and the minimum average current is 0.56 A. What is the value of the load resistance, and what is the angular firing control range? What is the range of dc load current? Assume that V_F is 2 V.

22-1.6 Solve Problem 22-1.5 if the source voltage is 10 V rms.

Section 22-2 Phase-Shift Circuits

A commonly used circuit arrangement to obtain a 180° delay range for the control signal uses a simple *LR* or *RC* series circuit. A voltage V_{AC} from a center-tapped transformer is applied to a network of *R* and *L* (Fig. 22-4). The current *I* lags V_{AC} by an angle θ. The *IR* drop is in phase with *I* and the current lags the IX_L drop by 90°. These phasors are sketched in the phasor diagram. Then *B* is the midpoint of V_{AC} since it is the center tap on the transformer. Angle *ADC* is the right angle of a right triangle whose hypotenuse is *AC*. If either *L* or *R* is varied, the lengths of the legs of the right triangle change, but the hypotenuse is fixed. By a theorem in plane geometry, the locus of the point *D* must follow a semicircle. The voltage V_{BD} represents the voltage from the center tap to the junction of the resistor and the inductance. In the phasor diagram, *BC* and *BD* are radii of the semicircle making triangle *BCD* isosceles. Then, angle *BCD* equals angle *BDC*. But, since triangle *ACD* is a right triangle,

$$\theta + \angle ACD = 90$$

then

$$\angle BCD = 90 - \theta$$

Since

$$\angle BCD = \angle BDC$$

and since

$$\angle BCD + \angle BDC + \phi = 180$$

then

$$(90 - \theta) + (90 - \theta) + \phi = 180$$

Simplifying, we have

$$\phi = 2\theta \qquad \text{where} \qquad \tan\theta = X_L/R \qquad (22\text{-}5a)$$

If V_{BC} (or another voltage in phase with V_{BC}) is used as the anode supply and if V_{BD} is used as the control voltage, a

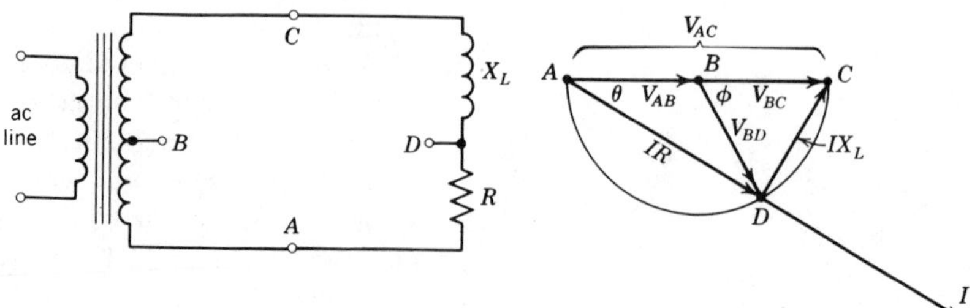

Figure 22-4 *LR* network producing a lagging phase angle.

variation in either L or R can produce almost a full range of lagging phase-shift control over 180°. The presence of resistance in the inductance prevents a full 180° control. The control voltage V_{BD} is a radius of the semicircle and will not change its magnitude as the phase angle is varied. When R equals X_L, point D is the midpoint on the arc between A and C, and the phase angle is 90°. If R is the variable element in the phase-shift circuit, an increase in R moves point D from A toward C, decreasing the phase angle. If L is variable, an increase in L moves point D toward A, increasing the phase angle.

The lagging phase-shift angle may also be obtained from an RC circuit, (Fig. 22-5). The current in the RC circuit leads the impressed voltage V_{AC} by θ degrees. The current leads the IX_C voltage phasor by 90° and is in phase with (parallel to) the IR voltage phasor. By the same logic as that in the LR circuit, the locus of point D traces out a semicircle. Then V_{BD} lags V_{BC} by the angle ϕ. The angle ϕ can vary from zero to 180° by a change in either R or C. Since $\angle ACD$ is θ, the angle by which V_{BD} lags V_{BC} is

$$\phi = 180 - 2\theta$$

where

$$\tan\theta = \frac{X_C}{R} \qquad (22\text{-}5b)$$

When the value of the resistor is increased, point D moves toward A, and the phase-shift angle ϕ increases. If the capacitance is increased, X_C decreases, and point D again moves toward A. When R equals X_C, point D is at the midpoint on the semicircle, and the phase shift is 90°.

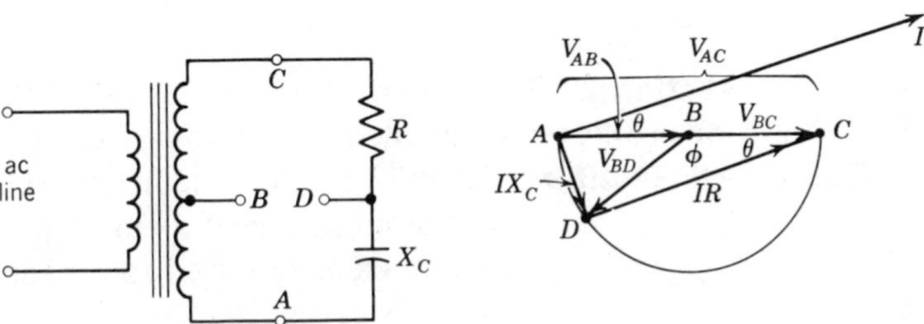

Figure 22-5 *RC* network producing a lagging phase angle.

There are many ways of obtaining variable elements for these phase-shift circuits. A direct means is to use a rheostat mounted on a panel as R. The pointer can be calibrated directly in terms of firing angle, conduction angle, load current, or load voltage. In other applications, the variable R is a temperature sensitive element. If the dc voltage from emitter to collector is fixed, the current through the transistor is determined by the emitter-to-base voltage, allowing the transistor to serve as the variable resistive element in a phase-shift circuit.

Problems

Note: The frequency is 60 Hz for all problems.

22-2.1 Interchange R and X_L in Fig. 22-4. Draw the phasor diagram. Determine the relationship between X_L and R to give a 180° phase control for a leading angle.

22-2.2 Interchange R and X_C in Fig. 22-5. Draw the phasor diagram. Determine the relationship between X_C and R to give a 180° phase control for a leading angle.

22-2.3 The phase-shift control circuit shown in Fig. 22-4 is used to control a controlled rectifier over a 5-to-1 current range. The inductor varies to 10 H. What is the value of R?

22-2.4 The phase-shift control circuit shown in Fig. 22-4 is used to control a controlled rectifier over a 3-to-1 current range. The inductor varies to 10 H. What is the value of R?

22-2.5 The phase-shift control circuit shown in Fig. 22-5 is used to control a controlled rectifier from 20% to maximum current. The capacitor is 0.05 μF. What size rheostat should be used for R?

22-2.6 The phase-shift control circuit shown in Fig. 22-5 is used to control a controlled rectifier from 10% to maximum current. The capacitor is 0.15 μF. What size rheostat should be used for R?

Section 22-3 The Firing of a Silicon Controlled Rectifier

A summary of the essential characteristics of a typical silicon controlled rectifier, the Texas Instrument 2N1604, is listed in Table 22-1. The silicon controlled rectifier is fired or triggered by the initiation of a suitable current in the gate. The value of gate current required to guarantee triggering is specified in the data sheet as 10 mA. The gate current must be maintained

Table 22-1
Texas Instrument 2N1604 Characteristics

Anode Characteristics		
$V_{F(off)}$	Forward voltage in the OFF condition	400 V
V_R	Peak inverse voltage	400 V
I_F	Average rectified forward current	3 A
i_f	Recurrent peak forward current	10 A
$i_{f(surge)}$	Surge current, 1 cycle at 60 cps	25 A
BV_F	Min forward breakover voltage	480 V
BV_R	Min reverse breakdown voltage	480 V
I_R	Max dc reverse current at rated V_R	0.25 mA
V_F	Max forward voltage drop at $I_F = 3$ A	2 V
I_H	Max holding current	25 mA
$I_{F(OFF)}$	Max dc forward current at $V_{F(OFF)}$	0.25 mA
Gate Characteristics		
I_G	Forward gate current, maximum	100 mA
V_{GR}	Gate peak inverse voltage	5 V
I_{GT}	Max gate current to trigger[a]	10 mA
BV_G	Min gate breakdown voltage	6 V
V_G	Max forward gate voltage drop at I_G = 25 mA	3 V

[a]Positive triggering is assured with a 10-mA gate current.

until the anode current has sufficient time to increase to at least the holding current value I_H. The relation between the gate current and the turn-on time is shown in Fig. 22-6.

When the switch is closed in the series switch (Fig. 22-7*a*), current flows in the gate circuit to trigger the anode on the half of the cycle on which the anode is positive. As soon as

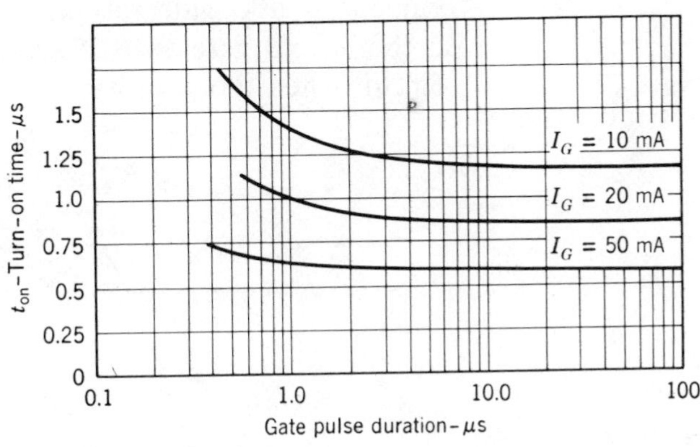

Figure 22-6 Turn-on characteristic for 2N1604.

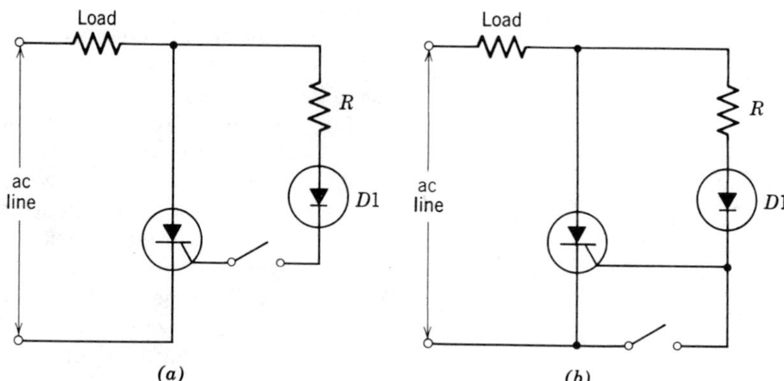

Figure 22-7 AC switches. (*a*) Series. (*b*) Shunt.

the SCR fires, the forward voltage drop from anode to cathode V_F is very small, of the order of one volt. Since the entire circuit, consisting of R, the diode $D1$, and the gate is a series string in parallel with the anode-cathode path of the rectifier, the gate current falls to a very low value. When the line voltage reverses its polarity, the cathode is positive and the anode is negative. Consequently, the anode current in the silicon controlled rectifier falls to zero. The diode $D1$ is required to prevent a reverse current flow in the gate. When the switch is opened, the rectifier will not conduct on any following cycle. The shunt switch (Fig. 22-7*b*) operates in a similar fashion except that it must be opened to turn on the rectifier.

In many applications, the gate circuit is operated from a sinusoidal source of emf (Fig. 22-8). As in the ac switch, a provision must be made to prevent a reverse breakdown current in the gate circuit. A diode is placed between the cathode and gate that effectively shunts or *clamps* the gate circuit when the polarity is reversed.

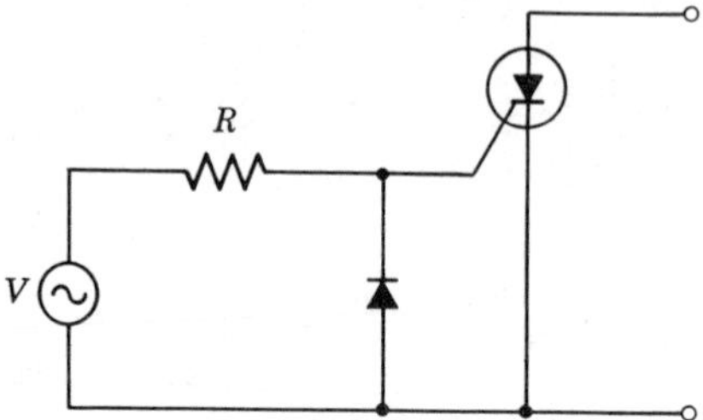

Figure 22-8 A gate clamp.

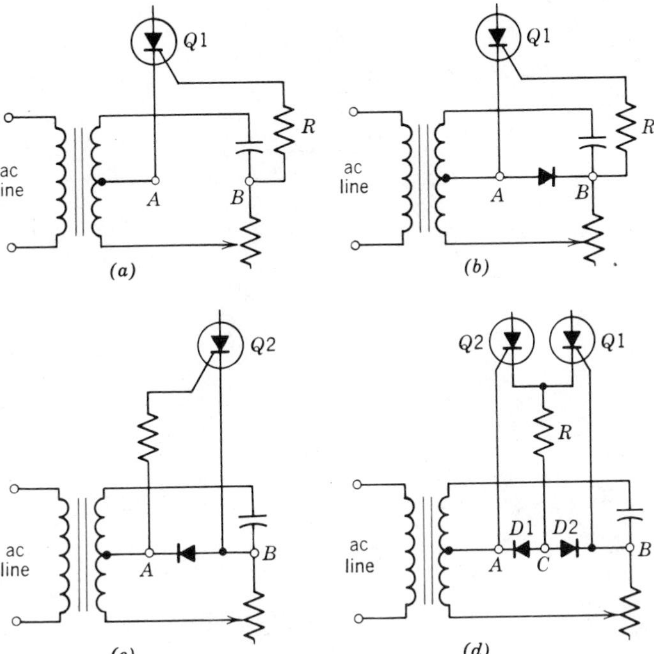

Figure 22-9 Phase-shift control circuits. (*a*) Basic control circuit. (*b*) With clamping diode. (*c*) Control circuit for second rectifier. (*d*) Control circuit for full-wave rectifier.

The phase-shift circuit shown in Fig. 22-9*a* produces a gate voltage that lags the main anode voltage to control the angle of the cycle at which firing occurs. Gate current is limited by a series resistor *R*. To prevent excessive leakage currents, reverse current flow in the gate is prevented by means of a clamping diode (Fig. 22-9*b*). If *Q*1 is one rectifier in a full-wave circuit and if the voltage from *A* to *B* provides the lagging control voltage to trigger *Q*1 (Fig. 22-9*b*), then the voltage from *B* to *A* is the proper lagging voltage required to trigger a second SCR, *Q*2 (Fig. 22-9*c*), which is the other rectifier of the full-wave circuit. This is true because the voltage from *B* to *A* is 180° out of phase with the voltage from *A* to *B* and the supply voltage on the anode of *Q*2 is 180° out of phase with the other anode supply voltage if full-wave rectification is to take place. These two circuits may be combined (Fig. 22-9*d*). The diodes *D*1 and *D*2 effectively switch point *C* to *A* for one half the ac cycle and then to *B* for the other half of the ac cycle. Thus one phase-shift circuit can serve both silicon controlled rectifiers. Complete rectifier circuits are shown in Fig. 22-10.

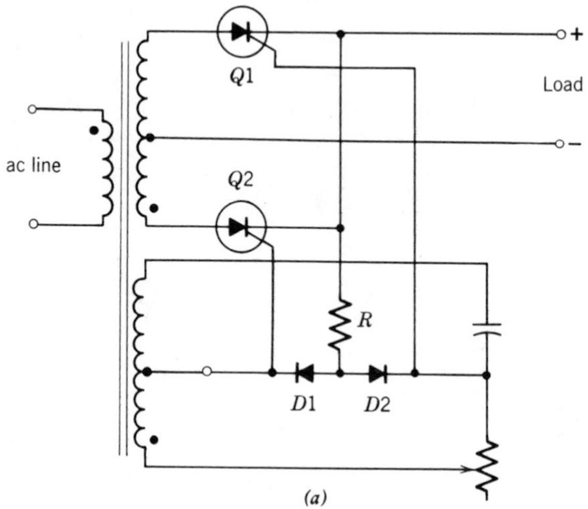

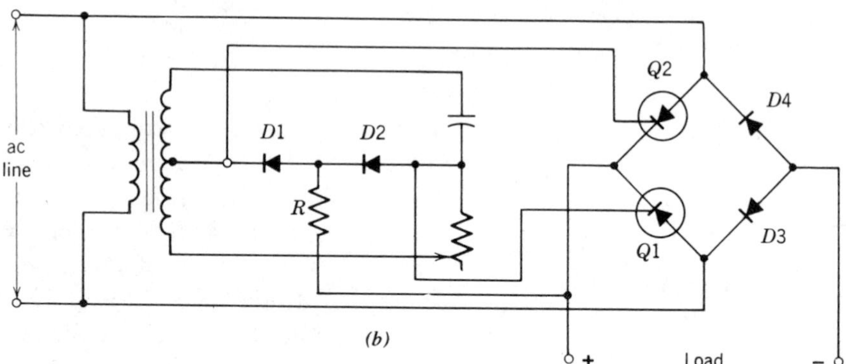

Figure 22-10 Complete rectifier circuits. (*a*) Full-wave. (*b*) Bridge.

When the SCR is used in a full-wave application, the values of V_L and I_L given by Eqs. 22-4, 22-5, and 22-6 are merely multiplied by two. The curve shown in Fig. 22-2 can be used, provided that the vertical-axis value is multiplied by two also.

The unijunction transistor relaxation oscillator explained in Section 21-2 is often used to trigger an SCR in a controlled rectifier circuit.

When the UJT is used to trigger an SCR, the source of power for the UJT is the same ac supply that is used for the

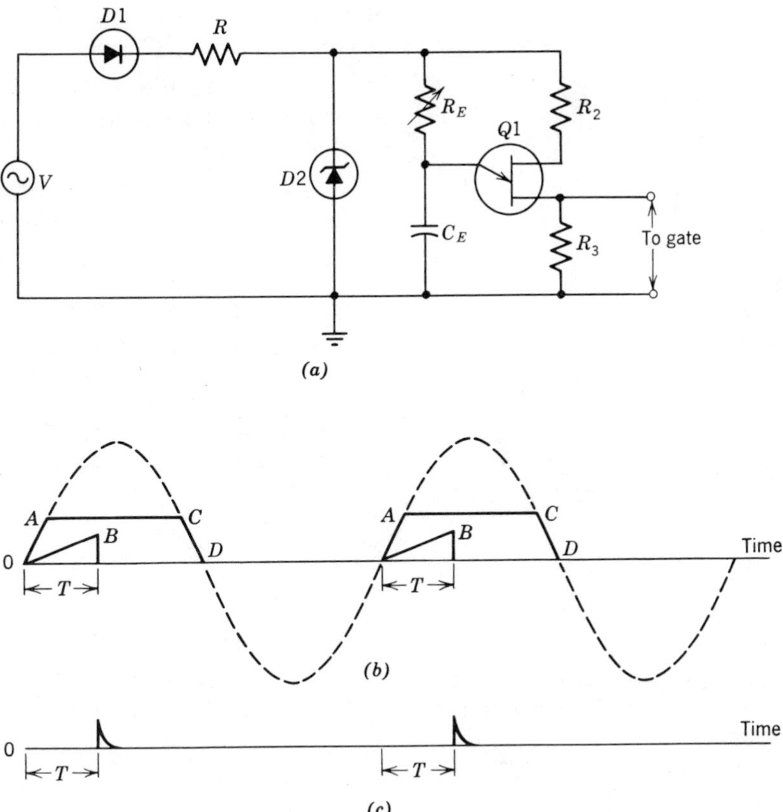

Figure 22-11 Firing on a unijunction transistor from an ac source. (*a*) Circuit. (*b*) Capacitor voltage. (*c*) Output pulse to gate.

SCR (Fig. 22-11). A diode *D*1 is used in conjunction with a Zener diode *D*2. The waveform of the voltage across the unijunction transistor circuit is the trapezoidal waveform, *O-A-C-D*, shown in Fig. 22-11*b*. On the negative half of the ac cycle, the diode *D*1 does not conduct. Capacitor C_E in the emitter circuit begins to charge as soon as the supply voltage goes positive. At point *B*, the voltage across the capacitor reaches the peak voltage V_P and the unijunction transistor conducts to discharge the capacitor. The time delay *T* between 0 and *B* can be converted in terms of degrees lag in the ac cycle. The angle of lag is controlled by varying the series charging resistor R_E. At *B* the firing of the unijunction transistor develops the firing pulse as the voltage drop across R_3 (Fig. 22-11*c*) for the silicon controlled rectifier gate.

The waveforms in Fig. 22-11*b* and Fig. 22-11*c* show only one oscillation and one output pulse per ac cycle. We show that, when the UJT breaks down at point B, the oscillation ceases. Actually, C_E would become discharged at B and then the sawtooth would start up again from zero. In order to use this circuit to trigger an SCR in a stable manner, the capacitor in the unijunction transistor circuit cannot be allowed to recharge once it has discharged in a particular positive half cycle. To accomplish this, the entire unijunction transistor

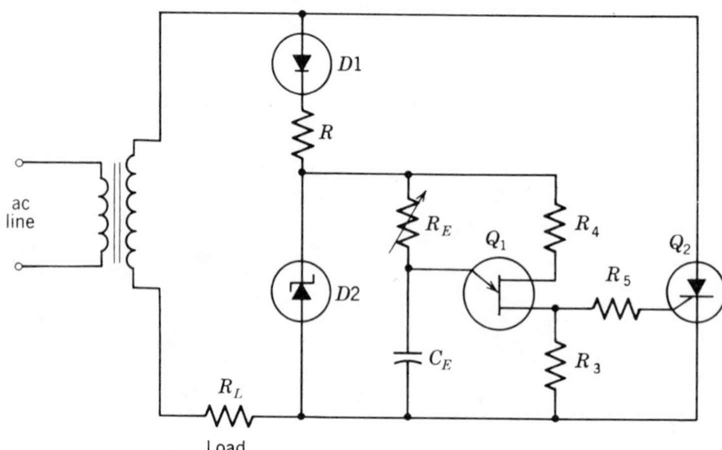

Figure 22-12 Half-wave silicon controlled rectifier circuit with phase-shift control.

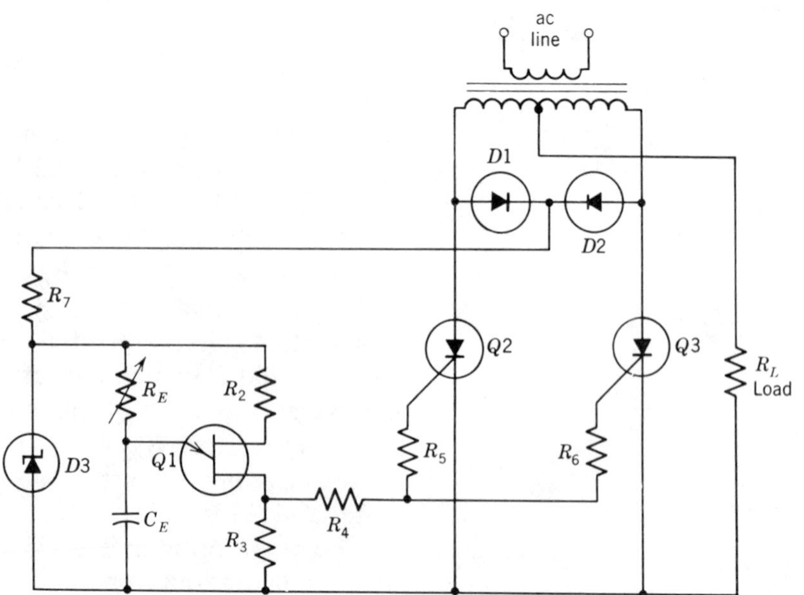

Figure 22-13 Full-wave silicon controlled rectifier circuit with phase-shift control.

pulse forming circuit is connected between the anode and the cathode of the silicon controlled rectifier it controls. Before the silicon controlled rectifier breaks down, the voltage across both the rectifier and the control circuit follows the supply voltage. After the silicon controlled rectifier fires, the forward voltage drop falls to the value of V_F, which is of the order of 1 V. Accordingly, the entire voltage available for the triggering circuit is insufficient to produce a recharge of the capacitor. Such an arrangement is shown in Fig. 22-12. This concept can be readily extended to the full-wave controlled rectifier circuit shown in Fig. 22-13. The waveforms for this circuit are given in Fig. 22-14. By a suitable choice of R_E and C_E, phase control can be obtained over nearly the full cycle from 0 to 180°.

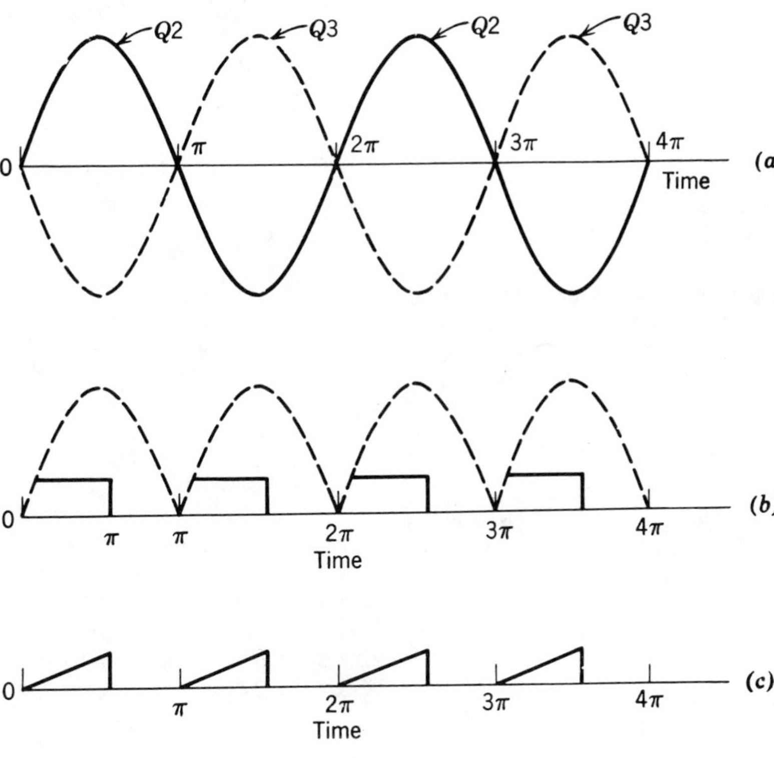

Figure 22-14 Waveforms in a full-wave controlled rectifier. (*a*) Anode supply voltage. (*b*) Voltage waveform across Zener diode *D*3. (*c*) Voltage waveform across capacitor. (*d*) Gate pulses.

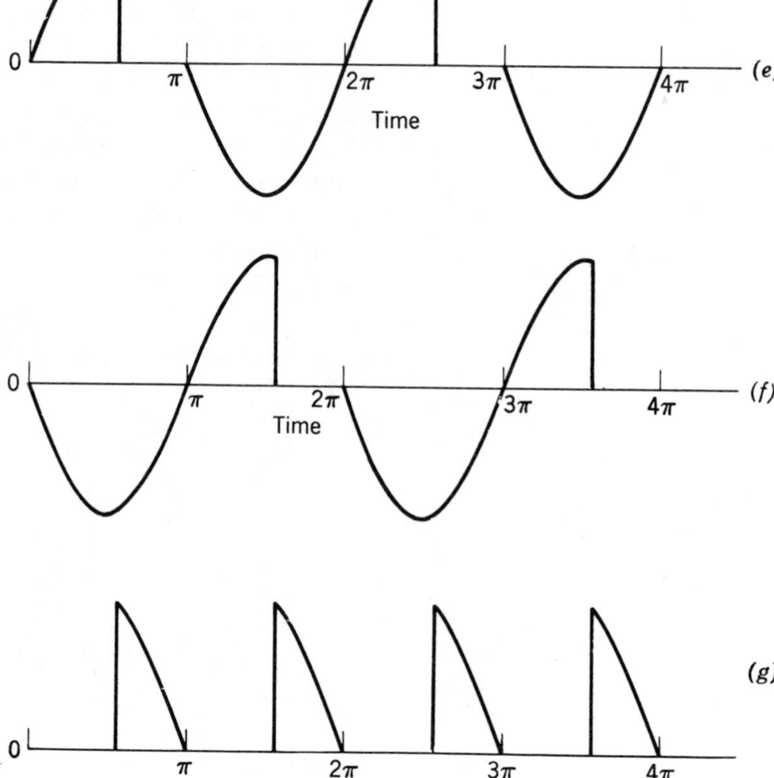

Figure 22-14 (Continued) (*e*) Waveform voltage across *Q*2. (*f*) Voltage waveform across *Q*3. (*g*) Load voltage or load current waveform.

Problems

Problem	R_1(kilohms)	X_{C1}(kilohms)	R_2(kilohms)	R_L(ohms)
22-3.1	2.5	1.2	100	10
22-3.2	5.0	5.0	100	20
22-3.3	5.0	20.0	100	60
22-3.4	2.0	0.5	100	60

For each of the above four problems, draw the phasor diagram. Draw waveforms for line voltage, for v_L, for the voltage across the SCR, and for load current. Calculate the firing angle, the load current, and the load voltage. Calculate the maximum gate current. Assume that the gate firing voltage is zero, V_F is 2.0 V for the SCR.

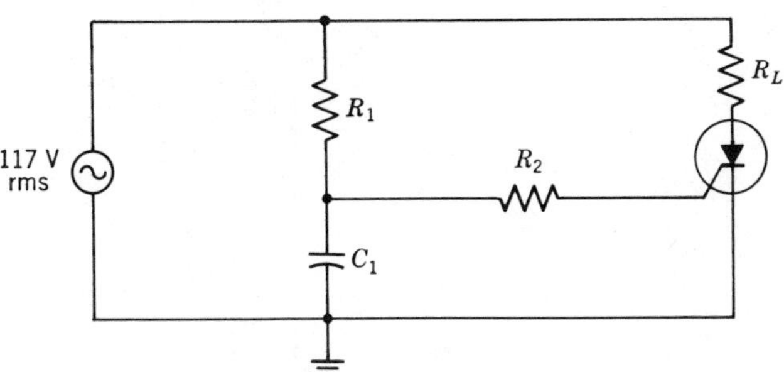

Circuit for Problems 22-3.1 to 22-3.4.

22-3.5 In Fig. 22-11*a*, *D*2 is a 20-V Zener, R_E is 1 MΩ, and η is 0.55. What values of C_E will fire the circuit at 30°, 60°, 90°, 120°, and 150° points on the positive half of a 220-V, 60-Hz supply?

22-3.6 In Problem 22-3.5, the capacitor C_E is fixed at 0.01 μF. What values of R_E provide for firing at the listed angles?

Section 22-4 The Firing of a Triac

A typical circuit used to vary load current in a triac is shown in Fig. 22-15*a*. When the voltage across the trigger diac is sufficiently high, the diac breaks down to provide gate current. The capacitor and resistor combination causes the voltage to the diac to lag the voltage to the triac. The amount of lag adjusted by the rheostat determines the point in the cycle at which the diac and triac turn on and, thus, controls the amount of load current. When the triac turns on, its low forward voltage drop effectively short circuits *a* and *c* and, thus, the diac current falls to zero and the diac recovers its off condition. This process repeats for each half cycle of the line supply (Fig. 22-15*d*).

A single *SCR* produces half-wave rectification. Two *SCR*s are used in a full-wave rectifier. Both circuits convert ac power into controlled dc power. The triac differs by not rectifying. The triac controls ac power in an ac load. Consequently, the equations developed for load current, load voltage, and load power for the SCR have no meaning for the triac. Now what is required is the rms or the effective value of the waveform of load current or load voltage of Fig. 22-15*d*.

The average value (the dc value) of the load current for the triac (Fig. 22-15*d*) is zero. Therefore, this waveform represents an ac value only. The useful result is in terms of an

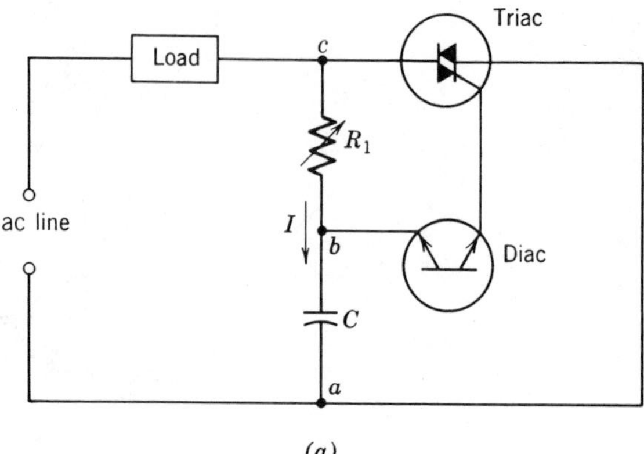

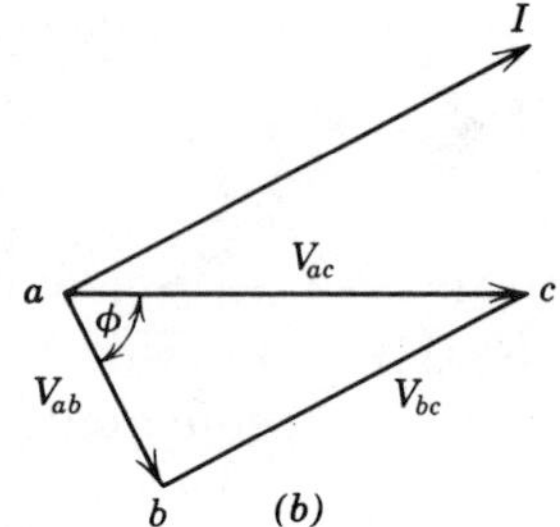

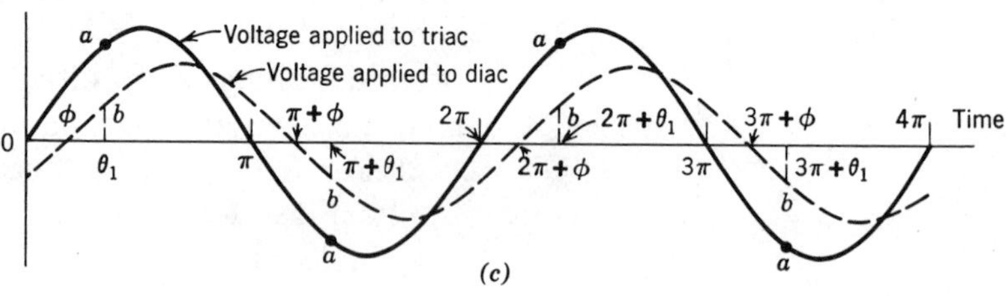

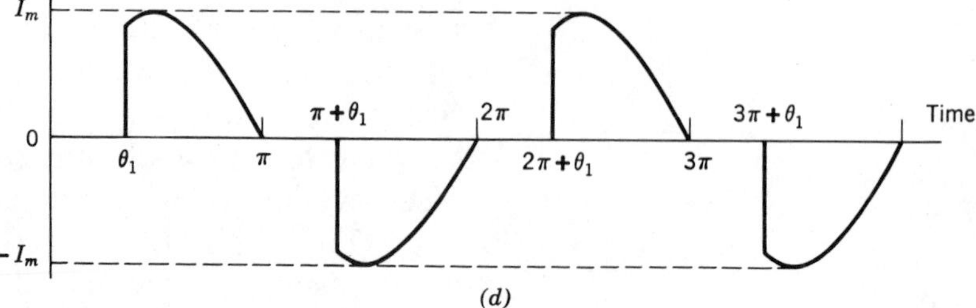

Figure 22-15 Triac application. (*a*) Circuit. (*b*) Phasor diagram. (*c*) Voltage waveforms. (*d*) Load current.

effective or rms value, which is defined by

$$I_{\text{rms}} \equiv \sqrt{\frac{1}{2\pi}\int_0^{2\pi} (I_m \sin\theta)^2\, d\theta} \tag{22-6}$$

If we use methods of calculus to evaluate Eq. 22-6 for the waveform shown in Fig. 22-15*d*, we find

$$\boxed{I_{\text{rms}} = \frac{I_m}{\sqrt{2}}\sqrt{\frac{180° - \theta_1}{180°} - \frac{\sin 2\theta_1}{2\pi}}} \tag{22-7}$$

where θ_1 is the angle in the cycle at which firing occurs.

An examination of Eq. 22-7 shows that, when the firing angle θ_1 is zero, the load current waveform is a complete ac cycle and Eq. 22-7 reduces to the expected value $I_m/\sqrt{2}$, which is the basic conversion from peak to rms in ac circuit analysis. When θ_1 is 180°, Eq. 22-7 yields zero, since the value of $\sin 2\theta_1$ is sin 360°, which is zero. Naturally, if the triac fires at 180°, there is zero load current. When θ_1 is 90°, $\sin 2\theta_1$ is sin 180°, which is zero. Now the rms value is $I_m/2$.

The power in a resistive load can be obtained from

$$P_L = I_{\text{rms}}^2 R_L$$

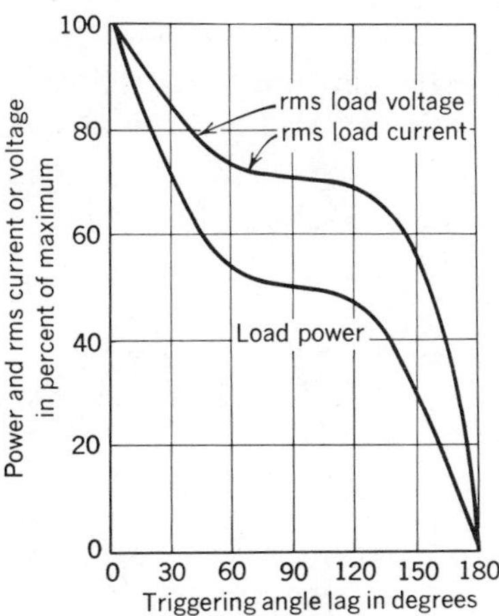

Figure 22-16 Variation of rms load voltage, of rms load current, and of load power with control angle for a triac.

Substituting Eq. 22-7 into this expression, we have

$$P_L = I_{\text{rms}}^2 R_L = \frac{I_m^2 R_L}{2}\left[\frac{180° - \theta_1}{180°} - \frac{\sin 2\theta_1}{2\pi}\right] \tag{22-8}$$

in which θ_1 is expressed in degrees.

Various values of θ_1 are substituted into Eq. 22-7 and into Eq. 22-8 to obtain the graphs given in Fig. 22-16. It should be noted that, when the triggering angle is 90°, the load power is 50% of its maximum value. Problem solutions can be obtained either from the equations or from the graphs.

Problems

22-4.1 In Fig. 22-15, the supply voltage is 117 V rms and the breakdown voltage of the diac is 20 V. What is the least value of θ_1 (Fig. 22-15*d*) that will fire the triac? If C is 1 μF, what value of R_L will fire the circuit at 20°, at 40°, and at 60°.

22-4.2 Show how the values for the curves of Fig. 22-16 were obtained at 45° and at 150°.

22-4.3 A triac is used to control the input power into a 1000-W heater element that operates from 117 V rms. What firing angles are required for $\frac{3}{4}$, $\frac{1}{2}$, and $\frac{1}{4}$ heat?

22-4.4 A triac controls the power to 100-Ω load. A switch reduces the power from maximum to 70% of full power when the device is operated from a 117-V supply. What firing angle should be inserted by the switch when the circuit uses a diac that has a 20-V breakdown voltage?

22-4.5 Using Eq. 22-8, plot the power in a 100-Ω resistor controlled by a triac for a 180° variation in phase shift. The supply voltage is 117 V rms. Obtain data at 30° intervals. Assume that the forward voltage drop across the triac is zero after breakdown. Also plot the rms load current.

Answers to Odd-Numbered Problems

Chapter 2

2-1.1 $r_B = 3.5\ \Omega$; $0.67\ \Omega$; $0.029\ \Omega$
2-2.1 $R_L = 80\ \Omega$; $50\ \Omega$
2-2.3 $V_{1,\max} = 19$ V; $V_{1,\min} = 13$ V
2-2.5 $R = 253\ \Omega$; $P_Z = 90.2$ mW
2-2.7 Change 150 mV

2-3.1

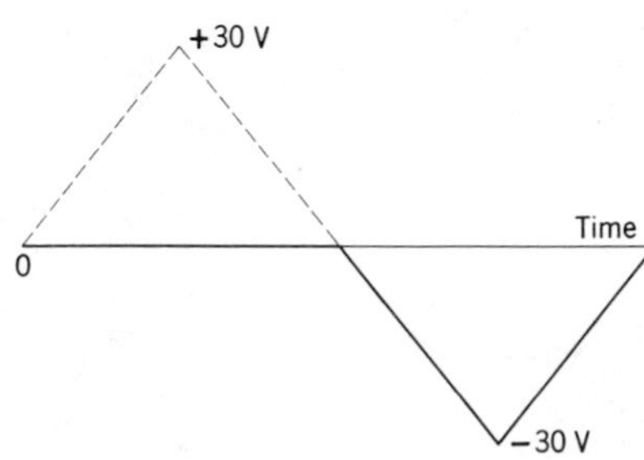

2-3.3

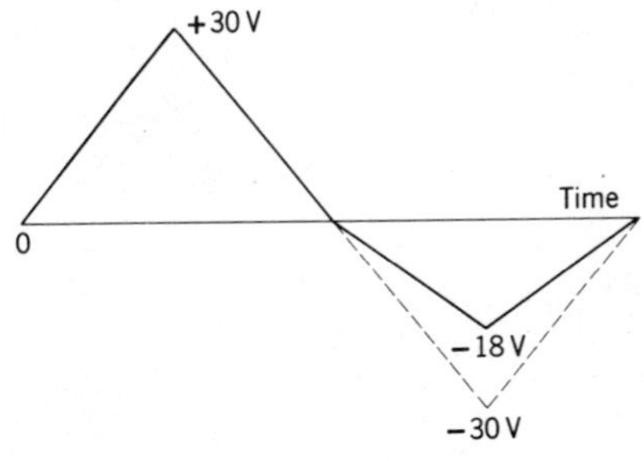

2-3.5

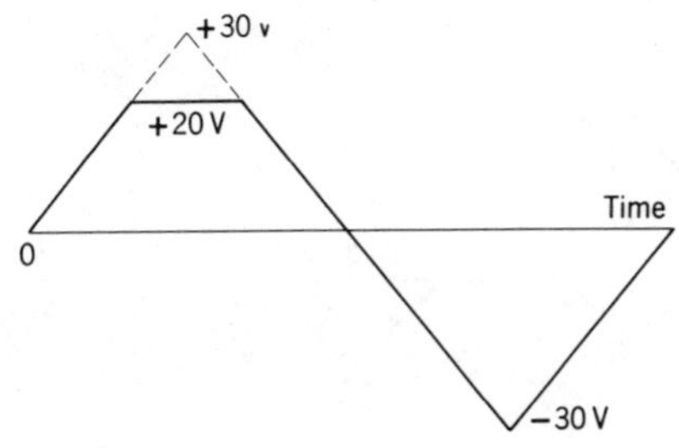

2-3.7

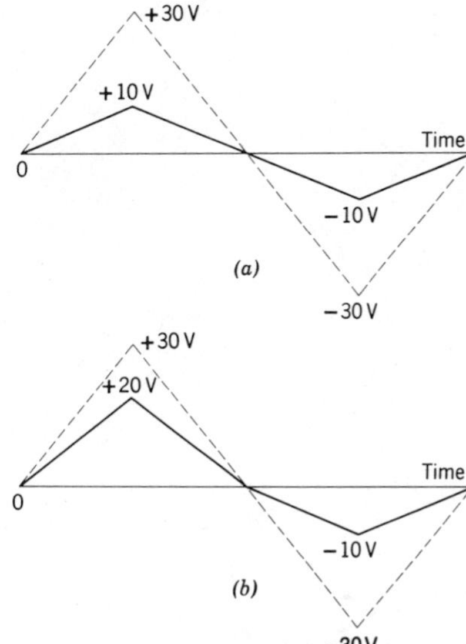

2-4.1

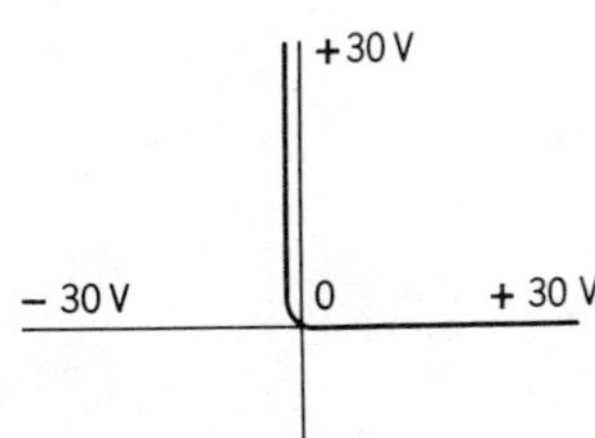

2-4.3

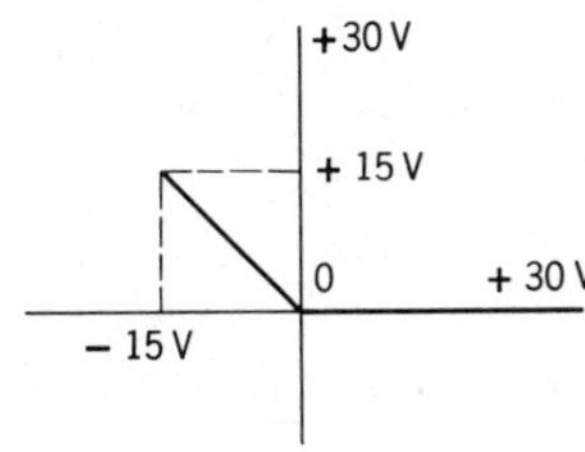

2-4.5

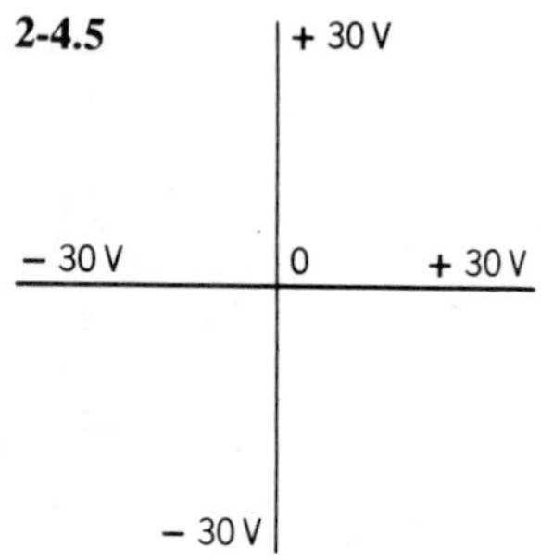

2-4.7

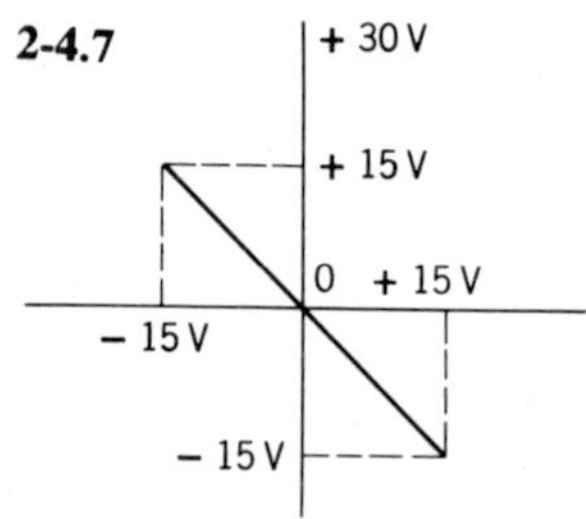

2-4.9

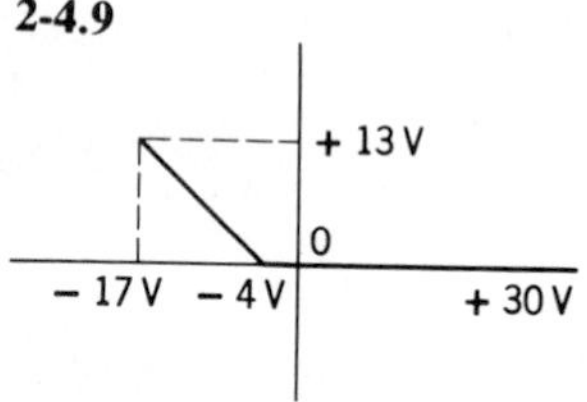

2-4.11

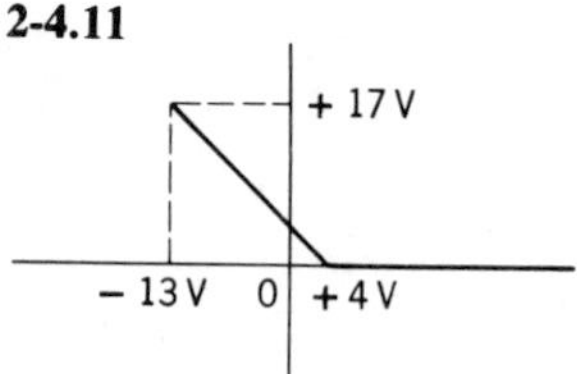

2-4.13

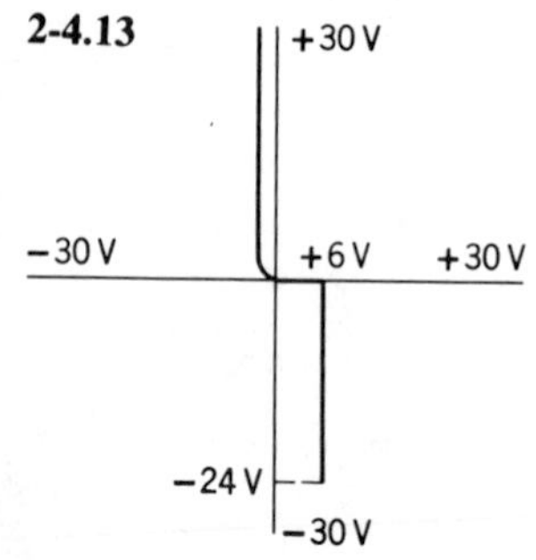

2-4.15

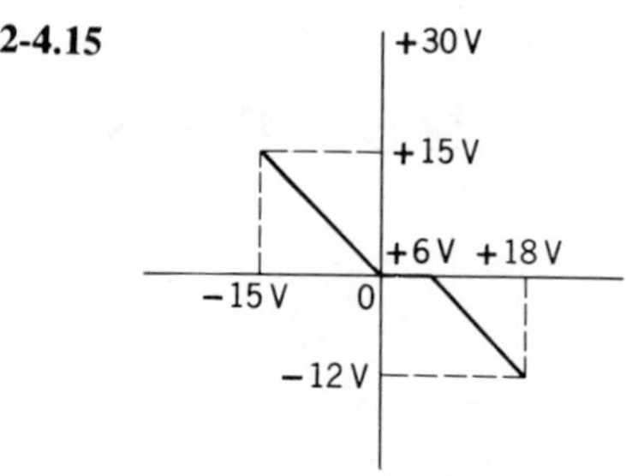

2-4.17

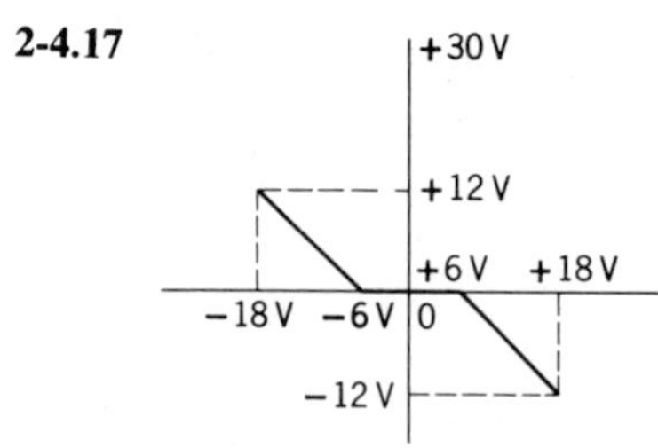

2-4.19

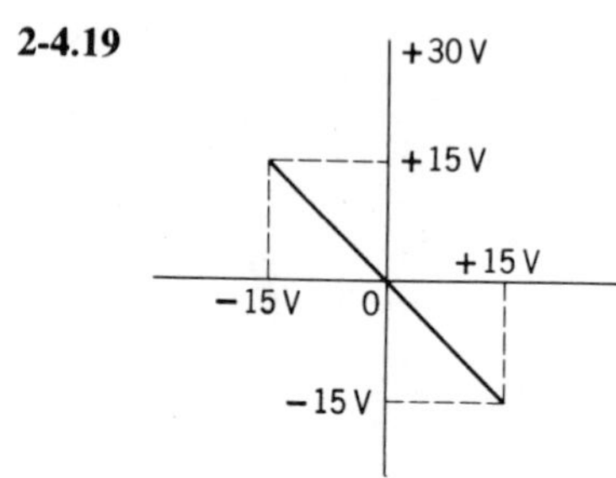

2-4.21

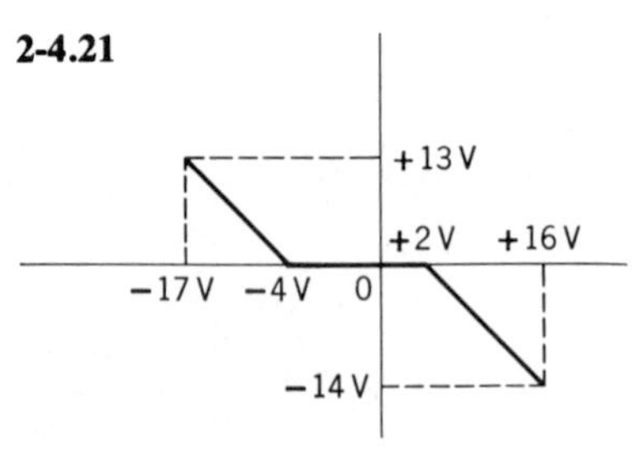

2-4.23

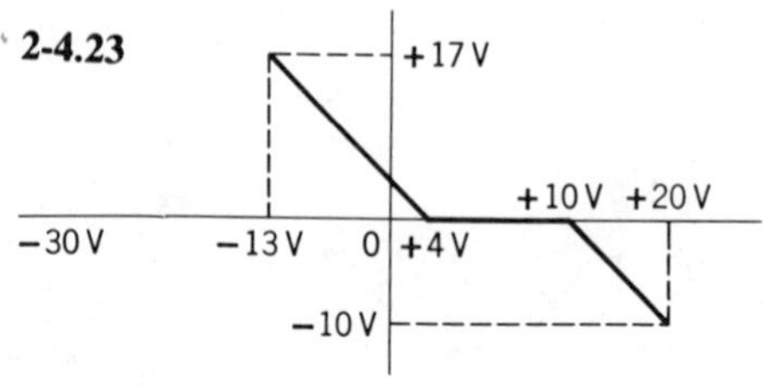

2-5.1 $V_{out} = 8.1$ mV
2-5.3 $V_{out} = 0.71$ mV
2-5.5 $V_{out} = 83\ \mu V$
2-1 $I_{L,max} = 4$ A; $I_{L,min} = 2$ A
2-3 $V_{Z,max} = 6.236$ V; $V_{Z,min} = 6.140$ V
2-5 $V_{in,max} = 16.817$ V; $V_{in,min} = 13.183$ V; $\pm 12\%$
2-7 $P_D = 584$ mW
2-9 When $R_L = \infty\ \Omega$; $I_L = 0$ mA; $P_Z = 0.94$ W
When $I_Z = 0$ mA; $I_L = 152$ mA; $R_L = 40.8\ \Omega$
2-11

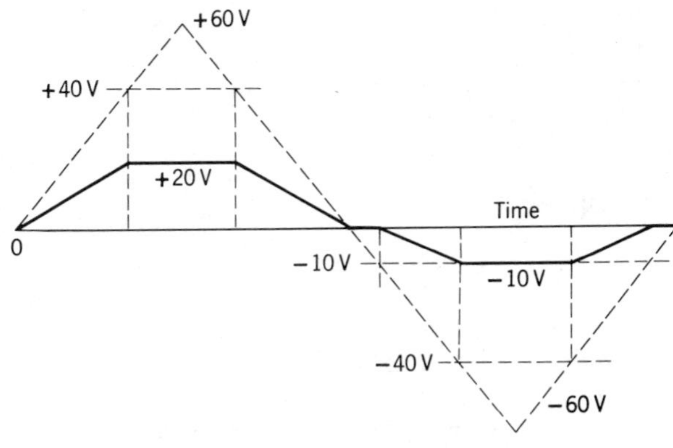

2-13 $V_{out} = 11.3$ mV
2-15 $V_{out} = 44.8$ mV

Chapter 3

3-1.1 $R_L = 13.6\ \Omega$; $I_{dc} = 0.38$ A
3.1.3 $I_{dc} = 5$ A; $R_L = 4\ \Omega$; $V_s = 44.4$ V
3-1.5 First quadrant 100 V; 100 mA
3-1.7 $I_L = 0$; $V_L = 0$
3-1.9 Polarities reversed
3-2.1 248 V; 248 mA
3-2.3 22.7 V − 0 − 22.7 V at 5 A dc
3-3.1 $V_{dc} = 225$ V; $I_{dc} = 22.5$ mA; $I_L = 54$ mA; $P_{ac} =$ 6.27 W
3-3.3 117 V and 4.2 A; 22.2 V and 22.2 A
3-3.5 Transformer shorted
3-3.7 Operates as half-wave rectifier
3-1 6.25 V; 62.5 mA
3-3 628 V_{p-p}
3-5 25 V; 250 mA
3-7 12.5 V; 125 mA
3-9 628 V_{p-p}

Chapter 4

4-6.1 $\alpha = 0.980$; 0.990; 0.992; 0.993; 0.995
4-6.3 $\beta = 199$; 99; 79; 42
4-6.5 $\alpha = 0.997$; $\beta = 319$

Chapter 5

5-2.1 $R_C = 539\ \Omega$
5-2.3 $R_C = 782\ \Omega$
5-2.5 $R_B = 6\ k\Omega$
5-2.7 $R_B = 3.3\ M\Omega$
5-2.9 $R_E = 5.3\ k\Omega$
5-2.11 $R_E = 5.3\ k\Omega$
5-3.1 $V_{CE} = 9.5$ V
5-3.3 $R_B = 1.46\ M\Omega$
5-3.5 $V_{CE} = 13$ V; $R_B = 10\ k\Omega$
5-3.7 $R_B = 165 + 81.6\ R_C$
5-3.9 $R_C = 2.63\ k\Omega$; $R_B = 69\ k\Omega$
5-3.11 $I_C = 1.16$ mA; $V_{CE} = 4.16$ V
5-3.13 $I_C = 2.220$ mA; $V_{CE} = 4.4$ V
5-3.15 $R_1 = 23.9\ k\Omega$
5-3.17 $R_1 = 18.3\ k\Omega$
5-3.19 $R_1 = 8.75\ k\Omega$
5-3.21 Measured to ground: $V_C = -1.12$ V; $V_B = -7.17$ V; $V_E = -7.87$ V
5-1 $R_B = 153\ k\Omega$
5-3 Measured to ground: $V_B = -0.84$ V; $V_C = -6.24$ V; $V_E = -0.15$ V
5-5 $V_{CE} = 7.83$ V
5-7 $R_1 = 151\ k\Omega$
5-9 $R_2 = 57\ k\Omega$
5-11 Measured to ground: $V_B = -1.8$ V; $V_C = -7.2$ V; $V_E = -0.15$ V

Chapter 6

6-1.1 1.7 A; 51 V; 69 V
6-1.3 $\Delta V_F = 0.08$ V; $\Delta I_F = 18$ mA
6-2.1 12.8 V_{p-p}
6-2.3 $R_B = 193\ k\Omega$; 20 V_{p-p}
6-3.1 $V_{out} = 22.5\ V_{p-p}$; $R_B = 1.25\ M\Omega$
6-3.3 $V_{out} = 26.7\ V_{p-p}$; $R_B = 1.47\ M\Omega$
6-3.5 $V_{out} = 20\ V_{p-p}$; $R_B = 1.24\ M\Omega$
6.1 $I_{CQ} = 1.46$ mA; $V_{CEQ} = 5.08$ V; $V_{out} = 5.84\ V_{p-p}$
6-3 $I_{CQ} = 1.21$ mA $V_{CEQ} = 5.5$ V; $V_{out} = 11.0\ V_{p-p}$
6-5 $R_B = 1.16\ M\Omega$
6-7 $V_{out} = 5.8\ V_{p-p}$
6-9 $V_{out} = 13.3\ V_{p-p}$; $R_B = 870\ k\Omega$
6-11 $V_{out} = 18.5\ V_{p-p}$; $R_B = 1.7\ M\Omega$

Chapter 7

7-3.1 $A_v = 150$; $A_e = 53$; $5570\ \Omega$
7-3.3 $A_e = 116$
7-4.1 $I_s = 17\ \mu A$; $V_{out} = 2\ V$
7-4.3 $I_s = 0.22\ mA$; $V_{out} = 0.72\ V$
7-4.5 $r_{out} = r'_{out} = 25\ \Omega$
7-4.7 $r_{out} = 62\ \Omega$; $r'_{out} = 38\ \Omega$
7-5.1 $V_{out} = 15.6\ V_{p-p}$; $A_v = 311$; $E_s = 50\ mV_{p-p}$
7-5.3 $I_{CQ} = 0.48\ mA$; $V_{CBQ} = -10.4\ V$; $A_v = 192$
7-5.5 $I_{CQ} = 0.066\ mA$; $V_{CBQ} = -2.7\ V$; $V_{out} =$ 251 mV; 87 mV
7-6.1 $r_{in} = r'_{in} = 6528\ \Omega$; $A_v = 78$
7-6.3 $A_e = 60$; $8528\ \Omega$
7-6.5 $A_e = 5.65$; $14.5\ k\Omega$
7-6.7 $R_B = 1.3\ m\Omega$; $E_s = 167\ mV_{p-p}$
7-7.1 $A_e = 143$; $3.5\ k\Omega$
7-7.3 $V_{out,max} = 16\ V_{p-p}$; $R_B = 913\ k\Omega$; $E_s =$ $130\ mV_{p-p}$
7-7.5 $R_B = 730\ k\Omega$; $E_s = 0.174\ V_{p-p}$
7-8.1 $V_{out} = 1208\ mV$
7-8.3 $V_{out} = 57\ mV$
7-1 $A_v = 100$; $A_e = 26.2$; $V_{in} = 21\ mV$; $V_{out} = 2.1\ V$
7-3 $A_v = 75$; $A_e = 51$; $V_{out} = 1.53\ V$; $V_{in} = 20.4\ mV$
7-5 $A_v = 70.6$; $A_e = 18.5$; $V_{out} = 1.48\ V$; $V_{in} =$ 21 mV
7-7 $A_v = 85.7$; $A_e = 20$; $V_{out} = 500\ mV$; $V_{in} =$ 5.8 mV
7-9 $A_v = 33.9$: $A_e = 14.5$; $V_{out} = 290\ mV$; $V_{in} =$ 8.6 mV
7-11 $A_v = 3.96$; $A_e = 3.08$; $V_{out} = 139\ mV$; $V_{in} =$ 35 mV
7.13 $V_{out} = 3.75\ V$; $V_{in} = 4.13\ V$
7-15 $R_E = 2.90\ \Omega$; $V_{out} = 2.4\ V$
7-17 $A_v = 80$; $A_e = 3.95$; $V_{out} = 395\ mV$; $V_{in} =$ 4.9 mV
7-19 $A_v = 16$; $A_e = 3.25$; $V_{out} = 325\ mV$; $V_{in} =$ 20 mV
7-21 $A_v = 7.69$; $A_e = 3.06$; $V_{out} = 184\ mV$; $V_{in} =$ 24 mV

Chapter 8

8-1.3 $V_P = 5\ V$; $I_{DSS} = 20\ mA$
8-1.5 $I_D = 28.8\ mA$
8-2.1 $g_{m0} = 5.6\ mS$
8-2.3 $V_P = -4\ V$; $I_{DSS} = 4\ mA$
8-3.3 $1250\ \Omega$; $176\ \Omega$
8-4.1 At $V_{GS} = 4.0\ V$; $I_D = 6.94\ mA$
8-4.3 At $V_{GS} = 4.0\ V$; $I_D = 4.08\ mA$
8-4.5 $\infty\Omega$; $431\ \Omega$
8-1 11.25 mA; 7.5 mS
8-3 $V_{GS} = +1.9\ V$
8-5 $R = 986\ \Omega$
8-7 $R = 19.8\ k\Omega$
8-9 Position A overloads; $1500\ \Omega$
8-11 $240\ \Omega$; $\infty\Omega$

Chapter 9

9-1.1 $I_{DQ} = 3.6\ mA$; $V_{DSQ} = 7.8\ V$
9-1.3 $V_{GSQ} = -1.5\ V$; $R_S = 750\ \Omega$; $V_{DSQ} = 22\ V$; $R_D = 3250\ \Omega$
9-1.5 $R_S = 2.5\ V/0.22\ mA = 11.4\ k\Omega$
9-1.7 $V_{GS} = -2.2\ V$; $I_{DQ} = 2.0\ mA$; $R_D = 9.0\ k\Omega$
9-2.1 End points: 20 V and 4 mA; $I_{DQ} = 1.8\ mA$; $V_{DSQ} = 10.9\ V$
9-2.3 End points: 20 V and 5 mA; $I_{DQ} = 1.8\ mA$; $V_{DSQ} = 12.8\ V$
9-2.5 $A_v = 3.75$
9-3.1 $R_D = 6.67\ k\Omega$; $V_{out} = 1.0\ V$
9-3.3 $R_S = 146\ \Omega$; $V_{out} = 1.13\ V$
9-3.5 $V_{DS} = 5\ V$; $g_m = 1\ mS$; $V_{out} = 2\ V$
9-3.7 $g_m = 2.4\ mS$; $A_v = 24$; $V_{DS} = 8.0\ V$
9-3.9 $g_m = 1.96\ mS$; $A_v = 19.6$; $V_{DS} = 12.0\ V$
9-4.1 $I_D = 6.1\ mA$; $A_v = 0.72$
9-4.3 $I_D = 3.6\ mA$; $A_v = 0.57$
9-1 $I_D = 8.9\ mA$; $V_{DS} = 18.6\ V$
9-3 $I_D = 3.2\ mA$; $V_{DS} = 29.1\ V$
9-5 $I_D = 6.8\ mA$; $V_{DS} = 3\ V$
9-7 $A_e = 8.6$
9-9 $A_e = 1.41$
9-11 $A_e = 9.4$

Chapter 10

10-1.1 $I_{CQ_1} = 3.2$ mA; $I_{CQ_2} = 6.4$ mA; $V_{CEQ_1} = 13.8$ V; $V_{CEQ_2} = 7.2$ V
10-2.1 $I_{CQ_2} = 6.4$ mA
10-2.3 $I_C = 2.75$ mA; $V_{CE} = 11.7$ V; $K_1 = 0.91$; $K_2 = 0.80$; $I_{C_1} = 1.75$ mA; $I_{C_2} = 3.85$ mA; $V_{CE_1} = 14.7$ V; $V_{CE_2} = 8.4$ V
10-2.5 $\beta_1 = 127$; $\beta_2 = 76$
10-2.7 $I'_C = 1.64$ mA
10-2.9 $I_C = 0.686$ mA; $I_{C_1} = 0.44$ mA; $I_{C_2} = 0.84$ mA
10-2.11 $K = 0.408$; $I_B = 2.06$ mA; $I'_C = 1.78$ mA
10-3.1 92°C
10-3.3 66°C
10-3.5 103°C
10-3.7 $I_{CEO} = 2.265$ pA; 3.615 pA
10-4.1 114.5°C
10-4.3 89°C
10-4.5 87°C
10-4.7 50°C
10-4.9 $I'_C = 0.42$ mA
10-5.1 +100%; −66%
10-5.3 +24%; −23%
10-5.5 $g_m = 4.5\,(1 + V_{GS}/4)$ mS
10-5.7 $A_v = 5.5$; 4.6; 2.4
10-5.9 $A_v = 1.68$; 1.76; 1.41
10-5.11 $A_v = 7.7$; 8.6; 11
10-1 $I_{CQ} = 0.99$ mA; $V_{CEQ} = 3.27$ V; $I'_{CQ} = 1.44$ mA; $V'_{CEQ} = 0.2$ V
10-3 $I_{CQ} = 0.729$ mA; $V_{CEQ} = 4.2$ V; $I'_{CQ} = 0.838$ mA; $V'_{CEQ} = 3.3$ V
10-5 $I_{CQ} = 0.639$ mA; $V_{CEQ} = 5.0$ V; $I'_{CQ} = 0.754$ mA; $V'_{CEQ} = 4.1$ V
10-7 0.16 of 1%
10-9 66°C
10-11 $S = 19.8$; saturated; $I_{CQ,\text{sat}} = 1.23$ mA
10-13 $S = 23.4$; $I'_{CQ} = 0.687$ mA
10-15 From graphs: (7.2 mA, 6.4 V): (10.7 mA, 4.7 V); (3 mA, 8.5 V)
10-17 From graphs: (2.26 mA, 19.7 V); (3.00 mA, 13.0 V); (2.16 mA, 20.6 V)

Chapter 11

11-3.1 *a* 3.42; *b* 2.12; *c* 5.89; *d* 0.16; *e* 18.47; *f* −2.64; *g* −0.058; *h* −1.20; *i* −1.33; *j* −4.076
11-4.1 3.36 V
11-4.3 80×10^{-9} W; 1.13 mV
11-4.5 1 dB/100 ft or 3.3 dB/100 m
11-4.7 Space loss 238 dB; $V_r = 1.58\ \mu$V
11-4.9 −40 dB; −26 dB; −18 dB; −13 dB; −6 dB

Chapter 12

12-1.1 $V_{\text{out,dc}} = 1.88$ V; $V_{\text{in,dc}} = 0.15$ V; $R = 52.5$ kΩ; $A_e = A_v = 85$; $V_{\text{out,max}} = 2.24\ V_{p-p}$
12-1.3 $R_{C_2} = 2000$ Ω; $R_{E_2} = 392$ Ω; $R_{E_1} = 1912$ Ω; $R_{C_1} = 450$ Ω; $R = 4545$ Ω; $A_v = A_e = 4930$
12-1.5 $R_{C_2} = 300$ Ω; $R = 9375$ Ω; $R_{C_1} = 1667$ Ω; $A_e = A_v = 55$
12-2.1 $R_1 = R_B = 1.875$ MΩ; $E_S = 7.2\ V_{p-p}$; $I_s = 6.5\ \mu A_{p-p}$
12-2.3 $R_{\text{ac}} = 104.4$ kΩ; $A_e = 0.52$; $A_i = 5416$
12-2.5 $R_1 = 145$ kΩ; 13.8 kΩ
12-3.1 $I_B = 4.35\ \mu$A; $I_C = 0.348$ mA; $I_E = 0.352$ mA; $V_B = 0.2$ V; $V_E = -0.5$ V; $V_C = +8.0$ V; $A_e = A_v = 282$; $V_{\text{out}} = 14\ V_{p-p,\max}$
12-3.3 $I_B = 5.14\ \mu$A; $I_C = 0.411$ mA; $I_E = 0.416$ mA; $V_B = 0.05$ V; $V_E = -0.65$ V; $V_C = 5.8$ V; $A_e = A_v = 167$; $V_{\text{out}} = 8.2\ V_{p-p,\max}$
12-4.1 $I_B = 13.8\ \mu$A; $I_C = 1.38$ mA; $I_E = 1.39$ mA; $V_B = +0.3$ V $V_E = -0.4$ V; $V_C = 10.6$ V; $A_e = A_v = 190$; $V_{\text{out,max}} = 18.8\ V_{p-p}$
12-4.3 $I_B = 2.15\ \mu$A; $I_C = 0.215$ mA; $I_E = 0.217$ mA; $V_B = 16$ mV; $V_E = -0.68$ V; $V_C = 5.7$ V; $A_e = A_v = 87$; $V_{\text{out,max}} = 8.6\ V_{p-p}$
12-5.1 106 dB
12-5.3 150 mV; 20 V
12-7.1 $V_{\text{out}} = 9.4\ V_{p-p}$; $A_v = 37.6$; $V_{\text{out}} = 752$ mV; $I_s = 1\ \mu$A
12-7.3 $V_{\text{out,max}} = 3.6\ V_{p-p}$; $A_e = A_v = 4900$

Chapter 13

13-2.1 $\theta_{JA} = 75°C/W$; $\theta_{JC} = 10°C/W$

13-2.3 2.5 mW/°C

13-2.5 $\theta_{CA} = 0.3°C/mW$

13-2.7 NC441 with a 5°C reserve

13-3.1 176.8 mA to 23.2 mA; $R_B = 6.78\ k\Omega$

13-4.1 $\alpha = 2.12/1$; $P_C = 4\ W$; $I_C = 333\ mA$; $BV_{CE} = 24\ V$

13-4.3 $\alpha = 1/1$; $P_C = 20\ W$; $I_C = 1\ A$; $BV_{CE} = 40\ V$

13-4.5 $\alpha = 1.36/1$

13-4.7 +26.3 dB

13-1 Use NC403 with safety of 7°C

13-3 $T_J = 125°C$

13-5 $P_C = 5.1\ W$

13-7 $P_C = 20\ W$; $\alpha = 1/1.26$

Chapter 14

14-2.1 $A_v = 101$; $R = 480\ k\Omega$

14-4.1 $\alpha = 2.12/1$; $P_C = 20\ W$; $I_C = 333\ mA$; $BV_{CE} = 120\ V$; $R_B = 10.7\ k\Omega$

14-4.3 $\alpha = 1.5/1$; $I_C = 333\ mA$; $P_C = 4\ W$; $BV_{CE} = 24\ V$; $R_B = 2034\ \Omega$

14-4.5 $E_{s,max} = 3.39\ V_{peak}$; +30.3 dB

14-5.1 $\alpha = 7.14/1$; $P_C = 40\ mW$; $I_C = 14.2\ mA$; $BV_{CE} = 18\ V$

14-5.3 $\alpha = 1.5/1$; $P_C = 0.8\ W$; $BV_{CE} = 24\ V$; $I_C = 212\ mA$

14-5.5 $E_s = 106\ mV_{peak}$; +38.3 dB

14-5.7 $P_L = 100\ W$; $E_{s,max} = 6.48\ V_{peak}$

14-6.1 $P_L = 200\ W$; $P_C = 40\ W$

14-6.3 $V_{CC} = +17.9\ V$; $-17.9\ V$; $I_C = 356\ mA$; $P_C = 2\ W$; $BV_{CE} = 35.8\ V$

14-6.5 $V_{CC} = +10\ V$; $-10\ V$; $I_C = 100\ mA$; $BV_{CE} = 20\ V$; $P_C = 1\ W$; $E_{s,max} = 10.9\ V_{peak}$; +16.6 dB

14-6.7 $P_L = 200\ W$; $P_C = 200\ W$; $R_B = 1269\ \Omega$

14-6.9 $P_L = 10\ W$; $P_C = 10\ W$; $R_B = 1229\ \Omega$

14-1 $\alpha = 4.24/1$; $P_C = 2\ W$; $I_C = 83\ mA$; $BV_{CE} = 48\ V$

14-3 $\alpha = 4.24/1$: $BV_{CE} = 48\ V$; $P_C = 0.4\ W$; $I_C = 53\ mA$

14-5 $V_{CC} = +5.66\ V$; $-5.66\ V$; $I_C = 353\ mA$; $P_C = 2\ W$; $BV_{CE} = 11.3\ V$

14-7 $V_{CC} = \pm 5.66\ V$; $I_C = 353\ mA$; $P_C = 2\ W$; $BV_{CE} = 11.3\ V$

Chapter 15

15-1.1 5 Hz – 0.031 V – 88.2°; 60 Hz – 0.353 V – 69.3°; 159 Hz – 0.707 V – 45°; 300 Hz – 0.883 V – 27.9°; 500 Hz – 0.923 V – 18°; 1 kHz – 0.988 V – 9°

15-1.3 5 kHz – 0.156 V – 81.1°; 20 kHz – 0.532 V – 58°; 31.8 kHz – 0.707 V – 45°; 50 kHz – 0.844 V – 32.5°; 100 kHz – 0.953 V – 17.7°

15-1.5 1 Hz – 0.62 V – 84.1°; 2 Hz – 0.122 V – 78.3°; 4 Hz – 0.230 V – 67.5°; 9.65 Hz – 0.424 V – 45°; 15 Hz – 0.505 V – 32.7°; 40 Hz – 0.583 V – 13.6°; 200 Hz – 0.599 V – 2.8°

15-2.1 0.2 Hz – 1.000 V – 1.4°; 2 Hz – 0.970 – 14.1°; 4 Hz – 0.893 V – 25.7°; 7.95 Hz – 0.707 V – 45°; 25 Hz – 0.303 V – 72.3°; 75 Hz – 0.106 V; 83.9°; 200 Hz – 0.040 V – 87.7°

15-2.3 2 Hz – 1.000 V – 1.4°; 20 Hz – 0.970 V – 14.1°; 40 Hz – 0.893 V – 26.7°; 79.6 Hz – 0.707 v – 45°; 200 Hz – 0.370 V – 68.3°; 500 Hz – 0.157 V – 81°; 1000 Hz – 0.079 V – 85.5°

15-2.5 1 kHz – 952 mV – 1.7°; 10 kHz – 912 mV – 16.7°; 20 kHz – 817 mV – 30.9°; 33.4 kHz – 673 mV – 45°; 100 kHz – 302 mV – 71.5°; 1 MHz – 32 mV – 88.1°

15-3.1 $f_1 = 159\ Hz$; $A_v = -4.0\ dB$; $\theta = 54°$

15-3.3 $f_1 = 31.8\ kHz$; $A_v = -4.0\ dB$; $\theta = 54°$

15-3.5 $f_1 = 9.65\ Hz$; $A_v = -12.0\ dB$; $\theta = 62.2°$

15-4.1 $f_2 = 7.96\ kHz$; $A_v = -9.9\ dB$; $\theta = 67.4°$ lag

15-4.3 $f_2 = 79.6\ kHz$; $A_v = -16\ dB$; $\theta = 81°$ lag

15-4.5 $f_2 = 33.4\ kHz$; $A_v = -0.4\ dB$; $\theta = 35°$ lag

15-6.1 $A_v = 29\ dB$; $f_1 = 47.5\ Hz$; $f_2 = 45.4\ kHz$. Cascaded: $A_v = +58\ dB$; $f_1 = 74\ Hz$; $f_2 = 29\ kHz$; on phase 47.5 Hz and 45.4 kHz are the 90° values

15-7.1 $A_e = 13.5$; $C_1 = 0.26\ \mu F$; $C_2 = 0.49\ \mu F$; $A_e = 6.75$; $\theta = 90°$ lead

15-7.3 $\omega_2 = 1.57 \times 10^6$ rad/s; $\omega_2' = 25 \times 10^6$ rad/s; $A_e = A_{e,MF} = 13.5$ at 200 kHz; $A_e = 1.7$ at 2 MHz

15-7.5 $A_{e,MF} = 39$; $f_2 = 36.7\ kHz$; $f_2' = 1.6\ MHz$; $A_e = 18.1$ at 62° lag at 70 kHz; $A_e = 13.4$ at 70° lag at 100 kHz

15-7.7 $f_\alpha = 15.9\ MHz$

15-1 $K_{LF} = 0.36$; $V_{out} = 0.265\ V$ at 69° lead

15-3 $K_{LF} = 0.838$: $V_{out} = 617\ mV$ at 33° lead

15-5 $V_{out} = 0.379\ V$ at 59° lag

15-7 $V_{out} = 44\ mV$ at 86.6° lag

15-9 $A_e = +27.7\ dB$; $f_1 = 10.5\ Hz$; $C_{in} = 3950\ pF$; $f_2 = 23.9\ kHz$

Chapter 16

16-2.1 Missing values: 0.6; 13.33 mV; 33.3 mV; 666.7 mV
16-2.3 Missing values: 267 mV; 0.4; 40 mV; 667 mV
16-2.5 Missing values: 0.2 of 1%; 0.25 V; 1.25 V; 125 V
16-3.1 Missing values: 15%; 4; 150 mV; 50 mV
16-3.3 Missing values: 250 mV; 2.5; 150 mV; 100 mV
16-3.5 Missing values: 20; 5; 400 mV; 100 mV
16-3.7 3%: 0.75 V; 0.25 V; 25 V
16-3.9 0.15 of 1%
16-3.11 $\beta_f = 12\%$; $A'_v = 8$; $f'_2 = 40.2$ MHz
16-3.13 Increase is 0.8 of 1%
16-6.1 941 mV; $V_{out} = 1.30$ V
16-7.1 $A_e = 292$; $A'_e = 27.2$
16-7.3 $R_F = 1.04$ MΩ
16-8.1 $V_{in} = 0.762$ mV; $V_{out} = 2.92$ V; $V'_{in} = 5.97$ mV; $V'_{out} = 884$ mV
16-8.3 $A_v = 14.46 \times 14.4 \times 18.8 = 3915$; $V_{in} = 1.52$ mV; $V_{out} = 5.97$ V; $A_e = 2.98$ V
16-9.1 $V_{out} = 192$ mV; $V_{in} = 1.52$ mV; $V'_{in} = 0.256$ mV; $V'_{out} = 32.2$ mV
16-1 $V_B = 0.4$ V; $V_D = 0.4$ V; $V_A = 0$ V; $\beta_f = 2\%$ positive
16-3 $V_B = 187.5$ mV; $V_D = 150$ mV; $V_A = 337.5$ mV
16-5 $\beta_f = 0.45$ of 1%; $D' = 1.55\%$
16-7 $A'_v = 49.2$ to 50.4
16-9 $A'_v = 7.5$; $r'_{in} = 1$ kΩ; $r_{out} = 50$ Ω; $D' = 3.75\%$

Chapter 17

17-2.1 Error = −0.05 of 1%; −0.5 of 1%
17-2.3 Error = +10.5%; −9.5%
17-2.5 Error = +0.2 of 1%; −0.2 of 1%
17-3.1 4.2 V at5 kΩ; 2.87 V at 7.5 kΩ; 2.2 V at 10 kΩ; 1.2 V at 20 kΩ; 0.7 V at 40 kΩ; 0.4 V at 100 kΩ
17-3.3 $R_1 = 55$ kΩ; $R_2 = 82.5$ kΩ; $R_3 = 100$ kΩ; $R_4 = 330$ kΩ

17-3.5

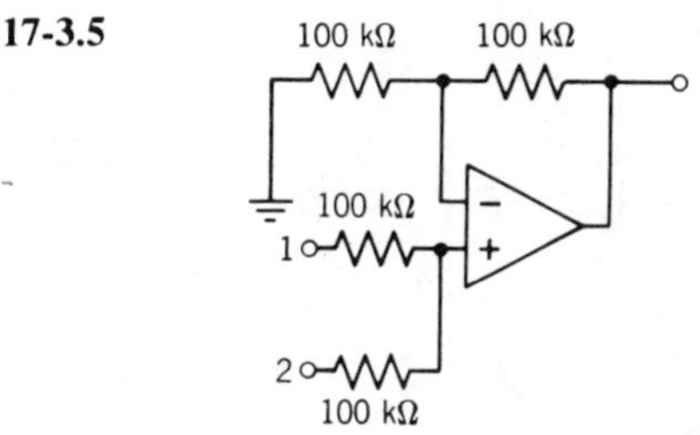

17-3.7

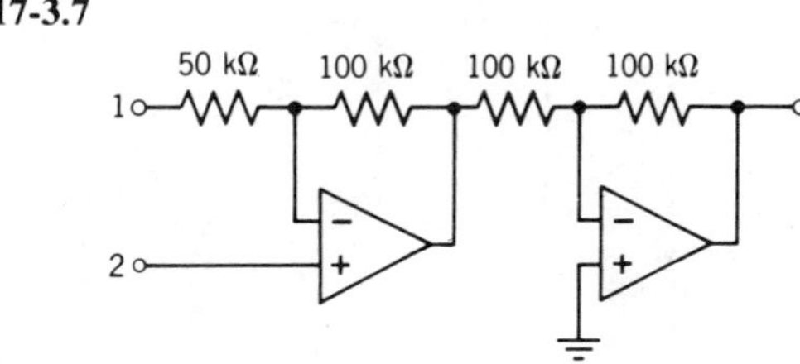

17-1 *a* −1.2 V; *b* −3.9 V; *c* −0.5 V; *d* −0.55 V; *e* −7 V; *f* −7.7 V
17-3 *a* triangular, +1.6 V, −1.6 V;

17-3*b*

+3.8 V
+1.8 V
v_{out}
+0.2 V
0
−0.2 V
Time
−1.8 V
−3.8 V

17-3*c*

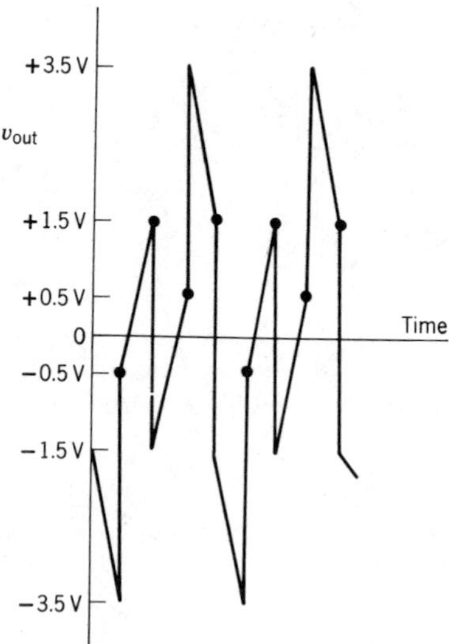

17-3*e*

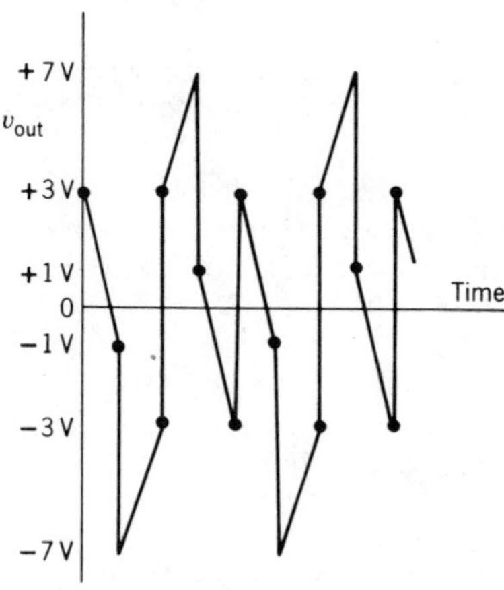

17-3*d*

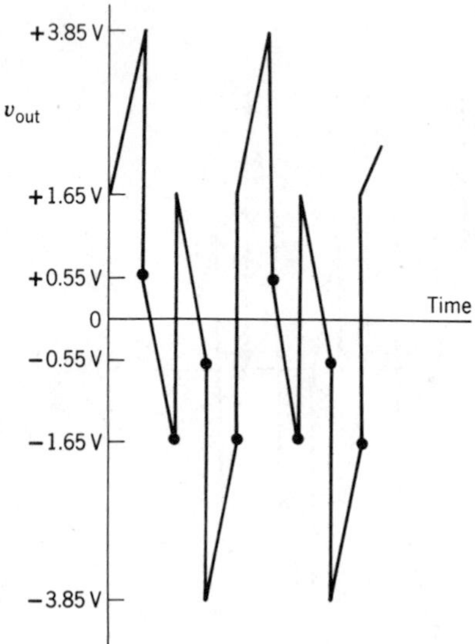

17-3*f*

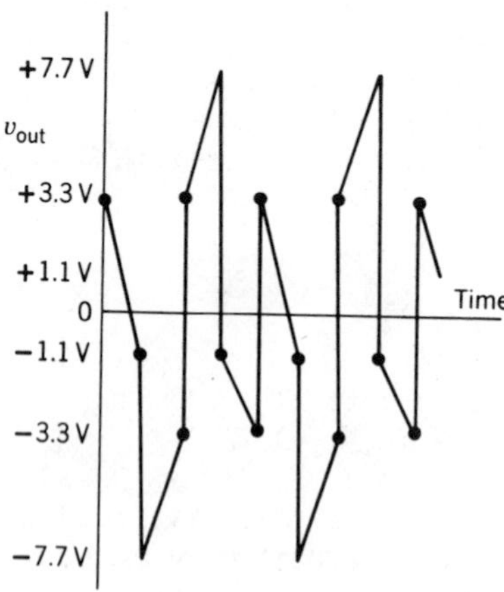

Chapter 18

18-2.1 All circuits stable
18-2.3 No. 2 and No. 3 stable
18-3.1 6121 Hz
18-3.3 No because $f_{max} = 7.96$ kHz

Chapter 19

19-1.1

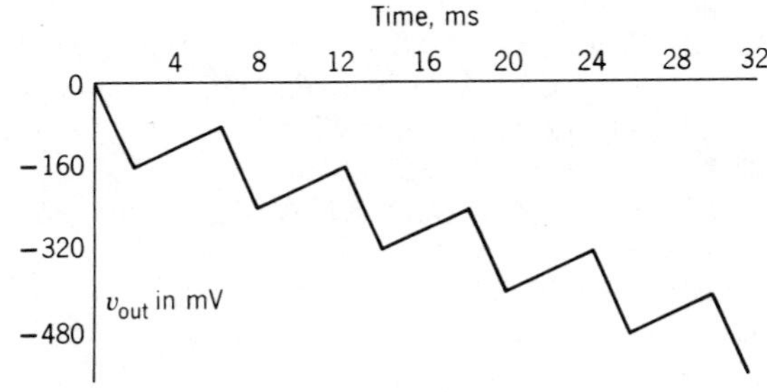

19-1.3

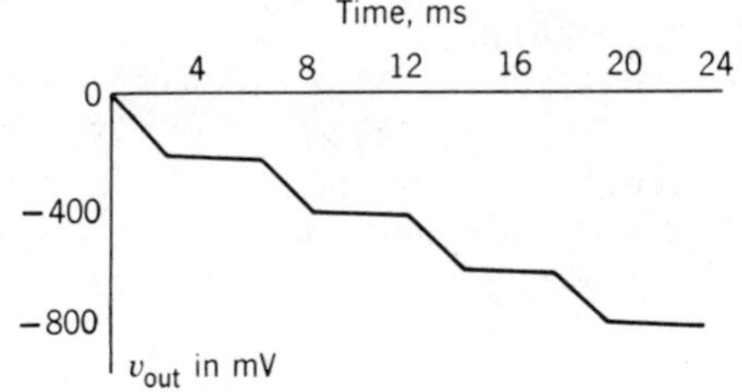

19-1.5 4 hrs, 48 min, 53 sec
19-2.1 Rectangular output: −136; +45.3; +63 mV
19-2.3 Rectangular output: −51 mV and +17 mV
19-3.1

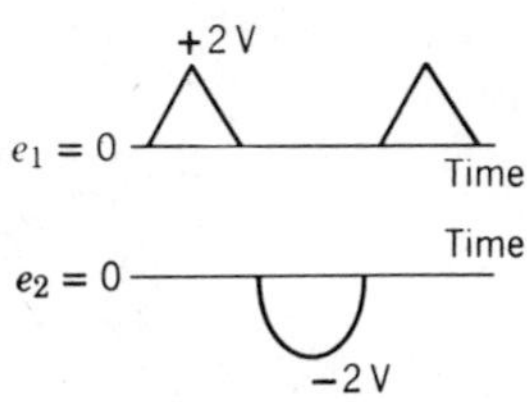

19-3.3

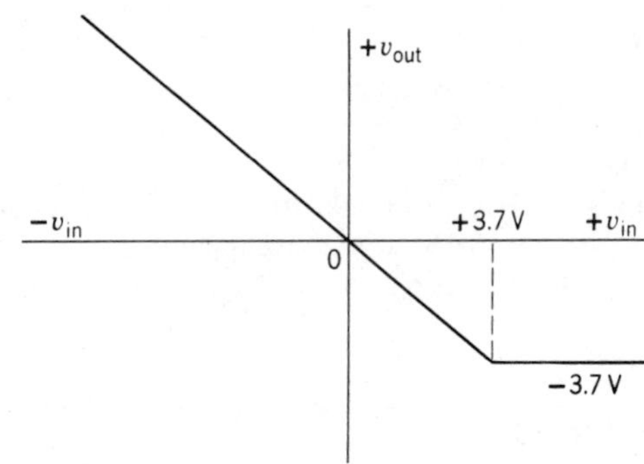

19-3.5

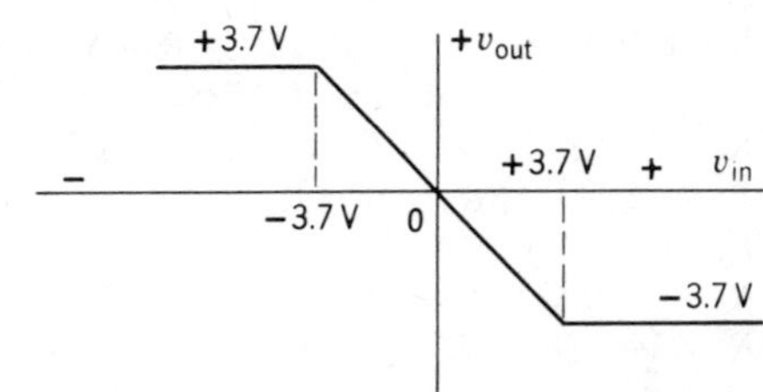

19-3.7 $V_A = +2.079$ V; $V_B = +1.881$ V; $V_H = 0.198$ V

19-3.9

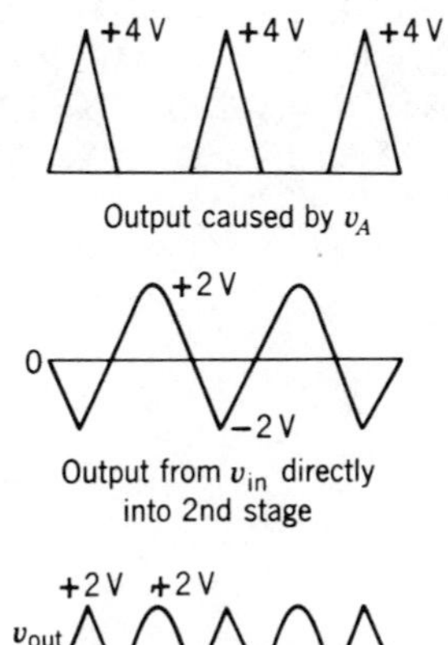

19-4.1 $P_{\text{out}} = 0.5$ W/channel; $I = 175$ mA
19-4.3 $\beta_f = 2\%$ negative
19-4.5 $f_2' = 400$ kHz
19-4.7 $f = 27.85$ kHz; slew rate limits
19-4.9 $I = 292$ mA
19-4.11 $A_v' = -51$
19-4.13 At 0 Hz, $A_v = +90$ dB; at midfrequencies, $A_v = +40$ dB; at high frequencies, $A_v = +34$ dB

Chapter 20

20-1.1 $I_{L,\min} = 19$ mA; $I_{L,\max} = 18$ mA
20-1.3 $V_{\text{in,max}} = 11.18$ V; $V_{\text{in,min}} = 10.22$ V
20-2.1 $I_L = 0$ at $V_L = 5.52$ V; $I_L = 610$ mA at $V_L = 5.50$ V
20-3.1 $R = 829\ \Omega$; $R_f = 15.48$ kΩ; $R_C = 6.08$ kΩ
20-3.3 $R = 854\ \Omega$; $V_{\text{out}} = 0$ V to $V_{\text{out}} = +6.2$ V
20-3.5 $R_f = 20$ kΩ; $R_2 = 625\ \Omega$; $I_L = 33.6$ mA
20-5.1 $R_2 = 14.24$ kΩ; $R_3 = 5875\ \Omega$: $I_{L,\max} = 80$ mA; $R_{sc} = 8.125\ \Omega$
20-5.3 $R_2 = 17.97$ kΩ; $I_L = 2.83$ A; $P_C = 42.5$ W
20-5.5 $R_2 = 3986\ \Omega$: $I_{L,\max} = 2.7$ A; $P_C = 40.5$ W; $R_{sc} = 0.241\ \Omega$
20-1 $R_2 = 23{,}256\ \Omega$; $I_{L,\max} = 80$ mA; $P_C = 400$ mW; $R_3 = 7.0$ kΩ
20-3 Circuit cannot be used
20-5 $I_{L,\max} = 150$ mA; $P_C = 750$ mW; $R_3 = 10$ kΩ $R_4 = 28.46$ kΩ; $R_{sc} = 17.57\ \Omega$
20-7 $I_{L,\max} = 40$ mA; $P_C = 800$ mW; $R_2 = 9108\ \Omega$; $R_3 = 4767\ \Omega$; $R_{sc} = 16.25\ \Omega$
20-9 $I_{L,\max} = 0.5$ A; $P_C = 10$ W; $P_C = 196$ mW; $R_2 = 9108\ \Omega$; $R_3 = 4767\ \Omega$; $R_{sc} = 1.3\ \Omega$
20-11 $I_L = 0.5$ A; $P_C = 153$ mW; $P_C = 10$ W; $R_1 = 10$ kΩ; $R_2 = 228.5$ kΩ

Chapter 21

21-2.1 $C = 0.040\ \mu$F
21-2.3 $C = 4420$ pF
21-2.5 $C = 0.048\ \mu$F; 20% error
21-2.7 $C = 5320$ pF; 20% error
21-3.1 $R_{L,\min} = 40\ \Omega$; $v_s = 17$ V; $r_{ac} = 1.11\ \Omega$
21-3.3 $\theta = 10.5°$
21-3.5 $\theta = 10.5°$; $\theta = 190.5°$

Chapter 22

22-1.1 When $\theta = 0°$, $I_L = 63.7$ mA; when $\theta = 45°$, $I_L = 54.3$ mA; when $\theta = 90°$, $I_L = 31.8$ mA; when $\theta = 135°$, $I_L = 9.3$ mA
22-1.3 At 18.2%, $\theta = 82°$; at 8.6%, $\theta = 117°$
22-1.5 $R_L = 41.4\ \Omega$: $I_{dc} = 1.27$ A; $I_{dc} = 0.56$ A; 0 to 97°
22-2.1 $\theta = 180° - 2\alpha$; $\alpha = \tan^{-1} X_L/R$
22-2.3 Range of R: ∞ to 1880 Ω
22-2.5 Range of R: 0 to 106.3 kΩ
22-3.1 $\alpha = 25.6°$; $\theta = 64.4°$; $V_L = 37.7$ V; $I_L = 3.77$ A; $I_{gm} = 716\ \mu$A
22-3.3 $\alpha = 76°$; $\theta = 14°$; $V_L = 51.9$ V; $I_L = 0.86$ A; $I_{gm} = 1.61$ mA
22-3.5 For 30°, $C = 1738$ pF; for 60°, $C = 3477$ pF; for 90°, $c = 5215$ pF; for 120°, $C = 6953$ pF; for 150°, $C = 8691$ pf
22-4.1 $\theta = 6.9°$; $R = 966\ \Omega$; $R = 2226\ \Omega$; $R = 4595\ \Omega$
22-4.3 3/4 heat, 25°; 1/2 heat, 90°; 1/4 heat 155°

Index

with unbalanced output:

$$A_v = \frac{R_C}{2r'_e} \quad (12\text{-}12) \qquad A_v = \frac{R_C}{2(r'_e + R_E)} \quad (12\text{-}13)$$

Common-mode rejection ratio:

$$A_{cm} \equiv \frac{V_{\text{out},cm}}{V_{\text{in},cm}} \quad (12\text{-}16) \qquad \text{CMRR} \equiv \frac{A_v}{A_{cm}} \quad (12\text{-}17)$$

$$V'_{\text{out}} = \left[1 + \frac{1}{\text{CMRR}} \times \frac{V_{\text{in},cm}}{(V_{\text{in}_1} - V_{\text{in}_2})}\right] V_{\text{out}} \quad (12\text{-}19)$$

Chapter 13 Single-Ended Power Amplifiers

Heat Sinks:

$$P_C = V_{CE} I_C \quad (13\text{-}1) \qquad \theta = \frac{T_2 - T_1}{P_C} \text{ °C/W or °C/mW} \quad (13\text{-}2a)$$

$$\theta_{JA} = \theta_{JC} + \theta_{CS} + \theta_{SA} \text{ °C/W or °C/mW} \quad (13\text{-}3a)$$

$$T_J = T_A + \theta_{JA} P_C \text{ °C} \quad (13\text{-}3b)$$

Transformers:

$$\alpha \equiv \frac{N_2}{N_1} = \frac{V_2}{V_1} \quad (13\text{-}5a) \qquad R_a = \frac{1}{\alpha^2} R_L \quad (13\text{-}5b)$$

For the load:

$$P_L = \frac{(V_{\max} - V_{\min})(I_{\max} - I_{\min})}{8} \quad (13\text{-}6)$$

$$V_m = \sqrt{2 P_L R_L} \quad (13\text{-}8)$$

Chapter 14 Push-Pull Amplifiers

For maximum conditions, class A:

$$\eta_{\text{overall}} = \eta_{\text{collector}} = 50\% \quad (14\text{-}1a)$$

$$P_C = P_L \quad \text{each transistor} \quad (14\text{-}1b)$$

For maximum conditions, class B:

$$I_C = \frac{I_m}{\pi} \quad (14\text{-}2a) \qquad \eta_{\text{overall}} = \eta_{\text{collector}} = \frac{\pi}{4} \times 100\% = 78.5\% \quad (14\text{-}5)$$

$$P_{C,\max} = 0.20 P_{L,\max} \quad \text{each transistor} \quad (14\text{-}11)$$

Chapter 15 Frequency Response

For low frequencies:

$$K_{LF} \equiv \frac{A_{LF}}{A_v} \quad (15\text{-}3) \qquad K_{LF} = \frac{1}{1 - j\dfrac{X_{C_1}}{R_1 + R_2}} \quad (15\text{-}4)$$

$$K_{LF} = \frac{1}{\sqrt{1 + \left(\dfrac{X_{C_1}}{R_1 + R_2}\right)^2}} \angle + \tan^{-1} \frac{X_{C_1}}{R_1 + R_2} \quad (15\text{-}5)$$

For high frequencies:

$$K_{HF} \equiv \frac{A_{HF}}{A_v} \quad (15\text{-}6) \qquad R_{\text{eq}} \equiv \frac{R_1 R_2}{R_1 + R_2} \quad (15\text{-}7)$$

$$K_{HF} = \frac{1}{1 + j\dfrac{R_{\text{eq}}}{X_{C_2}}} = \frac{1}{\sqrt{1 + \left(\dfrac{R_{\text{eq}}}{X_{C_2}}\right)^2}} \angle - \tan^{-1} \frac{R_{\text{eq}}}{X_{C_2}} \quad (15\text{-}8)$$

For low frequencies:

$$\tau_1 \equiv (R_1 + R_2) C_1 \quad (15\text{-}9) \qquad \omega_1 \equiv 2\pi f_1 \equiv \frac{1}{\tau_1} \quad (15\text{-}10)$$

$$f_1 = \frac{1}{2\pi\tau_1} = \frac{\omega_1}{2\pi} \quad (15\text{-}11)$$

$$K_{LF} = \frac{1}{1 - j\dfrac{f_1}{f}} = \frac{1}{1 - j\dfrac{\omega_1}{\omega}} = \frac{1}{\sqrt{1 + \left(\dfrac{f_1}{f}\right)^2}} \angle + \tan^{-1} \frac{f_1}{f} \quad (15\text{-}12)$$

For high frequencies:

$$\tau_2 \equiv R_{\text{eq}} C_2 \quad (15\text{-}15) \qquad \omega_2 \equiv \frac{1}{\tau_2} \quad (15\text{-}16)$$

$$f_2 = \frac{\omega_2}{2\pi} \quad (15\text{-}17)$$

$$K_{HF} = \frac{1}{1 + j\dfrac{f}{f_2}} = \frac{1}{1 + j\dfrac{\omega}{\omega_2}} = \frac{1}{\sqrt{1 + \left(\dfrac{f}{f_2}\right)^2}} \angle - \tan^{-1} \frac{f}{f_2} \quad (15\text{-}18)$$

$$C_{\text{out}} = C_{ob} \quad (15\text{-}24) \qquad C'_{\text{in}} = (1 + A_v) C_{ob} \quad (15\text{-}23)$$

$$C_{\text{in}} = C_{ob_1} + C_{b'e_2} + (1 + A_v) C_{ob_2} \quad (15\text{-}26)$$

Chapter 16 Feedback

$$A'_v = \frac{A_v}{1 - \beta_f A_v} \quad (16\text{-}1) \qquad A'_v = -\frac{1}{\beta_f} \quad (16\text{-}3)$$